EATCS
Monographs on Theoretical Computer Science

Ryszard Janicki Peter E. Lauer

Specification and Analysis of Concurrent Systems

The COSY Approach

With 128 Figures

Springer-Verlag
Berlin Heidelberg New York
London Paris Tokyo
Hong Kong Barcelona
Budapest

Authors

Prof. Dr. Ryszard Janicki
Prof. Dr. Peter E. Lauer

Department of Computer Science and Systems
McMaster University
1280 Main Street West
Hamilton, Ontario L8S 4K1, Canada

Editors

Prof. Dr. Wilfried Brauer
Institut für Informatik, Technische Universität München
Arcisstrasse 21, W-8000 München 2

Prof. Dr. Grzegorz Rozenberg
Institute of Applied Mathematics and Computer Science
University of Leiden, Niels-Bohr-Weg 1, P. O. Box 9512
2300 RA Leiden, The Netherlands

Prof. Dr. Arto Salomaa
The Academy of Finland
Department of Mathematics, University of Turku
SF-20 500 Turku, Finland

This book is the result of the combined efforts of the authors. Their names have been listed in alphabetical order with no implication that one is senior and the other junior.

ISBN 3-540-55204-9 Springer-Verlag Berlin Heidelberg New York
ISBN 0-387-55204-9 Springer-Verlag New York Berlin Heidelberg

Library of Congress Cataloging-in-Publication Data
Janicki, Ryszard. Specification and analysis of concurrent systems: the COSY approach / Ryszard Janicki, Peter E. Lauer. p. cm. - (EATCS monographs on theoretical computer science; v. 26) Includes bibliographical references and index. ISBN 0-387-55204-9 (U.S.) 1. Parallel processing (Electronic computers) I. Lauer, Peter E., 1934-. II. Title. III. Series.
QA76.58.J36 1992 005.1'2-dc20 92-7430

Printed in the United States of America

Typesetting: Camera ready by authors
45/3140-5 4 3 2 1 0 - Printed on acid-free paper

Preface

Complex systems of processes which can progress in parallel (or concurrently) and which involve critical periods of activity during which they may interfere with each other in some detrimental manner are becoming increasingly common in practice. Examples of such systems are air, train, or car traffic control, process control (such as a chemical or nuclear plant), communication systems and operating systems, etc. The complexity of the systems and the limitations of human capabilities for making complex decisions have led to attempts to reduce the complexity of the decisions required from the human operator by delegating large amounts of detailed sifting of intermediate information required for making decisions to automatic computers.

Most of the types of system listed above usually require a very high degree of validation before implementation because of the extremely critical nature of the processes being controlled, for example, the various traffic control systems whose malfunctioning could entail not only loss of property but loss of human life as well. The complexity of such validation can be reduced by initially validating the soundness of the intended control strategy independently from implementation details, and only subsequently validating that the implementation of the strategy by a specific computer system preserves the soundness of the strategy.

Concurrent systems are more difficult to specify and analyse than sequential ones, because they require the conceptualization not only of their sequential subsystems, but also of the complex interactions between them. It follows that the programmer's intuition is not enough, being unreliable in cases of high complexity. Here solution of the problem of verification of correct behaviour of the design becomes crucial, and a satisfactory conceptual apparatus for rigorous verification becomes essential.

In such an apparatus, reduction of complexity, by abstracting away from all irrelevant detail specific to some implementation of a concurrent and distributed system strategy, is desirable. But it must be easy to obtain implementations from the abstract specification of a strategy, given enough information about the synchronization mechanisms of the concrete system on which it is to be implemented. We therefore need a notion of "system" sufficiently abstract to allow analysis only of those aspects of systems arising from their concurrency and yet capable of being readily translated into practical terms.

The COSY (COncurrent SYstem) notation and theory presented in this book constitutes such a conceptual apparatus, and was developed by the authors and others in the last decade as one of a number of mathematical approaches for conceptualizing concurrent and reactive systems.

It is based on the recognition that formal language and automata theory have been very successful in the specification and analysis of sequential and centralized systems, including both software and hardware. Hence, the COSY approach extends the conventional notions of grammar and automaton to collections of "synchronized" grammars and automata, permitting system specification and analysis of "true" concurrency without reduction to non-determinism.

In the book the COSY theory is developed to a great level of detail and this constitutes the first uniform and self-contained presentation of all results about COSY published since 1974, as well as including many new results not published elsewhere.

The COSY theory is used to analyze a sufficient number of typical problems involving concurrency, synchronization and scheduling, to allow the reader to apply the techniques presented to similar problems of interest to him or her. Furthermore, a high-level notation supporting modular development and verification of specifications is presented. A number of computer based analysis systems are also presented briefly and their possible use indicated.

The COSY model is also related to a number of major alternative models of concurrency, particularly Petri Nets, which have since the beginning served as one standard truly concurrent semantics of COSY. Indeed, much of the central part of the chapter on COSY theory is concerned with a detailed study of the relationship between COSY and Petri Nets. Other major alternative models related to COSY are the Communicating Sequential Processes (CSP) of C.A.R. Hoare, and the Calculus of Communicating Systems (CCS) of R. Milner, both of which were developed after COSY had already been introduced.

Finally, although COSY incorporates Path notation, which is due to R. Campbell and A.N. Habermann, it should be understood that COSY Paths are a generalization of the original Paths first introduced by R. Campbell and P.E. Lauer in 1974. Most references to our work in the literature have missed this important point and hence published remarks about COSY paths tend to be wrong since they concern only the original Path notation.

Concerning the structure of the book, it consists of six chapters and eight appendices. The appendices present basic mathematical concepts used throughout the book as well as long, complex and tedious proofs. Chapter 1 provides detailed motivation, intentions of the book and acknowledgements. Chapter 2 is devoted to the formal theory of (basic) COSY. It contains the majority of formal definitions and results. Its major contents are: two kinds of semantics for COSY, one direct in terms of vector firing sequences; the other indirect in terms of Petri Nets; analysis of adequacy properties, execution strategies, and priorities. Chapter 3 deals with high-level COSY specifications including Process Notation, the Macro Notation and expansion theory. Chapter 4 analyzes 6 different applications of COSY theory. Chapter 5 compares COSY with other approaches including CCS and CSP. The last chapter gives a historical perspective and information about what has been excluded from the book.

R. Janicki
P. E. Lauer

Hamilton, Ontario, Canada, January 1992

Contents

Chapter 1

What COSY Is and What It Is For

1.1 Introduction

Concurrent systems abound in human experience but their fully adequate conceptualization as yet eludes our most able thinkers. Even the concurrent systems we construct on the basis of modern computer and communication technology continue to elude adequate conceptual analysis in general. However, our increasing dependence on ever more complex concurrent systems in the management and control of human affairs and activities increases the urgency for developing more adequate concepts to maintain reliable control over systems we have created.

Static systems are much easier to conceptualize than dynamic systems. The mathematical concepts for describing static systems were essentially in place during the time of Aristotle and Euclid, but the mathematical concepts required for describing dynamic systems were only obtained after Newton's time. Mathematical concepts for describing asynchronous dynamic systems interconnected by signals, without the assumption of absolute space and time, have appeared only since Einstein's time. But the mathematical concepts of physics and astronomy are not of much use to the average human, nor are they of much use in conceptualizing concurrent systems at the human level (rather than at the physically minute or at the astronomically large levels). At the human level one wishes to abstract from all irrelevant details of a system and focus on those properties of the system which are important to the human.

The COSY notation and theory presented in this book was developed in the last decade as one of a number of alternatives for conceptualizing concurrent systems. There are many possibilities for representing behaviours of dynamic systems. Since human knowledge is expressed in language, that is, by sequences of sentences involving names of entities, actions, events, etc., and names of relationships between them, behaviours of dynamic systems would be most readily comprehensible to humans if they could be directly identified with sentences of a language, that is, if a historical record of such a behaviour could just be read like a sentence in this language.

To explain in more detail, there are three basic questions which arise with regard to a dynamic system:

1. How does one characterize **all legal behaviours** of the system?
2. How does one **generate** a legal behaviour of the system?
3. How does one **recognize** a legal behaviour of the system?

Analogous questions arise with regard to a language:

1. How does one characterize **all legal sentences** in the language?
2. How does one **generate** a legal sentence in the language?
3. How does one **recognize** a legal sentence in the language?

In formal language theory, grammars are used to generate legal sentences in a language, automata are used to recognize legal sentences of a language, and both grammars and automata are used to characterize all the legal sentences of a language. In fact one usually uses both a grammar and automaton and demonstrates that they are equivalent in the sense that they both characterize the same language.

If one identifies a dynamic sequential system with a grammar and/or automaton, and identifies legal behaviours with legal sentences in the language they characterize, one can use language theory and automata theory to express knowledge about dynamic systems.

Concurrent systems consisting of parallel subsystems are in general more difficult to conceptualize than systems consisting of sequential subsystems. In the former case, additional complexity is introduced by the need to conceptualize all possible and actual interactions between concurrently active subsystems which may occur at arbitrary times. Given the behaviours of individual subsystems, the parallel system needs to ensure that the individual behaviours of all subsytems are preserved, and at the same time that the synchronization constraints of the system as a whole are satisfied.

Extending our reduction of dynamic sequential systems to grammars and/or automata above, we identify a concurrent system with a set of synchronized grammars and/or automata, and if concurrency is not reduced to interleaving, this results in a notion of sentence in the language where components of the sentence are no longer linear strings.

A very simple way to conceptualize a parallel system starts from the notion of an **atomic event (action, state transition, operation**, etc.), which is capable of occurring (being performed, taking place, being executed, etc., respectively) whenever the synchronization constraints of its immediate context allow, whereupon it will actually occur after a finite duration, provided the immediate context still permits its occurrence. A **sequential system** consists of a collection of atomic events whose occurrences exclude each other at any instant, that is, they may only occur one at a time, and whose mutual occurrences are further constrained by the internal synchronic structure of the system, that is one event may only occur after some other event has occurred in the immediate past, or some collection of events may occur repeatedly, etc. A **behaviour** of a sequential system consists of any sequence (total order) of event occurrences which satisfy this internal synchronic structure. If we think of the events of a system as an alphabet, then we can represent behaviours by sequences of letters (words) in the "formal language" of the system. This interpretation immediately allows us to use formal language theory to reason about the behaviours of sequential systems. A **parallel system** consists of a **parallel composition** of a collection of sequential subsystems, which ensures that an event common to more than one subsystem satisfies the synchronization constraints expressed by each of the subsystems sharing that event. Behaviours of parallel systems can be conceptualized as (partially ordered) traces (histories) of occurrences of events belonging to the system. Two events may occur in parallel if they do not both occur in the alphabet of a single subsystem. An event that is shared by several subsystems may only occur if permitted by the respective internal synchronic structure of all the subsystems which share it.

If one wishes to apply conventional formal language theory to analyze parallel systems one may proceed in two ways. The first approach assumes that parallel occurrences of events may be represented as single occurrences of both events but in either order. In this case, parallel behaviours (partially ordered event occurrences) are reduced to sequential behaviours (totally ordered event occurrences), and conventional formal language theory may be applied to study parallel systems. The second approach does not assume that all important properties of parallel systems can be dealt with adequately by reducing them to (nondeterministic) sequential systems. Hence, to apply formal language theory one must first generalize this theory to allow words to consist of partially ordered letters from some alphabet. Of course ordinary sequential words will be a special case of this more general kind of word.

The simple conceptual apparatus sketched above constitutes the basis of the COSY approach [LC75], [LTS79] which forms the subject matter of this book. It also underlies a number of other approaches which were subsequently developed. Examples that come to mind are the "Calculus of Communicating Systems" (CCS) of Robin Milner [Mil80] and the "Communicating Sequential Processes" of C. A. R. Hoare [Hoa85]. Milner and Hoare, however, originally chose to reduce parallelism to sequentiality, and the COSY approach has chosen the non reductionist way. In other words, Milner and Hoare use a total order trace theory and the axiomatic method as conceptual basis, and COSY uses partial order trace theory and formal language and automata theories as conceptual basis. The use of partial order semantics for COSY from its inception make it much more akin to "Petri Nets" of C. A. Petri [Pet62] which is essentially also based on the same simple model. A major part of the present book develops language theory to permit explicit representation of parallelism while preserving the applicability of as many useful tools and results from formal language theory and automata theory as possible. We believe both the reductionist and non-reductionist approaches are important and each is better suited to certain tasks than the other, as we shall try to point out in the course of the book.

Adequate representation of true concurrency is one of the aspects of systems which have only recently been stressed by other approaches to concurrency theory, but which were considered in the design of the COSY notation and theory from the very outset. In short, COSY was designed to:

1. represent true concurrency
2. be compositional
3. represent the distributed nature of systems
4. be reactive systems, in the sense of never terminating
5. stress intermediate states of computation
6. avoid global states, time or control as basis of the model
7. represent causal dependencies
8. permit refinements, that is, our sequentialization is more general than prefixing in CCS or CSP, and allows refinement, whereas prefixing does not.

Another distinguishing characteristic of our approach is that it is motivated by the desire to reduce as many global behavioural properties of concurrent systems to static structural properties of their specifications in the COSY notation. It is our view that the most extensively used verification tools available to the practicing programmer are the syntactically based portions of compilers concerned with debugging programs. Our concern is to extend these important tools to deal with problems arising in the context of concurrent systems. Although we give numerous and typical examples of the utility of our results for the practicing programmer, we have chosen in this book to concentrate on a thorough and self-contained presentation of the formal results themselves.

Two main types of reader should benefit from this book. The more mathematically inclined reader should be able to convince himself of the correctness and theoretical significance of our results solely on the basis of what we write, since our proofs are on the whole detailed and complete. Such readers may also find enough detail for them to develop our results further, possibly in directions we ourselves have not foreseen.

The more practically oriented reader should be able to develop correct specifications without requiring a complete understanding of their full mathematical characterization. This is possible because we have been careful to design the notation so as to map good behavioural structure onto syntactic structure. This means that syntactic structures have been designed to guarantee correctness as far as possible, and the less mathematically oriented programmer can develop his designs safely on the basis of informal arguments at the language level. It has been our experience that most designs developed in this informal manner turned out to be correct upon formal verification. This is not surprising since one largely constructs new solutions from earlier verified solutions, and detailed mathematical reasoning is only necessary to verify new components and novel interconnections.

The largest body of classical synchronization problems arise in the context of operating systems. Hence, the reader will find that the largest number of examples in this book treat these problems. However, COSY is applicable to a much wider class of problems which arise in any concurrent system from the hardware of computers and communication systems, to such real world systems as concurrent negotiations between management and labour to achieve the goal of mutually acceptable compromise on some question. If the system to be analyzed can be expressed in terms of classes of important events (state changes) COSY can be used both for formal analysis and simulation without having to use a "programming" language. Hence, COSY is also suited as a conceptual framework for non-programmers and non-mathematicians. This property of COSY also makes it a useful modelling and prototyping tool. As yet there has not been much development of examples of this usage, but here is a fruitful avenue of discovery of novel concurrent strategies for application in areas not specifically concerned with computers.

A computer based environment of the type we discuss in a later section of the book greatly enhances sound system design based on such a largely intuitive and informal approach. Where detailed mathematical reasoning is needed, the mathematical sophistication required is modest since it requires only elementary set theory, algebra of relations, graph theory, order theory, logic, language and automata theory. Furthermore, the computer based environment and its associated notion of system dossier permits the collaborative development of a design by programmers from both the orientations mentioned above. In fact, the dossier facility allows verification and design at numerous levels each of which can be more or less developed at any one time. The dossier manages all interactions, particularly those which fulfill management functions.

1.2 Concepts, Objectives and Design Decisions

From a practical standpoint concurrent systems are most usefully specified by means of an implemented parallel programming language since it allows immediate execution of the program. From the standpoint of concurrency theory a parallel programming language is not an appropriate vehicle for specifying concurrent systems. One reason is that here one is concerned with the specification of **what** behaviours a system should allow, and not **how** such behaviours are enforced by specific synchronization mechanisms of some specific programming language. Furthermore, conventional sequential programming language constructs have synchronization properties which may not be appropriate for combination in concurrent systems. Additionally, the built-in synchronization properties of such language constructs predispose or exclude certain decompositions of a system into concurrent subsystems. For these reasons we have decided to abstract from all semantics of systems except that which has to do with synchronization, and to design the notation at the level of systems of uninterpreted events which are constrained to occur as specified by means of explicit synchronization operators. These operators only state what behaviours are to constitute system activity. Any implementation of the system must then enforce the specified behaviours by means of the specific synchronization primitives of the concurrent programming language used.

Although our notation is at this level of abstraction it is possible to execute the specifications themselves in the COSY computer based environment introduced later in the book. Hence, we do not lose the ability for execution due to our abstraction from the non-synchronization semantic details of any specific programming language. Our notation also places no constraints on the decomposition of systems into concurrent subsystems.

One might wonder if our level of abstraction prevents us from using the COSY approach to study real existing systems since we abstract from all semantics not to do with synchronization in our concurrency theory. But this is not the case. In concurrency theory we formally characterize all possible behaviours of the system whatever the interpretation of the events might be. This activity is orthogonal to the activity of interpretation of the system. To interpret such an abstract specification all that is necessary is to interpret each event of the specification as the corresponding event of the system to be modelled. Such an initial interpretation immediately induces an interpretation of every behaviour of the uninterpreted system, and the structure of such an interpreted behaviour represents the structure of the corresponding behaviour of the system modelled. However, the correctness of the specification with regard to synchronization properties has been factored out of the correctness of the interpretation in our approach. This is analogous to and brings the same advantages as factoring out functional (partial) correctness of sequential programs from termination (total) correctness of sequential programs. Such factorization usually makes the task of verification of correctness considerably simpler.

We base our approach on a simple but very general concept of a dynamic system utilizing the notions of event and state, introduced as independent notions in the most general case, as for example in Petri nets. The dynamism is represented by the fact that an event may occur (is enabled) in certain states and not in other states, and when an event occurs it results in a change of state. In COSY we define the notion of state in terms of histories of event occurrences rather than introducing it as an independent concept. This book presents a formal treatment of synchronization issues arising from highly parallel asynchronous systems in the absence of global states and universal time.

A **sequential system** consists of a number of events which are constrained to occur as required by the **synchronic structure** of the system. A **behaviour** of the system is any sequence (total order) of event occurrences as required by the synchronic structure of the system.

A **concurrent system** consists of events which are constrained to occur as required by the synchronic structure of **all** its constituent sequential subsystems involving that event. This means that:

> if an event belongs to only one subsystem it may occur whenever the synchronic structure of that subsystem permits it,

and

> if an event belongs to more than one subsystem it may occur whenever the synchronic structure of every subsystem involving the event permits it.

Hence, subsystems are synchronized on common events. Note that the synchronic structure of one subsystem may contradict the synchronic structure of another subsystem, and hence, exclude some of the possible behaviours of the subsystems considered independently. This can lead to **partial or total system deadlock**, or **starvation** of some subsystem in the sense that it may not be able to perform some task represented by a behaviour which is excluded by constraints in other subsystems. A **behaviour** of a concurrent system is any partial order of event occurences as required by the synchronic structures of all its constituent subsystems. If two events belong to no single sequential subsystem they may occur simultaneously or both in either order if both are simultaneously enabled. This is expressed by the fact that there is no order between occurrences of these events in any partial order representing a system behaviour.

From the above discussion we see that there are two types of synchronic structure which are important in conceptualizing concurrent systems, the local synchronic structure of component sequential subsystems, and the global synchronic structure induced by the sharing of events by several sequential subsystems. In the **basic COSY notation** the local synchronic structure of a sequential subsystem is stated explicitly by embedding its event names in synchronization expressions called paths[1]. The global synchronic structure is implicit and has to be deduced from the sharing patterns of the entire concurrent system.

In general, a concurrent system naturally decomposes into a collection of separate (possibly concurrent) subsystems, which can be thought of as separate entities or objects. For example, an entity might represent an abstract data type thought of as a collection of operations capable of being used by sequential processes executing in parallel. Concurrently pending requests for the execution of operations will be granted to processes in the order specified by the internal synchronic structure of the abstact data type.

This decomposition of a system into separate subsystems (modules) greatly reduces the complexity of the conceptualization and verification of the system. One of its major consequences is that implicit sharing patterns can now be discovered by an analysis of smaller parts of the specification of the system. But within each separate subpart global synchronic structure is still implicit since a COSY specification consists of a collection of synchronization expressions, and all sharing patterns can in general only be deduced from the occurrences of the event names involved in all the separate expressions. There is no single expression which states the shared synchronic structure explicitly. In the COSY semantical theory the shared structure is made explicit by identifying shared event occurrences in their partial order representation. In other words, in basic COSY we cannot explicitly express shared synchronic structure at the level of system specification, but we can express it at the semantical or behavioural level. In the two dimensional representation of concurrent systems using Petri nets, the sharing patterns also become explicit at the system specification level. This is one reason we also use a standard semantics for basic COSY in terms of behaviourally equivalent Petri nets in this book.

Everything we have stated above about the inability of basic COSY to explicitly express shared structure is also true for Hoare's Communicating Sequential Processes (CSP) and Milner's Calculus of Communicating Systems (CCS).

Using synchronization expressions for specifying synchronic structure has another consequence which restricts their utility. For such expressions to be well defined they must be definite in the sense that they are stated in terms of a definite number of events. But in the conceptualization of systems one wishes to be more general and abstract to synchronization schemes which express the common aspects of whole classes of instances of specific systems. This is one of

[1] Readers should note that COSY paths [LC75] are a generalization of and should not be confused with the paths introduced by Campbell and Habermann [CH74] which represent concurrency in a radically different way.

the reasons for developing the high level COSY notations presented in the second half of the book.

Another aspect of the **high level COSY notation** also has to do with making synchronic structure more explicit at the specification level. For example, suppose one system consists of a three dimensional cube of events, e.g. a stack of core memory cells, and another system shares the events of one plane of this cube, e.g., the read/write control over the cells of that plane. This means that the shared structure of the two systems is precisely the intersection of the two systems on the plane in question. This structure is directly expressed in the high level COSY notation by associating the indices of the points in Cartesian space with the event names. In this way any structure in n-dimensional discrete Cartesian space can be expressed explicitly in COSY by generating indexed events using conventional mathematical notation to characterize the index sets. In fact index sets can even be characterized by predicates in the so called predicate generators. We sometimes allow indexes to range over arbitrary (not necessarily numerical) sets including strings.

The specification of such indexed event structures also permits the use of APL like aggregate processing by distributing synchronization operations over all or selected elements of some sub-structure. This ability allows very concise and yet natural formulations of synchronization constraints involving arbitrary numbers of events. The distributors also enhance modifiability of specifications. A specification involving the distribution of the mutual exclusion operator over all elements of a one dimensional array of operations specifies that all their executions mutually exclude each other during any system behaviour. This is specified for such arrays of any size. Hence, changing the size of the array does not require a rewriting of the specifications involving the distributor.

This generator and distributor notation still leaves implicit the conceptually important sharing structure between separate subparts discussed earlier, as well as any hierarchic ordering of systems among each other. The inability to express hierarchy also prevents the notation from explicitly representing any hierarchic structure of real systems modelled in the notation. The high level **system notation** introduced in the third chapter of the book allows the explicit expression of both these types of structure in COSY.

Our conceptual apparatus for specifying concurrent systems is based on formal language theory and partial order theory. Behaviours of systems will be expressed in terms of words (although partially ordered) representing possible or actual histories of occurrences of events. Conventionally, languages are formally specified by grammars and this is done in COSY. A specification in basic COSY notation is a collection of grammars defining all possible behaviours (words) of the system (language) specified. One of the objectives of the COSY approach is to allow the expression and verification of global behavioural structures in terms of textual structures in the corresponding specification. Hence it should be possible to directly perceive structural properties of behaviours from an understanding of the specification without its dynamic simulation. In particular, if behaviours have periodic sub-behaviours, these periods should be directly represented in the specification. Basic COSY was designed to support such direct representation of structure as far as possible, while avoiding excess complexity.

Since specifications are expressed as grammars in basic COSY, it seemed most natural to express high level and general specifications also as grammars, but at one level above the grammars used in the basic specification. That is, a high level specification is a grammar which generates or represents other grammars which specify the system at the basic COSY level. Using a higher level grammar to characterize the grammars of the lower level permits the explicit representation of much of the structure left implicit at the basic level. But the enhanced ability to express structure at the specification level should also enhance the ability to perceive more regularity of structure at the behavioural level. Our desire to achieve these enhancements have greatly determined our design of the high level notations, and given rise to a number of interesting properties higher level grammars must satisfy to achieve these goals. For this reason we go into considerable syntactical detail in the section on high level notation. But it must be remembered that in our approach, much that appears highly syntactical in nature also has great semantical importance, since we are concerned with reducing as many behavioural semantical properties of systems to syntactical (static, analytical) properties of their corresponding specifications.

1.3 Structure of Book

In Chap. 2, after a brief motivation for vectors of sequences as a model for concurrent system behaviours thought of as partially ordered event occurrences, we define them formally and demonstrate simple but useful properties they have. Subsequently we show how vectors of sequences formally relate to partial orders. On the basis of these concepts we correlate COSY specifications with the vectors of sequences they specify, and express important properties of COSY specifications in terms of these vectors of sequences. We also prove some elementary properties of vectors of sequences generated by such specifications. Next we characterize the descriptive power of basic COSY specifications using the notions of concurrent products of regular expressions and finite automata.

In the following section we give Petri net semantics of basic COSY, beginning with an introduction and adaptation of elementary net concepts required for our present purposes. We then give a vector of sequences semantics of state machine decomposable (smd) nets. But the general case arising from the full COSY notation requires the use of labelled smd nets for which we introduce a novel semantics in terms of vectors of sequences. We then demonstrate that closed labelled smd-nets define the same vector of sequences languages as finite asynchronous automata. Next we give Petri net semantics of COSY which is well founded for ambiguous specifications (whereas vector of sequences semantics is not). For this purpose we define a unique and ambiguity preserving mapping of regular expressions into labelled state machines.

Next we give a more detailed presentation of adequacy results for COSY specifications, with emphasis on verification techniques and the reduction of semantics to syntax. We also give algorithms for verifying that syntactically conflict free (SCF) specifications are adequate. Adequacy preserving transformations of specifications are introduced to extend the results obtainable by means of "static" criteria of adequacy. These transformations are insertions and deletions.

Adequacy preserving transformations and the results about SCF specifications are applied to a buffered reader/printer system with interrupt handler, to a mechanism for controlling access to a critical section for processes with the same priority, and to a mechanism for controlling access to a critical section for processes belonging to different priority classes.

Next we define a larger class than SCF specifications called free choice specifications, for which adequacy is equivalent to deadlock-freeness, and relate this class briefly to free choice Petri nets.

The subsequent section introduces a framework for specifying and analyzing various execution semantics for COSY, for example, interleaving semantics, full semantics, maximally concurrent semantics and optimal semantics. The framework is based on the notion of strategy which represents an arbitrary execution semantics. We define behavioural equivalence of strategies, which preserves adequacy and deadlock-freeness. It is demonstrated that for every strategy there is a behaviourally equivalent optimal strategy. Furthermore, we give sufficient and necessary conditions for equivalence of a full strategy and a maximal concurrent strategy. To obtain more structural criteria for this equivalence we also obtain a family of sufficient conditions.

The next major section introduces a formal semantics for COSY with static priorities. We point out that vector of sequences and partial orders do not suffice to express the semantics of this type of priority. Although priorities are often used in specific examples, they cause many problems for obtaining general results. We introduce a new semantics based on the notion of observation of behaviours (histories), and the corresponding formal objects called multiple sequences.

Subsequently we discuss the important problem of starvation which is often avoided in practical solutions by using priorities. In contradistinction to approaches based on Hoare logic [Hoa85] or Modal logic [Pnu86], we prevent starvation not by general fairness assumptions, but by explicit use of a **priority relation** denoted by "<" in specifications. Starvation-freeness is a stronger property than adequacy. First, we indicate why it is difficult to formally define starvation and to achieve absence of starvation in systems. But, as opposed to adequacy, there are no formal results concerning starvation, so the definition of starvation has to be used to verify absence of starvation, which we do for the second reader/writer problem.

Next we show how we can simulate certain dynamic priorities on the basis of the static priorities and their semantics introduced previously using so-called ghost events.

We also demonstrate that vector of sequences semantics can be used to model semantics of COSY with priorities if we forget about their usual interpretation as causality relations. That is, we can use this theory to reason about priority COSY, but to do so we need to distinguish between events being potentially enabled and being actually enabled. We go on to show that if we are only interested in proving sequentializability, deadlock-freeness, adequacy, and not in proving, e.g., atomicity of certain sequences of event occurrences, vector of sequences semantics of priority COSY can be used.

In Chap. 3 we formally develop and study several higher level COSY notations based on a number of design criteria which were chosen to obtain an extension of the basic notation which would preserve the semantics of the basic notation while extending its ability to express conceptual structure, to support modular development and verification of complex specifications. Not only should the semantics of the basic notation be preserved, but the semantics of the high level notation should reveal the semantics of the corresponding specification in basic notation directly in terms of the high level notation, as far as possible.

The introductory section of this chapter uses the more complex examples from Chap. 2 to motivate the introduction of high level notation and to introduce high level constructs in an intuitive setting. It begins by distinguishing between interpreted and uninterpreted specifications (programs). Uninterpreted specifications involve only uninterpreted events and they are used to define concepts arising in concurrency theory. They allow the formulation and proof of important laws relating to these concepts. Interpreted specifications involve interpreted events, and permit the results of concurrency theory to be applied to develop verified solutions to problems arising in the modelling of real world concurrent systems.

In the analysis of uninterpreted specifications in Chap. 2 one often uses meta-variables to state relations between specifications, and one uses ellipses to indicate "arbitrary context" or "regular repetitions of some pattern". These meta-linguistic features allow one to talk about indefinite aspects of basic COSY specifications which in themselves are always definite. The high level notation should incorporate features which allow one to express such indefiniteness at the object language level.

Interpreted specifications usually involve many interpreted events usually partitioned into subsets whose elements are distinguished by affixing indices to the common generic event name. The high level notation should include features for expressing this partitioning explicitly, and to facilitate the comprehension of large systems in terms of its sub-systems. Furthermore, the decomposition of the system into disjoint sub-systems should be explicit at the object language level. The interconnections between sub-systems constituting a connected (non-disjoint) system should also be explicitly expressible at the object level.

A component which is disjoint from other components of a system can be thought of as an abstract data type (adt) which grants execution of pending operations according to the internal synchronic structure of the adt. In general, a specific task to be performed by the system usually involves the performance of operations from any number of adt's. The high level notation should permit the formulation of such a task either as a sequential (possibly non-deterministic) system called a process, or alternatively as a compound operation consisting of events from the adt's involved. Processes and compound operations can progress only by operation executions in a way that does not conflict with the internal synchronic structures of all the adt's involved.

To this end we discuss how the power of generators is extended to permit greater expressiveness with regard to index sets, such as the ability to specify index sets which do not form arithmetic progressions, or which are non- arithmetic, such as strings, etc. This extension is attained by introducing "predicate replicators", that is, replicators which implicitly characterize their index sets by means of first order predicate logic statements.

We next explain how the expressive power of generators may be further extended by the introduction of the "system notation" which permits concise representation of more regularities in basic programs than is possible in all the notations discussed above. The system notation permits one to express at a single place separate occurrences of more than one subsystem of identical structure, possibly differing in event names, or involving occurrences of compound expressions where another subsystem involves occurrences of event names only. Such regularities could not be represented by any of the replicators introduced before. The introduction of formal parameters in the system notation allows one to state more general specifications as well.

Throughout the introduction we also indicate how the high level notation supports natural readings of specifications, as well as modular and hierarchic incremental development of specifications and their verifications. The COSY system notation allows specifications to be developed within the object oriented paradigm which is very appropriate for directly modelling the object oriented view of the world which lies at the basis of our approach.

In Sect. 3.2 the process notation is formally defined and its semantics is given, first by reducing it to the semantics of a corresponding equivalent set of paths, and secondly, directly in terms of vector firing sequences for processes and paths. The latter approach is more appropriate if one is concerned with giving direct high level semantics for the high level notation, that is, a semantics which enhances one's ability to perceive structure at the behavioural level directly from the structure of the high level specification without requiring prior translation to the basic level. We prove a theorem which states that the two semantics are equivalent with respect to behaviours.

Sect. 3.3 begins with a discussion of the criteria for the design of high level notations, and derives properties of the notation which must be satisfied if the objectives stated are to be achieved. Throughout the subsequent formal development we will demonstrate that our design does satisfy these properties. The heart of this section formally defines the syntax of high level programs involving generators of the type described earlier. Throughout we indicate how the syntax satisfies such properties as expansion correctness, uniformity of high level notation with basic level notation, context, neighbourhood and range independence, and syntactic strength.

Subsequently, expansion rules are given which reduce the semantics of any high level specification involving generators to the semantics of their corresponding expanded basic COSY specifications. The main theorem of this subsection states that the expansion of a well-formed high level specification results in a well-formed basic specification.

Sect. 3.4 presents a brief overview of a computer based environment for specifying, analyzing and verifying concurrent systems using the general COSY notation and partial order semantics. This environment has been implemented in two prototypes using the SIMULA programming language in the context of two operating systems. Recently, a new prototype was developed in C^{++}. The environment permits incremental development of COSY specifications, including automatic or interactive simulation of the behaviours of the systems specified. Simulation may proceed sequentially, concurrently or maximally concurrently. Regardless as to which type of simulation was used, the resulting history may be displayed as if it were the result of any of the other types of simulation. Hence it is easy to determine the precise benefits of concurrent execution over sequential execution. The simulator also permits verification by exhaustive simulation of many solutions to synchronization problems stated in the literature. In cases where exhaustive simulation is not possible, the simulator is still useful in the search for counter-examples to conjectures, or in the search for important sub-behaviours, for example, periodic ones, which indicate how a complete verification may be obtained by additional reasoning by means of the formal COSY theory.

The notion of "system dossier" which serves to organize and represent the knowledge about a system developed in the environment is briefly described in

Sect. 3.5. This knowledge is not static but is continually developed and refined further due to additional analysis of the system by means of the tools of the environment. The system dossier manages this development which may in general be driven by any number of systems analysts over a long period of time.

Chap. 4 deals with COSY applications. In general we can find two classes of applications. The first kind is the application of COSY and its theory to specify and prove properties of some realistic non-sequential systems or their parts. Some examples of this class appear in Chapters 2 and 3. In Chap. 4 we additionally apply the COSY approach to analyze such problems as the Cigarette Smoker's Problem, where we show how the COSY simulator can be used to prove some specification property. We also specify and verify a Two-way Channel (part of a communication system), as well as a Hyperfast Banker Strategy (part of an operating system), and indicate some advantages of the COSY approach for this purpose. The final example of this kind of application is the specification of a solution to the Merlin/Randell Problem of Train Journeys which arose from the construction of a model train set. The problem is to find a synchronization among train movements which allows parallel movements where possible, and permits the completion of each train journey.

The second kind of application of COSY involves the use of the theory as a theoretical tool which aids in the development of another theory. We present two such applications. One is the introduction of an alternative methodology for representing non-sequential systems which is more suited for specifying systems in which concurrency is not the primary goal of specification, but where concurrency is only used to speed up the performance of algorithms which may also be solved sequentially. In this approach the programmer starts with a sequential solution together with a characterization of independent events, from which a concurrent implementation can be derived automatically. The second application in this latter class shows how COSY can be used in the theory of N-Modular Redundancy in the area of reliability of systems.

In Chap. 5 we analyze some relationships of the COSY model to other models such as Petri Nets [Rei85], CSP [Hoa78a], CCS [Mil80], Synchronized Behaviours of Processes [AN80] and Mazurkiewicz Traces [Maz86]. The basic relationship between Petri Nets and COSY was analyzed in Sect. 2.3. In this final chapter we present additional properties such as connections between COSY with priorities and inhibitor and priority nets, as well as relationships between some subclaues of COSY programs with processes and asymmetric-choice Petri nets. Although the intuitive principles are virtually the same for COSY, CCS and CSP, each model evolved in its own direction. In this final chapter we point out some major similarities and differences.

In the final chapter we compensate for omissions from the book to prevent its expansion beyond reasonable size. These omissions include results in COSY theory which have been published, but which are based on techniques whose introduction here would have lengthened the book unreasonably; discussions of other important formal approaches to concurrency; discussion of alternative path notations based on different models of concurrency; and other path or COSY like models developed independently of our approach.

Clearly, a uniform presentation of all major results obtained in the last 16 years, which were expressed in different notations and using various techniques, required the redevelopment of the meta-notation and the definition of new concepts. This can particularly be seen in Chapter 3, where the new version of macro-notation, its formal theory and the concept of system notation are presented.

Since many of the earlier results were obtained for subsets of the notation, or were given in a more informal manner, and since the book was to treat the most general notation and preferably in a formal way, new concepts, techniques and results were developed, chief of which are:

- vector firing sequence semantics for state machine decomposible nets,
- net semantics of COSY path programs which covers both ambiguity and the Kleene star without restrictions, i.e. the most general notation,
- formal proofs concerning the relationship between vector firing sequence semantics and net semantics of COSY path programs,
- formal proofs of properties of the Cigarette Smokers problem (Chapter 4.3),
- the formal relationship between COSY with priorities, priority nets and inhibitor nets in terms of multiple sequence semantics (Chapter 5.3.2).

1.4 Acknowledgements

Much credit for the original formulations of the foundations and results of some of the material in this book must go to the full-time members of the original COSY project (1974 -1981), Mike Shields and Eike Best, who have continued to be valued colleagues after their departure from the project. Furthermore, H. Genrich, P. Torrigiani, Y. Cotronis and R. Devillers have contributed at one time or another to the development of various high-level COSY notations. A number of recent theoretical results are due to M. Koutny. All their major contributions to our work are indicated where appropriate throughout the book, and discussed further in the Historical Perspective.

The authors would like to express special gratitude to Raymond Devillers and Maciej Koutny for their meticulous reading of an earlier version of this manuscript, and who not only found multitudes of annoying typographical errors but also a few technical errors which led to their correction in the final edition.

Peter Lauer would like to thank the Gesellschaft für Mathematik und Datenverarbeitung (West Germany), and Carl Adam Petri and all his colleagues at the GMD, for a number of invitations to spend extended periods at the GMD in order to work on aspects of the work presented in this book. These visits have always proved to be very inspiring and productive due to the great expertise in concurrency theory represented by Petri and the members of his group at the GMD. Particulars of contributions of specific members of the group to our work are discussed in more detail in the final chapter of this book. Peter Lauer would also like to acknowledge his general indebtedness to Tony Hoare and Robin Milner for their encouragement and inspiration over many years.

Ryszard Janicki would like to specially thank Antoni Mazurkiewicz of the Polish Academy of Science, Warsaw, Poland, for introducing him to the difficult but fascinating world of concurrency.

Thanks are due to Art Mulder for reformulating the manuscript to satisfy the requirements of the publisher, and organizing it in a form for flexible development of the bibliography and the index. Furthermore, we want to express our appreciation of Janie Lauer's readiness to learn Latex and to type much of the manuscript during a critical phase of the production of the text. Thanks are also due to Dr. Hans Wössner of Springer Verlag for his continuing support and excellent advice during the production of the book.

Both authors have been supported by grants from the Science and Engineering Research Council of the U.K., and the National Science and Engineering Research Council of Canada. Furthermore, both authors have been supported at various times by the University of Newcastle upon Tyne (U.K.), Aalborg University (Denmark), and McMaster University (Canada). Ryszard Janicki also received support from the Warsaw Technical University (Poland). We would like to thank all these establishments for their generous support during the past 16 years.

Chapter 2

Formal Theory of Basic COSY

2.1 Basic COSY Syntax and Semantics

A **basic COSY path program** is a collection of paths in <u>program</u> and <u>endprogram</u> parentheses. A path is an expression, similar to a regular expression (see Appendix B.3), built from event names, semicolons, commas, conventional parentheses and Kleene stars, enclosed by <u>path</u> and <u>end</u> parentheses. For instance:

$$Pr_1 = \begin{cases} & \underline{program} \\ P(1): & \underline{path}\ a; b, c\ \underline{end} \\ P(2): & \underline{path}\ (d; e)*; b\ \underline{end} \\ & \underline{endprogram} \end{cases}$$

In every expression as described above, the semicolon specifies sequential occurrences of the events named or subexpressions, and comma specifies a mutually exclusive occurrence of one of the events named or subexpressions. The comma binds more strongly than the semicolon, so that the expression "$a; b, c$" means "first event a must occur, after which exclusively either event b or event c must occur". An expression may be enclosed in conventional parentheses with Kleene star appended, as for instance "$(d; e)*$" which means that the enclosed specification applies zero or more times. In other words, an expression between <u>path</u> and <u>end</u> may be understood as an ordinary regular expression. The only difference is that "$\cup$" is replaced by ",", concatenation is replaced by ";", and mutually exclusive choice binds more strongly than concatenation[1]. Thus for instance "$a; b, c$" is equivalent to "$a(b \cup c)$" in the traditional notation for regular expressions. Moreover, by definition, the parentheses <u>path</u> and <u>end</u> correspond to "(" and ")*" respectively, so that a single path specifies repeated (or *cyclic*) sequences of event occurrences.

Formally, a basic COSY path program is a string derived from the production rules given below. The following meta-linguistic conventions have been used in stating the syntactic rules. The symbols "=","{","}","|", "*","+", are used as

[1] In traditional notation the opposite assumption is made about the strength of binding.

meta-symbols. The symbol "=" indicates replacement of its left hand side by the string on its right hand side. The braces "{ }" are used to group items together, "|" indicates alternative productions, "$\{item\}^*$" indicates production of "*item*" zero or more times, and "$\{item\}^+$" indicates production of "*item*" one or more times.

The syntax of a basic COSY path program is given by the following rules:

B1. *basicpathprogram* = <u>*program*</u> *programbody* <u>*endprogram*</u>

B2. *programbody* = $\{path\}^+$

B3. *path* = <u>*path*</u> *sequence* <u>*end*</u>

B4. *sequence* = *orelement* $\{;\ orelement\}^*$

B5. *orelement* = *starelement* $\{,\ starelement\}^*$

B6. *starelement* = *element* | *element*$*$

B7. *element* = *event* | (*sequence*)

B8. *event* = *name*

where "name" is a sequence of letters and digits beginning with a letter, with possible indices if necessary. Note that $*$ is used both as a terminal symbol of the basic notation where it appears at the same hight as the separators ";" and ",", and as an operator in the meta language used to state the production rules, where it is in the position of a superscript. Assume additionally that we may write some comments immediately before <u>program</u> and <u>path</u>. We shall use this feature to write names of programs and paths. In this book the names of programs will be either Pr or Qr with subscripts, superscripts and primes if necessary. The names of paths would usually be P_i and/or Q_i, where i is a constant or variable, with primes if necessary. In the examples of paths, instead of P_i or Q_i we prefer to write $P(i)$ or $Q(i)$, but it will be assumed that P_i and $P(i)$ in fact denote the same path.

Of course the above syntax is neither unique nor even the simplest one, but it indicates the precedence of "$*$" over "," and ";", it is easy to handle and seems to be easy to understand.

Intuitively, each path specifies a sequential subsystem and expresses a constraint on the order of occurrences of events belonging to the subsystem. During any period of system activity, the events belonging to a single path may only be active or occur one at a time, and the sequence of event occurrences constituting a history of this activity must be either a concatenation of cyclic behaviours specified by the path, or a prefix of such a concatenation. From this it follows that each occurrence of an event of the system specified by a program is constrained by every path in which it is named, that is, the **synchronization among paths is due to common events.** Let us consider the path program Pr_1 above. The program consists of two paths which share one event named "b". The execution of the program is achieved by executing these two paths simultaneously. Individually, the execution of P_1 proceeds by repeatedly executing "a" followed by the execution of either "b" or "c". The execution of P_2 proceeds by repeatedly executing the sequence "de" any (possibly zero) number of times, followed by the execution of "b". Since "b" is a shared event, in the combined viewpoint, the first path can execute "b" if and only when the "b" in the second path can be executed, and this achieves the synchronization between these paths. This kind of synchronization is often called by **handshake** or by **rendez-vous.**

The standard formal semantics of path programs is expressed by means of **vectors of strings.** Let us explain this idea briefly.

Consider the very simple example program Pr_2:

$$Pr_2 = \begin{cases} & \underline{program} \\ P(1): & \underline{path}\ a; c\ \underline{end} \\ P(2): & \underline{path}\ b; c\ \underline{end} \\ & \underline{endprogram} \end{cases}$$

At the beginning we are able to perform events a and b, both simultaneously or one after the other in some arbitrary order. After performing a alone, only event b is enabled; and after first performing b alone, only event a is enabled. After the performance of both a and b, simultaneously or in any order, c is enabled, and after its occurrence the same procedure starts again. What kind of mathematical objects can represent a possible behaviour of Pr_2 ? This object should represent the fact that every occurrence of c must be preceded by appropriate occurrences of a and b, but there are no direct connections between occurrences of a and b. Furthermore, the object should conceptually be as simple as possible. A **(labelled) partial order** (see Appendix A) **of event occurrences** as depicted in Fig. 2.1 seems to be a good candidate for such an object. This partial order simply represents a **causality relation** among event occurrences, and no order between two or more events means that they can occur in any order or simultaneously.

Unfortunately, pure partial orders are rather difficult to work with, and only some of them can be interpreted as behaviours of non-sequential systems. But one may look at the partial order of Fig. 2.1 from another viewpoint. Note that we can unambiguously represent this partial order (see Fig. 2.2) as a **vector of strings:** $(acac, bcbc)$, where the first coordinate represents a behaviour generated by the path $\underline{path}\ a; c\ \underline{end}$, while the second coordinate represents a behaviour generated by the path $\underline{path}\ b; c\ \underline{end}$.

Defining concatenation of string vectors by concatenation of their coordinates and slightly modifying the notation, we may write[2]:

[2] ε denotes the empty string, and clearly $\varepsilon x = x\varepsilon = x$ for every string x.

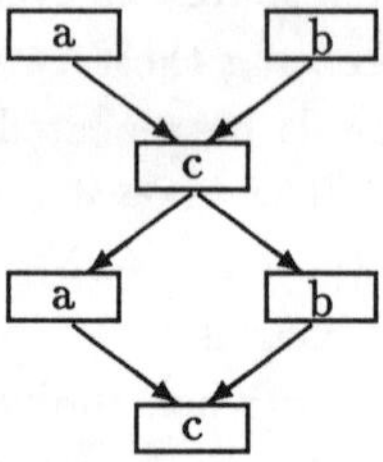

Figure 2.1. Example behaviour of Pr_2

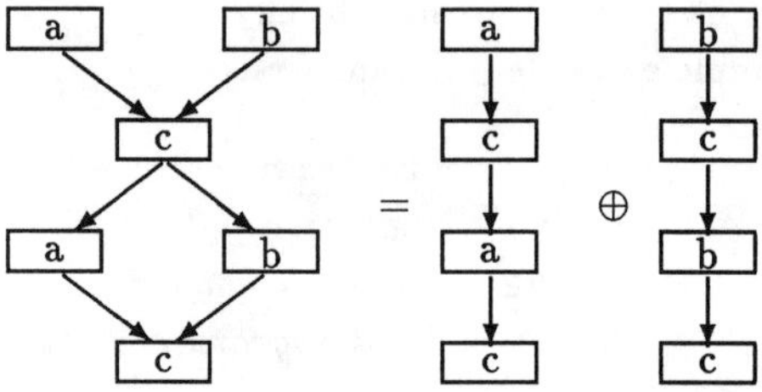

Figure 2.2. Decomposition of example behaviour of Pr_2 into sequential behaviours

$$(acac, bcbc) = \begin{bmatrix} acac \\ bcbc \end{bmatrix} = \begin{bmatrix} a\varepsilon ca\varepsilon c \\ \varepsilon bc\varepsilon bc \end{bmatrix} = \begin{bmatrix} a \\ \varepsilon \end{bmatrix}\begin{bmatrix} \varepsilon \\ b \end{bmatrix}\begin{bmatrix} c \\ c \end{bmatrix}\begin{bmatrix} a \\ \varepsilon \end{bmatrix}\begin{bmatrix} \varepsilon \\ b \end{bmatrix}\begin{bmatrix} c \\ c \end{bmatrix}.$$

Let us denote:

$$\underline{a} = \begin{bmatrix} a \\ \varepsilon \end{bmatrix}, \ \underline{b} = \begin{bmatrix} \varepsilon \\ b \end{bmatrix}, \ \underline{c} = \begin{bmatrix} c \\ c \end{bmatrix}.$$

The vectors $\underline{a}$, $\underline{b}$, $\underline{c}$ are called **vector events**, and we may now write:

$$(acac, bcbc) = \underline{abcabc} = \underline{bacabc} = \underline{abcbac} = \underline{bacbac}.$$

We formally define the vector sequences and their properties in the next section. The vector sequence approach results from the following reasoning. Single paths are nothing but regular expressions and the meaning (behaviour) of regular expressions can be defined in terms of strings generated by these. Path programs are compositions ("vectors") of single paths, so one might have expected that a behaviour of a path program should be characterizable as a composition (a "vector") of behaviours (i.e. strings) of its components. A different, although similar, reasoning is that the regular expressions correspond to finite automata (see Appendix B.4), hence path programs should correspond to appropriate compositions of finite automata. We shall discuss this approach in Sect. 2.3.

2.2 VFS Semantics of COSY

2.2.1 Simple Properties of Vector Sequences

Let $n \geq 1$ be an integer, and let $Ev_1, \ldots, Ev_n$ be non-empty finite sets of elements which we shall call **events**. Define $Ev = Ev_1 \cup \ldots \cup Ev_n$. Intuitively each Ev_i represents a set of events that must occur one at a time. Ev is the set of all events of the system. Consider a Cartesian product $Ev_1^* \times \ldots \times Ev_n^*$. If vectors $(x_1, \ldots, x_n)$ and $(y_1, \ldots, y_n)$ belong to this set, their **concatenation** is defined as:

$$(x_1, \ldots, x_n) \cdot (y_1, \ldots, y_n) = (x_1y_1, \ldots, x_ny_n).$$

For every set $A \subseteq Ev$, let $h_A : Ev^* \longrightarrow A^*$ be the erasing homomorphism defined as the natural extension of:

$$(\forall a \in Ev) h_A(a) = \begin{cases} a & if\ a \in A \\ \varepsilon & if\ a \notin A \end{cases}.$$

For $i = 1, \ldots, n$, we shall write h_i instead of h_{Ev_i}.
Let $\_ : Ev^* \longrightarrow Ev_1^* \times \ldots \times Ev_n^*$ be the mapping defined as follows:

$$(\forall x \in Ev^*)\ \underline{x} = (h_1(x), \ldots, h_n(x)).$$

In other words: $\_ = h_1 \times \ldots \times h_n$. Clearly, we may have $x \neq y$ and $\underline{x} = \underline{y}$.

Example 2.2.1 (See also Fig. 2.2 Let $Ev_1 = \{a, c\}, Ev_2 = \{b, c\}$. Then $\underline{abcabc} = (h_1(abcabc), h_2(abcabc)) = (acac, bcbc)$; and for instance $\underline{bacbac} = (acac, bcbc) = \underline{abcabc}$, while clearly $abcabc \neq bacbac$. ■

The set $Vev = \{\underline{a} | a \in Ev\}$ is called the set of **vector events defined by** $Ev_1, \ldots, Ev_n$. For Example 2.2.1 $Vev = \{\underline{a}, \underline{b}, \underline{c}\}$, where $\underline{a} = (a, \varepsilon), \underline{b} = (\varepsilon, b), \underline{c} = (c, c)$. One may easily show that for every $x, y \in Ev^*, \underline{x} \cdot \underline{y} = \underline{xy}$, thus in the sequel we shall replace "·" by juxtaposition and simply write $(x_1, \ldots, x_n)(y_1, \ldots, y_n)$ or $\underline{x}\underline{y}$ or $\underline{xy}$ instead of $(x_1, \ldots, x_n) \cdot (y_1, \ldots, y_n)$ or $\underline{x} \cdot \underline{y}$. Let $\underline{\varepsilon} = (\varepsilon, \ldots, \varepsilon)$ and let Vev^* denote the submonoid of $Ev_1^* \times \ldots \times Ev_n^*$ generated by Vev and $\underline{\varepsilon}$ under vector concatenation. Formally:

$$Vev^* = \{\underline{a}_1 \ldots \underline{a}_k | k \geq 1, \underline{a}_i \in Vev, i = 1, \ldots, k\} \cup \{\underline{\varepsilon}\}.$$

It is easy to prove that

$$Vev^* = \{\underline{x} | x \in Ev^*\} \subseteq Ev_1^* \times \ldots \times Ev_n^*$$

and, in general, $Vev^* \neq Ev_1^* \times \ldots \times Ev_n^*$. For Example 2.2.1, we have $(cc, c) \in Ev_1^* \times Ev_2^*$ and clearly $(cc, c) \notin Vev^*$.

$ind =$ 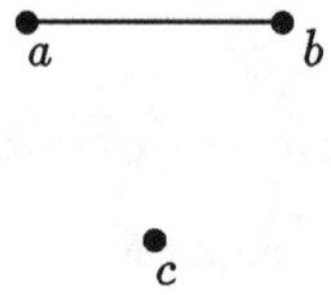

$$Ind = \{\{a\},\{b\},\{c\},\{a,b\}\}$$

Figure 2.3. An example of *ind* and *Ind* for $Ev_1 = \{a,c\}, Ev_2 = \{b,c\}$ (Ex 2.2.1)

For every $i = 1,\ldots,n$, let $pr_i : Ev_1^* \times \ldots \times Ev_n^* \longrightarrow Ev_i^*$ be the projection defined by:

$$(\forall(x_1,\ldots,x_n) \in Ev_1^* \times \ldots \times Ev_n^*)\ pr_i((x_1,\ldots,x_n)) = x_i.$$

Note that for every $\underline{x} \in Ev^*, pr_i(\underline{x}) = h_i(x)$.
Let $ind \subseteq Ev \times Ev$ be the following relation:

$$(\forall a,b \in Ev)(a,b) \in ind :\Leftrightarrow ((\forall i = 1,\ldots,n) a \notin Ev_i \vee b \notin Ev_i).$$

The relation *ind* is called **the independency**, and intuitively, only independent events *can* be executed concurrently.

For every $a \in Ev$, let $i(a) = \{j | a \in Ev_j\}$.

Corollary 2.1 *Let $a,b \in Ev$. The following are equivalent:*
(a) $(a,b) \in ind$
(b) $\underline{ab} = \underline{ba}$ *and* $a \neq b$
(c) $(\forall i = 1,\ldots,n) pr_i(\underline{a}) \neq \varepsilon \Rightarrow pr_i(\underline{b}) = \varepsilon$
(d) $i(a) \cap i(b) = \emptyset$ □

For Example 2.2.1, $i(a) = \{1\}$, $i(b) = \{2\}$, $i(c) = \{1,2\}$ and the independency relation is depicted in Fig. 2.3.

Let $Ind \subseteq 2^{Ev} \setminus \emptyset$ be the following family of sets of events:

$$A \in Ind :\Leftrightarrow ((\forall a,b \in A) a = b \vee (a,b) \in ind).$$

The elements of *Ind* are sets of *pairwise independent events*. If $A = \{a_1,\ldots,a_k\} \in Ind$, then $\underline{a}_1 \ldots \underline{a}_k = \underline{a}_{i_1} \ldots \underline{a}_{i_k}$ for every permutation $i_1,\ldots,i_k$ of $1,\ldots,k$ so that we may write $\underline{A} = \underline{a}_1 \ldots \underline{a}_k$ and $\underline{xA} = \underline{xa}_1 \ldots \underline{a}_k$. For Example 2.2.1, $Ind = \{\{a\},\{b\},\{c\},\{a,b\}\}$.
For every set of vector sequences $X \subseteq Vev^*$,

$$Pref(X) = \{\underline{x} \in Vev^* | (\exists \underline{y} \in Vev^*) \underline{xy} \in X\}.$$

For Example 2.2.1, if $X = \{\underline{abca}, \underline{ab}\}$, then $Pref(X) = \{\underline{\varepsilon}, \underline{a}, \underline{b}, \underline{ab}, \underline{abc}, \underline{abca}\}$.

For every $\underline{x} \in Vev^*$ we shall write $Pref(\underline{x})$ instead of $Pref(\{\underline{x}\})$.
Every element of Vev^* can be uniquely decomposed in a special manner.
For every $\underline{x} \in Vev^* \setminus \{\underline{\varepsilon}\}$, a sequence of sets $(A_1, \ldots, A_k), k \geq 1$, is said to be a **left-normal form** (abbreviated **ln-form**) of $\underline{x}$ if and only if the following conditions are satisfied:

(1) $(\forall i = 1, \ldots, k) A_i \in Ind$

(2) $\underline{x} = \underline{A_1} \ldots \underline{A_k}$

(3) *if* $k > 1$, *then* $(\forall i = 2, \ldots, k)\ (\forall a \in A_i)(\exists b \in A_{i-1})\ \ (a, b) \notin ind.$

The last condition expresses the fact that the decomposition of $\underline{x}$ is "left maximal", in the sense that no event may be pushed to the left, from A_i to A_{i-1}, while preserving the first two conditions, or else that, for any i, A_i is a maximal independent prefix of $\underline{A_i}\underline{A_{i+1}} \ldots \underline{A_k}$.
If $(A_1, \ldots, A_k)$ is a ln-form of $\underline{x}$, we shall write

$$\underline{x} \stackrel{\bullet}{=} \underline{A_1} \ldots \underline{A_k}.$$

For Example 2.2.1, we have $\underline{abac} \stackrel{\bullet}{=} \underline{\{a,b\}\{a\}\{c\}}$, but *not* $\underline{abac} \stackrel{\bullet}{=} \underline{\{a\}\{a,b\}\{c\}}$.
We will prove that every $\underline{x} \in Vev^* \setminus \{\underline{\varepsilon}\}$ has exactly one ln-form. Notice that, if $\underline{x} \stackrel{\bullet}{=} \underline{A_1} \ldots \underline{A_k}$, $\underline{y} \stackrel{\bullet}{=} \underline{B_1} \ldots \underline{B_m}$ and $(\forall b \in B_1)(\exists a \in A_k)(a, b) \notin ind$, then $\underline{xy} \stackrel{\bullet}{=} \underline{A_1} \ldots \underline{A_k}\underline{B_1} \ldots \underline{B_m}$. Conversely, if $\underline{x} \stackrel{\bullet}{=} \underline{A_1} \ldots \underline{A_k}$, then $(\forall i < k)\underline{A_1} \ldots \underline{A_i} \stackrel{\bullet}{=} \underline{A_1} \ldots \underline{A_i}$ and $\underline{A_{i+1}} \ldots \underline{A_k} \stackrel{\bullet}{=} \underline{A_{i+1}} \ldots \underline{A_k}$. We can now show that in a ln-form of $\underline{x}$ the first set is uniquely determined by $\underline{x}$.

Lemma 2.1 *If* $\underline{x} \stackrel{\bullet}{=} \underline{A_1} \ldots \underline{A_k}$, *then*
$A_1 = \{a \in Ev | (\forall i \in i(a))(\exists y \in Ev_i^*) pr_i(\underline{x}) = ay\}$
$= \{a \in Ev | (\exists \underline{y} \in Vev^*) \underline{x} = \underline{ay}\}.$

Proof. Let $A = \{a \in Ev | (\forall i \in i(a))(\exists y \in Ev^*) pr_i(\underline{x}) = ay\}$. Suppose that $a \in A_1$. Since $\underline{x} = \underline{A_1}\underline{x}'$ and $A_1 \in Ind$, we have $\underline{x} = \underline{az}$ for some $\underline{z}$. Hence, $a \in A$ and $A_1 \subseteq A$. If $A_1 \neq A$, let $a \in A \setminus A_1$ and let j be the index of the first A_i containing a, i.e., $a \in A_j$ and $a \notin A_i$ for $i = 1, \ldots, j-1$ $(2 \leq j \leq k)$. This means that there exists $b \in A_{j-1}$, such that $b \neq a$ and $(a, b) \notin ind$. Thus there is $i \in i(a) \cap i(b)$ with $a \in A$, and then $pr_i(\underline{x}) = ybaw$, where for some $i \leq n$, $w \in Ev_i^*$ and $y \in (Ev_i \setminus \{a\})^*$, a contradiction. Hence $A = A_1$. The second equality is a simple consequence of the first one. □

Our existence and unicity result may now be proven by induction on the "length" of the vector considered. Let us define the function $length : Vev^* \longrightarrow \{0, 1, 2, \ldots\}$, by: $length(\underline{\varepsilon}) = 0$, and if $\underline{y} = \underline{a_1} \ldots \underline{a_m}$ where $(\forall i = 1, \ldots, m) a_i \in Ev$, then $length(\underline{y}) = m$.

The definition of length is sound since any other expression of $\underline{y}$ is a permutation of the same a_i's.

Theorem 2.1 (Existence and unicity of left-normal form)
Each $\underline{x} \in Vev^* \setminus \{\underline{\varepsilon}\}$ has exactly one ln-form.

Proof. By induction on $length(\underline{x})$. Let $\underline{x} \in Vev^* \setminus \{\underline{\varepsilon}\}$. If $length(\underline{x}) = 1$, then, of course, it has exactly one ln-form $\{a\}$, where $a \in Ev$ and $\underline{x} = \underline{a}$. Assume that $length(\underline{x}) \geq 2$ and that each $\underline{y} \in Vev^*$ such that $1 \leq length(\underline{y}) < length(\underline{x})$ has exactly one ln-form. We first need to prove that $\underline{x}$ has at least one ln-form.
Let $A = \{a \in Ev | (\exists \underline{y} \in Vev^*)\underline{x} = \underline{ay}\}$. Clearly, $A \in Ind$, and $\underline{x} = \underline{Ay}$ for some $\underline{y} \in Vev^*$. Since $length(\underline{y}) < length(\underline{x})$, by the induction hypothesis, we have $\underline{y} = \underline{\varepsilon}$ or $\underline{y} \doteq \underline{B_1} \ldots \underline{B_k}$. If $\underline{y} = \underline{\varepsilon}$ then $\underline{x} \doteq \underline{A}$.
Suppose $\underline{y} \neq \underline{\varepsilon}$. Let $b \in B_1$. Suppose that $(\exists a \in A)(a, b) \in ind$. Then $\underline{x} = \underline{b}\ \underline{z}$ for some $\underline{z} \in Vev^*$, a contradiction with $b \notin A$. Consequently, we have $\underline{x} \doteq \underline{AB_1} \ldots \underline{B_k}$.
We still have to prove that the ln-form is unique. Suppose that $\underline{x} \doteq \underline{A_1} \ldots \underline{A_p}$ and $\underline{x} \doteq \underline{B_1} \ldots \underline{B_m}$. By Lemma 2.1 we have $A_1 = B_1$. If $p = 1$, this ends the proof. Otherwise, let $\underline{y} = \underline{A_2} \ldots \underline{A_p}$; thus, $\underline{y} \doteq \underline{A_2} \ldots \underline{A_p}$, and $\underline{y} \doteq \underline{B_2} \ldots \underline{B_m}$. Since $length(\underline{y}) < length(\underline{x})$, by the induction hypothesis, we obtain $A_i = B_i$ for $i = 2, \ldots, p$ and $p = m$. □

Now we shall show that for ln-forms the prefix relation corresponds to an inclusion property.

Theorem 2.2 *If $\underline{y} \in Vev^*$, $\underline{x} \in Pref(\underline{y})$, $\underline{x} \doteq \underline{A_1} \ldots \underline{A_k}$ and $\underline{y} \doteq \underline{B_1} \ldots \underline{B_m}$, then $k \leq m$ and $A_i \subseteq B_i$ for $i = 1, \ldots, k$.*

Proof. Let us first consider the simple case where $\underline{y} = \underline{xa}$ for some $a \in Ev$. We have to consider three cases.
Case 1: $(\forall b \in A_i)(a, b) \in ind$ for $i = 1, \ldots, k$. Then $\underline{y} \doteq \underline{B_1} \ldots \underline{B_k}$, where $B_1 = A_1 \cup \{a\}, B_i = A_i$ for $i \neq 1$.
Case 2: $(\forall b \in A_k)(a, b) \in ind$ and $(\exists c \in A_j)(a, c) \notin ind$ for some $j < k$. Let j be the last such index, i.e., $(\exists c \in A_j)(a, c) \notin ind$ and $(\forall b \in A_i)(a, b) \in ind$ for $i = j + 1, \ldots, k$. Then $\underline{y} \doteq \underline{B_1} \ldots \underline{B_k}$, where $B_{j+1} = A_{j+1} \cup \{a\}$ and $B_i = A_i$ for $i \neq j + 1$.
Case 3: $(\exists c \in A_k)(a, c) \notin ind$. Then $\underline{y} \doteq \underline{A_1} \ldots \underline{A_k}\underline{\{a\}}$.
The general case $\underline{y} = \underline{x}\ \underline{a_1} \ldots \underline{a_r}$, where $a_i \in Ev$, results from r successive applications of the simple case. □

The last two theorems play an important part in analyzing different execution strategies, the problem we shall discuss in Section 2.5.
Now we prove some simple but frequently used properties of Vev^*.
For every string $x \in Ev^*$ and every $a \in Ev$, let $\#_a(x)$ denote the number of occurrences of a in x. For instance, $\#_a(cbcabac) = 2$ and $\#_c(cbcabac) = 3$. See also Appendix B.

Lemma 2.2
(1) $(\forall a \in Ev)(\forall j \in i(a))(\forall x \in Ev^*))\ \#_a(x) = \#_a(h_j(x))$.
(2) $(\forall a \in Ev)(\forall x, y \in Ev^*)\underline{x} = \underline{y} \Rightarrow \#_a(x) = \#_a(y)$.

Proof. (1) If $j \in i(a)$ then $a \in Ev_j$ and h_j does not erase any a from x.
(2) If $\underline{x} = \underline{y}$, then $h_j(x) = h_j(y)$ for every $j = 1, \ldots, n$. Thus, by (1), $\#_a(x) = \#_a(h_j(x))$ where $j \in i(a)$, and $\#_a(x) = \#_a(h_j(x)) = \#_a(h_j(y)) = \#_a(y)$. □

For every vector sequence $\underline{x} \in Ev^*$, and every $a \in Ev$, let $\#_a(\underline{x}) = \#_a(x)$. The Lemma 2.2 (2) guarantees that the definition is sound.

Lemma 2.3
(1) Let $y_1, \ldots, y_n$, $x_1, \ldots, x_n \in Ev^*$ *satisfy the following:*

(a) $(\forall i = 1, \ldots, n)x_i \in Pref(y_i)$

(b) $(\forall a \in Ev)(\forall i, j = 1, \ldots, n)a \in Ev_i \cap Ev_j \Rightarrow \#_a(x_i) = \#_a(x_j)$.

Then: $(y_1, \ldots, y_n) \in Vev^* \Rightarrow (x_1, \ldots, x_n) \in Vev^*$.

(2) Let $x_1, \ldots, x_n \in Ev^*$, $a \in Ev$, $i(a) = \{j\}$, $x_j = uv$ *where* $u, v \in Ev^*$.

Then: $(x_1, \ldots, x_n) \in Vev^* \Leftrightarrow (x_1, \ldots, x_{j-1}, uav, x_{j+1}, \ldots, x_n) \in Vev^*$.

Proof.
(1) For every $a \in Ev$, let $n_a = \#_a(x_i)$, some $i \in i(a)$. The condition (b) guarantees that the definition of n_a is sound. Let $\underline{y} = (y_1, \ldots, y_n)$ and let x be a string obtained from y by deleting the last $\#_a(y) - n_a$ instances of a for each $a \in Ev$. Clearly $\underline{x} \in Vev^*$ and $h_i(x) = x_i$ for $i = 1, \ldots, n$. Hence $\underline{x} = (x_1, \ldots, x_n) \in Vev^*$.
(2) Assume that $uav = b_1, \ldots, b_k a b_{k+1}, \ldots, b_m$, $b_i \in Ev$ with $k = 0 \Rightarrow u = \varepsilon$ and $k = m \Rightarrow v = \varepsilon$.

($\Rightarrow$) $(x_1, \ldots, x_n) \in Vev^*$ implies that $(\exists x \in Ev^*)\underline{x} = (x_1, \ldots, x_n)$.
Since $x_i = uv$ then $x = z_0 b_1 z_1 b_2 \ldots z_m b_m z_{m+1}$, where $z_i \in Ev^*$.
Define $x' = z_0 b_1 z_1 \ldots z_k b_k a z_k b_{k+1} \ldots b_m z_{m+1}$. Clearly $\underline{x'} \in Vev^*$ and $\underline{x}' = (x_1, \ldots, x_{j-1}, uav, x_{j+1}, \ldots, x_n)$.

($\Leftarrow$) If $(x_1, \ldots, x_{j-1}, uav, x_{j+1}, \ldots, x_n) \in Vev^*$ then
$(x_1, \ldots, x_{j-1}, uav, x_{j+1}, \ldots, x_n) = \underline{x}'$, and clearly $\underline{x} = (x_1, \ldots, x_n)$.

□

For every $x \in Ev^*$, let $Ev(x)$ denote the set of all events occurring in x, i.e., $Ev(aba) = \{a, b\}$, and similarly for every $\underline{x} \in Vev^*$, let $Ev(\underline{x})$ denote the set of all events occurring in $\underline{x}$. Clearly $Ev(x) = Ev(\underline{x})$ for every $x \in Ev^*$.

We mentioned at the beginning of this chapter that we may have $\underline{x} = \underline{y}$ and $x \neq y$. We now show what conditions should x and y satisfy to ensure the equality $\underline{x} = \underline{y}$.
For every $\underline{x} \in Vev^*$, let $lin(\underline{x}) = \{y \in Ev^* | \underline{x} = \underline{y}\}$. Elements of $lin(\underline{x})$ are called **linearizations** of $\underline{x}$.
Let $\hat{-} \subseteq Ev^* \times Ev^*$ be the following relation:

$$(\forall x, y \in Ev^*) x \hat{-} y \quad :\Leftrightarrow (\exists x_1, x_2 \in Ev^*)(\exists a, b \in Ev)(a, b) \in ind \\ \wedge\, x = x_1 ab x_2 \wedge y = x_1 ba x_2.$$

Let $\hat{=}$ be the transitive closure of $\hat{-}$ ($\hat{=} = (\hat{-})^*$), i.e.,

$$x \hat{=} y \Leftrightarrow x = y \vee ((\exists x_1, \ldots, x_k) x \hat{-} x_1 \hat{-} x_2 \hat{-} \cdots \hat{-} x_k = y).$$

Clearly $\hat{=}$ is an equivalence relation. Let $[x]_{\hat{=}}$ denote the equivalence class of $\hat{=}$ containing x (see Definitions A.3 and A.4 in Appendix A).

Theorem 2.3 (Equality of Vector Sequences)
Let $x, y \in Ev^*$. The following are equivalent:
(1) $\underline{x} = \underline{y}$
(2) $x \hat{=} y$
(3) $[(\forall a \in Ev)\ \#_a(x) = \#_a(y)] \wedge [(\forall (a,b) \notin ind) h_{\{a,b\}}(x) = h_{\{a,b\}}(y)]$.

Proof. (2 $\Rightarrow$ 1): Let $x \hat{-} y$, i.e., $x = x_1 ab x_2$, $y = x_1 ba x_2$, $(a,b) \in ind$. Since $(a,b) \in ind$ then for every $i = 1, \ldots, n$, either $a \notin Ev_i$ or $b \notin Ev_i$. If $a \notin Ev_i$ then $h_i(x) = h_i(x_1 ab x_2) = h_i(x_1 b x_2) = h_i(x_1 ba x_2) = h_i(y)$, and similarly for $b \notin Ev_i$. Thus $x \hat{-} y$ implies $\underline{x} = \underline{y}$. Hence, since $\hat{=} = (\hat{-})^*$ we obtain (2) $\Rightarrow$ (1).
(1 $\Rightarrow$ 3): The equality $\#_a(x) = \#_a(y)$ follows from Lemma 2.2 (2). Let $\underline{x} = \underline{y}$ and $(a,b) \notin ind$. This means: $(\exists i)\{a,b\} \subseteq Ev_i$. Thus: $h_{\{a,b\}}(x) = h_{\{a,b\}}(h_i(x)) = h_{\{a,b\}}(h_i(y)) = h_{\{a,b\}}(y)$.
(3 $\Rightarrow$ 2): By induction on x. For $x = \varepsilon$ and $x = a \in Ev$ the implication is obviously true. Suppose that this implication is true for x' satisfying $x = x'a$ for $a \in Ev$, i.e.,
$(\forall y' \in Ev^*)[(\forall a \in Ev)\ \#_a(x') = \#_a(y')] \wedge [(\forall (a,b) \notin ind) h_{\{a,b\}}(x') = h_{\{a,b\}}(y')] \Rightarrow x' \hat{=} y'$.
We have $\#_a(x) = \#_a(y)$ which means that $\#_a(y) \geq 1$, i.e., $y = y_1 a y_2$ where $y_1, y_2 \in Ev^*$ and $a \notin Ev(y_2)$.
Then Case 1: $y_2 = \varepsilon$, i.e., $y = y_1 a$, and clearly $y \hat{=} y_1 y_2 a$.
Case 2: $y_2 = a_1 \ldots a_r$. If $(a, a_k) \notin ind$ for some $k \leq r$, then $h_{\{a,a_k\}}(x) = ta$, $h_{\{a,a_k\}}(y) = u a_k \ldots a_k$, where $t, u \in \{a, a_k\}^*$, i.e., $h_{\{a,a_k\}}(x) \neq h_{\{a,a_k\}}(y)$. Thus $(\forall k = 1, \ldots, r)(a, a_k) \in ind$, but this means that $y \hat{=} y_1 y_2 a$.
Hence in both cases, $y \hat{=} y'a$ for some $y' \in Ev^*$. Since we have already proved (2) $\Rightarrow$ (1) $\Rightarrow$ (3), then $y \hat{=} y'a$ implies that (3) holds also for y and $y'a$. Thus

we have $(\forall b \in Ev)\ \#_b(x'a) = \#_b(y) = \#_b(y'a)$, which implies $\#_b(x') = \#_b(y')$, and $(\forall(b,c) \notin ind)h_{\{b,c\}}(x'a) = h_{\{b,c\}}(y) = h_{\{b,c\}}(y'a)$, which implies $h_{\{b,c\}}(x') = h_{\{b,c\}}(y')$, i.e., the condition (3) is satisfied by x' and y'. Hence, by the induction hypothesis, $x' \hat{=} y'$. From the definition of $\hat{-}$ and the fact that $\hat{=} = (\hat{-})^*$ we have $x'a \hat{=} y'a$. This and $y'a \hat{=} y$, yields $x \hat{=} y$ since $\hat{=}$ is transitive. □

Corollary 2.2 $(\forall x \in Ev^*)[x]_{\hat{=}} = lin(\underline{x})$ □

2.2.2 Vector Sequences and Partial Orders

Let $Ev_1, \ldots, Ev_n$ be sets of events, $Ev = Ev_1 \cup \ldots \cup Ev_n$, Vev be defined by $Ev_1, \ldots, Ev_n$, and $\underline{x} \in Vev^*$. In this chapter we define the partial order $PO_{\underline{x}}$ corresponding to $\underline{x}$. For the definition and some properties of partial and total orders see Appendix A.

Let us define $EO = Ev \times \{1, 2, \ldots\}$. The set EO will be called the set of **event occurrences.** Usually we shall write $a^{(i)}$ to denote an element (a, i) for elements of EO.

One can observe that every string $x \in Ev^*$ generates, in a very natural way, a unique total order $TO_x \subseteq EO \times EO$. For instance, $x = abacba$ generates $TO_x = a^{(1)} \longrightarrow b^{(1)} \longrightarrow a^{(2)} \longrightarrow c^{(1)} \longrightarrow b^{(2)} \longrightarrow a^{(3)}$.

Formally, the **total order** TO_x may be defined as follows:

$$a^{(i)} TO_x b^{(j)} :\Leftrightarrow (\#_a(x) \geq i \wedge a^{(i)} = b^{(j)})$$

$$\vee((\exists x_1, x_2, x_3 \in Ev^*) x = x_1 a x_2 b x_3 \wedge \#_a(x_1 a) = i \wedge \#_b(x_1 a x_2 b) = j).$$

We left to the reader an easy proof that TO_x is indeed a total order on its domain, as well as the proof that $x = y \Leftrightarrow TO_x = TO_y$.

We now are going to apply the construction presented above to sequences of vector events, i.e., elements of Vev^*. Since it might happen that $\underline{ab} = \underline{ba}$, the orders associated with the $\underline{x}$'s from Vev^* shall rather be *partial* than total. On the other hand, the string x may be interpreted as a linear representation of $\underline{x}$, thus the partial order defined by $\underline{x}$ should be extensible to TO_x. This leads to the following definition.

For every $\underline{x} \in Vev^*$, its **canonical partial order** $PO_{\underline{x}} \subseteq EO \times EO$ is the least partial order $PO \subseteq EO \times EO$ satisfying:

$$a^{(i)} PO b^{(j)} \Leftrightarrow (\forall y \in lin(\underline{x})) a^{(i)} TO_y b^{(j)}$$

or, in other words:

$$PO_{\underline{x}} = \bigcap_{\underline{y} = \underline{x}} TO_y.$$

The partial order $PO_{\underline{x}}$ represents a causality relation among event occurrences, no order means that events may occur in any order or simultaneously. We show that every $PO_{\underline{x}}$ is consistent with *ind*, i.e., no order between event occurrences holds only for independent events. Recall from Appendix A (Definition A.6) that $\overline{PO_{\underline{x}}} = Dom(PO_{\underline{x}}) \times Dom(PO_{\underline{x}}) \setminus (PO_{\underline{x}} \cup PO_{\underline{x}}^{-1})$.

Lemma 2.4 *For every* $\underline{x} \in Vev^* : a^{(i)}\overline{PO_{\underline{x}}}b^{(j)} \Rightarrow (a,b) \in ind$.

Proof. $a^{(i)}\overline{PO_{\underline{x}}}b^{(j)}$ means that $\neg(a^{(i)}PO_{\underline{x}}b^{(j)}) \wedge \neg(b^{(j)}PO_{\underline{x}}a^{(i)})$, hence there are $y,z \in lin(\underline{x})$ such that $a^{(i)}TO_y b^{(j)}$ and $b^{(j)}TO_z a^{(i)}$. Assume that $(a,b) \notin ind$. Since $\underline{x} = \underline{y} = \underline{z}$, from Theorem 2.3 it follows that $h_{\{a,b\}}(y) = h_{\{a,b\}}(z)$. But from $a^{(i)}TO_y b^{(j)}$ we have that in y the i-th instance of a occurs before the j-th instance of b, while from $b^{(j)}TO_z a^{(i)}$ we have that in z the i-th instance of a occurs after the j-th instance of b, thus clearly $h_{\{a,b\}}(y) \neq h_{\{a,b\}}(z)$, a contradiction. Hence $(a,b) \in ind$. □

The opposite of the lemma is not true. In Example 2.2.1 we have $(a,b) \in ind$. Consider $\underline{x} = \underline{acb} = (ac, cb)$. In this case $PO_{\underline{x}} = TO_x$ and clearly $a^{(1)}PO_{\underline{x}}b^{(1)}$. Now we show that $PO_{\underline{x}}$, where $\underline{x} = (x_1, \ldots, x_n)$ can be obtained as a composition of total orders $TO_{x_1}, \ldots, TO_{x_n}$, i.e. $PO_{\underline{x}}$ is exactly the partial order intuitively depicted in Fig.2. Recall that for any partial (i.e. also total) order R, $Dom(R) = \{x|(\exists y)\ xRy\} \cup \{y|(\exists x)\ xRy\}$ (Definition A.6 in Appendix A).

Lemma 2.5

(1) If $x = yabz, w = ybaz$ *and* $i(a) \cap i(b) = \emptyset$*, where* $y,z \in Ev^*$ *and* $a,b \in Ev$*, then* $\underline{x} = \underline{w}$.

(2) If $x = yavbz$ *and* $i(a) \cap i(b) \neq \emptyset$ *then* $a^{(i)}PO_{\underline{x}}b^{(j)}$ *where* $i = \#_a(ya)$ *and* $j = \#_b(yavb)$.

(3) If $a^{(i)}\widehat{PO}_{\underline{x}}b^{(j)}$ *then* $i(a) \cap i(b) \neq \emptyset$ *(see Definition A.9 in Appendix A).*

Proof. (1) Follows from (2⇒1) in the proof of Theorem 2.3.
(2) Clearly, $\neg(b^{(j)}PO_{\underline{x}}a^{(i)})$. If also $\neg(a^{(i)}PO_{\underline{x}}b^{(j)})$ then, by Lemma 2.4, $(a,b) \in ind$, contradicting $i(a) \cap i(b) \neq \emptyset$.
(3) Suppose $i(a) \cap i(b) = \emptyset$. Clearly $a \neq b$. From $a^{(i)}PO_{\underline{x}}b^{(j)}$ it follows that each $y \in lin(\underline{x})$ can be represented as $y = zaa_1 \ldots a_{k(y)}bw$ and $\#_a(za) = i$ and $\#_b(zaa_1 \ldots a_{k(y)}b) = j$. Let $Y = \{y \in lin(\underline{x}) | k(y) \leq k(v) \text{ for all } v \in lin(\underline{x})\}$. Clearly for all $y \in Y$, $k(y)$ is the same so let us set $k = k(y)$, $y \in Y$. We have $k \geq 1$. Indeed, if $k = 0$, i.e. $y = zabw \in Y$, then by Lemma 2.5(1), $y' = zbaw \in lin(\underline{x})$, which together with $\#_a(za) = i$ and $\#_b(zab) = j$, contradicts $a^{(i)}PO_{\underline{x}}b^{(j)}$ (notice $a \neq b$). Hence $k \geq 1$.
Let $y = zaa_1 \ldots a_k bw \in Y$ and $\#_a(za) = i$ and $\#_b(zaa_1 \ldots a_k b) = j$.
Denote $y = za_0a_1 \ldots a_k a_{k+1}w$ (i.e. $a_0 = a$ and $a_{k+1} = b$) and $\#_l = \#_{a_l}(za_0 \ldots a_l)$

for $l = 0, 1, \ldots, k+1$. We observe that for every $1 \leq s \leq k$ there is $t \in \{s+1, \ldots, k+1\}$ such that $i(a_s) \cap i(a_t) \neq \emptyset$. Indeed, otherwise, by Lemma 2.5(1), $y' = zaa_1 \ldots a_{s-1}a_{s+1}a_{s+2} \ldots a_k ba_s w \in lin(\underline{x})$. This, however, contradicts the minimality of k. Furthermore, $i(a) \cap i(a_1) \neq \emptyset$ for the same reason (one just takes $y' = za_1 aa_2 \ldots a_k bw$).
Hence there is a sequence $0 = i_1 < i_2 < \ldots < i_m = k+1, m \geq 3$, such that $i(a_{i_\gamma}) \cap i(a_{i_{\gamma+1}}) \neq \emptyset$ for $\gamma = 1, \ldots, m-1$.
Hence, by Lemma 2.5(2), $a_{i_\gamma}^{(\#i_\gamma)} PO_{\underline{x}} a_{i_{\gamma+1}}^{(\#i_{\gamma+1})}$, for $\gamma = 1, \ldots, m-1$.
Hence we have $a^{(i)} PO_{\underline{x}} a_{i_2}^{(\#i_2)} PO_{\underline{x}} b^{(j)}$.
Clearly $a^{(i)} \neq a_{i_2}^{(\#i_2)} \neq b^{(j)}$. This means that we obtained a contradiction with $a^{(i)} \widehat{PO}_{\underline{x}} b^{(j)}$. □

Theorem 2.4 (Representation of $PO_{\underline{x}}$)
$PO_{\underline{x}} = (TO_{x_1} \cup \ldots \cup TO_{x_n})^*$, for every $\underline{x} = (x_1, \ldots, x_n) \in Vev^*$.

Proof. From Lemma 2.5(3), $\widehat{PO}_{\underline{x}} \subseteq TO_{x_1} \cup \ldots \cup TO_{x_n}$. Hence $PO_{\underline{x}} \subseteq (TO_{x_1} \cup \ldots \cup TO_{x_n})^*$ since $PO_{\underline{x}}$ is finite (by Lemma A.1 in Appendix A).
From Lemma 2.5(2), $TO_{x_1} \cup \ldots \cup TO_{x_n} \subseteq PO_{\underline{x}}$. Hence $(TO_{x_1} \cup \ldots \cup TO_{x_n})^* \subseteq PO_{\underline{x}}$ since $PO_{\underline{x}}$ is transitive and reflexive. □

A proof of a unique representation of $\underline{x}$ by $PO_{\underline{x}}$ can easily be derived from Theorem 2.4.

Theorem 2.5 *$\underline{x} = \underline{y} \Leftrightarrow PO_{\underline{x}} = PO_{\underline{y}}$, for all $\underline{x}, \underline{y} \in Vev^*$.*

Proof. Let $\underline{x} = (x_1, \ldots, x_n)$ and $\underline{y} = (y_1, \ldots, y_n)$.
($\Rightarrow$) Obvious, from $PO_{\underline{x}}$ definition.
($\Leftarrow$) If $x_l \neq y_l$ for some l then there are $a, b \in Ev_l$ such that $a^{(i)} TO_{x_l} b^{(j)}$ and $b^{(j)} TO_{y_l} a^{(i)}$ and $a^{(i)} \neq b^{(j)}$. This and Theorem 2.4 yields $a^{(i)} PO_{\underline{x}} b^{(j)}$ and $b^{(j)} PO_{\underline{y}} a^{(i)}$, a contradiction with $PO_{\underline{x}} = PO_{\underline{y}}$. □

The last result of this section characterizes the membership of Vev^* in terms of partial orders.

Theorem 2.6 (Membership of Vev^*)
Let $(x_1, \ldots, x_n) \in Ev_1^* \times \ldots \times Ev_n^*$.
$(x_1, \ldots, x_n) \in Vev^* \Leftrightarrow$
(1) $(\forall i, j = 1, \ldots, n)(\forall a \in Ev_i \cap Ev_j)\#_a(x_i) = \#_a(x_j)$
(2) $(TO_{x_1} \cup \ldots \cup TO_{x_n})^*$ is a partial order.

Proof. ($\Rightarrow$) Let $\underline{x} = (x_1, \ldots, x_n) \in Vev^*$. Then for every $i = 1, \ldots, n$, $a \in Ev_i \Rightarrow \#_a(x) = \#_a(x_i)$, which implies (1). The condition (2) follows directly from Theorem 2.4.

($\Leftarrow$) From Lemma A.3 (Appendix A) it follows that there is a total order TO such that $Dom(TO) = Dom((TO_{x_1} \cup \ldots \cup TO_{x_n})^*)$ and $(TO_{x_1} \cup \ldots \cup TO_{x_n})^* \subseteq TO$. Clearly there is $x \in Ev^*$ such that $TO = TO_x$. Thus $(TO_{x_1} \cup \ldots \cup TO_{x_n})^* \subseteq TO_x$. Let $EO_i = Ev_i \times \{1, 2, \ldots\}$ for $i = 1, \ldots, n$. We have

$$(\forall i = 1, \ldots, n)\ TO_{x_i} = (TO_{x_1} \cup \ldots \cup TO_{x_n})^* \cap EO_i \times EO_i$$

$$\subseteq TO_x \cap EO_i \times EO_i = TO_{h_i(x)}.$$

Hence $TO_{x_i} \subseteq TO_{h_i(x)}$, for each i. Assume that $TO_{x_i} \neq TO_{h_i(x)}$. This means that $Dom(TO_{x_i}) \subset Dom(TO_{h_i(x)})$ and $Dom(TO_{x_i}) \neq Dom(TO_{h_i(x)})$. Let $a^{(k)} \in Dom(TO_{h_i(x)}) \setminus Dom(TO_{x_i})$. Since $a^{(k)} \in Dom(TO_x)$, and $Dom(TO_x) = Dom(TO_{x_1} \cup \ldots \cup TO_{x_n}) = Dom(TO_{x_1}) \cup \ldots \cup Dom(TO_{x_n})$, then there is $j \neq i$ such that $a^{(k)} \in Dom(TO_{x_j})$, which means that $a \in Ev_i \cap Ev_j$. But $a^{(k)} \in Dom(TO_{x_j}) \Rightarrow \#_a(x_j) \geq k$, while $a^{(k)} \notin Dom(TO_{x_i}) \Rightarrow \#_a(x_i) < k$, a contradiction with (1). Thus $TO_{x_i} = TO_{h_i(x)}$, i.e., $x_i = h_i(x)$, or in other words $\underline{x} = (x_1, \ldots, x_n)$. □

2.2.3 Vector Sequences of Path Programs

In this section we show how path programs can generate vector sequences and how certain properties of path programs can be defined in terms of such vector sequences. Furthermore, we shall prove some elementary properties of vector sequences generated by path programs.

A path P = <u>path</u> *sequence* <u>end</u> determines a set of *sequences* of occurrences of events, according to the interpretation of the symbols ";",",","*", "<u>path</u>" and "<u>end</u>", similarly to the way a classical regular expression determines a regular language (see Appendix B.3).

Let us say that a string is a "path", "sequence", "orelement", "starelement","element" or "event", if it may be generated from a non-terminal *path*, *sequence*, *orelement*, *starelement*, *element* or *event* by the production rules of the basic COSY path program grammar from Sect. 2.1. Thus "$(a; b, c), (d; e)$" is an "orelement" (and also a "sequence"), "$a; b, c$" is a "sequence", "$(d; e)*$" is a "starelement" (and also an "orelement" and a "sequence").

We may now consider a path P = <u>path</u> *sequence* <u>end</u> to define a set of strings whose elements are events of the path.

To a string E being either a path, sequence, starelement, element or event, respectively, we associate a set of strings $Cyc(E)$, defined recursively by the following rules:

1. $Cyc(\underline{path}\ E\ \underline{end}) = Cyc(E)$
2. $Cyc(E_1; \ldots; E_k) = Cyc(E_1) \ldots Cyc(E_k)$
3. $Cyc(E_1, \ldots, E_k) = Cyc(E_1) \cup \ldots \cup Cyc(E_k)$

4. $Cyc(E*) = Cyc(E)^*$
5. $Cyc((E)) = Cyc(E)$
6. $Cyc(a) = \{a\}$, if a is an event.

For a path P = <u>path</u> *sequence* <u>end</u>, $Cyc(P)$ represents the set of all full sequences of event occurrences permitted by the sequencing constraints of P ignoring the cyclicity rule. In other words, $Cyc(P)$ is nothing but a regular language defined by "*sequence*" from P treated as a regular expression. The set of strings $Cyc(P)$ is called the set of **cycles** of P. For instance for the path program Pr_1 from Sect. 2.1 we have:

$$Cyc(P_1) = \{ab, ac\},\ Cyc(P_2) = \{de\}^*\{b\}.$$

From the set $Cyc(P)$ we construct the set of **finite firing sequences of** P, denoted by $FS(P)$, as follows (see Appendix B for the definition of prefix $Pref$):

$$FS(P) = Pref(Cyc(P)^*).$$

The set $FS(P)$ is the set of sequences of event occurrences permitted by the path P. For the path program Pr_1, we have:

$$FS(P_1) = \{ab, ac\}^*\{a, \varepsilon\}$$

$$FS(P_2) = (\{de\}^*\{b\})^*\{de\}^*\{d, \varepsilon\}.$$

Let Pr = <u>program</u> $P_1 \ldots P_n$ <u>endprogram</u> (or simply $Pr = P_1 \ldots P_n$) be a path program, fixed for the rest of this section except for examples, where the P_i's are paths.

With every path P_i, we associate its set of events denoted by $Ev(P_i)$, and define $Ev(Pr) = Ev(P_1) \cup \ldots \cup Ev(P_n)$, $Vev(Pr)$ is the set of vector events defined by $Ev(P_1), \ldots, Ev(P_n)$.

For example Pr_1 we have $Ev(P_1) = \{a, b, c\}$, $Ev(P_2) = \{b, d, e\}$, $Ev(Pr_1) = \{a, b, c, d, e\}$, $Vev(Pr_1) = \{\underline{a}, \underline{b}, \underline{c}, \underline{d}, \underline{e}\} = \{(a, \varepsilon), (b, b), (c, \varepsilon), (\varepsilon, d), (\varepsilon, e)\}$.

In the sequel, if it does not lead to misunderstanding, we shall write Ev_i, Cyc_i, FS_i, Ev, Vev instead of $Ev(P_i)$, $Cyc(P_i)$, $FS(P_i)$, $Ev(Pr)$, $Vev(Pr)$, respectively.

A vector of strings $(x_1, \ldots, x_n)$ is a possible behaviour of Pr if each x_i for $i = 1, \ldots, n$ is a possible firing sequence of P_i and, furthermore, if the x_i's agree about the number and order of occurrences of events they share.

Let us denote by VFS (or $VFS(Pr)$ if the explicit name Pr is necessary) the set of all vectors $(x_1, \ldots, x_n)$ that might be produced by Pr. Intuitively, we may regard elements of VFS as describing histories as follows. There are two kinds of constraint imposed on a system specified by Pr. First, in any history of the system, for every i, the events mentioned in P_i must occur in an order permitted by P_i, and the P_i's together define the only sequential constraints. A history may then be completely described by listing the sequences, one for each P_i, of event occurrences from Ev_i, that is, it may be completely described by some element of $FS_1 \times \ldots \times FS_n$. Secondly, the sequences must be consistent with each other. Thus if $a, b \in Ev_i \cap Ev_j$ and $(x_1, \ldots, x_n)$ is a history, then the k-th occurrence of a in x_i precedes the m-th occurrence of b in x_i if and only if the k-th occurrence of a in x_j precedes the m-th occurrence of b in x_j. Additionally there must be no *cycle* in the graph specified by $TO_{x_1} \cup \ldots \cup TO_{x_n}$, which means that $(TO_{x_1} \cup \ldots \cup TO_{x_n})^*$ can be interpreted as causal relationships among event occurrences. Theorem 2.6 says that it happens only if $(x_1, \ldots, x_n) \in Vev^*$.

Thus we define:

$$VFS = (FS_1 \times \ldots \times FS_n) \cap Vev^*.$$

The set VFS is called the set of **vector firing sequences of** Pr and it *represents the behaviour of* Pr.

For example Pr_2 from Sect. 2.1, $VFS(Pr_2) = \{\underline{abc}\}^*\{\underline{a}, \underline{b}, \underline{ab}, \underline{\varepsilon}\}$, where $\underline{a} = (a, \varepsilon)$, $\underline{b} = (\varepsilon, b)$, $\underline{c} = (c, c)$.

Since $VFS \subseteq Vev^*$ we may apply the whole theory of Sections 2.2.1 and 2.2.2, in particular we can use *ind*, the independency relation defined by $Ev_1, \ldots, Ev_n$; *Ind*, the set of sets of independent events; and for every $\underline{x} \in VFS$ we can define $PO_{\underline{x}}$. Sometimes we shall write *ind*(Pr), *Ind*(Pr) if the explicit name Pr is necessary.

Note that $(a, b) \in ind$ *does not* imply that a and b may ever occur simultaneosly, i.e. there are $\underline{x} \in VFS$ and $i, j \geq 1$ such that $(a^{(i)}, b^{(j)}) \notin PO_{\underline{x}}$. Consider the path program Pr_3 defined as:

$$Pr_3 = \begin{cases} P(1): & \underline{path}\ a; b\ \underline{end} \\ P(2): & \underline{path}\ b; c\ \underline{end} \end{cases}$$

Clearly $(a, c) \in ind(Pr_3)$, but $VFS(Pr_3) = \{\underline{abc}\}^*\{\underline{a}, \underline{ab}, \underline{aba}, \underline{\varepsilon}\}$, where $\underline{a} = (a, \varepsilon), \underline{b} = (b, b), \underline{c} = (\varepsilon, c)$. Thus a and c will *never* be performed simultaneously by Pr_3, since for every $\underline{x} \in VFS(Pr_3)$, $PO_{\underline{x}}$ is a total order.
Now, we describe some simple properties of VFS.

Lemma 2.6
(1) $Pref(VFS) = VFS$.
(2) $VFS = \{\underline{x} \in Vev^* | pr_i(\underline{x}) \in FS_i, i = 1, \ldots, n\}$.

Proof. (1) Clearly $VFS \subseteq Pref(VFS)$. Let $\underline{x} \in Pref(VFS)$, i.e., $(\exists \underline{y} \in Vev^*)\underline{xy} \in VFS$. Let $\underline{x} = (x_1, \ldots, x_n)$, $\underline{y} = (y_1, \ldots, y_n)$. From the definition of VFS it follows that $x_i y_i \in FS_i$, for $i = 1, \ldots, n$. Since $FS_i = Pref(Cyc_i^*)$ then $Pref(FS_i) = FS_i$ and $x_i y_i \in FS_i$ implies $x_i \in FS_i$ for $i = 1, \ldots, n$. Hence

$(x_1, \ldots, x_n) \in FS_1 \times \ldots \times FS_n$. On the other hand, from the definition of $Pref$, $\underline{x} \in Pref(VFS) \Rightarrow \underline{x} \in Vev^*$, which means that $\underline{x} \in VFS = (FS_1 \times \ldots \times FS_n) \cap Vev^*$.
(2) $\underline{x} \in VFS \Leftrightarrow \underline{x} \in (FS_1 \times \ldots \times FS_n) \cap Vev^* \Leftrightarrow (\forall i) pr_i(\underline{x}) \in FS_i \wedge \underline{x} \in Vev^* \Leftrightarrow \underline{x} \in \{\underline{y} \in Vev^* | pr_i(\underline{y}) \in FS_i, i = 1, \ldots, n\}$. □

Another way of describing the set VFS is by a *transition* relation, i.e., more or less as one defines behaviours of Petri nets, or finite automata.

For every $\underline{x} \in VFS$, an event $a \in Ev$ is said to be **enabled at** $\underline{x}$ if and only if:

$$(\forall i = 1, \ldots, n) a \in Ev_i \Rightarrow pr_i(\underline{x}) a \in FS_i.$$

In other words, a is enabled at $\underline{x}$ if every sequential subsystem involving a permits its occurrence.

For every $\underline{x} \in VFS$, a set of independent events $A \in Ind$ is said to be **simultaneously enabled at** $\underline{x}$ if and only if every $a \in A$ is enabled at $\underline{x}$.

Corollary 2.3 *For every* $\underline{x} \in VFS$*:*
(1) $A \in Ind$ *is simultaneously enabled at* $\underline{x} \Rightarrow (\forall B \subseteq A) B = \emptyset$ *or* B *is simultaneously enabled at* $\underline{x}$.
(2) $a \in Ev$ *is enabled at* $\underline{x} \Leftrightarrow \underline{x}\ \underline{a} \in VFS$.
(3) $A \in Ind$ *is simultaneously enabled at* $\underline{x} \Leftrightarrow \underline{x}\ \underline{A} \in VFS$. □

For every $\underline{x} \in VFS$, let $enabled(\underline{x})$ (or $enabled_{Pr}(\underline{x})$ if the explicit name Pr is necessary) denote the family of all sets of events simultaneously enabled at $\underline{x}$. Assume also that for $\underline{x} \in Vev^* \backslash VFS$, $enabled(\underline{x}) = \emptyset$.

Let $SEQ, CON \subseteq Vev^* \times Vev^*$ be the following relations:

$$\underline{x}\ SEQ\ \underline{y} \Leftrightarrow (\exists a \in Ev)\{a\} \in enabled(\underline{x}) \wedge \underline{y} = \underline{x}\ \underline{a}$$

$$\underline{x}\ CON\ \underline{y} \Leftrightarrow (\exists A \in enabled(\underline{x}))\underline{y} = \underline{x}\ \underline{A}.$$

The relations SEQ and CON are called respectively: **the sequential reachability in one step**, and **the concurrent reachability in one step.**

Lemma 2.7 (VFS and Reachabilities)
$VFS = \{\underline{x} | \underline{\varepsilon} SEQ^* \underline{x}\} = \{\underline{x} | \underline{\varepsilon} CON^* \underline{x}\}$.

Proof. By the definition, we have $SEQ^* = CON^*$; hence $\{\underline{x} | \underline{\varepsilon} SEQ^* \underline{x}\} = \{\underline{x} | \underline{\varepsilon} CON^* \underline{x}\}$. Furthermore, $VFS = FS_1 \times \ldots \times FS_n \cap Vev^*$ and from the definition of SEQ and Corollary 2.3(2) we obtain:

(1) $\underline{\varepsilon} \in \{\underline{x} | \underline{\varepsilon} SEQ^* \underline{x}\} \cap (FS_1 \times \ldots \times FS_n) \cap Vev^*$

(2) $\underline{y} \in \{\underline{x} | \underline{\varepsilon} SEQ^* \underline{x}\} \cap Vev^* \Rightarrow$

$\underline{y}\underline{a} \in \{\underline{x} | \underline{\varepsilon} SEQ^* \underline{x}\} \Leftrightarrow \underline{y}\underline{a} \in (FS_1 \times \ldots \times FS_n) \cap Vev^*$.

Thus, $\{\underline{x} | \underline{\varepsilon} SEQ^* \underline{x}\} = FS_1 \times \ldots \times FS_n \cap Vev^*$, which ends the proof. □

For the example path program Pr_1 we have:

$$\begin{aligned} enabled(\underline{\varepsilon}) &= \{\{a\},\{d\},\{a,d\}\} \\ enabled(\underline{a}) &= \{\{d\},\{b\},\{c\},\{c,d\}\} \\ enabled(\underline{ad}) &= \{\{e\},\{c\},\{c,e\}\} \\ enabled(\underline{ade}) &= \{\{d\},\{b\},\{c\},\{c,d\}\} \\ enabled(\underline{adc}) &= \{\{a\},\{e\},\{a,e\}\},\ etc. \end{aligned}$$

and for instance:

$$\underline{ad} SEQ \underline{ade},\ \underline{a} CON \underline{adc},\ \underline{a} CON \underline{ad},\ etc.$$

The formal model of behaviour defined above allows us to speak formally of dynamic properties of systems specified by a path program $Pr = P_1 \ldots P_n$.

We say that $Pr = P_1 \ldots P_n$ is **deadlock-free** if and only if

$$(\forall \underline{x} \in VFS)\ enabled(\underline{x}) \neq \emptyset.$$

that is every history $\underline{x}$ may be continued.

The simplest example of a deadlocked path program is:

$$Pr_4 = \begin{cases} P(1) : \underline{path}\ a; b\ \underline{end} \\ P(2) : \underline{path}\ b; a\ \underline{end} \end{cases}$$

for which $VFS(Pr_4) = \{\underline{\varepsilon}\}$ and $enabled(\underline{\varepsilon}) = \emptyset$. However, path programs generating infinite sets of vector sequences may deadlock as well. For example:

$$Pr_5 = \begin{cases} P(1) : \underline{path}\ a, (c; d; e)\ \underline{end} \\ P(2) : \underline{path}\ b, (c; e; d)\ \underline{end} \end{cases}$$

where $VFS(Pr_5) = \{\underline{a}, \underline{b}\}^*\{\underline{a}, \underline{b}, \underline{c}, \underline{\varepsilon}\}$, and Pr_5 deadlocks after $\underline{c}$, $\underline{ac}$, $\underline{bc}$, $\underline{abc}$, etc.

For every $\underline{x} \in VFS$, the set $CON^*(\underline{x}) = \{\underline{y} | \underline{x} CON^* \underline{y}\}$ is called the set of **continuations of** $\underline{x}$. Note that $\underline{y} \in CON(\underline{x}) \Rightarrow \underline{x} \in Pref(\underline{y})$.

We say that $Pr = P_1 \ldots P_n$ is **adequate** if and only if:

$$(\forall \underline{x} \in VFS)(\forall a \in Ev)(\exists \underline{y} \in CON^*(\underline{x}))\{a\} \in enabled(\underline{y}),$$

that is, every history $\underline{x}$ of Pr may be continued, eventually enabling every event in Pr. Adequacy is a property akin to absence of partial system deadlock, and roughly corresponds to the concept of **liveness** for Petri nets. For a path program to be adequate there must be no event which is prevented from ever occurring again.

Adequacy is not always **a priori** of practical interest. In a program there may well be events which become blocked after a while, for example those involved in the initialization phase. Adequacy of a program (or a part of a program), asserts that all events of the program (or the part), retain a chance of occurring at some future, thus, in a sense, adequacy asserts that the system can be considered to be cyclic.

Corollary 2.4

(1) Pr is deadlock-free $\Leftrightarrow (\forall \underline{x} \in VFS)(\exists a \in Ev)\underline{x}\,\underline{a} \in VFS$.

(2) Pr is adequate $\Leftrightarrow (\forall \underline{x} \in VFS)(\forall a \in Ev)(\exists \underline{y} \in Vev^*)\underline{x}\,\underline{y}\,\underline{a} \in VFS$. □

We say that $Pr = P_1 \dots P_n$ is **periodic** if and only if enabled$(\underline{\varepsilon}) \neq \emptyset$ and

$$(\forall \underline{x} \in VFS)(\exists \underline{y} \in CON^*(\underline{x}))\underline{y} \in Cyc_1^* \times \dots \times Cyc_n^*$$

The structure of the set of behaviours of periodic programs is extremely simple. All the system does is to "cycle" non-sequentially forever through a number of alternatives.

Corollary 2.5

(1) Pr is periodic $\Rightarrow$ *Pr is deadlock-free.*

(2) Pr is adequate $\Rightarrow$ *Pr is deadlock-free.* □

The implications from the above corollary do not, in general, hold in the reverse direction; although they do for certain classes of path programs, as we shall see in Sect. 2.4. The following path program is deadlock-free but not adequate:

$$Pr_6 = \begin{cases} P(1) : \underline{path}\ a, (b; c)\ \underline{end} \\ P(2) : \underline{path}\ a, (c; b)\ \underline{end} \end{cases}$$

One may show that $VFS(Pr_6) = \{a\}^*$, Pr_6 is thus periodic (and hence deadlock-free) but not adequate, as b and c will never be performed. As an example of an adequate but not periodic path program we may take the following:

$$Pr_7 = \begin{cases} P(1) : \underline{path}\ a; b*; c\ \underline{end} \\ P(2) : \underline{path}\ c; b\ \underline{end} \end{cases}$$

It may be shown that $VFS(Pr_7) = Pref(\{\underline{a}\}\{\underline{cab}\}^*)$, thus Pr_7 is adequate (and hence deadlock-free) but not periodic. To prove the non-periodicity of Pr_7, let us observe that Cyc_1 ends with c, Cyc_2 ends with b, so no $\underline{y} \neq \underline{\varepsilon}$ may be such that $pr_i(\underline{y}) \in Cyc_i^*$, $i = 1, 2$.
Checking and analyzing the adequacy and periodicity properties of path programs is in general a very difficult task. We shall devote the whole of Sect. 2.4 to this problem.

The final elementary definitions used in the sequel are as follows.
Let $conn \subseteq \{P_1, \ldots, P_n\} \times \{P_1, \ldots, P_n\}$ be the following relation:

$$(P_i, P_j) \in conn \Leftrightarrow Ev_i \cap Ev_j \neq \emptyset.$$

We say that $Pr = P_1 \ldots P_n$ is **connected** if and only if the graph of *conn* is connected, i.e., if and only if:

$$(\forall i, j = 1, \ldots, n)(\exists i_1, \ldots, i_k) i_1 = i, i_k = j$$

$$\wedge (\forall j = 1, \ldots, k-1)(P_{i_j}, P_{i_{j+1}}) \in conn.$$

For path program $Pr = P_1 \ldots P_n$, let $Vcyc$ (or $Vcyc(Pr)$ if the name Pr is necessary) denote the following set of vector sequences:

$$Vcyc = \{\underline{x} \in VFS | (\forall i = 1, \ldots, n) pr_i(\underline{x}) \in Cyc_i\}.$$

2.2.4 Concurrent Product of Regular Expressions

From the formal viewpoint every path program $Pr = P_1 \ldots P_n$ generates a certain subset of Vev^*.

Each path P_i can be interpreted as a kind of (cyclic) regular expression. We show that path program Pr can also be interpreted as some restricted version of concurrent product of regular expressions. In regular expressions written in this book we shall use ";" to denote concatenation, "," to denote mutual exclusive choice, and assume that "," binds more strongly than ";", so we shall write for instance "$a; b, c, \varepsilon$" instead of the traditional "$a(b \cup c \cup \varepsilon)$". For every regular expression R, $L(R)$ will denote the (regular) language generated by R (see Appendix B, Definition B.3). Note that the only difference between regular expressions and strings derived from the nonterminal *sequence* of the basic COSY grammar is that the regular expressions may contain ε's and $\emptyset$'s while strings derived from *sequence* may not. Thus, if R does not contain ε or $\emptyset$, then $L(R) = Cyc(R)$. Of course there is a difference in the interpretation, $L(R)$ is just the language defined by R, $Cyc(R)$ defines one full cycle of the sequential behaviour of the path $P = \underline{path}\ R\ \underline{end}$, and the whole (sequential) behaviour of P is given by $Pref(Cyc(R)^*)$.
We shall say that the term:

$$CR = R_1 || \ldots || R_n$$

where $R_1, \ldots, R_n$ are regular expressions, is a **concurrent product of regular expressions.**

For every $CR = R_1 || \ldots || R_n$, let $VL(CR)$ be the following set of vector sequences:

$$VL(CR) = L(R_1) \times \ldots \times L(R_n) \cap Vev^*,$$

where Vev^* is defined by $Ev_1, \ldots, Ev_n$, and Ev_i is the alphabet of R_i for $i = 1, \ldots, n$.
The set $VL(CR)$ is called the **vector language** of $CR = R_1 || \ldots || R_n$.

Corollary 2.6 (Path Programs and Concurrent Products)
For every path program $Pr = P_1 \ldots P_n$:
$VFS(Pr) = VL(R_1 || \ldots || R_n)$ *where R_i is a regular expression such that $L(R_i) = FS(P_i), i = 1, \ldots, n$.* □

For example $VFS(Pr_2) = VL((a;c)^*; a, \varepsilon || (b;c)^*; b, \varepsilon) = \{\underline{abc}\}^* \{\underline{ab}, \underline{a}, \underline{b}, \underline{\varepsilon}\}$.

Corollary 2.7

$$VL(R_1 || \ldots || R_n) = \{\underline{x} \in Vev^* | (\forall i = 1, \ldots, n) \ pr_i(\underline{x}) \in L(R_i)\} = \\ = \{\underline{x} \in Vev^* | (\forall i = 1, \ldots, n) h_i(x) \in L(R_i)\}.$$

Proof. Identically to proof of Lemma 2.6 (2). □

2.3 Petri Net Semantics of COSY

Petri nets constitute a widely accepted and well established model of non-sequential systems. There are a variety of different types of Petri nets, but in principle a Petri net can be regarded as a bipartite graph with two types of nodes called places (denoted by circles) and transitions (denoted by boxes). The nodes are connected by directed arcs from places to transitions, and from transitions to places. The dynamic properties of a Petri net result from its execution, namely, tokens (sometimes with different colours indicating some additional information) reside on places of a net, and they can move along arcs according to the transition firing rules for Petri nets.

Petri nets have been used to model a wide range of systems, including hardware systems, software projects, production systems, etc. One of the reasons for the utility of Petri nets in modelling systems follows from the fact that different interpretations may be given to places, transitions and tokens. Places may represent conditions, transitions may represent actions or events, and tokens may indicate that the conditions on which they reside are true; places may also represent states, transitions may represent state changes and tokens may represent objects, such as jobs or resources that change state. On the other hand, Petri nets are rather far removed from any current programming notation, and because of their generality, the appropriate interpretation is not always clear for practical applications.

COSY is much closer to ordinary programming notation, and furthermore, it is sufficiently general to make formal reasoning possible. In a sense, COSY occupies a middle position "between" Petri nets and non-sequential programs , see [Bes87].

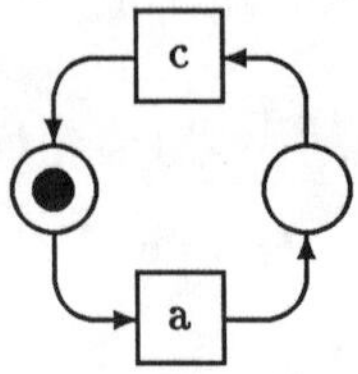

(a): Finite automaton for $\underline{path}\ a; c\ \underline{end}$

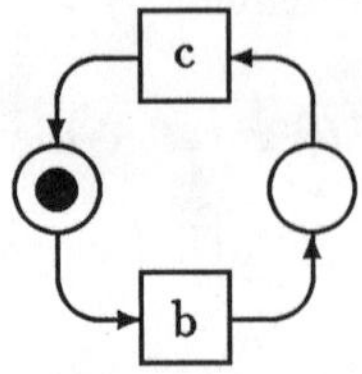

(b): Finite automaton for $\underline{path}\ b; c\ \underline{end}$

(c):$Pr_2 = \begin{cases} \underline{path}\ a; c\ \underline{end} \\ \underline{path}\ b; c\ \underline{end} \end{cases}$

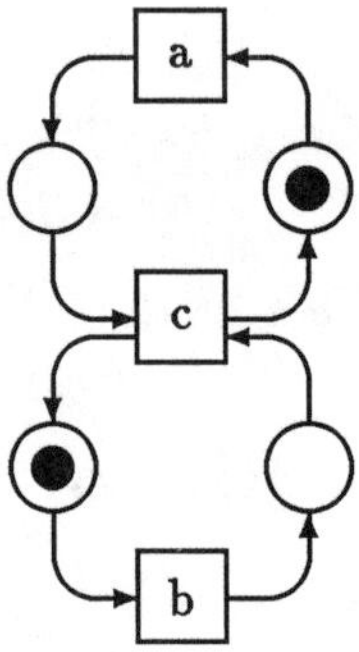

Figure 2.4. PN for Pr_2

In this section we will only consider a special class of Petri net, for more details the reader is advised to refer to any popular book about Petri nets such as [Rei85, Pet81]. The nets we will be using are a kind of 1-safe Place/Transition net. However, for our purposes the general definition of Place/Transition nets is unnecessarily general and complex, so we will introduce some different simplified concepts.

To explain informally what kind of Petri nets we will consider, and what their relationship to COSY is, recall the non-sequential system specified by the path program Pr_2, which we repeat here for convenience:

$$Pr_2 = \begin{cases} P(1): & \underline{path}\ a; c\ \underline{end} \\ P(2): & \underline{path}\ b; c\ \underline{end} \end{cases}$$

Since both P_1 and P_2 are a sort of regular expression, they may equivalently be represented by finite automata. The transition graph representing an automaton corresponding to $P(1) : \underline{path}\ a; c\ \underline{end}$, and the transition graph representing an automaton corresponding to $P(2) : \underline{path}\ b; c\ \underline{end}$ are depicted in Fig. 2.4 (a) and Fig. 2.4 (b), respectively. The automata have no final states (or *all* states are final), and the solid dot (token) indicates the initial state.

The graphs of Fig. 2.4 (a) and Fig. 2.4 (b) can be regarded as examples of a

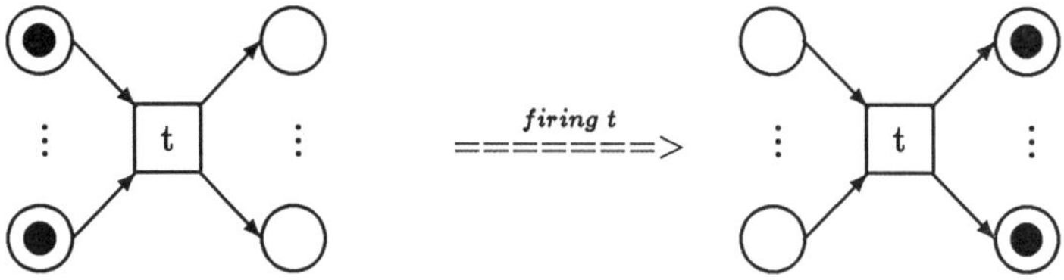

Figure 2.5. Firing Rule.

class of Petri net called **state-machines**. From the graphs depicted in the two figures we can obtain the graph depicted in Fig. 2.4 (c) by identifying common transitions, i.e., c in this case. The resulting graph represents the Petri net corresponding to the path program Pr_2. This net is an example of the class of **state-machine decomposible nets**. The path program Pr_2 is very simple. For instance, each single path contains no repeated event, so the transformation into an equivalent Petri net is rather straightforward. In the general case the rules are slightly more complex and we shall define them later on.

As mentioned above, the dynamic properties of nets are described by the motion of tokens along arcs according to the transition firing rules. The firing rule for the kind of Petri net we will use is illustrated in Fig. 2.5, and says that:

> if all the places connected to the input arcs of a transition contain one token, and all the places connected to the output arcs of this transition contain no token, then the tokens are removed from all input places and one token is placed on each output place.

An example of applying this firing rule to the net from Fig. 2.4 (c) is depicted in Fig. 2.6.

2.3.1 Elements of Net Theory

In this section we use some elements of Petri net theory, which is of course much more extensive than might appear from our presentation. The first restriction we will impose is that *all* nets defined below are *finite*.

A triple $N = (S, T, F)$ where S and T are sets, is called a **general net** if and only if

(1) $S \cap T = \emptyset$, $S \cup T$ is finite

(2) $F \subseteq (S \times T) \cup (T \times S)$

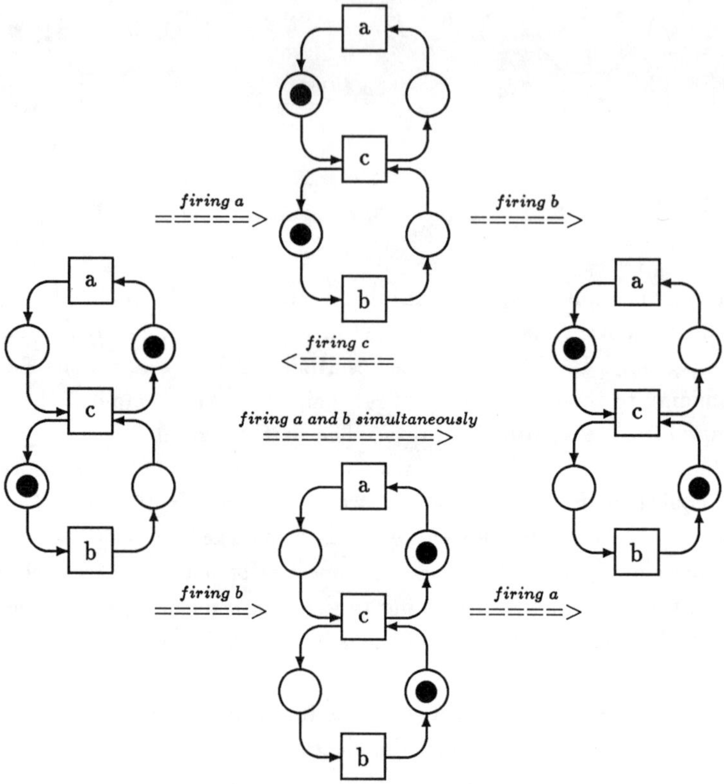

Figure 2.6. Moving tokens in net for Pr_2

The elements of S are called **places**, the elements of T are called **transitions**, F is the **flow-relation** of N, and the set $S \cup T$ is called the set of **elements** of the general net N. The places are represented by circles ○, the transitions are represented by boxes □, and the flow relation F is represented by arrows between elements of $S \cup T$ in the way shown in Fig. 2.7.

Let $N = (S, T, F)$ be a general net. For every $a \in S \cup T$, we define ${}^\bullet a = \{b \in S \cup T | (b, a) \in F\}$ and $a^\bullet = \{b \in S \cup T | (a, b) \in F\}$. The set ${}^\bullet a$ is called the **pre-set** of a, and $a^\bullet$ is called the **post-set** of a. For $Y \subseteq S \cup T$, we define ${}^\bullet Y = \bigcup_{a \in Y} {}^\bullet a$ and $Y^\bullet = \bigcup_{a \in Y} a^\bullet$.

We say that $N = (S, T, F)$ is a ***T*-restricted net** if and only if it is a general net and $T = {}^\bullet S \cap S^\bullet$. *Throughout the rest of this book we shall only be concerned with T-restricted nets, which will simply be called nets.* The condition $T = {}^\bullet S \cap S^\bullet$ means that each transition has at least one pre-place and at least one post-place. For example the graphs of Fig. 2.4 (a), (b), (c) and Fig. 2.6 considered without tokens represent (T-restricted) nets. The two simple graphs from Fig. 2.7 represent general nets which are not T-restricted nets.

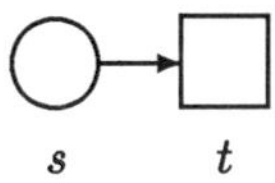

$$\Leftrightarrow \qquad (s,t) \in F \cap (S \times T)$$

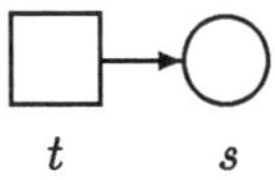

$$\Leftrightarrow \qquad (t,s) \in F \cap (T \times S)$$

Figure 2.7. Flow Relation F

From a formal point of view we can interpret the relation F as a function $F : (S \times T) \cup (T \times S) \longrightarrow \{0,1\}$, with

$$(a,b) \in F \Leftrightarrow F(a,b) = 1$$

Note that such an interpretation allows the extension of the whole theory to nets with weighted arrows. One simply has to replace $\{0,1\}$ by the set of non-negative integers. We mention this because it is a frequently used net definition. However, we will not use such an extension in this section.
We say that M is a **marking** of (S,T,F) if and only if M is a mapping

$$M : S \longrightarrow \{0,1\}.$$

Graphically, markings are denoted by putting tokens inside circles, i.e., $M(s) = 1$ is denoted by $\odot$, and $M(s) = 0$ is denoted by $\bigcirc$. From a formal viewpoint M can be interpreted as a subset of S, with $s \in M \Leftrightarrow M(s) = 1$, and we will frequently use this form of representation. In general net theory M is usually defined as $M : S \longrightarrow \{0,1,2,\ldots\}$, but there is no need for such a general definition in this section.
Let $N = (S,T,F)$ be a net, M a marking of N, and $t \in T$. We say that M **enables** t if and only if

$${}^\bullet t \subseteq M \wedge M \cap (t^\bullet \setminus {}^\bullet t) = \emptyset.$$

We say that M' **is produced from** M **by the firing of** $t \in T$ if and only if:

(1) M enables t,

(2) $M' = (M \setminus {}^\bullet t) \cup t^\bullet$.

This rule is represented graphically in Fig. 2.5.
We say that $t_1, \ldots, t_k \in T$ where $k \geq 1$ are **simultaneously enabled** by a marking M (or M enables $\{t_1, \ldots, t_k\}$) if and only if:

(1) M enables t_i for every $i = 1, \ldots, k$,

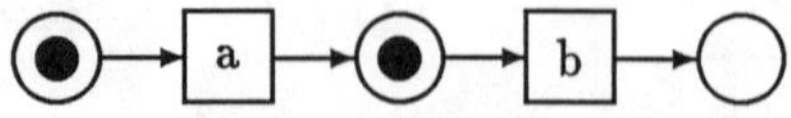

Figure 2.8. Marked Net with a Contact

(2) $({}^{\bullet}t_i \cup t_i^{\bullet}) \cap ({}^{\bullet}t_j \cup t_j^{\bullet}) = \emptyset$ if $i \neq j$ for all $i, j = 1, \ldots, k$.

We say that **M' is produced from M by the simultaneous firing of a non-empty $A \subseteq T$** if and only if:

(1) M enables A,

(2) $M' = (M \setminus {}^{\bullet}A) \cup A^{\bullet}$.

If M' is produced from M by the firing of $t \in T$ we will write $M[t\rangle M'$ and if M' is produced by the simultaneous firing of $A \subseteq T$ we will write $M[A\rangle M'$. We will also write $M[t_1 \ldots t_r\rangle M'$ if $M[t_1\rangle M_1[t_2\rangle \ldots M_{r-1}[t_r\rangle M'$ for some $M_1, \ldots, M_{r-1}$, and $M[A_1 \ldots A_r\rangle M'$ if $M[A_1\rangle M^1[A_2\rangle \ldots M^{r-1}[A_r\rangle M'$ for some $M^1, \ldots, M^{r-1}$. Directly from the definition we have the following result.

Corollary 2.8
If $A = \{t_1, \ldots, t_k\} \subseteq T$ and $M[A\rangle M'$,
then $M[t_{i_1} \ldots t_{i_k}\rangle M'$ for any permutation $i_1, \ldots, i_k$ of $1, \ldots, k$. □

We say that **M' is reachable from M** if either $M' = M$ or there is a sequence $A_1, \ldots, A_k \subseteq T$ such that $M[A_1 \ldots A_k\rangle M'$.
Note that we may replace sets of simultaneously fired transitions by single transitions in the above definition. Let $[M\rangle$ denote the **set of all markings reachable from M**.
A quadruple $N = (S, T, F, M_0)$, where (S, T, F) is a net, M_0 is a marking of (S, T, F), is called a **marked net**, and M_0 is called **the initial marking of N**. In graphical representation, tokens denote the initial marking. The nets of Fig. 2.4 (a), (b), (c) and Fig. 2.6 are marked nets.
A marked net $N = (S, T, F, M_0)$ is said to be **contact-free** if and only if:

$$(\forall M \in [M_0\rangle)(\forall t \in T)\, {}^{\bullet}t \subseteq M \Rightarrow M \cap (t^{\bullet} \setminus {}^{\bullet}t) = \emptyset.$$

Contact-freeness means that a transition can fail to be enabled only because some of its pre-places fail to hold, rather than because some of its post-places hold. An example of a marked net which is not contact-free is given in Fig. 2.8.

Lemma 2.8 *Let $N = (S, T, F, M_0)$ be a contact-free marked net, $M \in [M_0\rangle$, and $t_1, \ldots, t_k \in T$, $k \geq 1$. Then the following are equivalent:*

(1) M enables $\{t_1, \ldots, t_k\}$,
(2) M enables t_i *for every* $i = 1, \ldots, k$ *and*
$^\bullet t_i \cap {}^\bullet t_j = \emptyset$ *if* $i \neq j$ *for all* $i, j = 1, \ldots, k$.

Proof. $(1 \Rightarrow 2)$: Obvious.
$(2 \Rightarrow 1)$ Let $i \neq j$. We have the following:

(a) $^\bullet t_i \cap {}^\bullet t_j = \emptyset$ (follows from (2)).

(b) $^\bullet t_i \cap t_j^\bullet = \emptyset = t_i^\bullet \cap {}^\bullet t_j$ (follows from the fact that both t_i and t_j are enabled).

(c) $t_i^\bullet \cap t_j^\bullet = \emptyset$ (Let $M[t_i\rangle M'$. We have $^\bullet t_j \subseteq M'$ (see (b)) and $t_i^\bullet \subseteq M'$. Thus, since N is contact-free, $t_j^\bullet \cap t_i^\bullet = \emptyset$)

Hence, $(t_i^\bullet \cup {}^\bullet t_i) \cap (t_j^\bullet \cup {}^\bullet t_j) = \emptyset$ which means that M enables $\{t_i, \ldots, t_k\}$. □

Usually the concurrent semantics of contact-free Petri nets is given in terms of special nets called occurrence nets.

A net (S, T, F) is an **occurrence net** if and only if:

(1) $(\forall s \in S) card({}^\bullet s) \leq 1 \wedge card(s^\bullet) \leq 1$,

(2) $(\forall a, b \in S \cup T)(a, b) \in F^* \wedge (b, a) \in F^* \Rightarrow a = b$,
i.e. the graph of F is acyclic.

The places of an occurrence net are usually denoted by B and are used to represent state holdings. The transitions of an occurrence net are usually denoted by E and are used to represent event occurrences. The condition (1) of the definition means that an occurrence net contains no non-deterministic choice, the idea being that all choices are resolved at the behaviour level. The condition (2) means that an occurrence net contains no cycle, the idea being that all loops are unfolded at the behaviour level. Because of (2), one may easily check that F^* is a partial order on $B \cup E$ (i.e., on $S \cup T$). Occurrence nets will usually be denoted by ON.

Let $ON = (B, E, F)$ be an occurrence net. We define the relation $co \subseteq (B \cup E) \times (B \cup E)$ (or co_{ON} if the net is not known from the context) as: $co = (\overline{F^*})$, and we call co the **concurrency relation on** ON.

A set $C \subseteq B \cup E$ is called a **cut** of ON if and only if:

$$(\forall a, b \in C)((a, b) \in co \wedge (\forall c \in (B \cup E) \setminus C)(\exists a \in C)(a, c) \notin co) \text{ or } a = b$$

In other words, a cut is a maximal antichain of F^* (see Appendix A for the definition of antichain).

A set $C \subseteq B \cup E$ is called a B-**cut** if and only if C is a cut and $C \subseteq B$. The set of all B-cuts on ON will be denoted by $bcuts(ON)$. Finally, let $init$ and $final$ (or $init(ON)$, $final(ON)$ if necessary) be subsets of $B \cup E$ defined by:

$$init = \{a \in B \cup E | {}^\bullet a = \emptyset\},$$
$$final = \{a \in B \cup E | a^\bullet = \emptyset\}.$$

Since ON is T-restricted and finite, one may easily show that $init \in bcuts(ON)$ and $final \in bcuts(ON)$. This property may not hold for infinite occurrence nets considered in general net theory.
Now we can define the concept of **process** (not to be confused with the *process notation* from Chapter 3) for marked contact-free nets.
Let $N = (S,T,F,M_0)$ be a marked contact-free net, $ON = (B,E,F')$ be an occurrence net, and $\varphi : B \cup E \longrightarrow S \cup T$ be a mapping such that:

(1) $\varphi(init(ON)) = M_0$,

(2) $(\forall e \in E)\varphi({}^\bullet e) = {}^\bullet\varphi(e) \wedge \varphi(e^\bullet) = \varphi(e)^\bullet$,

(3) $(\forall C \in bcuts(ON))\varphi(C) \in [M_0\rangle \wedge$
$((\forall c_1, c_2 \in C)\varphi(c_1) = \varphi(c_2) \Leftrightarrow c_1 = c_2)$.

The pair (ON, φ) is called a **process** of N.
The condition (2) says that φ respects transition environments. The set ${}^\bullet e$ models the tokens consumed by the occurrence of the transition $\varphi(e)$, and $e^\bullet$ models the tokens produced by the occurrence of the transition $\varphi(e)$. The condition (3) says that each B-cut is mapped injectively into a marking reachable from M_0. The condition (1) specifies that the initial cut corresponds to the initial marking. *This definition is only sound for contact-free nets.* For example, for the net shown in Fig. 2.8, the transition rule specified that a can occur only strictly after b, but it is impossible to express this order by means of the above definition. Fig. 2.9 shows an example of a process for a given contact-free net, while the theorem below characterizes the basic properties of processes.

Theorem 2.7 *Let $N = (S,T,F,M_0)$ be a contact-free marked net, $ON = (B,E,F')$ be an occurrence net such that (ON,φ) is a process of N for some φ. Then:*
(1) $\varphi(B) \subseteq S \wedge \varphi(E) \subseteq T$.
(2) $(\forall a,b \in B \cup E)(a,b) \in F' \Rightarrow (\varphi(a),\varphi(b)) \in F$.
(3) $(\forall a,b \in B \cup E)\varphi(a) = \varphi(b) \Rightarrow (a,b) \in (F')^* \vee (b,a) \in (F')^*$.
(4) For every cut D of ON: $(\forall d_1,d_2 \in D)\varphi(d_1) = \varphi(d_2) \Leftrightarrow d_1 = d_2$.
(5) $(\forall e_1,e_2 \in E)(e_1,e_2) \in co \Rightarrow (\exists M \in [M_0\rangle)M$ enables $\{\varphi(e_1),\varphi(e_2)\}$.

Proof. (1) If $b \in B$ then clearly there is $C \in bcuts(ON)$ such that $b \in C$. Hence from (3) of the process definition: $\varphi(b) \in S$. Let $e \in E$. Since ON is T-restricted, there is $b \in B$ such that $e \in {}^\bullet b$. From (2) of the process definition: $\varphi(b) \in \varphi(e)^\bullet$. We have already proven $\varphi(b) \in S$. Hence $\varphi(e) \in T$.
(2) Let $a,b \in B \cup E$ and $(a,b) \in F'$, i.e. $a \in B$, $b \in E$ or vice versa. Let $a \in B, b \in E$. Thus: $(a,b) \in F' \Rightarrow a \in {}^\bullet b \Rightarrow \varphi(a) \in {}^\bullet\varphi(b) \Rightarrow (\varphi(a),\varphi(b)) \in F$. Similarly for $a \in E, b \in B$.

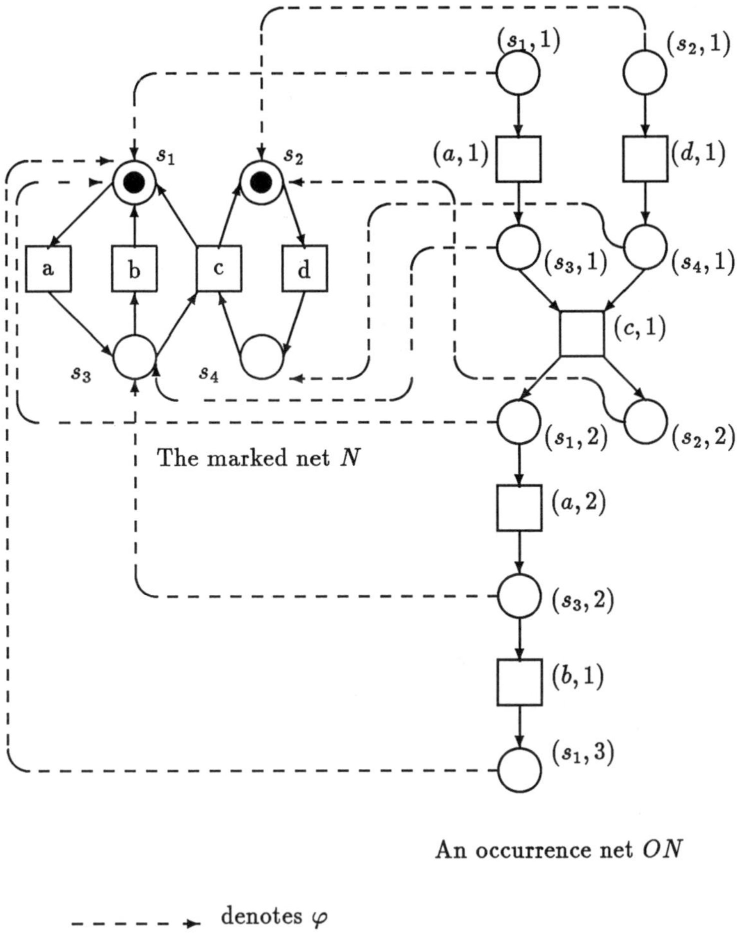

Figure 2.9. Marked net and one of its processes

(3) From (1) we know that either $a, b \in B$ or $a, b \in E$. If $a, b \in B$ then, from (3) of the process definition it follows that there is no B-cut containing both a and b, thus either $(a, b) \in (F')^*$ or $(b, a) \in (F')^*$. Let $a, b \in E$. The condition is clearly true if $a = b$, so assume $a \neq b$. Clearly $\varphi(a) = \varphi(b)$ implies ${}^\bullet\varphi(a) = {}^\bullet\varphi(b)$ so using (2) of the process definition we find $\varphi({}^\bullet a) = \varphi({}^\bullet b)$. Suppose $(a, b) \in co$, then from (1) in the definition of ON there is B-cuts $C \supseteq {}^\bullet a \cup {}^\bullet b$. Since ON is an occurrence net: ${}^\bullet a \cap {}^\bullet b = \emptyset$, and since ON is T-restricted, we have ${}^\bullet a \neq \emptyset$ and ${}^\bullet b \neq \emptyset$. Let $b_1 \in {}^\bullet a$. Since $\varphi({}^\bullet a) = \varphi({}^\bullet b)$, there is $b_2 \in {}^\bullet b$ such that $\varphi(b_1) = \varphi(b_2)$. This and ${}^\bullet a \cap {}^\bullet b = \emptyset$ implies $b_1 \neq b_2$, therefore φ restricted to the B-cut C is not injective, a contradiction with (3) of the process definition. Hence (3) holds.
(4) Directly from (3) and (3) of the process definition.
(5) Let $(e_1, e_2) \in co$. This means there is a B-cut $C \supseteq {}^\bullet e_1 \cup {}^\bullet e_2$, and clearly ${}^\bullet e_1 \cap {}^\bullet e_2 = \emptyset$. By (3) of the process definition, $\varphi(C) \in [M_0\rangle$ and clearly $\varphi(C)$ enables $\{\varphi(e_1), \varphi(e_2)\}$. □

We say that two occurrence nets $ON = (B, E, F)$ and $ON' = (B', E', F')$ are **isomorphic** if and only if there is a mapping $is : B \cup E \longrightarrow B' \cup E'$ such that:

(1) $(\forall a, b \in B \cup E) is(a) = is(b) \Leftrightarrow a = b$.

(2) $is(B \cup E) = B' \cup E'$.

(3) $(\forall a, b \in B \cup E)(a, b) \in F \Leftrightarrow (is(a), is(b)) \in F'$.

From here on, we will identify isomorphic occurrence nets. We have introduced the processes in a non-algorithmic (non-operational) manner. We now will present a method for generating processes for a given contact free marked net.
Let $N = (S, T, F, M_0)$ be a marked contact-free net, $M_1, \ldots, M_m \in [M_0\rangle$ and $A_1, \ldots, A_m \subseteq T$ be such that: $M_{i-1}[A_i\rangle M_i$ for $i = 1, \ldots, m$. We can define the occurrence net $ON = (B, E, F)$ recursively using the following rules:

(1) $ON_0 = (B_0, E_0, F_0)$, where $B_0 = M_0 \times \{1\}$, $E_0 = \emptyset, F_0 = \emptyset$.

(2) for $i = 1, \ldots, m, ON_i = (B_i, E_i, F_i)$ is constructed from A_i and $ON_{i-1} = (B_{i-1}, E_{i-1}, F_{i-1})$ in the following way:
$B_i = B_{i-1} \cup \{(s, k_s) | s \in A_i^\bullet \wedge k_s = num_s(ON_{i-1}) + 1\}$,
$E_i = E_{i-1} \cup \{(t, k_t) | t \in A_i \wedge k_t = num_t(ON_{i-1}) + 1\}$,
where $num_a(ON_i) = max(\{k | (a, k) \in E_i \cup B_i\})$, for $a \in T \cup S$,
$F_i = F_{i-1}$
$\cup\{((s, k_s), (t, k_t)) | (s, t) \in F \wedge (s, k_s) \in final(ON_{i-1}) \wedge (t, k_t) \in E_i \setminus E_{i-1}\}$
$\cup\{((t, k_t), (s, k_s)) | (t, s) \in F \wedge (s, k_s) \in B_i \setminus B_{i-1} \wedge (t, k_t) \in E_i \setminus E_{i-1}\}$

(3) $ON = (B, E, F')$, where $B = B_m, E = E_m, F' = F_m$.

One may easily check that ON is an occurrence net, and that (ON, φ) where $\varphi((a, k_a)) = a$ is a process of N. Fig. 2.9 illustrates this construction. We show that *every process* of a contact-free marked net can be derived by the above construction (up to isomorphism). Let (ON', φ'), where $ON' = (B', E', F')$, be *any* process of N. Recall that $card(X)$ denotes the cardinality of X, for any set X (see Appendix A). We define the mapping

$$iz : B' \cup E' \longrightarrow S \times \{1, 2, \ldots\} \cup T \times \{1, 2, \ldots\}$$

in the following manner:

$$(\forall b \in B') iz(b) = (\varphi'(b), k)$$
$$\text{where } k = card(\{b' \in B' | (b', b) \in (F')^* \wedge \varphi'(b') = \varphi'(b)\})$$
$$(\forall e \in E') iz(e) = (\varphi'(e), k)$$
$$\text{where } k = card(\{e' \in E' | (e', e) \in (F')^* \wedge \varphi'(e') = \varphi'(e)\}).$$

Now, let $B = iz(B'), E = iz(E')$ and $F \subseteq (B \times E) \cup (E \times B)$ be defined as: $(iz(e), iz(b)) \in F \Leftrightarrow (e, b) \in F'$ and $(iz(b), iz(e)) \in F \Leftrightarrow (b, e) \in F'$, for $b \in B', e \in E'$.
We leave to the reader the straightforward proof that $iz : B' \cup E' \longrightarrow B \cup E$ is one-to-one, $ON = (B, E, F)$ is isomorphic to $ON' = (B', E', F')$, and that ON can be derived by the above algorithm. Hence, we may define $process(N)$ as the set of all processes of N obtainable by the construction presented above and we may formulate the following result.

Theorem 2.8 *For every contact-free marked net N, $process(N)$ is the set of all its processes (up to isomorphism).* □

Let $(ON, \varphi) \in process(N)$, where $N = (S, T, F, M_0)$, and $A_1, \ldots, A_m \subseteq T$ be sets of transitions, such that $ON = ON_m$ where ON_m is constructed from M_0 and $A_1, \ldots, A_m$ using the procedure described above. Then $A_1, \ldots, A_m$ is called a **derivation** of ON in N.
If a mapping φ is understood, we shall identify a process (ON, φ) with an occurrence net ON.
A marked net $N = (S, T, F, M_0)$ is **deadlock-free** if and only if:

$$(\forall M \in [M_0\rangle)(\exists t \in T) M \textit{ enables } t.$$

A marked net $N = (S, T, F, M_0)$ is **live** if and only if:

$$(\forall M \in [M_0\rangle)(\forall t \in T)(\exists M' \in [M\rangle) M' \textit{ enables } t.$$

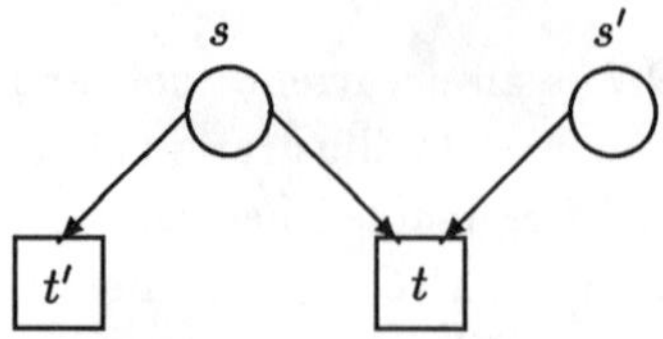

Figure 2.10. Pattern absent in free-choice nets

The property of liveness for Petri nets corresponds to the property of adequacy for COSY path programs. Both properties have the same "grass roots", i.e., absence of partial deadlock. However there are some formal differences between the two concepts which will be analyzed in detail at the end of Sect. 2.3.7. Liveness defined as above is a dynamic property. An important question is: "Can it be fully characterized by a static structure of a net?". Unfortunately, so far we can answer this question positively only for a subclass of Petri nets, namely for so called *Free-choice Petri nets* (and partially for some of their extensions). The major source of problems in the analysis of liveness is the pattern shown in Fig. 2.10. Free-choice nets are nets which exclude this kind of situation by their static structure, i.e. without regard for the marking classes of the nets of this type of net.

A marked net $N = (S,T,F,M_0)$ is called **free-choice** if and only if

$$(\forall(s,t) \in F \cap S \times T)\ s^\bullet = \{t\} \vee {}^\bullet t = \{s\}.$$

Since $(s,t) \in F$ always implies $t \in s^\bullet$ and $s \in {}^\bullet t$, N is not a free-choice net if and only if

$$(\exists(s,t) \in F \cap S \times T)\ s^\bullet \setminus \{t\} \neq \emptyset \wedge {}^\bullet t \setminus \{s\} \neq \emptyset$$

i.e. if N *contains* the net from Fig. 2.10.

The liveness property for Free-choice nets can be characterized statically in terms of H-deadlocks and H-traps (H for <u>H</u>ack who first used these concepts). Let $N = (S,T,F,M_0)$ be a marked net, and let $S' \subseteq S$:

(1) S' is called an **H-deadlock** if and only if ${}^\bullet S' \subseteq S'^\bullet$

(2) S' is called an **H-trap** if and only if $S'^\bullet \subseteq {}^\bullet S'$.

Theorem 2.9 (Liveness of Free-choice Nets)
A free-choice net $N = (S,T,F,M_0)$ is live if and only if every non-empty deadlock $S' \subseteq S$ contains a trap S'' such that $S'' \cap M_0 \neq \emptyset$. □

For more details including the proof of Theorem 2.9 the reader is referred to [Rei85, Pet81]. In Sect. 2.4.4 we show how the concept of free-choice may be applied to COSY path programs.

2.3.2 VFS semantics of SMD nets

In this section we further narrow the family of nets we are interested in.
A marked net $N = (S,T,F,M_0)$ is called a (finite) **state machine** if and only if:

(1) $(\forall t \in T)card({}^\bullet t) = card(t^\bullet) = 1$.

(2) $card(M_0) = 1$.

(3) $(\forall s \in S)\{s\} \in [M_0\rangle$

The nets from Fig. 2.4 (a) and Fig. 2.4 (b) are both examples of state machines. The condition (3) guarantees state machines do not possess useless places and transitions. Formally, state machines are equivalent to connected finite state automata without final states and with all states reachable from the initial one.
A marked net $N = (S,T,F,M_0)$ is called **state machine decomposable** if and only if there are state machines $N_i = (S_i,T_i,F_i,M_0^i)$, $i = 1,\ldots,n$, $n \geq 1$, such that:

(1) $S = \bigcup_{i=1}^{n} S_i$,

(2) $(\forall i,j = 1,\ldots,n) i \neq j \Rightarrow S_i \cap S_j = \emptyset$,

(3) $T = \bigcup_{i=1}^{n} T_i$,

(4) $F = \bigcup_{i=1}^{n} F_i$,

(5) $(\forall i = 1,\ldots,n) M_0 \cap S_i = M_0^i$.

By a **marking** of a state machine decomposable net we mean a set $M \subseteq S$ such that $card(M \cap S_i) = 1$ for all $i = 1,\ldots,n$.
The set $\{N_1,\ldots,N_n\}$ of state machines will be called a **state machine decomposition** of N, and will be denoted by $SMD(N)$. For example, the net from Fig. 2.4 (c) is state machine decomposable, and its state machine decomposition consists of the nets from Fig. 2.4 (a) and Fig. 2.4 (b). In general, $SMD(N)$ *is not* a unique set, as Fig. 2.11 shows, but its cardinality is always the same, equal to $card(M_0)$.

In the sequel we will use the abbreviation: **smd-net** instead of the full name: state machine decomposable net, and furthermore, when we talk about an smd-net we mean a pair $(N, \{N_1,\ldots,N_n\})$ where $\{N_1,\ldots,N_n\} = SMD(N)$, however, if it does not lead to confusion, we will denote this net by N itself.

The basic properties of smd-nets are described in the following lemma.

Lemma 2.9 *Let* $N = (S,T,F,M_0)$ *be a smd-net with* $SMD(N) = \{N_1,\ldots,N_n\}$ *where* $N_i = (S_i,T_i,F_i,M_0^i)$ *for* $i = 1,\ldots,n$. *Then:*
(1) $(\forall t \in T)card({}^\bullet t) = card(t^\bullet) = card(\{N_i | t \in T_i\})$.

$\{N_1, N_2\}$ is a state machine decomposition of N, and

$\{N_3, N_4\}$ is also a state machine decomposition of N.

Note that for instance $\{N_1, N_2, N_3, N_4\}$ **is not** a state machine decomposition of N because of (2) of the definition.

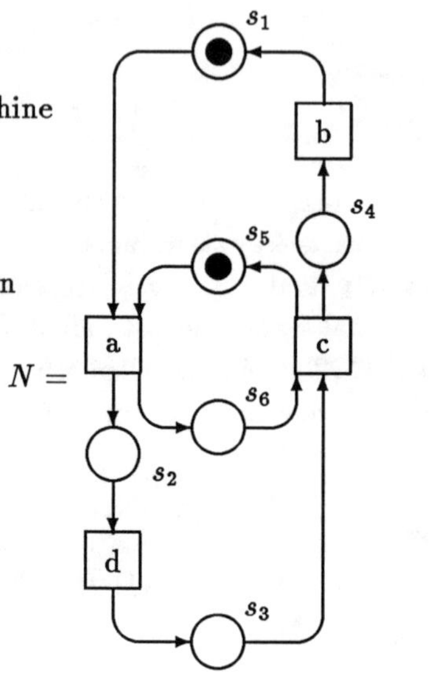

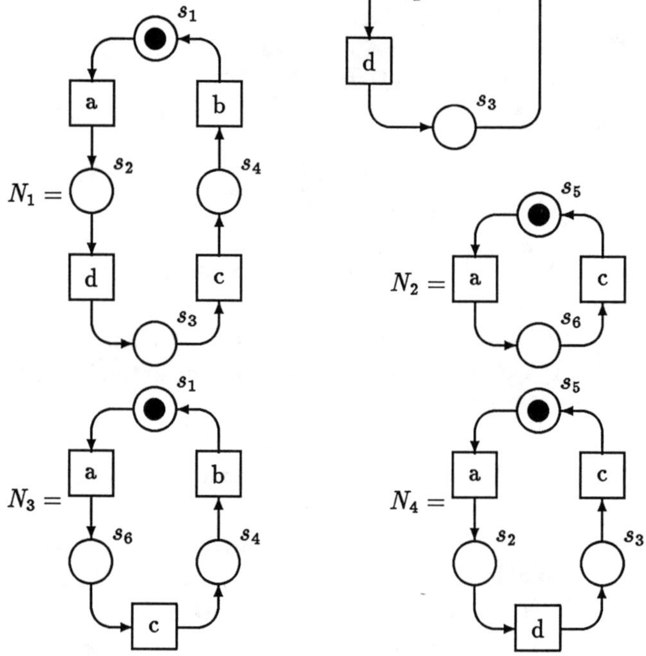

Figure 2.11. smd-net with many state machine decompositions

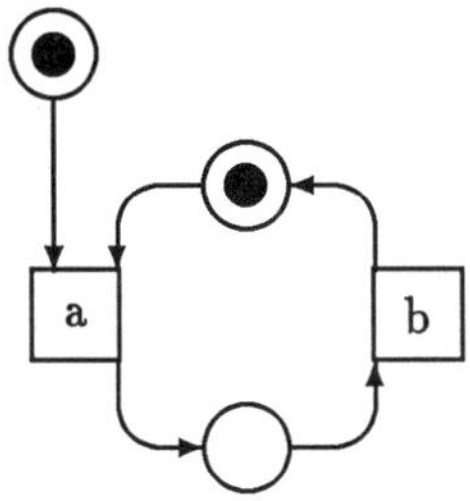

Figure 2.12. Contact free non smd-net

(2) $(\forall M \in [M_0\rangle) card(M) = n \wedge (\forall i = 1,\ldots,n) card(M \cap S_i) = 1$.
(3) N *is contact-free.*

Proof. (1) From (2) of the definition of smd-nets and (1) of the definition of state machine.
(2) This is clearly true for M_0. Assume that it holds for $M \in [M_0\rangle$ and consider M' such that $M[A\rangle M'$ for some $A \subseteq T$. This means that $M' = (M \setminus {}^\bullet A) \cup A^\bullet$, thus from (1) we get $card(M') = n$. Since obviously:
$(\forall t \in T) T_i \cap {}^\bullet t \neq \emptyset \Leftrightarrow T_i \cap t^\bullet \neq \emptyset$, we have $card(M' \cap S_i) = 1$.
(3) Assume that N is not contact-free, i.e. $(\exists M \in [M_0\rangle)(\exists t) {}^\bullet t \subseteq M \wedge (t^\bullet \setminus {}^\bullet t) \cap M \neq \emptyset$. Hence $(\exists i) {}^\bullet t \cap S_i \neq \emptyset$ and $(t^\bullet \setminus {}^\bullet t) \cap S_i \neq \emptyset$, which means that $card(M \cap S_i) = 2$, a contradiction. □

A net which is contact-free but not state machine decomposable is shown in Fig. 2.12. Let $N = (S, T, F, M_0)$ be a smd-net, fixed for the rest of this section, with $SMD(N) = \{N_1, \ldots, N_n\}$ where $N_i = (S_i, T_i, F_i, M_0^i)$ for $i = 1, \ldots, n$. In Sect. 2.2 we defined the set $Vev^* \subseteq Ev_1^* \times \ldots \times Ev_n^*$ on the basis of the sets $Ev_1, \ldots, Ev_n$. The only assumption about the Ev_i's was that they were finite, so we can repeat this construction using the sets $T_1, \ldots, T_n$. The only difference is that in this case, the set of vector events denoted by Vev in Sect. 2.2, will be denoted by Vtr and called the set of **vector transitions**. Otherwise we will preserve the notation from Sect. 2.2, so in particular, we have the relation $ind \subseteq T \times T$ defined as $(t_1, t_2) \in ind \Leftrightarrow (\forall i) t_1 \notin T_i \vee t_2 \notin T_i$, called the **independency relation** of N, and the set $Ind \subseteq 2^T$ defined as:

$$A \in Ind \Leftrightarrow (t_1, t_2 \in A \Rightarrow (t_1 = t_2 \vee (t_1, t_2) \in ind)) \wedge A \neq \emptyset$$

and called the set of **independent transitions** of N. $Vtr^* \subseteq T_1^* \times \ldots \times T_n^*$ corresponds to Vev^* of Sect. 2.2. We shall write $Vev^*(N)$, $ind(N)$ and $Ind(N)$, if necessary.
Directly from the definitions we have the following result.

Corollary 2.9
(1) $(t_1, t_2) \in ind \Rightarrow ({}^\bullet t_1 \cup t_1^\bullet) \cap ({}^\bullet t_2 \cup t_2^\bullet) = \emptyset$.
(2) $M[A\rangle M' \Rightarrow A \in Ind$.
(3) ${}^\bullet A \subseteq M \in [M_0\rangle \wedge A \in Ind \Rightarrow M$ *enables* A. □

For every N_i, let FT_i be the set of transition sequences given by

$$FT_i = \{x \in T_i^* | (\exists M \in [M_0^i\rangle) M_0^i[x\rangle_i M\}$$

where the index i in $[x\rangle_i$ indicates that the derivation was done in N_i. The set FT_i is called the set of **firing transition sequences** of N_i. Let VTS be the set of vector transition sequences defined as follows:

$$VTS = \{\underline{x} \in Vtr^* | (\exists M \in [M_0\rangle)(\exists A_1, \ldots, A_m \subseteq T)$$
$$M_0[A_1 \ldots A_m\rangle M \wedge \underline{x} = \underline{A_1} \ldots \underline{A_m}\}.$$

The set VTS is called the set of **vector transition sequences** of N.

Theorem 2.10 (Vector Transition Sequences of Nets)
$VTS = (FT_1 \times \ldots \times FT_n) \cap Vtr^*$.

Proof. Clearly $\underline{\varepsilon} \in VTS$ and $\underline{\varepsilon} \in (FT_1 \times \ldots \times FT_n) \cap Vtr^*$. Suppose that $\underline{x} \in VTS$ and $\underline{x} \in (FT_1 \times \ldots \times FT_n) \cap VTr^*$. Let us consider $\underline{xt}$ where $t \in T$. Suppose first that $\underline{xt} \in VTS$. $\underline{x} \in VTS$ and $\underline{xt} \in VTS$ means that: $M_0[t_1\rangle \ldots [t_m\rangle M_m$, $\underline{x} = \underline{t_1} \ldots \underline{t_m}$ for some $t_1, \ldots, t_m \in T$, and $M_m[t\rangle M_{m+1}$. Let $\underline{x} = (x_1, \ldots, x_n)$, $\underline{xt} = (x_1', \ldots, x_n')$. Clearly $x_i = x_i'$ if $t \notin T_i$, thus $x_i' \in FT_i$ if $t \notin T_i$. Suppose that $t \in T_i$ and consider $x_i' = x_i t$. Let us set: $[\varepsilon\rangle$ =identity relation. Thus we have $(M_{j-1} \cap S_i)[h_i(t_j)\rangle_i(M_j \cap S_i)$ for $j = 1, \ldots, m$. Furthermore, $M_m \cap S_i$ enables t in N_i, hence $(M_m \cap S_i)[t\rangle_i(M_{m+1} \cap S_i)$, i.e. $x_i t \in FT_i$. Thus $VTS \subseteq (FT_1 \times \ldots \times FT_n) \cap Vtr^*$.
Let $\underline{xt} \in (FT_1 \times \ldots \times FT_n) \cap Vtr^*$, $\underline{x} \in VTS \cap ((FT_1 \times \ldots \times FT_n) \cap Vtr^*)$, $\underline{x} = \underline{t_1} \ldots \underline{t_m}$, $t_1, \ldots, t_m \in T$, $M_0[t_1\rangle \ldots [t_m\rangle M_m$. Clearly, for every $i = 1, \ldots, n$: $(M_{j-1} \cap S_i)[h_i(t_j)\rangle_i(M_j \cap S_i)$ for $j = 1, \ldots, m$, and every sequence $(M_1 \cap S_i), \ldots, (M_m \cap S_i)$ is described unambiguously (up to identical consecutive elements) by $h_i(x)$. This means that: $(t \in T_i \wedge h_i(x)t \in FT_i)$ implies $M_m \cap S_i$ enables t in N_i, i.e. ${}^\bullet t \cap S_i \subseteq M_m \cap S_i$ and $(t^\bullet \setminus {}^\bullet t) \cap M_m \cap S_i = \emptyset$. Hence M_m enables t since the S_i' are disjoint. Thus $\underline{xt} \in VTS$, and as a consequence $VTS = (FT_1 \times \ldots \times FT_n) \cap Vtr^*$. □
The above shows certain similarities between path programs and smd-nets. We will discuss these relationships in detail later; now we show the connection between the sets *process*(N) and $VTS(N)$.

Lemma 2.10 *Let* $(ON, \varphi) \in process(N)$ *and* $A_1, \ldots, A_m$ *and* $A_1', \ldots, A_k'$ *be derivations of* ON *in* N. *Then:* $\underline{A_1} \ldots \underline{A_m} = \underline{A'_1} \ldots \underline{A'_k}$.

Proof. Suppose that $ON = (B, E, F')$. Corollary 2.8 enables us to reduce our considerations to the case when $A_i = \{t_i\}$, $t_i \in T$ for $i = 1, \ldots, m$, and $A'_j = \{t'_j\}$, $t'_j \in T$ for $j = 1, \ldots, k$. From the construction of elements of $process(N)$ we have that in such a case $m = card(E) = k$, i.e. $m = k$. Recall that $SMD(N) = \{N_1, \ldots, N_n\}$ with $N_i = (S_i, T_i, F_i, M_0^i)$, $i = 1, \ldots, n$. For every $i = 1, \ldots, n$, let $ON^i = (B^i, E^i, F'_i)$ be occurrence nets given by: $B^i = \varphi^{-1}(S_i)$, $E^i = \varphi^{-1}(T_i)$, $F'_i = F' \cap (B^i \times E^i \cup E^i \times B^i)$. Clearly $ON^i \in process(N_i)$, and the sequences $\{h_i(t_1)\}, \ldots, \{h_i(t_m)\}$, and $\{h_i(t'_1)\}, \ldots, \{h_i(t'_m)\}$ after removing all $\{\varepsilon\}$'s, are both derivations of ON^i in N_i, for $i = 1, \ldots, n$. But $(F'_i)^*$ is a total order, which means that all derivations of ON^i are identical, hence $h_i(t_1 \ldots t_m) = h_i(t'_1 \ldots t'_m)$ for all $i = 1, \ldots, n$. Consequently $\underline{t}_1 \ldots \underline{t}_m = \underline{t'}_1 \ldots \underline{t'}_m$. □

Lemma 2.10 makes it possible to introduce the following transformations. For every process $ON \in process(N)$, let $vts(ON) \in Vtr^*$ be the vector transition sequence defined as follows: if $A_1, \ldots, A_m$ is a derivation of ON in N, then $vts(ON) = \underline{A}_1 \ldots \underline{A}_m$.

Lemma 2.10 ensures that the above definition is independent of the choice of a derivation.

For every $\underline{x} \in VTS$, let $proc(\underline{x}) \in process(N)$ be the process such that its every derivation $A_1, \ldots, A_m$ satisfies $\underline{x} = \underline{A}_1 \ldots \underline{A}_m$.

Corollary 2.10

(1) $(\forall \underline{x} \in VTS) vts(proc(\underline{x})) = \underline{x}$.

(2) $(\forall ON \in process(N)) proc(vts(ON)) = ON$.

(3) $VTS(N) = vts(process(N))$.

(4) $process(N) = proc(VTS(N))$. □

As one might suspect, a process $ON \in process(N)$, and $vts(ON)$ define similar partial orders.

Theorem 2.11 *For every* $\underline{x} \in VTS$, *if* $proc(\underline{x}) = (B, E, F')$, *then:* $PO_{\underline{x}} = (F')^* \cap E \times E$.

Proof. Note that: $(F')^* \cap E \times E = ((F')^2 \cap E \times E)^*$. Define $F_E = (F')^2 \cap E \times E$. Since F' is acyclic, F_E^* is a partial order. For every $i = 1, \ldots, n$, let $B_i = \varphi^{-1}(S_i)$, $E_i = \varphi^{-1}(T_i)$, $F'_i = F' \cap (B_i \times E_i \cup E_i \times B_i)$. From the proof of Lemma 2.10 we have that $((F')^2 \cap E_i \times E_i)^* = TO_{h_i(x)}$, $i = 1, \ldots, n$. Define $F_{E_i} = (F')^2 \cap E_i \times E_i$, i.e. $F_{E_i}^* = TO_{h_i(x)}$ for $i = 1, \ldots, n$.

Let $(e_1, e_2) \in (F')^2 \cap E \times E$. From Theorem 2.7 (2) it follows that $(\varphi(e_1), \varphi(e_2)) \in F^2$, i.e. $(\exists s \in S) s \in \varphi(e_1)^\bullet \cap {}^\bullet\varphi(e_2)$. Since N is a smd-net, this means that $(\exists i \in \{1, \ldots, n\}) \varphi(e_1) \in T_i \wedge \varphi(e_2) \in T_i$. Thus, $(\exists i \in \{1, \ldots, n\})\{e_1, e_2\} \subseteq E_i$, so $F_E = \bigcup_{i=1}^n ((F')^2 \cap E_i \times E_i) = \bigcup_{i=1}^n F_{E_i}$. Hence, by Lemma A.2 (see Appendix A), $F_E^* = (F_{E_1}^* \cup \ldots \cup F_{E_n}^*)^*$, and since we have already shown $F_{E_i}^* = ((F'_i)^2 \cap$

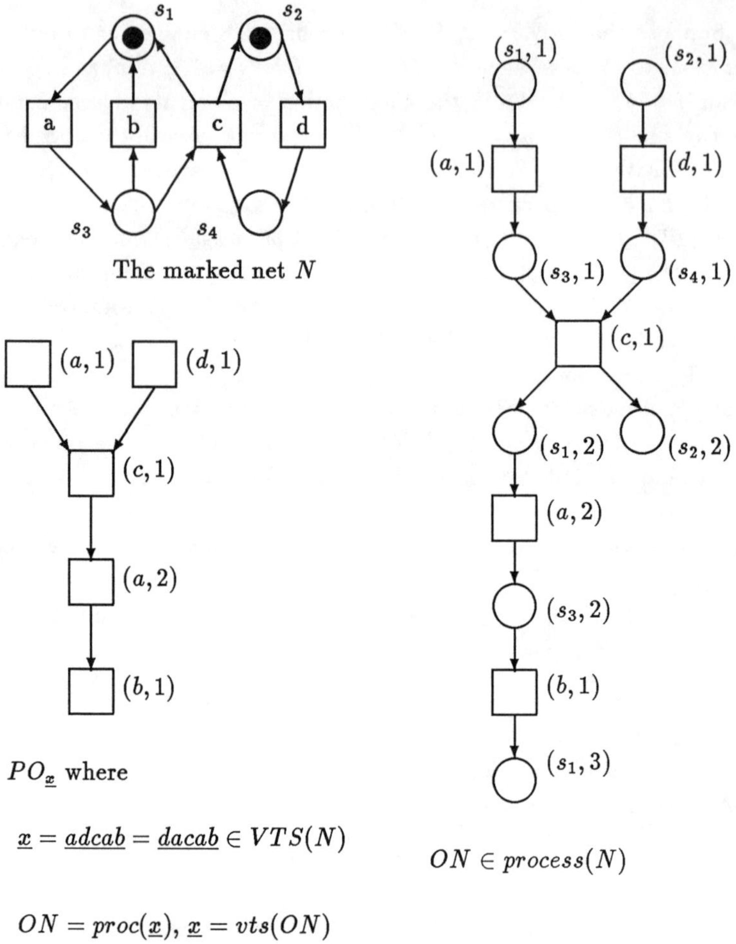

Figure 2.13. Relationship between processes, vts and partial orders

$E_i \times E_i)^* = TO_{h_i(x)}$ and $F_E^* = (F')^* \cap E \times E$, we can write $(F')^* \cap E \times E = (TO_{h_1(x)} \cup \ldots \cup TO_{h_n(x)})^*$, and by Theorem 2.4, we have $(F')^* \cap E \times E = PO_{\underline{x}}$. □

Since $vts = proc^{-1}$, we may also formulate Theorem 2.11 as follows.

Corollary 2.11 *For every* $ON = (B, E, F') \in process(N)$*:*
$(F')^* \cap E \times E = PO_{vts(ON)}$. □

Theorem 2.11 and Corollary 2.10 are illustrated in Fig. 2.13.

At this point the general idea of defining the meaning of path programs in terms of Petri nets can be grasped. This idea allows us to give meaning to some simple cases, such as those illustrated in Fig. 2.4 (a), Fig. 2.4 (b), and Fig. 2.4 (c), but for the general case some major obstacles still remain. First, consider the path path $a; a$ end. Intuitively this path corresponds to the structure depicted in Fig. 2.14 (a), but this structure *is not a Petri net*, but rather a *labelled Petri net*, i.e. a net which has labels associated with transitions, and different transitions may have the same label. The path path $a, b*; c$ end shows that even if we restrict our concern to paths with no repeated events, unlabelled Petri nets are not sufficient. A labelled Petri net corresponding to the latter path is depicted in Fig. 2.14 (b), and one can show that there is no (unlabelled) state machine Petri net which corresponds to this relatively simple path. Hence, we are required to introduce the concept of labelled Petri nets.

2.3.3 Semantics of Labelled Petri Nets

We will restrict our concern to labelled smd-nets and labelled occurrence nets since this is sufficient for our purposes, however, the concept of labeling can be applied to any kind of Petri net.

By a **labelled smd-net** we mean a 6-tuple

$$LN = (S, T, F, M_0, \nu, Ev)$$

where: $N = (S, T, F, M_0)$ is a smd-net with $SMD(N) = \{N_1, \ldots, N_n\}$ and

$N_i = (S_i, T_i, F_i, M_0^i)$ for $i = 1, \ldots, n$,

Ev is a finite set of **labels** or **events**,

$\nu : T \longrightarrow Ev$ is a mapping satisfying:

(1) $\nu(T) = Ev$

(2) $(\forall t_1, t_2 \in T)\nu(t_1) = \nu(t_2) \Rightarrow ((\forall i = 1, \ldots, n)t_1 \in T_i \Leftrightarrow t_2 \in T_i)$

The condition (2) says that **transitions with identical labels must belong to the same state machines.** Of course every smd-net can be interpreted as a labelled smd-net with ν being the identity.

By a **labelled occurrence net** we mean a 5-tuple

$$LON = (B, E, F, \nu, Ev)$$

where: $ON = (B, E, F)$ is an occurrence net,

Ev is a finite set of **labels** or **events**,

$\nu : E \longrightarrow Ev$ is a mapping satisfying:

(1) $\nu(E) = Ev$

(2) $(\forall e_1, e_2 \in E)\nu(e_1) = \nu(e_2) \Rightarrow (e_1, e_2) \notin co_{ON}$.

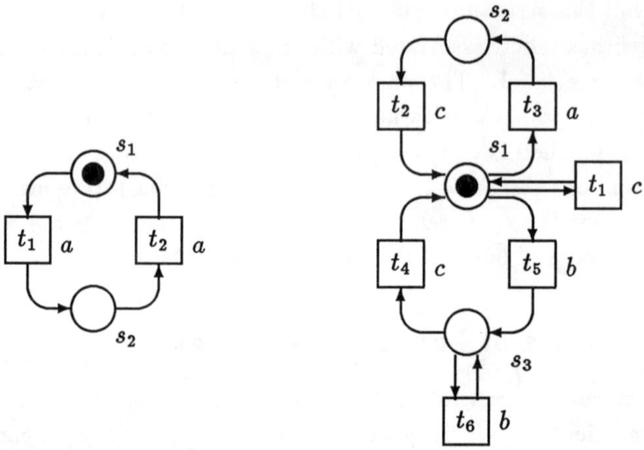

(a) Labelled Petri net which corresponds to *path* $a; a$ *end*

(b) Labelled Petri net which corresponds to *path* $a, b*; c$ *end*

Figure 2.14. Labelled Petri nets

Here the second condition of ν says that **transition occurrences with the same label cannot be concurrent.**
Let $LN = (S, T, F, M_0, \nu, Ev)$ be a labelled smd-net, $N = (S, T, F, M_0)$ be the underlying smd-net, $LON = (B, E, F', \nu', Ev')$ be a labelled occurrence net, and let $\varphi : B \cup E \longrightarrow S \cup T$.
We say that (LON, φ) is a **labelled process of** LN if

(1) $((B, E, F'), \varphi)$ is a process of (S, T, F, M_0)

(2) $(\forall e \in E)\ \nu'(e) = \nu(\varphi(e))$

Let $lprocess(LN)$ be the following set of labelled processes of LN:
$(LON, \varphi) \in lprocess(LN) \Leftrightarrow$
$((B, E, F'), \varphi) \in process(N) \wedge (\forall e = (t, k) \in E)\nu'(e) = \nu(t)$.
Proceeding as in Sect. 2.3.1 we can show that $lprocess(LN)$ contains (up to isomorphism) all labelled processes of LN.
Furthermore, every $\underline{x} = \underline{t}_1, \ldots, \underline{t}_m \in Vtr(N)^*$ defines unambiguously a *labelled partial order* $LPO_{\underline{x}} = (PO_{\underline{x}}, \nu', \Sigma')$, where:

(1) $\Sigma' = \nu(\{t_1, \ldots, t_m\})$

(2) $(\forall (t_i, k) \in Dom(PO_{\underline{x}}))\ \nu'((t_i, k)) = \nu(t_i)$.

For the definition of labelled partial order see Appendix A.
We left it to the reader to extend the result stated in Theorem 2.11 to the labelled partial orders of LN. Fig. 2.15 illustrates all the concepts introduced above.

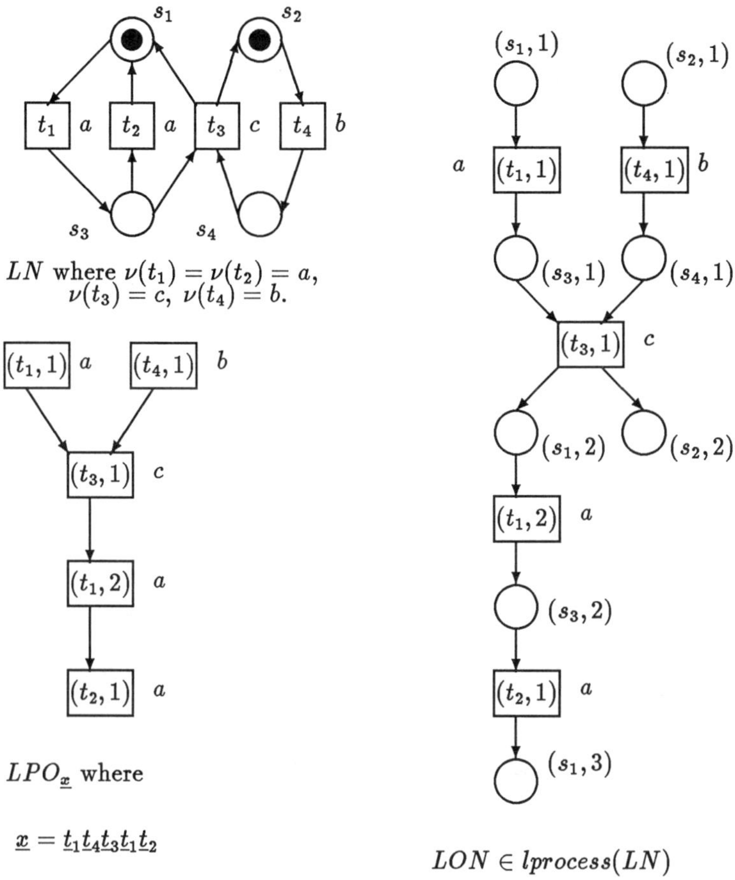

Figure 2.15. Labelled smd-net, labelled process and labelled partial order

For labelled nets we can also imagine another kind of semantics. When net transitions are interpreted as system events, it frequently happens that identical events appear several times, for instance "$x := x + 1$" may appear in different places of a program. Using standard unlabelled Petri nets we have to model such different appearances of identical events by different transitions, which sometimes is an advantage, but usually we do not want to distinguish between different appearances of the same event. So why not define the meaning of labelled smd-nets in terms of vector sequences of elements of Ev instead of elements of T?
Let $LN = (S, T, F, M_0, \nu, Ev)$ be a labelled smd-net with $N = (S, T, F, M_0)$, $SMD(N) = \{N_1, \ldots, N_n\}$, $N_i = (S_i, T_i, F_i, M_0^i)$ for $i = 1, \ldots, n$. Let us define $Ev_i = \nu(T_i)$ for $i = 1, \ldots, n$. Clearly $Ev = Ev_1 \cup \ldots \cup Ev_n$. We can apply the method from Sect. 2.2.1 and create $Vev^* \subseteq Ev_1^* \times \ldots \times Ev_n^*$, as the set of all vector event sequences generated by sets $Ev_1, \ldots, Ev_n$. Similarly, we can define $ind \subseteq Ev \times Ev$ by $(a, b) \in ind \Leftrightarrow (\forall i = 1, \ldots, n) a \notin Ev_i \vee b \notin Ev_i$, and $Ind \subseteq 2^{Ev}$ as $A \in Ind \Leftrightarrow A \neq \emptyset \wedge ((\forall a, b \in A) a = b \vee (a, b) \in ind)$, and $PO_{\underline{x}}$ for $\underline{x} \in Vev^*$, as in Sect. 2.2.2.
If necessary we write $ind(LN), Ind(LN), Vev(LN)$, etc.
We can also extend ν to $\underline{\nu} : Vtr(N)^* \longrightarrow Vev(LN)^*$ by:

(1) $\underline{\nu}(\underline{\varepsilon}) = \underline{\varepsilon}$

(2) $\underline{\nu}(\underline{x}\underline{t}) = \underline{\nu}(\underline{x})\underline{\nu}(\underline{t})$ for $\underline{x} \in Vtr(N)^*, \underline{t} \in Vtr(N)$.

Similarly we can extend ν to $\hat{\nu} : T^* \longrightarrow Ev^*$.
In the sequel we will use the same symbol ν to denote $\nu, \underline{\nu}$ and $\hat{\nu}$. Some simple relationships between $Vtr(N)^*$ and $Vev(LN)^*$ are characterized by the following lemma.

Lemma 2.11
(1) $(\forall t_1, t_2 \in T)(t_1, t_2) \in ind(N) \Leftrightarrow (\nu(t_1), \nu(t_2)) \in ind(LN)$
(2) $(\forall \underline{x} \in Vtr(N)^*) PO_{\underline{x}}$ *and* $PO_{\nu(\underline{x})}$ *are isomorphic partial orders.*

Proof. (1) Follows from (2) of the definition of ν for labelled smd-nets.
(2) Let $\underline{x} = \underline{t}_1 \ldots \underline{t}_m$, $\nu(\underline{x}) = \underline{a}_1 \ldots \underline{a}_m$, $\nu(t_i) = a_i$ for $i = 1, \ldots, m$. Let $TO_x = (t_1, i_1) \longrightarrow (t_2, i_2) \longrightarrow \ldots \longrightarrow (t_m, i_m)$, and let $TO_{a_1 \ldots a_m} = (a_1, j_1) \longrightarrow (a_2, j_2) \longrightarrow \ldots \longrightarrow (a_m, j_m)$.
Define $f : Dom(PO_{\underline{x}}) \longrightarrow Dom(PO_{\nu(\underline{x})})$ by: $f((t_r, i_r)) = (a_r, j_r)$ for $r = 1, \ldots, m$. One can show that f is an isomorphism. □
The second part (2) of Lemma 2.11 is illustrated in Fig. 2.16.

The set of **vector event sequences of** LN, denoted by $VES(LN)$ or VES, is defined as: $VES(LN) = \nu(VTS(N))$, i.e.

$$VES(LN) = \{\underline{x} \in Vev^* | (\exists \underline{y} \in VTS(N)) \underline{x} = \nu(\underline{y})\}.$$

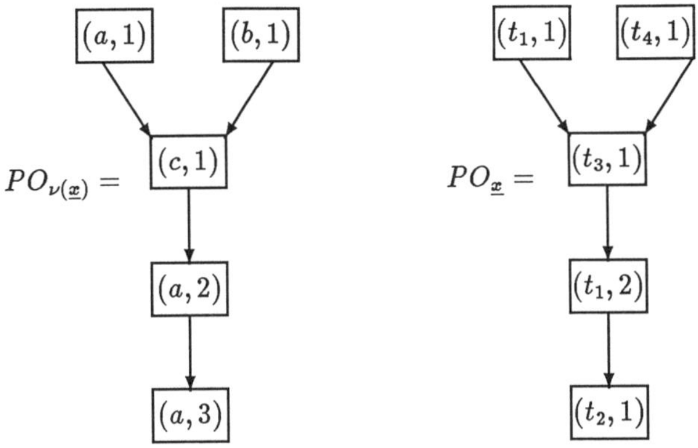

$\nu(t_1) = \nu(t_2) = a$ $\qquad$ $\underline{x} = \underline{t}_1\underline{t}_4\underline{t}_3\underline{t}_1\underline{t}_2$

$\nu(t_3) = c,\ \nu(t_4) = b$ $\qquad$ LN is depicted in Fig. 2.15.

Figure 2.16. $PO_{\underline{x}}$ and $PO_{\nu(\underline{x})}$

For every $i = 1, \ldots, n$, let $LN_i = (S_i, T_i, F_i, M_0^i, \nu_i, Ev_i)$, where ν_i is ν restricted to T_i. Every LN_i is called a **labelled state machine**, and the set of **event sequences of** LN_i, denoted by $ES(LN_i)$ or ES_i, is defined similarly as: $ES(LN_i) = \nu(FT_i)$, i.e.

$$ES_i = \{x \in Ev_i^* | (\exists y \in FT_i) x = \nu(y)\}.$$

The nets from Fig. 2.17 show that Theorem 2.10 cannot be extended to the case of labelled nets.

Clearly, VES semantics and VTS semantics are not equivalent. It may sometimes happen that $\nu(\underline{x}) = \nu(\underline{y})$ but $\underline{x} \neq \underline{y}$ and, for example, $\underline{x}$ may lead to deadlock, while $\underline{y}$ may not. This case is illustrated in Fig. 2.18, where $\underline{t}_1 \neq \underline{t}_2$ and $\nu(\underline{t}_1) = \nu(\underline{t}_2) = \underline{a}$, $\{s_1, s_2\}[t_1\rangle\{s_3, s_4\}$, $\{s_1, s_2\}[t_2\rangle\{s_1, s_4\}$, the marking $\{s_3, s_4\}$ enables t_3 while the marking $\{s_1, s_4\}$ is dead, i.e. there is no transition it enables. The vector of events $\underline{a} = (a, a)$ may lead either to $\{s_3, s_4\}$ or to $\{s_1, s_4\}$. Knowing $\underline{a}$ alone, we do not know if b is enabled after $\underline{a}$ or not!

It turns out, however, that because of condition (2) of ν, the case $\underline{x} \neq \underline{y}$ and $\nu(\underline{x}) = \nu(\underline{y})$ *may occur only for some special kind of labelled smd-nets.*

Let (S, T, F) be a net. We say that $s \in S$ is a **forward conflict place** for $t_1, t_2 \in T$ if and only if $t_1 \neq t_2$ and $s \in {}^\bullet t_1 \cap {}^\bullet t_2$.

Let (S, T, F, M_0) be a contact-free marked net, and let $M \in [M_0\rangle$.

We say that the transitions t_1, t_2 are in **conflict under** M if and only if:

(1) M enables t_1 and M enables t_2.

(2) M contains a forward conflict place for t_1 and t_2.

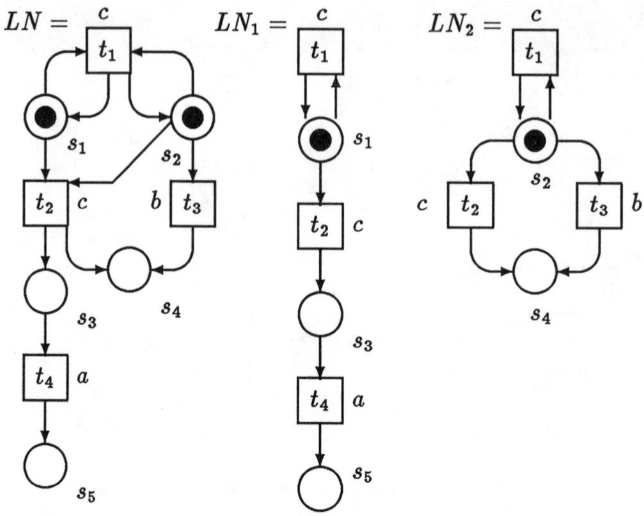

$VES = Pref(\underline{c}^*\underline{ca} \cup \underline{c}^*\underline{b}), ES_1 = Pref(c^*ca), ES_2 = Pref(c^*(c \cup b))$,
and $\underline{cab} \in (ES_1 \times ES_2 \cap Vev^*) \setminus VES$.

Figure 2.17. Theorem 2.10 does not hold for labelled nets

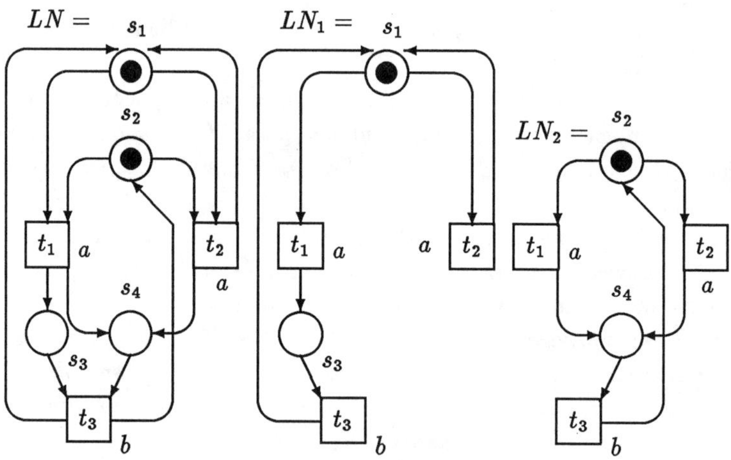

$VES(LN) = \{\underline{ab}\}^*\{\underline{a}, \underline{\varepsilon}\}, VTS(N) = \{\underline{t_1 t_3}\}^*\{\underline{t}_1, \underline{t}_2, \underline{\varepsilon}\}$.

Figure 2.18. Non-equivalence of VES and VTS Semantics

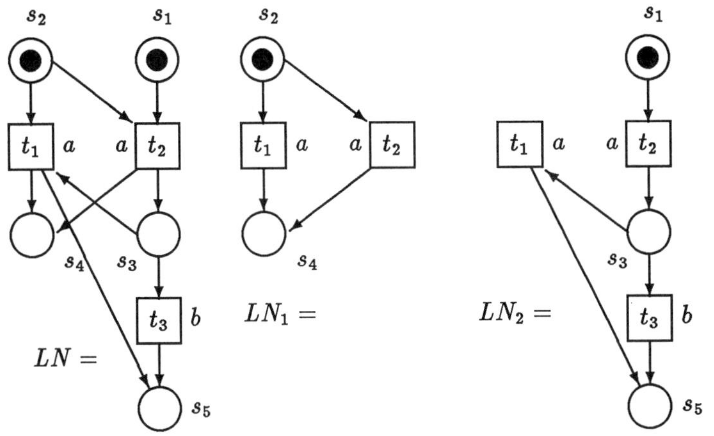

Figure 2.19. LN_1 ambiguous for a, but LN unambiguous.

The same definition holds for labelled smd-nets.
A labelled smd-net $LN = (S, T, F, M_0, \nu, Ev)$ is said to be **ambiguous for** $a \in Ev$ if and only if there are $M \in [M_0\rangle$, $t_1, t_2 \in T$ such that:

(1) t_1, t_2 are in conflict under M,

(2) $\nu(t_1) = \nu(t_2) = a$.

A labelled smd-net $LN = (S, T, F, M_0, \nu, Ev)$ is said to be **ambiguous** if there is $a \in Ev$ such that LN is ambiguous for a, and it is said to be **unambiguous** if it is not ambiguous. Clearly the nets from Fig. 2.18 are all ambiguous.

Corollary 2.12 *LN is ambiguous $\Rightarrow (\exists i)LN_i$ is ambiguous.* □

Fig. 2.19 shows that *the opposite implication is not true.*

Theorem 2.12 (Ambiguity of Labelled Nets)
Let $LN = (S, T, F, M_0, \nu, Ev)$ be a labelled smd-net, with $N = (S, T, F, M_0)$. The following are equivalent:
(1) LN is ambiguous.
(2) $(\exists \underline{x}, \underline{y} \in VTS(N))\underline{x} \neq \underline{y} \wedge \nu(\underline{x}) = \nu(\underline{y})$.

Proof. (2) $\Rightarrow$ (1) Let $SMD(N) = \{N_1, \ldots, N_n\}$ with $N_i = (S_i, T_i, F_i, M_0^i)$, $i = 1, \ldots, n$. Let $Q = \{(\underline{x}, \underline{y}) \in VTS(N) \times VTS(N) | \underline{x} \neq \underline{y} \wedge \nu(\underline{x}) = \nu(\underline{y})\}$. From (2) it follows that $Q \neq \emptyset$. Hence there is $(\underline{x}, \underline{y}) \in Q$ such that if $(\underline{z}, \underline{u}) \in Q$, then length$(\underline{x}) \leq$ length$(\underline{z})$. Clearly $\underline{x} \neq \underline{\varepsilon}$. Hence we may write $\underline{x} = \underline{x}'\underline{t}$ for some $\underline{x}' \in VTS(N)$ and $t \in T$. We have $\nu(\underline{x}) = \nu(\underline{x}')\nu(\underline{t})$. hence, by $\nu(\underline{x}) = \nu(\underline{y})$, there is $\underline{y}' \in VFS(N)$ and $t' \in T$ such that $\underline{y} = \underline{y}'\underline{t}'$ and $\nu(\underline{y}') = \nu(\underline{x}'), \nu(\underline{t}) = \nu(\underline{t}')$. If

$\underline{x}' \neq \underline{y}'$ then $(\underline{x}', \underline{y}') \in Q$, contradicting length$(\underline{x}) \leq$ length$(\underline{x}')$. Hence $\underline{x}' = \underline{y}'$. Let $\underline{x}' = \underline{t}_1 \ldots \underline{t}_m = \underline{y}'$. Since $\underline{x} = \underline{t}_1 \ldots \underline{t}_m \underline{t} \in VTS(N)$, then $(\exists M \in [M_0\rangle) M_0[t_1 \ldots t_m\rangle M$. It suffices to show that t and t' are in conflict under M. Since $\underline{x}, \underline{y} \in VTS(N)$ and $\underline{x}' = \underline{y}'$, t and t' must both be enabled by M, so (1) of the conflict definition is satisfied. If ${}^\bullet t \cap {}^\bullet t' = \emptyset$, then $\{t, t'\}$ is enabled by M, which by Corollary 2.9(2) means that $(\forall i) t \notin T_i \vee t' \notin T_i$, and hence $\nu(t) \neq \nu(t')$ by (2) of the definition of ν, which is a contradiction, so ${}^\bullet t \cap {}^\bullet t' \neq \emptyset$, so (2) of the conflict definition is also satisfied.
(1) $\Rightarrow$ (2) Let $M_0[t_1 \ldots t_m\rangle M$, t and t' are in conflict under M and $\nu(t) = \nu(t')$. Hence $t \neq t'$ and clearly $\underline{t}_1 \ldots \underline{t}_m \underline{t} \neq \underline{t}_1 \ldots \underline{t}_m \underline{t}'$, $\nu(\underline{t}_1 \ldots \underline{t}_m \underline{t}) = \nu(\underline{t}_1 \ldots \underline{t}_m \underline{t}')$. □

We end this section defining labelled sets with final markings. This allows nets to generate sets of vector sequences which are not necessary prefix-closed. By a **closed labelled smd-net** we mean a 7-tuple:

$$CLN = (S, T, F, M_0, FMar, \nu, Ev)$$

where (S, T, F, M_0, ν, Ev) is a labelled smd-net, and $FMar$ is a set of **final markings** of it.
The set of **vector event sequences generated by** CLN is defined as:

$$VES(CLN) = \{\nu(\underline{x}) | (\exists M \in FMar) M_0[\underline{x}\rangle M\}.$$

If (S, T, F, M_0) is a state machine net itself, i.e., if CLN is a **closed labelled state machine net**, then the set of **event sequences generated by** CLN is defined as:

$$ES(CLN) = \{\nu(x) | (\exists M \in FMar) M_0[x\rangle M\}.$$

Fig. 2.20 shows some examples of closed labelled smd-nets.
A closed labelled smd-net is said to be **ambiguous** if and only if there are $a \in Ev$, $M \in [M_0\rangle$, $t_1, t_2 \in T$ such that:

(1) t_1, t_2 are in conflict under M,

(2) $\nu(t_1) = \nu(t_2) = a$,

A closed labelled smd-net is called **unambiguous** if it is not ambiguous.
It is intuitively obvious that closed labelled state machine nets and finite automata are isomorphic objects, and now we define this relationship formally.
Let $\mathcal{M} = (Q, A, \delta, q_0, \mathcal{F})$ be a finite automaton (see Appendix B). Define the following closed labelled net:

$$CLN(\mathcal{M}) = (S, T, F, M_0, FMar, \nu, Ev),$$

where

$$S = \{q | (\exists x \in A^*) q \in \hat{\delta}(q_0, x)\} \cap \{q | (\exists y \in A^*) \hat{\delta}(q, y) \cap \mathcal{F} = \emptyset\},$$

i.e., the set of all states reachable from the initial state and which lead to a final state,

$$T = \{(q_1, a, q_2) | q_1, q_2 \in S \wedge q_2 \in \delta(q_1, a)\},$$

$$F = \{(q_1, (q_1, a, q_2)) | q_1, q_2 \in S \wedge (q_1, a, q_2) \in T\} \cup$$
$$\{((q_1, a, q_2), q_2) | q_1, q_2 \in S \wedge (q_1, a, q_2) \in T\},$$

$$M_0 = \{s_0\},$$

$$FMar = \{s | s \in \mathcal{F} \cap S\},$$

$$Ev = A,$$

$$(\forall (q_1, a, q_2) \in T) \;\; \nu((q_1, a, q_2)) = a.$$

The net $CLN(\mathcal{M})$ is a closed labelled state machine net, and clearly $ES(CLN(\mathcal{M})) = \{x | M_0[x\rangle M \wedge M \in FMar\} = L(\mathcal{M}) = \{x | \hat{\delta}(q_0, x) \in \mathcal{F}\}$.
Let $CLN(\mathcal{M}) = (S, T, F, M_0, FMar, \nu, Ev)$ be a closed labelled finite state machine. Because δ is a total function we have to consider two cases. If $(\exists a \in Ev)(\exists s \in S) a \notin \nu(s^\bullet)$ then we define the following finite automaton:

$$\mathcal{M}(CLN) = (Q, A, \delta, q_0, \mathcal{F})$$

where:

$$Q = S \cup \{q_{garbage}\}$$

$$A = Ev,$$

$$q_0 = s_0, \; where \; M_0 = \{s_0\},$$

$$\mathcal{F} = \{s | \{s\} \in FMar\},$$

$$\delta(q, a) = \begin{cases} \bigcup_{\{q\} = {}^\bullet t \wedge \nu(t) = a} t^\bullet & if \; (\exists t \in T)\{q\} = {}^\bullet t \wedge \nu(t) = a \\ \{q_{garbage}\} & otherwise. \end{cases}$$

If $(\forall a \in Ev)(\forall s \in S) a \in \nu(s^\bullet)$, then we define $Q = S, \delta(q, a) = \bigcup_{\{q\} = {}^\bullet t \wedge \nu(t) = a} t^\bullet$, and $A, q_0, \mathcal{F}$ as above. One may easily show that $L(\mathcal{M}(CLN)) = ES(CLN)$ in both cases.

2.3.4 Labelled Petri Nets defined by Path Programs

The general strategy for defining Petri net semantics of COSY path programs may be formulated as follows:
(1) construct labelled finite state machines equivalent to single paths in the sense that they both generate the same sets of firing sequences (behaviours).
(2) compose a labelled net from these finite state machines.
The construction should preserve not only vector firing sequences generated but also the other behavioural properties as deadlock-freeness, ambiguity, etc. However, when we come to details, the implementation of the above general rules

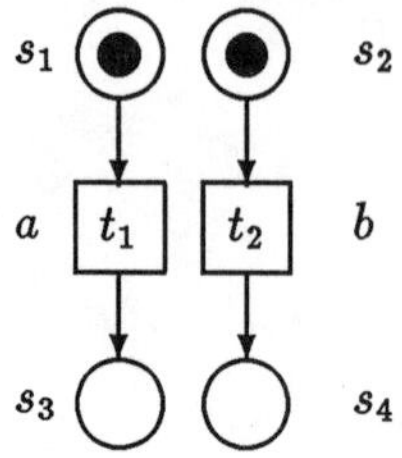

$Vev = \{(a, \varepsilon), (\varepsilon, b)\}$ $FMar = \{\{s_1, s_4\}, \{s_2, s_3\}\}$

$VES = \{\underline{a}, \underline{b}\}$

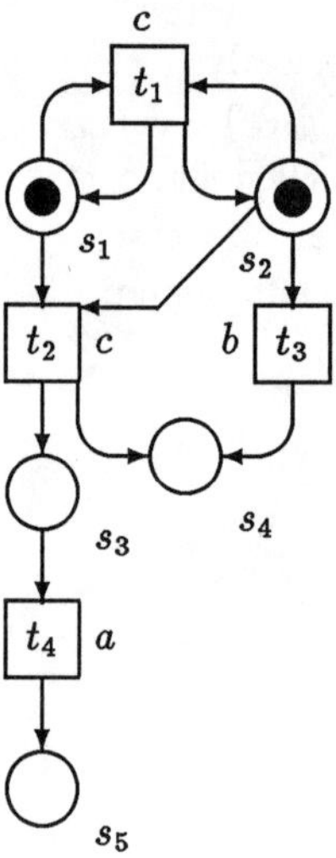

$Vev = \{(a, \varepsilon), (\varepsilon, b), (c, c)\}$ $FMar = \{\{s_4, s_5\}\}$

$VES = \underline{c}^*\underline{ca} \cup \underline{c}^*\underline{b}$

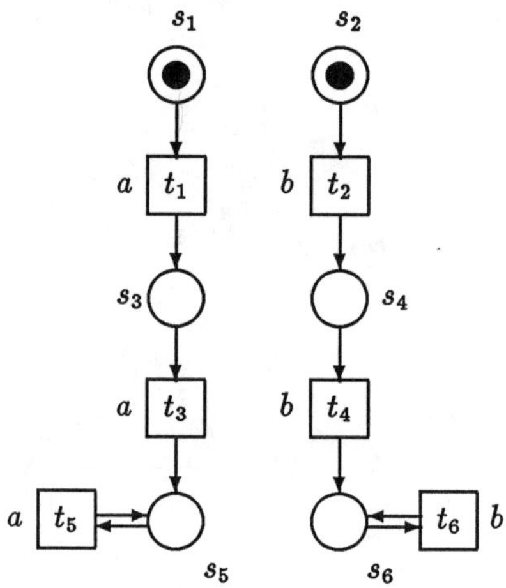

$Vev = \{(a, \varepsilon), (\varepsilon, b)\}$

$FMar = \{\{s_1, s_2\}, \{s_1, s_4\}, \{s_1, s_6\}, \{s_3, s_2\}, \{s_3, s_6\}, \{s_5, s_4\}, \{s_5, s_2\}, \{s_5, s_6\}\}$

$VES = \{\underline{a}, \underline{b}\}^* \setminus \{\underline{a}, \underline{b}\}$

Figure 2.20. Closed labelled smd-nets

is not obvious. There are two main sources of difficulties. The first is that the equivalence between regular expressions (and so simple paths) and /or finite automata (and so labelled state machines) defined as the equality of the languages they generate is *not sufficient.* The languages alone do not reflect for instance the ambiguity which may sometimes cause some problems for Petri nets. Let us consider the labelled smd-nets depicted in Fig. 2.21. The labelled state machines LN_1 and LN_1' treated as finite automata are equivalent, both generate the same event sequences, namely $(\{a\}\{b,c\})^*\{a,\varepsilon\}$, however LN_1 composed with LN_2 gives the net LN while LN_1' composed with the same LN_2 gives LN', and LN produces completely different behaviours than does LN'. For instance LN' deadlocks after an occurrence of the a on the right, i.e., the a which labels the transition t_1, while LN is live.
Note that $FT(N_1)$ differs from $FT(N_1')$ where N_1 and N_1' denote respectively LN_1 and LN_1' but without labels. These kinds of problems have been discussed in detail for instance in [Mil89], and they are one of the motivations behind the concept of *bisimulation* (see [Mil89] and Section 5.1). We think that the importance of the problems like those discussed above is often exaggerated in the (particularly CCS-related) literature, however we can not deny such problems do exist. The vector firing sequence semantics of COSY path programs cannot deal properly with them, but Petri net semantics, if chosen correctly, can.

To explain this point let us consider once more the labelled nets from Fig. 2.21, three simple paths:

$$P(1) : \underline{path}\ a; b, c\ \underline{end}$$

$$P'(1) : \underline{path}\ (a; b), (a; c)\ \underline{end}$$

$$P(2) : \underline{path}\ b; c\ \underline{end}$$

and two path programs: $Pr_8 = P_1P_2$, $Pr_8' = P_1'P_2$, i.e.:

$$Pr_8 = \begin{cases} P(1): & \underline{path}\ a; b, c\ \underline{end} \\ P(2): & \underline{path}\ b; c\ \underline{end} \end{cases}$$

$$Pr_8' = \begin{cases} P'(1): & \underline{path}\ (a; b), (a; c)\ \underline{end} \\ P(2): & \underline{path}\ b; c\ \underline{end} \end{cases}$$

First observe that

$$FS(P_1) = FS(P_1') = ES(LN_1) = ES(LN_1') = (\{a\}\{b,c\})^*\{a,\varepsilon\},$$

$$VFS(Pr_8) = VFS(Pr_8') = VES(LN) = VES(LN') =$$

$$\{\underline{abac}\}^*\{\underline{aba}, \underline{ab}, \underline{a}, \underline{\varepsilon}\}.$$

The net LN_1 is isomorphic to the minimal deterministic automaton accepting $(\{a\}\{b,c\})^*\{a,\varepsilon\}$, while the automaton represented by LN_1' is nondeterministic.

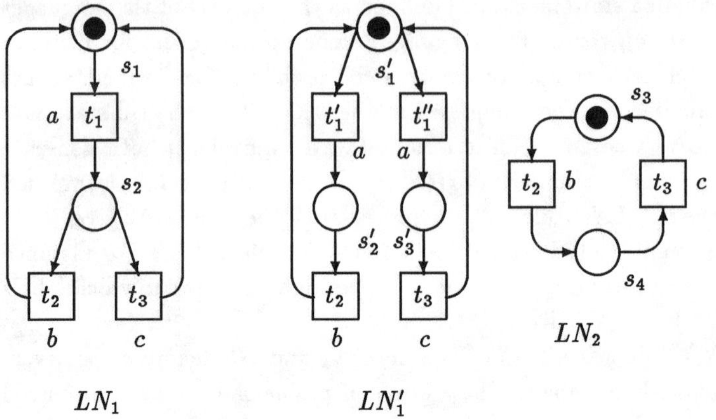

$$ES(LN_1) = ES(LN_1') = (\{a\}\{b,c\})^*\{a,\varepsilon\}$$

$$FT(N_1) = (\{t_1\}\{t_2,t_3\})^*\{t_1,\varepsilon\} \qquad FT(N_1') = (\{t_1't_2, t_1''t_3\})^*\{t_1', t_1'', \varepsilon\}$$

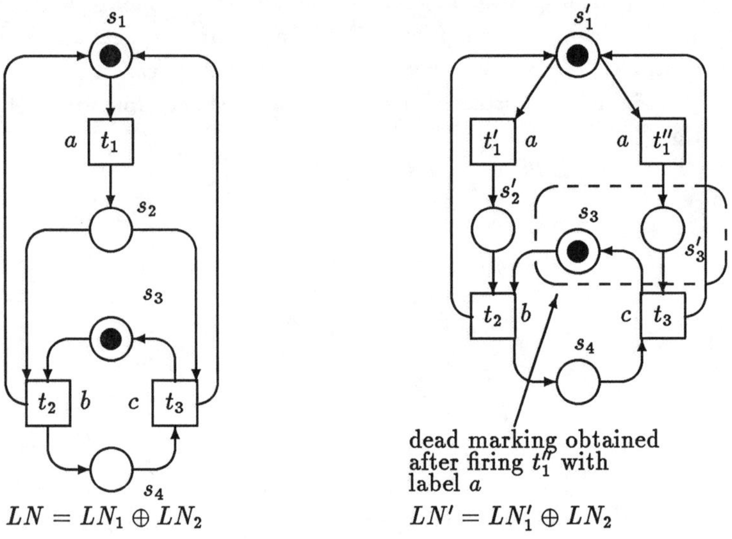

Figure 2.21. Illustration of a problem caused by use of language equivalence alone. $\oplus$ denotes "merging" of labelled state machines by identifying the same transitions (which must have the same labels). N_1 and N_1' denote LN_1 and LN_1' without labelling respectively.

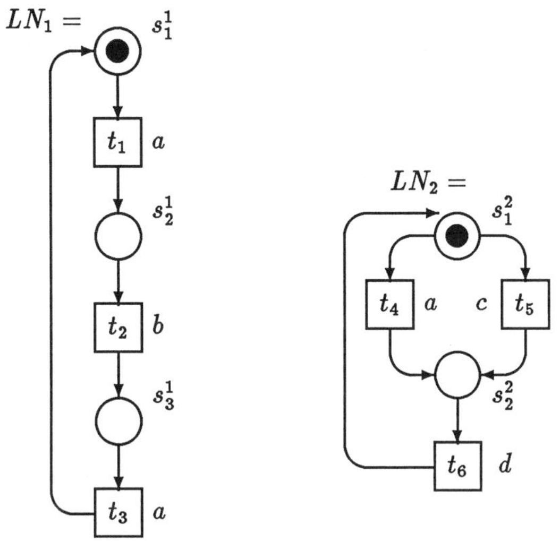

$P_1 = \underline{path}\ a; b; a\ \underline{end}$ $\qquad P_2 = \underline{path}\ a, c; d\ \underline{end}$

Figure 2.22. Labelled state machines for P_1, P_2

Thus, as far as the vector sequence semantics is concerned, the path programs Pr and Pr' are equivalent, on the other hand the net LN_1' is much more "close" to P_1' than LN_1, which is rather "close" to P_1, so LN is rather a net model of Pr_8, while LN' is a net model of Pr_8'.

The nets LN and LN' are clearly *not equivalent*, although they generate the same vector event sequences (but *different* vector transition sequences). What is the essential difference between P_1 and P_1', or rather, what is the essential difference between expressions $a; b, c$ and $(a; b), (a; c)$? In $(a; b), (a; c)$ we have a choice between a and a, the meaning of this expression treated as the description of a system may be defined as "perform either a sequence ab or a sequence ac", in other words, the choice between b and c must be made *before* performing a. We shall call such expressions "ambiguous" and we will give their formal definition later. For path programs with ambiguous subexpressions, a naive use of the vector sequence semantics may *sometimes be inconsistent* with the underlying intuition. The vector sequences are a very natural generalization of string sequences, and in terms of string sequences only, we cannot distinguish between $a; b, c$ and $(a; b), (a, c)$, since both expressions generate the same language $\{ab, ac\}$. The set of prefixes of $\{ab, ac\}$ is $\{\varepsilon, a, ab, ac\}$ and suppose that this set of prefixes describes the behaviour of a certain sequential system. What can we say about this system if we see such a behaviour? Well: "first a may be performed, and next either b or c may be performed", – this seems to be the most natural description of the system, and this corresponds directly to the expression

$a; b, c$. It is hardly possible to suspect that a history a might be ambiguous, and that the decision about b and c might be made before performing a. In this sense, strings generated by regular expressions create a more natural semantics for unambiguous expressions than for ambiguous ones.
This means that the vector sequence semantics of COSY path programs is intuitively well-founded *only* for unambiguous path programs. This restriction is however more theoretical than practical, since if P is an ambiguous path, we can always construct from it an unambiguous and intuitively equivalent path P' by differentiating appropriate occurrences of the same event. For instance, we can replace $(a; b), (a; c)$ by $(a_1; b), (a_2; c)$. We do not know any real example showing that we could not do this, in fact the a before b in $(a; b), (a; c)$ differs semantically from the a before c, since the first a enforces the performance of b, while the second enforces the performance of c, so differentiating them is justified.
Nevertheless, the Petri net approach gives us an opportunity to define a well-founded semantics for ambiguous paths as well, provided that we find a well-founded and ambiguity preserving transformation of ambiguous expressions into non-deterministic automata. Finding such a transformation is slightly more complex than it might seem after analyzing some simple cases, because we do not allow transitions to be labelled by empty string symbol ($\nu : T \longrightarrow Ev \wedge \varepsilon \notin Ev$), and in such a case, as Fig. 2.14 b shows, there is no straightforward transformation of regular expressions with Kleene stars into finite automata.
The second problem is that even if we finally find suitable transformations of regular expressions into automata, i.e., labelled finite state machine nets, when those labelled state machine nets have transitions with identical labels, it is still not always clear how to construct a labelled net corresponding to the whole path program. To illustrate this case let us consider the following path program:

$$Pr_9 = \begin{cases} P(1): & \underline{path}\ a; b; a\ \underline{end} \\ P(2): & \underline{path}\ a, c; d\ \underline{end} \end{cases}$$

The labelled state machines corresponding to P_1 and P_2 are depicted in Fig. 2.22, but how should they be composed to obtain one labelled smd-net equivalent to Pr_9?

Could we identify the transition t_1 of LN_1 with t_4 of LN_2, and if so, what should we do with t_3 of LN_1? Recall that the basic rule on which the whole theory of COSY path programs is based says that an event belonging to many paths can be performed only if *all* paths containing it enable the event. This means that we must identify *both* t_1 and t_3 of LN_1 with t_4 of LN_2, which leads to the construction in Fig. 2.23. Note that this type of construction will lead to a quite complex net even for simple path programs which have a number of repeated events. On the other hand, for path programs with no repeated events the construction is quite obvious and simple. The final conclusion is that although often the Petri net semantics of COSY path programs is rather straightforward, we cannot expect such straightforwardness in the general case.

2.3.5 Uniquely Named and Labelled Uniquely Named Regular Expressions

In this section we will introduce a tool used then to find a unique and ambiguity preserving mapping of regular expressions into labelled state machines (automata).

For every regular expression R, let Σ_R denote the alphabet of R. Two regular expressions R_1 and R_2 are called *isomorphic* if and only if there is an onto injection $\gamma : \Sigma_{R_1} \longrightarrow \Sigma_{R_2}$ such that R_2 can be obtained from R_1 by replacing every occurrence of a in R_1 by $\gamma(a)$ for all $a \in \Sigma_{R_1}$. For instance $a; b, a; c$ and $b; d, b; c$ are isomorphic with $\gamma(a) = b, \gamma(b) = d, \gamma(c) = c$. Isomorphic regular expressions can be treated as identical up to isomorphism of their alphabets.

A regular expression R is said to be **uniquely named** (abbreviated unr-expression) if and only if there are no repeated event names in R.

For example $a; b, c$ is uniquely named, while $a; b; a$ is not.

By a **labelled uniquely named regular expression** we mean a triple:

$$LR = (R, \psi, Ev)$$

where:

> R is an unr-expression with the alphabet Σ_R,
>
> $\psi : \Sigma_R \longrightarrow Ev$ is a mapping such that $\psi(\Sigma_R) = Ev$,
>
> Ev is a set of labels (events).

Clearly we can extend ψ to the homomorphism $\psi : \Sigma_R^* \longrightarrow Ev^*$ by $\psi(\varepsilon) = \varepsilon$, $\psi(xa) = \psi(x)\psi(a)$ for $x \in \Sigma_R^*$, $a \in \Sigma_R$ and in the sequel we will always regard ψ as such a homomorphism.

For every labelled unr-expression $LR = (R, \psi, Ev)$, let $L(LR)$, the language defined by LR, be defined as: $L(LR) = \psi(L(R))$, where $L(R)$ is the language defined by R (see Appendix B).

By $\psi(R)$ we denote an expression obtained from R by replacing every $a \in \Sigma_R$ by $\psi(a)$. Clearly $\psi(R)$ is a *regular expression*. On the other hand, *for every regular expression* R we can easily find a *labelled unr-expression* $LR = (R', \psi, \Sigma_R)$ such that $R = \psi(R')$. In fact we have infinitely many such unr-expressions for every regular expression R, but as one may easily observe if $LR_i = (R'_i, \psi_i.\Sigma_R)$, for $i = 1, 2$, and $R = \psi_1(R'_1) = \psi_2(R'_2)$, then R'_1 and R'_2 are *isomorphic*. For example, take $R = a_1; b; a$. Then for instance $LR_1 = (R'_1, \psi_1, \{a, b\})$, $LR_2 = (R'_2, \psi_2, \{a, b\})$, where $R'_1 = a_1; b; a_2$, $R'_2 = c; d; e$, $\psi_1(a_1) = \psi_2(c) = a$, $\psi_1(b) = \psi_2(d) = b$, $\psi_1(a_2) = \psi_2(e) = a$, satisfy $\psi_1(R'_1) = \psi_1(R_2) = R$. Thus, we

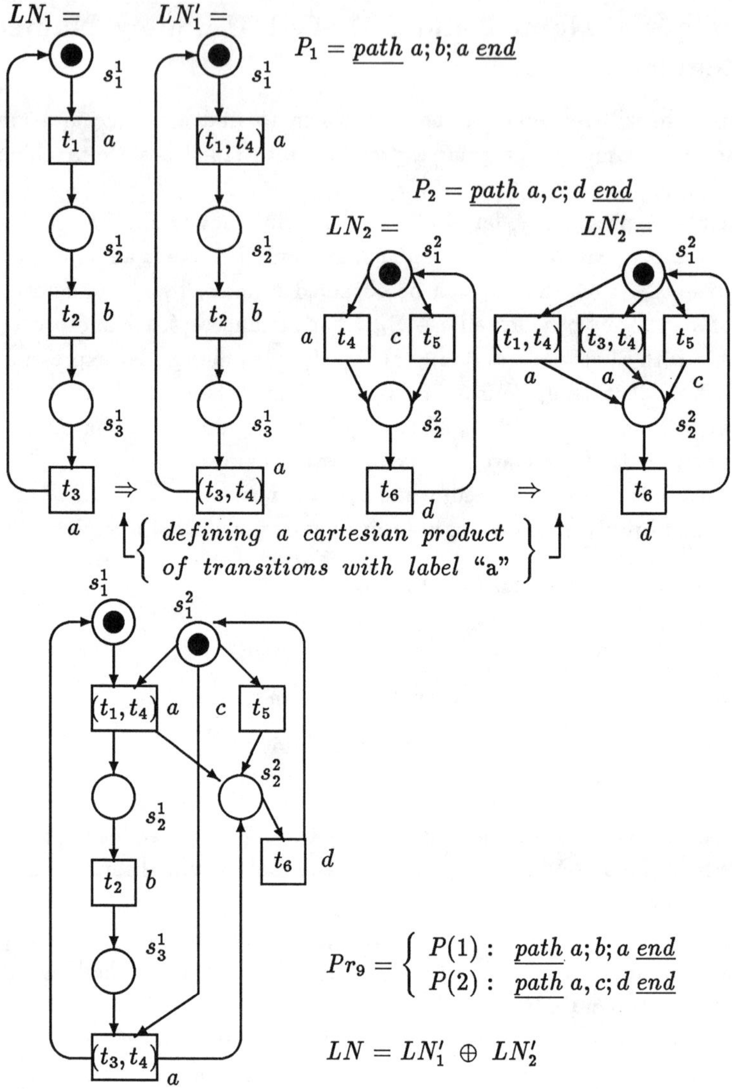

Figure 2.23. Illustration of the transformation of path programs into labelled smd-nets

can identify *regular expressions* with *labelled uniquely named regular expressions*! The relationship between regular expressions and labelled unr-expressions is one-to-one up to isomorphism of names in unr-expressions.
A labelled unr-expression $LR = (R, \psi, Ev)$ is said to be **ambiguous** if and only if:

$$(\exists x_1, x_2 \in Pref(L(R)))x_1 \neq x_2 \wedge \psi(x_1) = \psi(x_2).$$

Note that if $LR_1 = (R_1, \psi_1, Ev), LR_2 = (R_2, \psi_2, Ev)$ and $\psi(R_1 = \psi_2(R_2)$, then R_1 is ambiguous if and only if R_2 is. Hence we can define ambiguity for regular expressions in the following way.
A regular expression R is said to be **ambiguous** if any labelled unr-expression corresponding to R is ambiguous.
A labelled unr-expression (i.e. also regular expression) is called **unambiguous** if it is not ambiguous. In particular, unr-expressions are unambiguous.
Let $LR = (R', \psi, Ev)$ be a labelled unr-expression, and let $CLN(LR)$ be the closed labelled finite state machine defined in the following two steps:

Step 1. Let $CLN(R')$ be a closed *labelled* finite state machine representing the *minimal* finite automaton accepting the language $L(R')$[3].

Step 2. Suppose that $CLN(R') = (S, T, F, M_0, FMar, \nu', \Sigma_{R'})$. Define $CLN(LR) = (S, T, F, M_0, FMar, \nu, Ev)$, where ν is given by: $(\forall t \in T)\nu(t) = \psi(\nu'(t))$.

Fig. 2.14 (b) shows that in Step 1 we *cannot* replace labelled finite state machines by finite state machines although R' is a unr-expression.
The net $CLN(LR)$ will be called the **closed labelled finite state machine defined by LR**. Because of a one-to-one relationship (up to isomorphism of names in unr-expressions) between regular expressions and labelled unr-expressions, we can apply the above definition to regular expressions, simply by saying that for every regular expression R, the closed labelled finite state machine defined by R, $CLN(R)$, is $CLN(LR)$ where LR is any labelled unr-expression corresponding to R. The construction of $CLN(LR)$ is unique up to isomorphism of transition names. An example of this construction is depicted in Fig. 2.24.

Corollary 2.13
(1) For every regular expression R, $L(R) = ES(CLN(R))$.
(2) R is ambiguous $\Leftrightarrow$ $CLN(R)$ is ambiguous $\Leftrightarrow$ $\mathcal{M}(CLN(R))$ is non-deterministic. □

[3] A fast algorithm for finding the minimal state automaton can be found in [BS86].

$R = (a, b*; c)*; a \quad \Leftrightarrow \quad LR = (R', \psi, \{a, b, c\})$, where

$R' = (a_1, b*; c)*; a_2$ and $\psi(a_1) = \psi(a_2) = a, \psi(b) = b, \psi(c) = c$

$CLN(R') =$

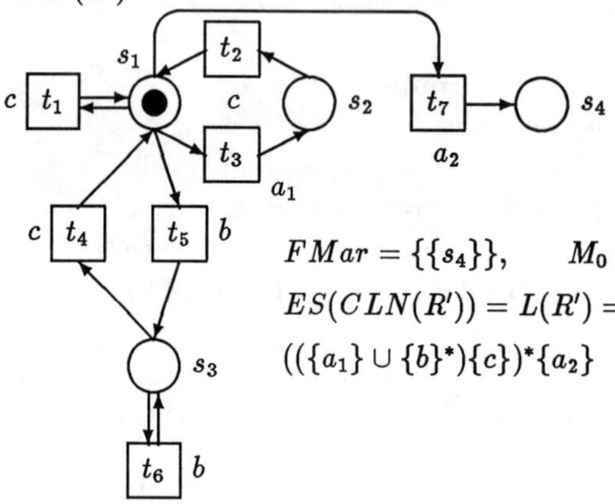

$FMar = \{\{s_4\}\}, \qquad M_0 = \{s_1\},$

$ES(CLN(R')) = L(R') =$

$((\{a_1\} \cup \{b\}^*)\{c\})^*\{a_2\}$

$CLN(R) = CLN(LR) =$

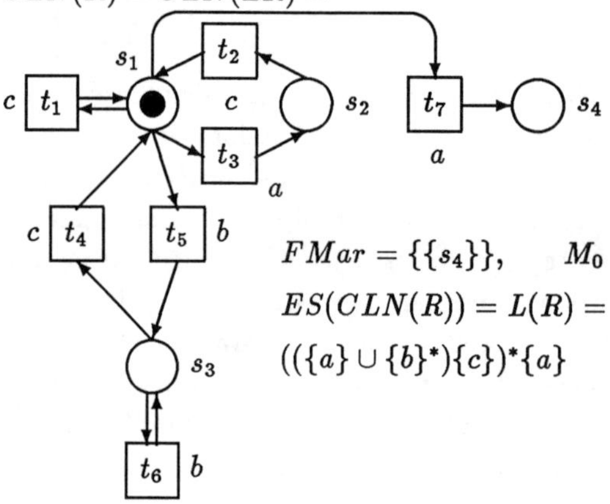

$FMar = \{\{s_4\}\}, \qquad M_0 = \{s_1\},$

$ES(CLN(R)) = L(R) = L(LR) =$

$((\{a\} \cup \{b\}^*)\{c\})^*\{a\}$

Figure 2.24. Closed labelled finite state machine net for a regular expression

2.3.6 Labelled Nets and Products of Regular Expressions

We shall now construct a closed labelled net modelling the behaviour of a product of regular expressions. The essence of this construct has been presented in Fig. 2.23.

Let $R_1||\dots||R_n$ be a product of regular expressions,
and let $CLN(R_1),\dots,CLN(R_n)$ be closed labelled state machine nets defined by $R_1,\dots,R_n$. Assume that $CLN(R_i)=(S_i,T_i,F_i,M_0^i,FMar_i,\nu_i,Ev_i)$ for $i=1,\dots,n$, and that $S_i\cap S_j=\emptyset$ if $i\neq j$. If for every i the mapping ν_i is a one-to-one function, then we can build the closed labelled smd-net that models the behaviour of $R_1||\dots||R_n$ directly from the $CLN(R_i)$'s. We can simply merge all $CLN(R_i)$'s by identifying transitions with identical labels, a construction similar to composing finite state machine nets to smd-nets. However, if at least one mapping ν_i is not a one-to-one function, we must adjust the $CLN(R_i)$'s to the "ready-to-merge" form. The idea of such an adjustment was illustrated in Fig. 2.23, where, for instance, we replaced the substructure of LN_2:

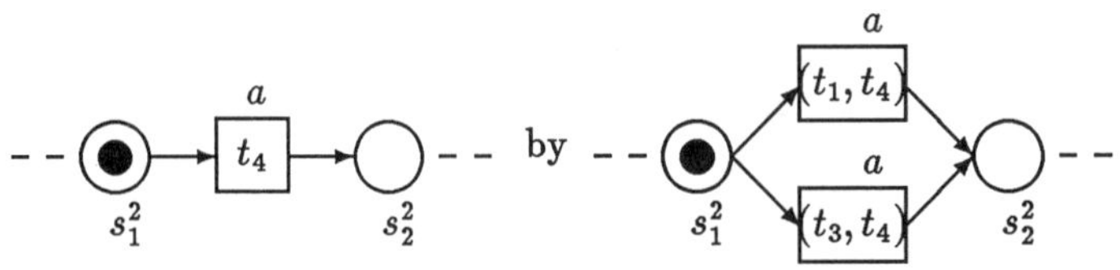

We will now define this construct in a formal way.
Let $t\in T_i$ for some i, with $\nu_i(t)=a$, $i(a)=\{i_1,\dots,i_k\}$ and $i_1<\dots<i_k$, with $i=i_r$. Define $\hat{t}$ as the following Cartesian product:

$$\hat{t}=\nu_{i_1}^{-1}(a)\times\dots\times\nu_{i_{r-1}}^{-1}(a)\times\{t\}\times\nu_{i_{r+1}}^{-1}(a)\times\dots\times\nu_{i_k}^{-1}(a).$$

Note that $t',t''\in T_i$ and $t'\neq t''$ implies $\hat{t}'\cap\hat{t}''=\emptyset$. Now, for every $i=1,\dots,n$, let

$$T_i'=\bigcup_{t\in T_i}\hat{t}$$

and $F_i'=\{(s,t')|(\exists t\in T_i)(s,t)\in F_i\wedge t'\in\hat{t}\}\cup\{(t',s)|(\exists t\in T_i)(t,s)\in F_i\wedge t'\in\hat{t}\}$. Clearly $F_i'\subseteq S_i\times T_i'\cup T_i'\times S_i$ for $i=1,\dots,n$. Finally, for every $i=1,\dots,n$, let CLN_i be the following closed labelled state machine net:

$$CLN_i=(S_i,T_i',F_i',M_0^i,FMar_i,\nu_i',Ev_i)$$

where $\nu_i':T_i'\longrightarrow Ev_i$ is defined by $(\forall t'\in T_i')\nu_i'(t')=\nu_i(t)$ where $t'\in\hat{t}$. The definition of ν_i' is correct since $t'\neq t''$ and $t',t''\in T_i$ implies $\hat{t}'\cap\hat{t}''=\emptyset$. Let $CLN(R_1||\dots||R_n)=(S,T,F,M_0,FMar,\nu,Ev)$ be the closed labelled smd-net defined by:

$S = S_1 \cup \ldots \cup S_n$.

$T = T_1' \cup \ldots \cup T_n'$.

$F = F_1' \cup \ldots \cup F_n'$.

$M_0 = M_0^1 \cup \ldots \cup M_0^n$.

$FMar = \{M \subseteq S | (\forall i) M \cap S_i \subseteq FMar_i\}$.

$Ev = Ev_1 \cup \ldots \cup Ev_n$.

$\nu : T \longrightarrow Ev$ and $(\forall t' \in T)\nu(t') = \nu_i(t')$ where $t' \in T_i$ for some i.

From the definition of $\hat{t}$ it follows that $(\forall t' \in T_i)(\forall t'' \in T_j)\hat{t}' \cap \hat{t}'' \neq \emptyset \Rightarrow \nu_i(t') = \nu_j(t'')$, which assures the correctness of the definition of ν. Note that if just one ν_i is not one-to-one, then the corresponding CLN_i is ambiguous, even if all $CLN(R_i)$'s are unambiguous. However, as we shall prove below, the whole of $CLN(R_1||\ldots||R_n)$ is unambiguous if all R_i's are unambiguous. The above construction is illustrated in Fig. 2.25 through Fig. 2.28. Fig. 2.25 shows the case where the same event occurs many times in more than one expression, while Fig. 2.26 shows how ambiguity of expressions is preserved. The case of a uniquely named concurrent product is shown in Fig. 2.27, and Fig. 2.28 presents the case when a uniquely named regular expression corresponds to a closed labelled state machine net with a many-to-one labeling function. The two theorems below show the formal equivalence between $R_1||\ldots||R_n$ and $CLN(R_1||\ldots||R_n)$.
Let $CLN(R_1||\ldots||R_n) = (S, T, F, M_0, FMar, \nu, Ev)$, $LN = (S, T, F, M_0, \nu, Ev)$, $N = (S, T, F, M_0)$, and for $i = 1, \ldots, n$, let $CLN(R_i) = (S_i, T_i, F_i, M_0^i, FMar_i, \nu_i, Ev_i)$, $LN(R_i) = (S_i, T_i, F_i, M_0^i, \nu_i, Ev_i)$, $CLN_i = (S_i, T_i', F_i', M_0^i, FMar_i, \nu_i, Ev_i)$, $LN_i = (S_i, T_i', F_i', M_0^i, \nu_i, Ev_i)$, $N_i = (S_i, T_i', F_i', M_0^i)$ where CLN_i is created from $CLN(R_i)$ in the construction of $CLN(R_1||\ldots||R_n)$. Recall also that Vev denotes the set of vector events defined by $Ev_1, \ldots, Ev_n$, while Vtr denotes the set of vector transitions defined by $T_1, \ldots, T_n$.

Lemma 2.12 *Let $(x_1, x_2, \ldots, x_n) \in FT(N_1) \times \ldots \times FT(N_n)$ and $M^1, \ldots, M^n$ be such that $(\nu(x_1), \ldots, \nu(x_n)) \in Vev^*$ and $M_0^i[x_i\rangle M^i$ for $i = 1, \ldots, n$. Then there is $\underline{y} \in VTS(N)$ such that $\nu(\underline{y}) = (\nu(x_1), \ldots, \nu(x_n))$ and $M_0[\underline{y}\rangle(M^1 \cup \ldots \cup M^n)$.*

Proof. Let $(\nu(x_1), \ldots, \nu(x_n)) = \underline{a}_1 \ldots \underline{a}_m$. Consider a_j and let $r \in i(a_j)$. Since $\nu(x_r) = pr_r(\underline{a}_1 \ldots \underline{a}_m)$, then there is $\tau_r^j \in T_r'$ such that $x_r = x_r' \tau_r^j x_r''$, $\nu(x_r) = \nu(x_r') a_j \nu(x_r'')$. Let M_1^r, M_2^r be markings in CLN_r such that $M_0^r[x_r'\rangle M_1^r[\tau_r^j\rangle M_2^r$. Suppose that $i(a_j) = \{i_1, \ldots, i_k\}$ with $i_1 < \ldots < i_k$. From the contruction of CLN_r we have that $\tau_r^j = (t_{ri_1}^j, \ldots, t_{ri_k}^j)$, where $t_{ri}^j \in T_i$ for all $i \in i(a_j)$. In particular $t_{rr}^j \in T_r$, and $M_1^r[t_{rr}^j\rangle M_2^r$ in $CLN(R_r)$. For every $r \in i(a_j)$ we can find such t_{rr}^j. Define $\tau^j = (t_{i_1 i_1}^j, \ldots, t_{i_k i_k}^j)$. Note that for every $r \in i(a_j)$, $M_1^r[\tau^j\rangle M_2^r$ in CLN_r. Now define $y = \tau^1 \tau^2 \ldots \tau^m$. Since $M_1^r[\tau^j\rangle M_2^r$ for every j and every

$r \in i(a_j)$, then $pr_i(\underline{y}) \in FS(N_i)$ and $M_0^i[pr_i(\underline{y})\rangle M^i$ for $i = 1,\dots,n$. Hence, by Theorem 2.10, $\underline{y} \in VTS(N)$, $\nu(\underline{y}) = \underline{a}_1 \dots \underline{a}_m = (\nu(x_1),\dots,\nu(x_n))$, and $M_0[\underline{y}\rangle(M^1 \cup \dots \cup M^n)$. □

Note: It may happen that there is another $\underline{y}' \in VFS(N)$, such that $\nu(\underline{y}) = \nu(\underline{y}')$, $\underline{y}' \neq \underline{y}$, $M_0[\underline{y}'\rangle M'$ and $M' \neq M^1 \cup \dots \cup M^n$, but this is not a problem!

Theorem 2.13 (On Preserving Set of Vector Sequences) $VL(R_1||\dots||R_n) = VES(CLN(R_1||\dots||R_n))$.

Proof. By Corollary 2.13, $(\forall i)L(R_i) = ES(CLN(R_i))$. Furthermore, from the definition of CLN_i it follows that $ES(CLN_i) = ES(CLN(R_i))$ for $i = 1,\dots,n$. Hence: $VL(R_1||\dots||R_n) = \times_{i=1}^{n} L(R_i) \cap Vev^* = \times_{i=1}^{n} ES(CLN_i) \cap Vev^*$. Thus it suffices to show: $VES(CLN(R_1||\dots||R_n) = \times_{i=1}^{n} ES(CLN_i) \cap Vev^*$. Suppose $\underline{x} \in VES(CLN(R_1||\dots||R_n))$. Then there is $\underline{y} \in VTS(N)$ and $M \in FMar$ such that $\nu(\underline{y}) = \underline{x}$ and $M_0[\underline{y}\rangle M$. Let $M^i = M \cap S_i$ and $x_i = pr_i(\underline{y})$. From Theorem 2.10, $x_i \in FT(N_i)$. Moreover, $M_0^i[x_i\rangle M^i \subseteq FMar_i$ and $\nu(x_i) = pr_i(\underline{x})$. Hence $\underline{x} \in \times_{i=1}^{n} ES(CLN_i) \cap Vev^*$. Suppose now $\underline{x} \in \times_{i=1}^{n} ES(CLN_i) \cap Vev^*$. Then there are $(x_1,\dots,x_n) \in FT(N_1) \times \dots \times FT(N_n)$ and $M^1,\dots,M^n$, each $M^i \subseteq FMar_i$ such that $(\nu(x_1),\dots,\nu(x_n)) = \underline{x}$ and $M_0^i[x_i\rangle M^i$. Hence, by Lemma 2.12, there is $\underline{y} \in VTS(N)$ such that $\nu(\underline{y}) = \underline{x}$ and $M_0[\underline{y}\rangle(M^1 \cup \dots \cup M^n) \subseteq FMar$. Hence $\underline{x} \in VES(CLN(R_1||\dots||R_n))$. □

The formal definition of ambiguity for products of regular expressions is more complicated than one might have expected. The natural definition: "$R_1||\dots||R_n$ is ambiguous if there is at least one ambiguous R_i" does not work properly as the example below shows. Let $CR = a;b||(a;b),(b;a;(c;d),(c;e))$. Clearly $(a;b),(b;a;(c;d),(c;e))$ is ambiguous, there is a choice between c and c after a sequence ba, but in CR c will never be enabled, so as far as behaviours of CR are concerned, CR is unambiguous with $VL(CR) = \{\underline{ab}\}$.

Let R be a regular expression and $LR = (R', \psi, \Sigma_R)$ be a labelled uniquely named regular expression associated with R (recall all such LR's for a given R are identical up to isomorphism of alphabets).

A sequence $x \in Pref(L(R))$ is said to be an **ambiguous sequence** of R if there are $x_1, x_2 \in Pref(L(R'))$ such that $x_1 \neq x_2$ and $\psi(x_1) = \psi(x_2) = x$.

Clearly R is ambiguous if there is at least one ambiguous sequence in $L(R)$.

A product of regular expressions $CR = R_1||\dots||R_n$ is said to be **ambiguous** if and only if there is $\underline{x} \in Pref(VL(CR))$ such that for some $i \in \{1,\dots,n\}$ $pr_i(\underline{x})$ is an ambiguous sequence of R_i.

Note that now the ambiguity of CR implies the existence of at least one ambiguous R_i but not vice versa. We prove that although the nets CLN_i, $i = 1,\dots,n$, are usually ambiguous (except some cases when all R_i's are uniquely named), the whole net $CLN(CR)$ is ambiguous only if CR is ambiguous.

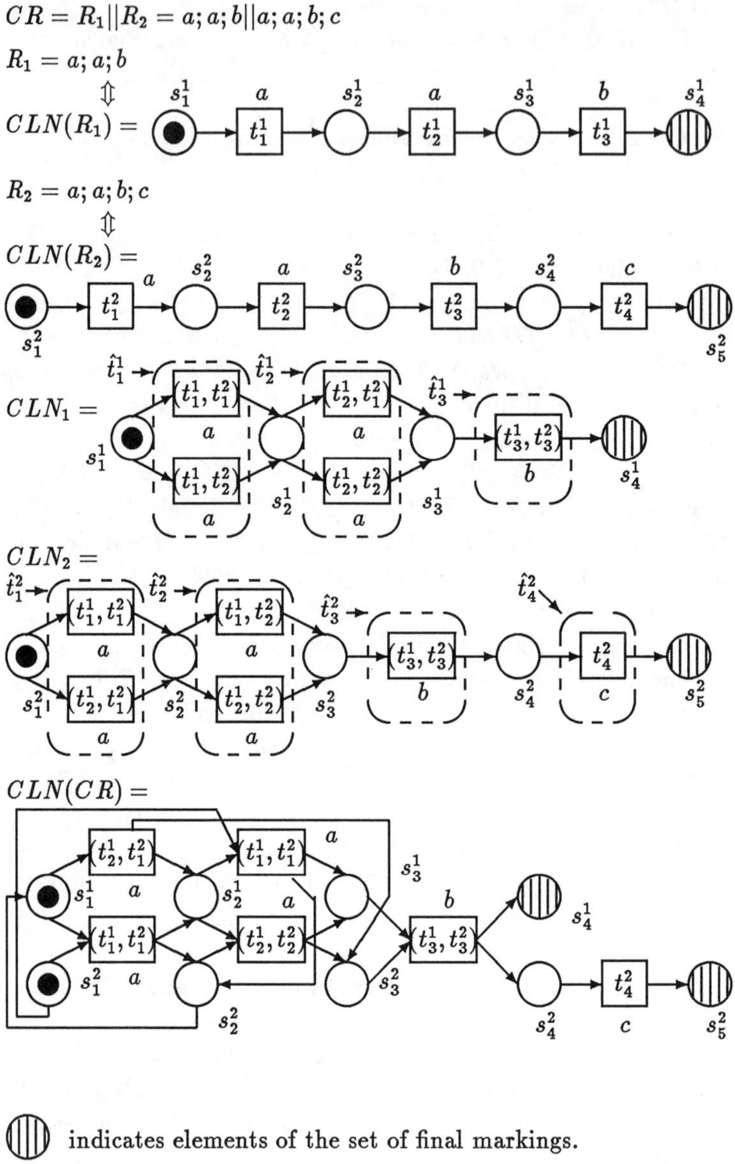

Figure 2.25. Illustration of the case when the same event (a here) occurs many times in more than one expression

Lemma 2.13
$VES(LN) = \times_{i=1}^{n} ES(LN_i) \cap Vev^* = \times_{i=1}^{n} ES(LN(R_i)) \cap Vev^*$.

Proof. Suppose $\underline{x} \in VES(LN)$. Then there is $\underline{y} \in VTS(N)$ such that $\nu(\underline{y}) = \underline{x}$. Hence, by Theorem 2.10, $pr_i(\underline{y}) \in FT(N_i)$ for all $i = 1, \ldots, n$. Thus, $pr_i(\underline{x}) = \nu(pr_i(\underline{y})) \in ES(LN_i)$ for $i = 1, \ldots, n$. Hence $\underline{x} \in \times_{i=1}^{n} ES(LN_i) \cap Vev^*$. Suppose $\underline{x} \in \times_{i=1}^{n} ES(LN_i) \cap Vev^*$. Then there is $(x_1, \ldots, x_n) \in FT(N_i) \times \ldots \times FT(N_n)$ such that $\nu(x_i) = pr_i(\underline{x})$ for $i = 1, \ldots, n$. Hence by Lemma 2.12, there is $\underline{y} \in VTS(N)$ such that $\nu(\underline{y}) = \underline{x}$. Hence $\underline{x} \in VES(LN)$, so we proved the first equality. The second follows from: $ES(LN_i) = ES(LN(R_i))$, $i = 1, \ldots, n$. □

Theorem 2.14 (On Preserving Ambiguity)
$R_1||\ldots||R_n$ is ambiguous $\Leftrightarrow$ $CLN(R_1||\ldots||R_n)$ is ambiguous.

Proof. ($\Rightarrow$) Suppose that $CR = R_1||\ldots||R_n$ is ambiguous, i.e. there is $\underline{x} = (x_1, \ldots, x_n) \in Pref(VL(CR))$ and $x_i = pr_i(\underline{x})$ is an ambiguous sequence for some R_i, $i \in \{1, \ldots, n\}$. Without loss of generality we can assume that $\underline{x} = \underline{x'a}$ and for every i, $pr_i(\underline{x'})$ is not an ambiguous sequence of R_i. By Lemma 2.13 and the construction of $LN(R_i)$ we have $Pref(VL(CR)) \subseteq VES(LN)$, so $\underline{x'}, \underline{x} \in VES(LN)$. From the construction of $CLN(R_i)$ it follows that: $(\exists y'_i \in T_i^*)(\exists t_1, t_2 \in T_i)t_1 \neq t_2$, $y'_it_1 \in FT(N_i)$, $y'_it_2 \in FT(N_i)$, $\nu_i(y'_it_1) = \nu_i(y'_it_2) = x_i$ $\wedge \nu_i(t_1) = \nu_i(t_2) = a$. Let $\underline{y'} \in VTS(N)$ be such that $\nu(\underline{y'}) = \underline{x'}$ (at this point we do not assume the unicity of $\underline{y'}$, although in fact $\underline{y'}$ is unique), and let M be a marking of N such that $M_0[\underline{y'}\rangle M$. Now, let $\tau \in T$ be such that: $\nu(\tau) = a$, M enables τ in LN. The existence of such a τ follows from: $M_0[\underline{y'}\rangle M$ and $\underline{x} = \underline{x'a} \in Pref(VL(CR)) \subseteq VES(LN)$. Suppose that $\tau = (t'_{i_1}, \ldots, t'_{i_k})$, $i(a) = \{i_1, \ldots, i_k\}$, $i_1 < \ldots < i_k$ and $i = i_r$. Let us define:
$\tau_1 = (t'_{i_1}, \ldots, t'_{i_{r-1}}, t_1, t'_{i_{r+1}}, \ldots, t'_{i_k}), \tau_2 = (t'_{i_1}, \ldots, t'_{i_{r-1}}, t_2, t'_{i_{r+1}}, \ldots, t'_{i_k})$.
Clearly $\tau_1 \neq \tau_2$, $\nu(\tau_1) = \nu(\tau_2) = \nu_i(t_1) = \nu_i(t_2) = a$. From the construction of $CLN(CR)$ we have that M enables both τ_1 and τ_2 in $CLN(CR)$, so $CLN(CR)$ is ambiguous. Hence, the ambiguity of CR implies the ambiguity of $CLN(CR)$.
($\Leftarrow$) Suppose that $CLN(CR)$ is ambiguous. This means that there is $\underline{x} = \underline{x'a} \in VES(LN)$ such that $\underline{y'} \in VTS(N)$, $\nu(\underline{y'}) = \underline{x'}$, and there are $\tau_1, \tau_2 \in T$ such that $\tau_1 \neq \tau_2$, $\underline{y'\tau_1}, \underline{y'\tau_2} \in VTS(N)$, $\nu(\tau_1) = \nu(\tau_2) = a$. Let, as usual, $i(a) = \{i_1, \ldots, i_k\}$, $i_1 < \ldots < i_k$, and suppose that $\tau_1 = (t^1_{i_1}, \ldots, t^1_{i_k})$, $\tau_2 = (t^2_{i_1}, \ldots, t^2_{i_k})$. Since $\tau_1 \neq \tau_2$ there is at least one i such that $t^1_i \neq t^2_i$. Suppose that $i = i_r \in \{i_1, \ldots, i_k\}$, and let $\tau'_2 = (t^1_{i_1}, \ldots, t^1_{i_{r-1}}, t^2_i, t^1_{i_{r+1}}, \ldots, t^1_{i_k})$. From the construction of $CLN(CR)$, we have that if $M_0[\underline{y'}\rangle M$ then M enables all of τ_1, τ_2 and τ'_2, hence $M \cap S_i$ enables both t^1_i and t^2_i in $CLN(R_i)$, so $pr_i(\underline{x})$ is an ambiguous sequence of R_i, or in other words $R_1||\ldots||R_n$ is ambiguous, which completes the proof. □

In this way we established a very close relationship between products of regular expressions and Petri nets. We have also shown that the Petri nets can adequately model ambiguity of products of regular expressions. However, as we can see from the examples, the net representations are usually much more complex than the original products of regular expressions. This seems to be the rule since the graphical representation of Petri nets has always been a mixed blessing. Clearly, for small examples a graphical representation has immediate and undeniable appeal, however, for larger examples, graphical representations are hard to comprehend[4]. One may then reason in the opposite way and use products of regular expressions to model properties of Petri nets. This is always possible for smd-nets (i.e. labelled smd-nets with one-to-one labelling function), but only in special cases for labelled smd-nets with many-to-one labelling function. For example, the nets from Fig. 2.20 *cannot* be modelled by products of regular expressions. Moreover, the standard semantics of products of regular expressions, vector sequence semantics, does not reflect ambiguity (this problem disappears for unlabelled smd-nets).

[4] A partial solution solution is to use some abbreviated ways to describe Petri nets, such as Predicate/Transition nets [GL81] or Coloured Petri nets [Jen81], however, the formal theory of these complex structures is still not completed.

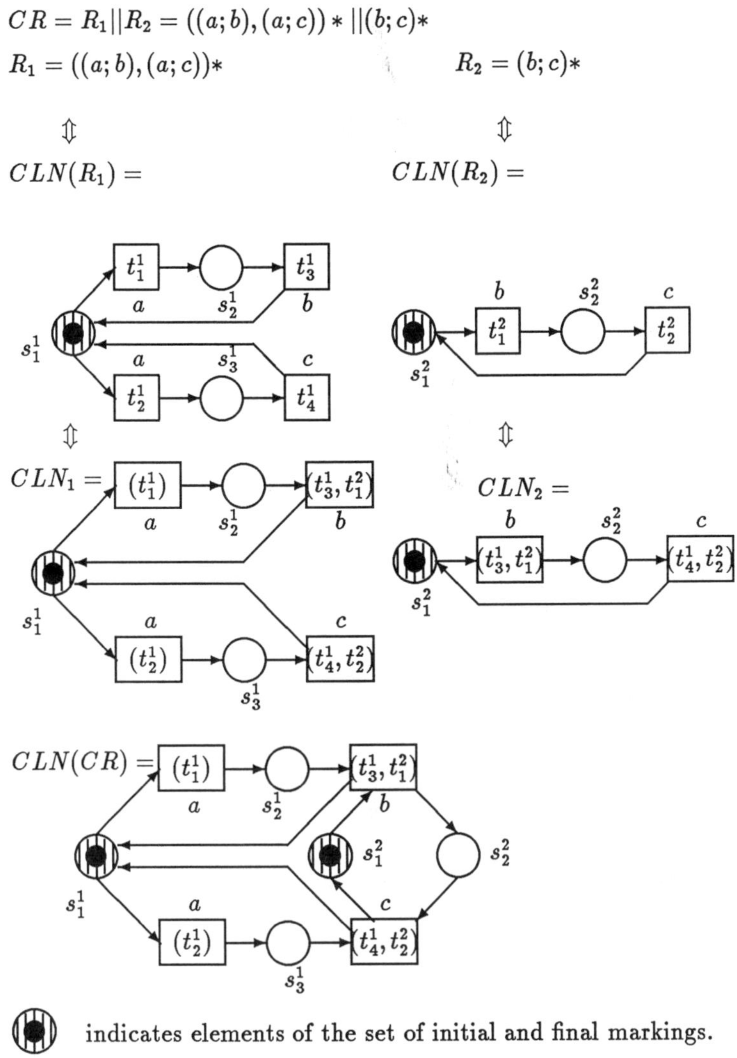

Figure 2.26. Illustration of preservation of ambiguity

$CR = R_1||R_2 = (a; b, c) * ||(b; c)*$

$R_1 = (a; b, c)*$ $R_2 = (b; c)*$

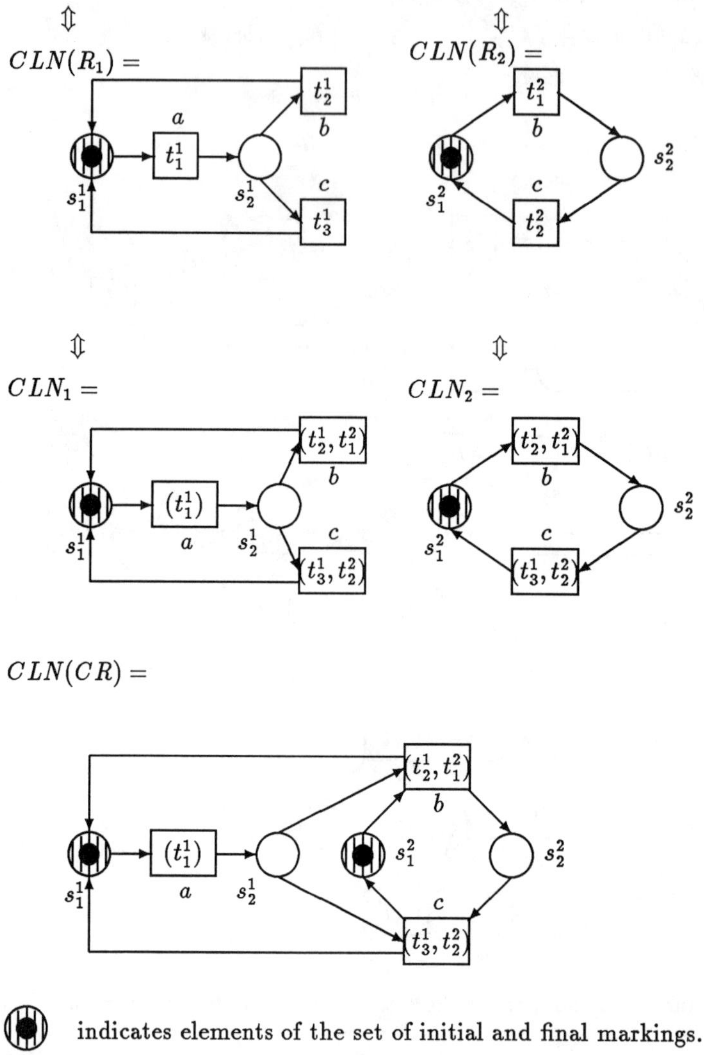

Figure 2.27. Illustration of the simplest case: all expressions in CR are uniquely named, and their associated nets have a one-to-one labelling function

2.3.7 Labelled Nets and COSY Path Programs

Because COSY path programs can be regarded as a special case of products of regular expressions, we might directly use the construction from the previous section. However, it is not clear which regular expression should correspond to a single path. Let us consider the path $P = \underline{path}\ a; a\ \underline{end}$ where $FS(P) = \{a\}^*$. Is an equivalent regular expression equal to: $(a; a)*; (\varepsilon, a)$ or rather simply $a*$? Of course $L((a; a)*; (\varepsilon, a)) = L(a*) = FS(P) = \{a\}^*$, but the nets defined by these expressions according to the construction from the previous section are different. The nets defined by $(a; a)*; (\varepsilon, a)$ and $a*$ are depicted in Fig. 2.29, but what we intuitively want is the net from Fig. 2.30.

To solve this problem let us observe that there is a semantical difference between $P = \underline{path}\ a; a\ \underline{end}$ and $P' = \underline{path}\ a\ \underline{end}$. Surely, $FS(P) = FS(P') = \{a\}^*$, but P specifies a sequential cyclic system with a cyclic behaviour "aa", while P' specifies a cyclic system with a cycle "a". In other words, P returns to its initial state after two occurrences of a, whereas P' returns to its initial state after one occurrence of a. Note that $Cyc(P)$, or $Cyc(P)^*$ capture more basic semantical properties of P than $FS(P)$. The firing sequences of P, $FS(P)$ is defined as the prefixes of $Cyc(P)^*$, and $Cyc(P)^*$ contains more semantical information about P than $FS(P)$, in particular, $Cyc(P)^*$ describes $FS(P)$, but not vice versa. So why not define a Petri net generating $Cyc(P)^*$? Thus the algorithm for transforming basic COSY path programs into labelled Petri nets can be defined as follows.

Let $Pr = P_1 \ldots P_n$ be a basic COSY path program with $P_i = \underline{path}\ seq_i\ \underline{end}$ for every $i = 1, \ldots, n$. Define $CR(Pr)$ to be the following product of regular expressions:

$$CR(Pr) = (seq_1)^* || \ldots || (seq_n)^*.$$

Suppose that $CLN(CR(Pr)) = (S, T, F, M_0, FMar, \nu, Ev)$, and define

$$LN(Pr) = (S, T, F, M_0, \nu, Ev).$$

The labelled smd-net $LN(Pr)$ is called a **net defined by** Pr. In fact, if we directly follow the construction of $CLN(CR(Pr))$ from $CR(Pr)$ and *omit all steps concerning final states of automata and final markings of nets*, we get exactly $LN(Pr)$ instead of $CLN(CR(Pr))$.

Theorem 2.15 (On Preserving Vector Firing Sequences)
$VFS(Pr) = VES(LN(Pr))$.

Proof. The proof is not immediate since it might happen that $VFS(Pr) \neq Pref(VL(CR(Pr)))$ as we have in the following case:

$$Pr_{10} = \begin{cases} P(1): & \underline{path}\ a*; b; c\ \underline{end} \\ P(2): & \underline{path}\ a*; c; b\ \underline{end} \end{cases}$$

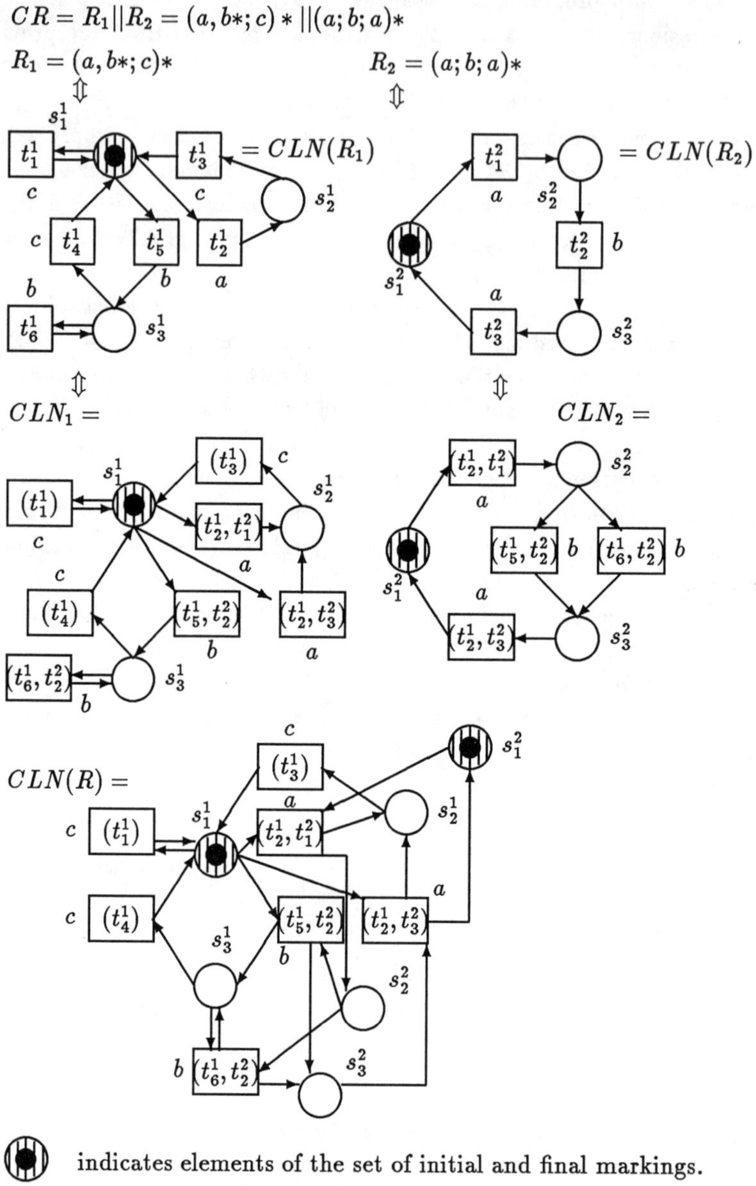

Figure 2.28. Illustration of the case when a uniquely named regular expression corresponds to a closed labelled net with a many-to-one labelling function (R_1 and $CLN(R_1)$)

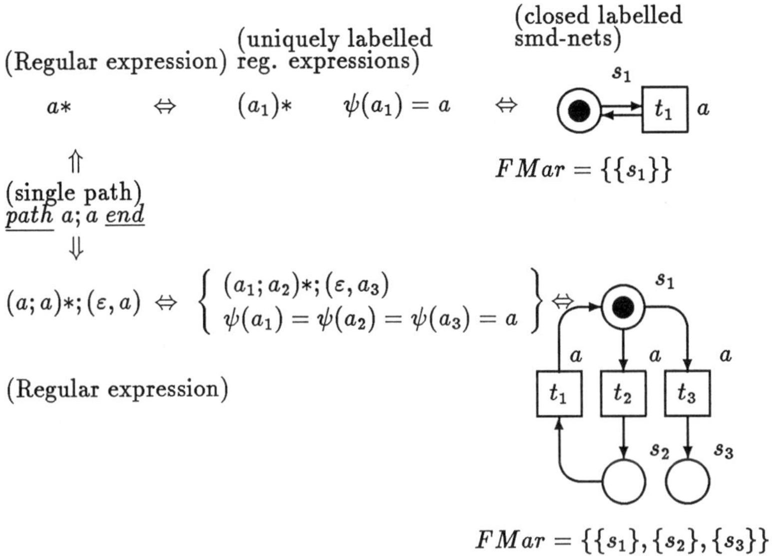

Figure 2.29. Different regular expressions and nets for same single path

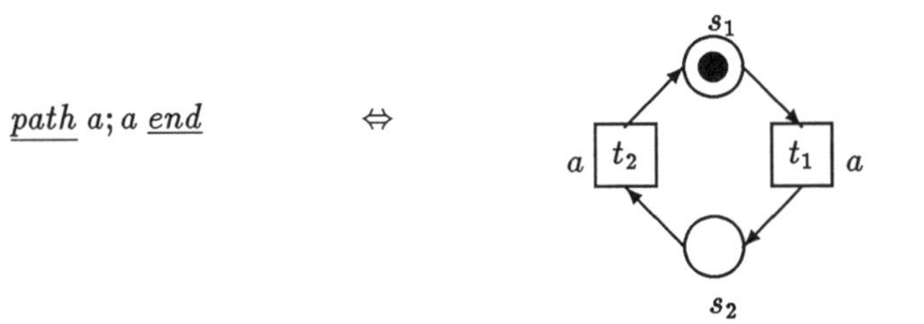

Figure 2.30. Labelled smd-net intuitively corresponding to path $a; a$ end

where $CR(Pr_{10}) = (a^*; b; c)^* || (a^*; c; b)^*$, $VS(CR(Pr_{10})) = \emptyset$, $VFS(Pr) = \{\underline{a}\}^*$. Let us set $R_i = (seq_i)^*$ for $i = 1, \ldots, n$. By Lemma 2.13,

$$VES(LN(Pr)) = (ES(LN_1) \times \ldots \times ES(LN_n)) \cap Vev^*,$$

where LN_i is a labelled smd-net defined by $R_i = (seq_i)^*$, for $i = 1, \ldots, n$ see Section 2.3.6. Thus, $ES(LN_i) = Pref(L(R_i)) = Pref(Cyc(seq_i)^*) = FS(P_i)$, for $i = 1, \ldots, n$. Hence: $VES(LN(Pr)) = (FS(P_1) \times \ldots \times FS(P_n)) \cap Vev^* = VFS(Pr)$. □

We can also define formally the concept of ambiguity for basic COSY path programs.

A path program $Pr = P_1 \ldots P_n$, where $P_i = \underline{path}\ seq_i\ \underline{end}$ for $i = 1, \ldots, n$, is said to be **ambiguous** if and only if there is $\underline{x} \in VFS(Pr)$ such that for some $i \in \{1, \ldots, n\}$, $pr_i(\underline{x})$ is an ambiguous sequence of $(seq_i)^*$. A path program Pr is called **unambiguous** if it is not ambiguous. We can show that in this case the construction also preserves ambiguity.

Theorem 2.16 (Ambiguity of Nets and Path Programs)
Pr is ambiguous $\Leftrightarrow$ $LN(Pr)$ is ambiguous.

Proof. From Theorem 2.14 it follows that: $CR(Pr)$ is ambiguous $\Leftrightarrow$ $CLN(CR(Pr))$ is ambiguous. By the definition of ambiguity for concurrent products of regular expressions and path programs we have: $CR(Pr)$ is ambiguous $\Leftrightarrow$ Pr is ambiguous. From the definition of ambiguity for labelled nets we have: $CLN(CR(Pr))$ is ambiguous $\Leftrightarrow$ $LN(Pr))$ is ambiguous. □

The construction presented above is illustrated in Fig. 2.23 of Sect. 2.3.4. Since the relationship between basic COSY path programs and labelled smd-nets is exactly the same as the relationship between products of regular expressions and closed labelled smd-nets, the comment from the end of the last section is valid also for path programs.

At the end of this section we shall discuss some relationships between liveness of nets and adequacy of COSY path programs. Since the intuition behind these two concepts is identical (i.e. absence of even a partial deadlock) one might have expected that the adequacy of a path program Pr implies the liveness of $LN(Pr)$ and vice versa . However, this is not the case. Clearly liveness of $LN(Pr)$ implies adequacy of Pr, but the opposite implication is false. To prove this let us consider the following path program:

$$Pr_{11} = \begin{cases} P(1): & \underline{path}\ a; a\ \underline{end} \\ P(2): & \underline{path}\ a; a\ \underline{end} \end{cases}$$

which is clearly adequate. The labelled smd-net $LN(Pr_{11})$ is presented in Fig. 2.31. (The Fig. 2.31 also illustrates how $LN(Pr_{11})$ is derived from Pr_{11}.)

$$Pr_{11} = \begin{cases} P(1): \underline{path}\ a; a\ \underline{end} \\ P(2): \underline{path}\ a; a\ \underline{end} \end{cases}$$

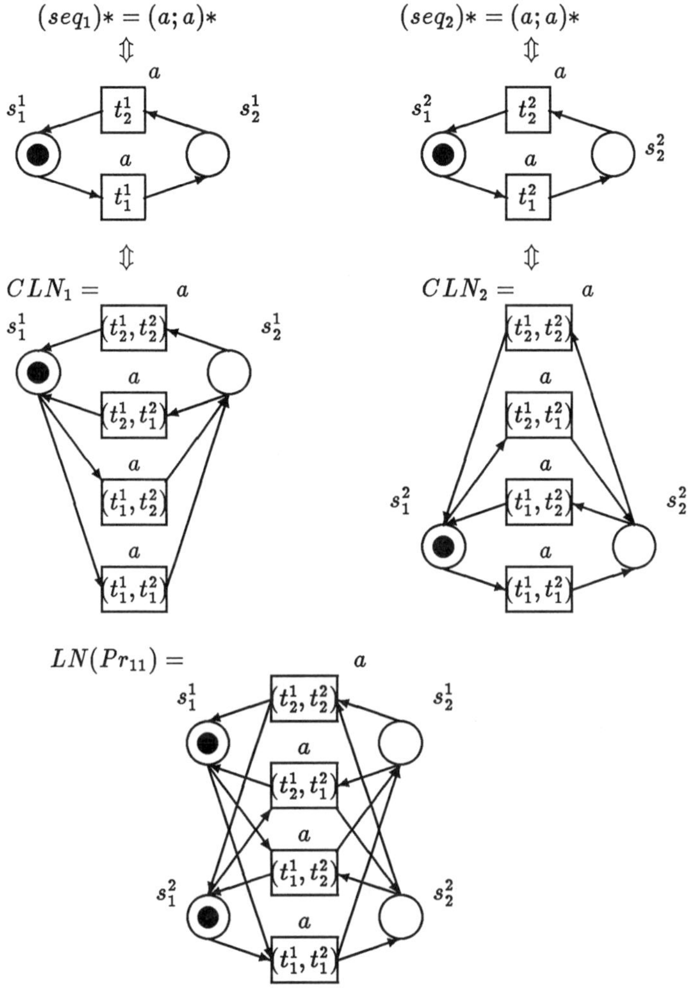

Figure 2.31. Adequate Pr s.t. $LN(Pr)$ is not live

The reader may quickly verify that transitions $t' = (t_2^1, t_1^2)$ and $t'' = (t_1^1, t_2^2)$ *never fire at all*, so $LN(Pr_{11})$ is not live.

The path program Pr_{11} above is unambiguous. For ambiguous path programs it may even happen that Pr is adequate while $LN(Pr)$ may deadlock! Consider the following path program:

$$Pr_{12} = \begin{cases} P(1): & \underline{path}\ (a;b),(a;c)\ \underline{end} \\ P(2): & \underline{path}\ b;c\ \underline{end} \end{cases}$$

Clearly Pr_{12} is adequate as $VFS(Pr_{12}) = \{\underline{abac}\}^*\{\underline{aba}, \underline{ab}, \underline{a}, \underline{\varepsilon}\}, LN(Pr_{12})$ is the labelled net LN' from Fig. 2.21 (up to isomorphism of transition names), and LN' deadlocks after firing t_1'' with label a.

This divergence follows from the fact that adequacy and liveness are defined for *different objects at different levels of abstraction*. Liveness is defined as a property of unlabelled marked nets. Its definition uses the notions of transitions and markings (i.e., system states) which do not occur in the COSY approach in an explicit manner.

Adequacy is a property of behaviours generated by path programs, and, as Corollary 2.4 (2) shows, it can entirely be defined in terms of $VFS(Pr)$, $Ev(Pr)$ and $Vev(Pr)$.

In the definition of VFS we relate only to the existence of a string vector, not to the way its string components are derived from single paths (regular expressions).

The liveness property, by using markings, relates to the way a string vector is derived. To narrow this divergence we shall try to define formally the concept of adequacy in the case of Petri nets.

Intuitively, adequacy means that in any system configuration there must be no event which is prevented from ever occurring again. In the case of nets, the events, as meant in COSY approach, are represented by *labels* of transitions of nets, so the adequacy should rather be defined as a property of *labelled* nets than of unlabelled ones.

It also seems to be reasonable to assume that *adequate labelled nets cannot deadlock*. The path program Pr_{12} analysed above shows that under such assumption it is impossible to find the definition of adequacy satisfying: Pr is adequate $\Leftrightarrow$ $LN(Pr)$ is adequate, without some restrictions imposed on Pr. In fact we might have expected this as Petri net semantics of COSY has more expressive power than the vector firing sequences have. The definition of adequacy for nets we propose is the following.

The labelled Petri net $LN = (S, T, F, M_0, \nu, Ev)$ is said to be **adequate** if and only if:

$$(\forall M \in [M_0\rangle)(\forall a \in Ev)(\exists M' \in [M\rangle)(\exists t \in T)\ M'\ enables\ t \wedge \nu(t) = a.$$

If ν is an injective function then net adequacy is equivalent to liveness, otherwise liveness always implies adequacy but not necessarily vice versa. Probably the simplest example of a labelled smd-net (a labelled finite state machine in fact) which is adequate but not live is given in Fig. 2.32. The net is not live since the transition t_1 can be fired only once. It is adequate since $\nu(t_1) = \nu(t_2) = a$.

The labelled net $LN(Pr)$ from Fig. 2.31 is adequate, while LN' from Fig. 2.21 (i.e., $LN(Pr_{12})$ up to isomorphism of transition names) is not. Clearly, for every path program Pr, adequacy of the labelled net $LN(Pr)$ implies adequacy of Pr. The opposite implication may not be true as the path program Pr_{12} and the labelled net $LN(Pr_{12})$ analysed above show. The formal relationship between adequacy of path programs and adequacy of labelled nets is characterised by the following theorem.

Theorem 2.17 (Adequacy of Nets and Path Programs)
If Pr is unambiguous, then : Pr is adequate $\Leftrightarrow$ $LN(Pr)$ is adequate.

Proof.
($\Leftarrow$) obvious.
($\Rightarrow$) Suppose that Pr is adequate but $LN(Pr) = (S,T,F,M_0,\nu,Ev)$ is not, i.e.,

$$(\exists M_1 \in [M_0\rangle)(\exists a \in Ev)(\forall M' \in [M_1\rangle)(\forall t \in T)\ M'\ \mathit{enables}\ t \Rightarrow \nu(t) \neq a.$$

Suppose that $M_1 = M_0$. Then $(\forall \underline{x} \in VES(LN(Pr)))\ \#_a(\underline{x}) = 0$. By Theorem 2.15 we have $VES(LN(Pr)) = VFS(Pr)$, i.e., $\underline{x} \in VFS(Pr) \Rightarrow \#_a(\underline{x}) = 0$, a contradiction with adequacy of Pr. Hence $M_1 \neq M_0$, which means that there is $\underline{v} \in VTS(N)$, where $N = (S,T,F,M_0)$, such that $M_0[\underline{v}\rangle M_1$ and $\underline{v} \neq \underline{\varepsilon}$. Let $\underline{x} = \nu(\underline{v})$. By Theorem 2.15, $\underline{x} \in VFS(Pr)$. By adequacy of Pr and Theorem 2.15 again, $(\exists \underline{y} \in Vev^*)\underline{x}\underline{y}\underline{a} \in VFS(Pr) = VES(LN(Pr))$. Thus, there is $\underline{v}_1\underline{v}_2\underline{t} \in VTS(N)$ such that $\nu(\underline{v}_1) = \underline{x}, \nu(\underline{v}_2) = \underline{y}, \nu(t) = a$.
Let $M_2, M_3 \subseteq S$ be markings defined by:$M_0[\underline{v}_1\rangle M_2$ and $M_2[\underline{v}_2\rangle M_3$. Since $M_3 \in [M_2\rangle \wedge M_3$ enables $t \wedge \nu(t) = a$, then $M_2 \neq M_1$. Hence $\underline{v} \neq \underline{v}_1$ as $M_0[\underline{v}\rangle M_1$ and $M_0[\underline{v}_1\rangle M_2$. On the other hand $\nu(\underline{v}) = \nu(\underline{v}_1) = \underline{x}$, which by Theorem 2.12 means that $LN(Pr)$ is ambiguous. From Theorem 2.16 it follows that Pr is also ambiguous, a contradiction. □

Theorem 2.15,Theorem 2.16 and Theorem 2.17 provide a formal support for our claim that for unambiguous path programs both the vector firing sequence semantics and Petri net semantics are equivalent, while for ambiguous path programs the Petri net semantics has greater expressive power.

2.3.8 Other Constructions.

Let $Pr = P_1 \ldots P_n$ be a path program with $P_i = \underline{path}\ seq\ \underline{end}$ for $i = 1, \ldots, n$. The construction presented in the previous chapter consists of two main steps:

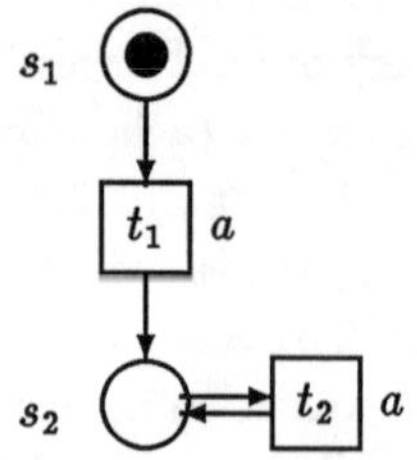

Figure 2.32. Labelled smd-net which is adequate but not live

(1) Transforming each P_i into an equivalent labelled state machine $LN(P_i)$,

(2) Transforming the set $\{LN(P_1), \ldots, LN(P_n)\}$ into one labelled net $LN(Pr)$.

For uniquely named $P_i = \underline{path}\ seq\ \underline{end}$ the first step is reduced to finding the minimal state (deterministic) automaton generating the language $Cyc(seq_i)^*$, in general case the concept of labelled unr-expressions is used. The construction preserves ambiguity. If all $LN(P_i)$ are uniquely labelled, the second step is reduced to "merging" $LN(P_i)$ by identifying transitions with identical labels. Fig. 2.14 (b) shows that even if P_i is uniquely named, $LN(P_i)$ may not be uniquely labelled.

A question one may ask is: "What would be the behavioural difference if we always transform P_i directly into an equivalent minimal state (deterministic) automaton without going through labelled unr-expressions?" So let us consider the following procedure:

(1a) for each P_i construct a (closed) labelled state machine $CLN^{msa}(P_i)$ as a net corresponding to the minimal state automaton which accepts the language $Cyc(seq_i)^*$.

(2a) transform the set $\{CLN^{msa}(P_1), \ldots, CLN^{msa}(P_n)\}$ into one labelled net, following the same procedure as defined in previous sections, but starting with $CLN^{msa}(P_i)$ instead of $CLN(seq_i^*)$.

Let $LN^{msa}(Pr)$ denote the result of the above procedure.
For the path program:

$$Pr_{11} = \begin{cases} P(1): & \underline{path}\ a; a\ \underline{end} \\ P(2): & \underline{path}\ a; a\ \underline{end} \end{cases}$$

from Fig. 2.31, the nets $LN^{msa}(Pr_{11})$ and $LN(Pr_{11})$ are identical. For the path program:

$$Pr_{12} = \begin{cases} P(1): & \underline{path}\ (a; b), (a; c)\ \underline{end} \\ P(2): & \underline{path}\ b; c\ \underline{end} \end{cases}$$

the nets $CLN^{msa}(P_1)$ and $CLN^{msa}(P_2)$ correspond respectively to LN_1 and LN_2 from Fig. 2.21, and $LN^{msa}(Pr_{12})$ is identical to LN from Fig. 2.21 (up to isomorphism of transition names). The formal relationship between Pr and $LN^{msa}(Pr)$ is characterised by the following theorem.

Theorem 2.18

(1) $VFS(Pr) = VES(LN^{msa}(Pr))$

(2) $LN^{msa}(Pr)$ *is always unambiguous labelled net*

(3) Pr *is adequate* $\Leftrightarrow$ $LN^{msa}(Pr)$ *is adequate*

Proof. (1) In the same way as Theorem 2.15. In the proof of Theorem 2.15 (and Lemma 2.13 on which the proof is based) we did not use the fact that $CLN((seq_i)^*)$ are built using labelled unr-expressions. The only relationship between $CLN((seq_i)^*)$ and P_i we used is that $VES(CLN((seq_i)^*)) = Cyc(seq_i)^*$, but this holds also for $CLN^{msa}(P_i)$.
(2) Suppose that $LN^{msa}(Pr)$ is ambiguous. Reasoning in the same way as in the proof of the implication ($\Leftarrow$) of Theorem 2.14, we can show that $CLN^{msa}(P_i)$ is ambiguous for some i, a contradiction, as $CLN^{msa}(P_i)$ corresponds to deterministic (minimal state) automaton.
(3) ($\Leftarrow$) Obvious.
($\Rightarrow$) Suppose that Pr is adequate but $LN^{msa}(pr)$ is not. Reasoning in the same way as in the proof of implication ($\Rightarrow$) of Theorem 2.17, we can show that $LN^{msa}(Pr)$ is ambiguous, a contradicion to (2). □
Thus, when using $LN^{msa}(Pr)$ instead of $LN(Pr)$, a Petri net semantics of COSY path programs does not reflect ambiguity but can be treated as equivalent to vector firing sequence semantics.

To construct both $LN(Pr)$ and $LN^{msa}(Pr)$ we need to use an algorithm constructing the minimal state (deterministic) automaton from a given regular expression. Although efficient algorithms can build such an automaton in linear time (see [BS86]), in general use, as Fig.. 2.14 (b) shows, the shape of such automaton is not obvious to predict. It turns out that for a restricted, but still quite general, class of path programs, we can skip both the concepts of unr-expression and minimal state automaton, and define an equivalent labelled state machine directly from a regular expression (single path) by using only a few recursive rules.

A path P is said to be *star-free* if it does not contain the symbol "*". A path program $Pr = P_1 \ldots P_n$ is said to be *star-free* if every P_i is star-free. Formally, the syntax of star-free paths is given by the following rules (compare Chapter 2.1):

$$\begin{aligned}
starfreepath &= \underline{path}\ sequence\ \underline{end}\\
sequence &= orelement\{;orelement\}^*\\
orelement &= element\{,element\}^*\\
element &= event \mid (sequence)\\
event &= name
\end{aligned}$$

We say that two labelled nets $LN_i = (S_i, T_i, F_i, M_{0i}, \nu_i, Ev_i)$, $i = 1, 2$, are *distinct* if $S_1 \cap S_2 = \emptyset$ and $T_1 \cap T_2 = \emptyset$.
A labelled finite state machine $LN = (S, T, F, M_0, \nu, Ev)$ is said to be **neat** if there are $s_b, s_e \in S$ such that $\{s \in S \mid {}^\bullet s = \emptyset\} = \{s_b\}, \{s \in S \mid s^\bullet = \emptyset\} = \{s_e\}$ and $M_0 = \{s_b\}$. If LN is a neat labelled finite state machine we define $first(LN) = s_b, last(LN) = s_e$.

Let $LN_i = (S_i, T_i, F_i, M_{0i}, \nu_i, Ev_i)$, $i = 1, 2$, be neat and distinct labelled finite state machines. We define the labelled finite state machines $LN_1 \textcircled{;} LN_2$ and $LN_1 \textcircled{,} LN_2$ in the following way:

- $LN_1 \textcircled{;} LN_2 = (S, T, F, M_0, \nu, Ev)$ is obtained from LN_1 and LN_2 by identifying $last(LN_1)$ with $first(LN_2)$ and defining $M_0 = M_{01}$, i.e., $S = S_1 \cup S_2 \setminus \{first(LN_2)\}$, $T = T_1 \cup T_2$, $F = F_1 \cup F_2 \setminus \{(first(LN_2), t) \mid t \in T_2\} \cup \{(last(LN_1), t) \mid (first(LN_2), t) \in F_2\}$, $Ev = Ev_1 \cup Ev_2$, $\nu(t) = \nu_i(t)$ if $t \in T_i, i = 1, 2$, $M_0 = \{first(LN_1)\}$.
- $LN_1 \textcircled{,} LN_2 = (S, T, F, M_0, \nu, Ev)$ is obtained from LN_1 and LN_2 by identifying $first(LN_1)$ with $first(LN_2)$, $last(LN_1)$ with $last(LN_2)$, and defining $M_0 = M_{01}$, i.e., $S = S_1 \cup S_2 \setminus \{first(LN_2), last(LN_2\}$, $T = T_1 \cup T_2$, $Ev = Ev_1 \cup Ev_2$, $\nu(t) = \nu_i(t)$, if $t \in T_i$, $i = 1, 2$, $F = F_1 \cup F_2 \setminus \{(first(LN_2), t) \mid t \in T_2\} \setminus \{(t, last(LN_2)) \mid t \in T_2\} \cup \{(first(LN_1), t) \mid (first(LN_2), t) \in F_2\} \cup \{(t, last(LN_1)) \mid (t, last(LN_2)) \in F_2\}$, $M_0 = \{first(LN_1)\}$.

The operations $\textcircled{;}$ and $\textcircled{,}$ can be extended for arbitrary neat labelled finite state machines by assuming that if LN_1 and LN_2 are not distinct we define $LN_1 \textcircled{;} LN_2 = LN_1 \textcircled{;} LN_2'$ and $LN_1 \textcircled{,} LN_2 = LN_1 \textcircled{,} LN_2'$ where $LN_2' = (S_2', T_2', F_2', M_{02}', \nu_2', Ev_2)$ is defined by:

$$\begin{aligned}
&S_2' = S_2 \times \{1\},\ T_2' = T_2 \times \{1\},\\
&F_2' = \{((s,1),(t,1)) \mid (s,t) \in F_2\} \cup \{((t,1),(s,1)) \mid (t, {}^\bullet s) \in F_2\}.\\
&\nu_2'((t,1)) = \nu_2(t).
\end{aligned}$$

The net LN_2' is clearly isomorphic to LN_2 and LN_1, LN_2' are always distinct. In general case $LN_1 \textcircled{;} LN_2$ and $LN_1 \textcircled{,} LN_2$ are defined uniquely up to isomorphism of transitions and places. One may also observe that $LN_1 \textcircled{;} LN_2$ and $LN_1 \textcircled{,} LN_2$ are always neat labelled state machines and

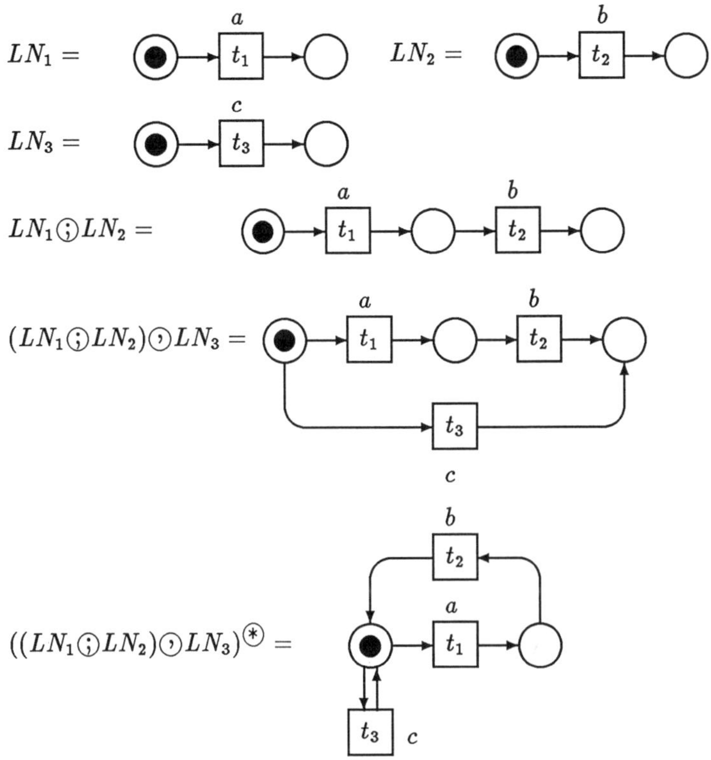

Figure 2.33. The semicolon, comma, and star operations for neat nets

$$
\begin{aligned}
(LN_1 \textcircled{;} LN_2) \textcircled{;} LN_3 &= LN_1 \textcircled{;} (LN_2 \textcircled{;} LN_3),\\
(LN_1 \textcircled{,} LN_2) \textcircled{,} LN_3 &= LN_1 \textcircled{,} (LN_2 \textcircled{,} LN_3).
\end{aligned}
$$

For every neat labelled state machine $LN = (S, T, F, M_0, \nu, Ev)$, let $LN^{\textcircled{*}} = (S', T', F', M_0', \nu', Ev)$ be the net obtained from LN by identifying $last(LN)$ with $first(LN)$, i.e., $S' = S \setminus \{last(LN)\}$, $T' = T$, $F' = F \setminus \{(t, last(LN)) \mid t \in T\} \cup \{(t, first(LN)) \mid (t, last(LN)) \in F\}$, $M_0' = \{first(LN)\}$. Note that $LN^{\textcircled{*}}$ is also a labelled state machine *but it is no longer neat*. The operations $\textcircled{;}$, $\textcircled{,}$ and $\textcircled{*}$, are illustrated in Fig. 2.33.

Let $P =$ path *seq* end be a star-free path. We define:

$$LN^{induc}(P) = LN(seq)^{\textcircled{*}}$$

where $LN(seq)$ is defined inductively:

- If $seq = a$ ($a \in Ev(P)$) then

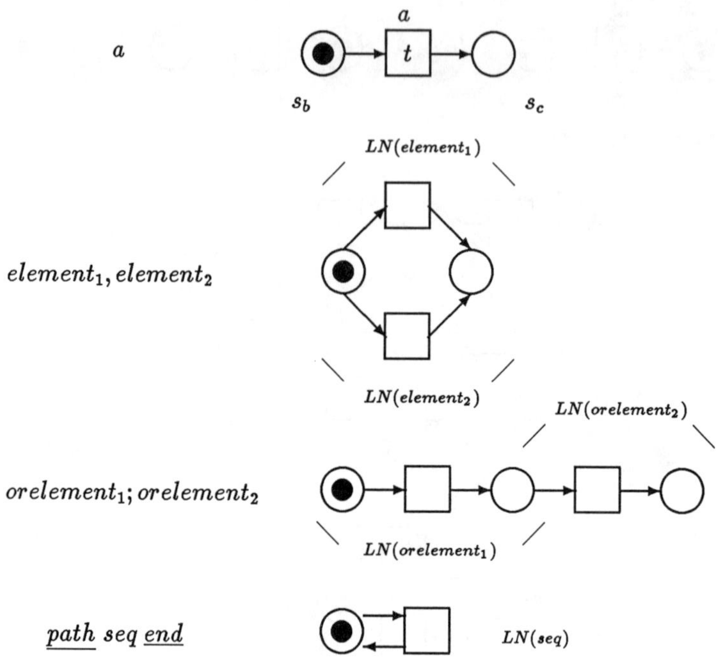

Figure 2.34. Translation of star-free path P into $LN^{induc}(P)$

$LN(a) = (\{s_b, s_e\}, \{t\}, \{(s_b, t), (t, s_e)\}, \{s_b\}, \nu, \{a\})$, where $\nu(t) = a$.

- If $seq = (sequence)$ then $LN(seq) = LN(sequence)$.
- If $seq = element_1, \ldots, element_k$ then

  $LN(seq) = LN(element_1)\textcircled{,}\ldots\textcircled{,}LN(element_k),\ k \geq 2$.
- If $seq = orelement_1; \ldots; orelemet_k$ then

  $LN(seq) = LN(orelement_1)\textcircled{;}\ldots\textcircled{;}LN(orelement_k),\ k \geq 2$.

This construction is informally illustrated in Fig. 2.34. Suppose that $LN^{induc}(P) = (S, T, F, M_0, \nu, Ev)$. We define $CLN^{induc}(P)$ by adding $FMar = \{M_0\}$ to $LN^{induc}(P)$, i.e., $CLN^{induc}(P) = (S, T, F, M_0, \{M_0\}, \nu, Ev)$.

Let $Pr = P_1 \ldots P_n$ with $P_i =$ <u>path</u> seq_i <u>end</u> be a *star-free* path program. Consider the following procedure:

(1b) for each P_i, construct $CLN^{induc}(P_i)$.

(2b) transform the set $\{CLN^{induc}(P_1), \ldots, CLN^{induc}(P_n)\}$ into one labelled net, following the same procedure as defined in previous sections, but starting with $CLN^{induc}(P_i)$ instead of $CLN((seq_i)^*)$.

Let $LN^{induc}(Pr)$ denote the result of the above procedure. We shall show that $LN^{induc}(Pr)$ and $LN(Pr)$ obtained by a general procedure are *identical* (up to isomorphism).

Lemma 2.14 *Let $CLN = (S, T, F, M_0, FMar, \nu, Ev)$ be a* uniquely *labelled finite state machine satisfying: $(\forall a \in Ev)(\exists x \in Ev^*) xa \in Pref(ES(CLN))$, and let $\mathcal{M}(CLN)$ be a finite automaton corresponding to CLN.*
The automaton $\mathcal{M}(CLN)$ is a minimal state deterministic automaton accepting $ES(CLN)$.

Proof. Obvious if $card(S) = 1$, so assume that $card(S) \geq 2$. Since CLN is uniquely labelled then $\nu^{-1} : Ev \longrightarrow T$ is correctly defined. For every $a \in Ev$, let ${}^\bullet a = {}^\bullet\nu^{-1}(a)$, $a^\bullet = \nu^{-1}(a)^\bullet$. Define $L = ES(CLN)$. For every $x \in Pref(L)$ we define ${}^\bullet x$ and $x^\bullet$ by: $\varepsilon = \varepsilon^\bullet = M_0$, ${}^\bullet(ax) = {}^\bullet a, (xa)^\bullet = a^\bullet$ for $xa \in Pref(L)$, $a \in Ev$.
Let $E \subseteq Ev^* \times Ev^*$ be the following relation:

$$xEy \Leftrightarrow (\{x,y\} \subseteq Pref(L) \wedge x^\bullet = y^\bullet) \vee (\{x,y\} \cap Pref(L) = \emptyset).$$

Clearly E is an equivalence relation. Recall that $[x]_E$ denote the equivalence class of E containing x, and Ev^*/E denote the set of all equivalence class of E (see Appendix A). Assume that $s_{garbage} \notin S$ and define the mapping $\varphi : Ev^*/E \longrightarrow S \cup \{s_{garbage}\}$ as follows:

$$(\forall x \in Ev^*)\ \varphi([x]_E) = \begin{cases} s & if x \in Pref(L) \wedge x^\bullet = \{s\} \\ s_{garbage} & if\ \ x \notin Pref(L) \end{cases}$$

By the definition of E, φ is injective. From the condition: $(\forall a \in Ev)(\exists x \in Ev^*)xa \in Pref(L)$, it follows that φ is onto. Thus $card(Ev^*/E) = card(S \cup \{s_{garbage}\}) = card(S) + 1$.
Let $R_L \subseteq Ev^* \times Ev^*$ be an equivalence relation defined by:

$$xR_Ly \Leftrightarrow (\forall z \in Ev^*)(xz \in L \Leftrightarrow yz \in L).$$

Let m be the number of states of the minimal state deterministic automaton accepting the language L. From Theorem B.3 of Appendix B it follows that $m = card(Ev^*/R_L)$. We show that $R_L \subseteq E$.
Let xR_Ly and $x \notin Perf(L)$. Then $y \notin Pref(L)$ and xEy.
Let xR_Ly and $x \in Pref(L)$. Suppose that $xz \in L$. Hence $yz \in L$. But $xz \in L \Rightarrow x^\bullet = {}^\bullet z$, and $yz \in L \Rightarrow y^\bullet = {}^\bullet z$. Thus $x^\bullet = y^\bullet$, which implies xEy. Hence $R_L \subseteq E$, which means that $m = card(Ev^*/R_L) \geq card(Ev^*/E) = card(S) + 1$. Let $\mathcal{M}(CLN) = (Q, A, S, q_0, \mathcal{F})$. Since $card(S) \geq 2$ then there are $s_0, s_1, \in S, t \in T$ such that $M_0 = \{s_0\} = {}^\bullet t, \{s_1\} = t^\bullet$ and $s_0 \neq s_1$. Suppose that $\nu(t) = a$. Because CLN is uniquely named then $a \notin \nu(s_1^\bullet)$ since $t \notin s_1^\bullet$. Hence,

by the definition of $\mathcal{M}(CLN)$, $Q = S \cup \{s_{garbage}\}$ and $card(Q) = card(S) + 1$. Thus $m \geq card(Q)$, but m is the minimal number of states, hence $m = card(Q)$ which ends the proof. □

Theorem 2.19 (Equivalences for Star-free Path Programs)
Let $Pr = P_1 \ldots P_n$ be a star-free path program with
$P_i = \underline{path}\ seq_i\ \underline{end}$, $i = 1, \ldots, n$.
(1) $CLN^{induc}(P_i) = CLN((seq_i)^*), i = 1, \ldots, n.$
(2) $LN(Pr) = LN^{induc}(Pr).$
The equalities in (1) and (2) are up to isomorphism of places and transitions.

Proof. (1) Let $LR_i = ((R_i)^*, \psi_i, Ev_i)$ be a labelled unr-expression such that $\psi_i(R_i) = seq_i$. Define $P_i' = \underline{path}\ R_i\ \underline{end}$. Clearly P_i' is uniquely named and star-free. Let $LN_j = (S_j, T_j, F_j, M_{0j}, \nu_j, Ev_j'), j = 1, 2$, be uniquely labelled neat state machines such that $Ev_1' \cap Ev_2' = \emptyset$. Then $LN_1 \textcircled{;} LN_2$ and $LN_1 \textcircled{,} LN_2$ are also uniquely labelled. Thus $CLN^{induc}(P_i')$ is uniquely labelled closed state machine. From the construction of $CLN^{induc}(P_i')$ it follows that $(\forall a \in Ev(P_i'))(\exists x \in Ev(P_i')^*) xa \in Pref(ES(CLN^{induc}(P_i')))$. Hence $CLN^{induc}(P_i')$ satisfies the conditions of Lemma 2.14, which means that $\mathcal{M}(CLN^{induc}(P_i'))$ is a minimal state automaton. Thus $CLN^{induc}(P_i')$ is the closed labelled state machine representing the minimal finite automaton accepting $L(R_i)$, which implies: $CLN^{induc}(P_i) = CLN(LR_i) = CLN((seq_i)^*)$.
(2) From (1) and the step (2b) of the construction. □

From Theorem 2.19 it follows that for star-free path programs we can use the more natural and easier inductive algorithm instead of going through unr-expressions and the result is the same. For both star-free and uniquely named path programs, all three procedures result in the same labelled net, i.e., in such a case $LN(Pr) = LN^{msa}(Pr) = LN^{induc}(Pr)$.

Another approach to define a Petri net semantics for COSY path programs is to use *Petri nets with silent transitions.* In principle, a **closed net with silent transitions,** CN^{st}, is a closed labelled net:

$$CN^{st} = (S, T, F, M_0, FMar, \nu, T_0 \cup \{\varepsilon\})$$

such that the set of transitions T, is divided into two sets T_0 and T_1, with $T = T_0 \cup T_1$, $T_0 \cap T_1 = \emptyset$, elements of T_1 are called **silent transitions,** and $\nu : T \longrightarrow T_0 \cup \{\varepsilon\}$ is defined as:

$$(\forall t \in T)\ \nu(t) = \begin{cases} t & if\ \ t \in T_0 \\ \varepsilon & if\ \ t \in T_1 \end{cases}$$

The set of event sequences generated by $CN^{st}, ES(CN^{st})$, is defined as:

$$ES(CN^{st}) = \{\nu(w) \mid (\exists M \in FMar) M_0[w\rangle M\}.$$

One may show that for every regular expression R there is (not necessarily unique) closed net with silent transitions $CN^{st} = (S, T, F, M_0, FMar, \nu, T_0 \cup \{\varepsilon\})$ such that $T_0 = \Sigma_R, L(R) = ES(CN^{st})$ and R is ambiguous if and only if CN^{st} is ambiguous. Unfortunately, if R is not uniquely named, none of the corresponding nets with silent transitions is a labelled state machine.

Let $Pr = P_1 \ldots P_n$ be a path program with $P_i =$ path seq_i end for $i = 1, \ldots, n$. Consider the following procedure:

(1c) For every P_i construct a closed net with silent transitions $CN_i^{st} = (S_i, T_i, F_i, M_{0i}, FMar, \nu, T_{0i} \cup \{\varepsilon\})$ such that $T_{0i} = Ev_i, ES(CN_i^{st}) = Cyc(seq_i)^*$, and CN_i^{st} is ambiguous if seq_i is ambiguous. Assume that for every i, $S_i, T_i \setminus T_{0i}$ are unique.

(2c) Construct $CN^{st}(Pr)$ by "merging" (i.e., identifying the same transitions from T_{0i}'s) $\mathcal{M}CN_i^{st}$.

Fig. 2.35 illustrates the above procedure. We shall not discuss this construction in great detail, since it would require additional formal concepts. The second step is usually easier in the last procedure than in the previous ones, but the price for it is quite heavy. We need silent actions which have no immediate interpretation in the COSY approach. If P_i contains many repeated events the net CN_i^{st} is complex and it is not a state machine. The latter fact makes the definition of vector event sequences for $CN^{st}(Pr)$ complex and problematic.

2.4 Adequacy Properties of Path Programs

Recall that adequacy is a strengthening of the notion of freedom from deadlock. In an adequate path program, no event ever becomes "dead", i.e., incapable of ever occurring again without infringing the constraints expressed by the path program. The problem we shall try to solve in this section is how to verify whether a path program is adequate or not. It would be interesting to answer this question solely from looking at the text of the path program. Since in general the program text tends to conceal that property to a great degree, it is necessary to obtain some formal results about adequacy. But unfortunately, in the general case, very little can be done, so we have to search for subclasses of programs which can reveal their adequacy syntactically, that is, for which we can decide about adequacy by examining the program text. Having defined such subclasses, we can look for larger classes derived from the previous ones by adequacy-preserving transformations.

2.4.1 Syntactically Conflict-Free Path Programs

A single path $P =$ path *sequence* end is said to be **conflict-free** if *sequence* contains no comma and no Kleene star. In other words, if P is either $P =$

$$Pr_{12} = \begin{cases} P(1): & \underline{path}\ (a;b),(a;c)\ \underline{end} \\ P(2): & \underline{path}\ b;c\ \underline{end} \end{cases}$$

$CN^{st}(P_1) =$

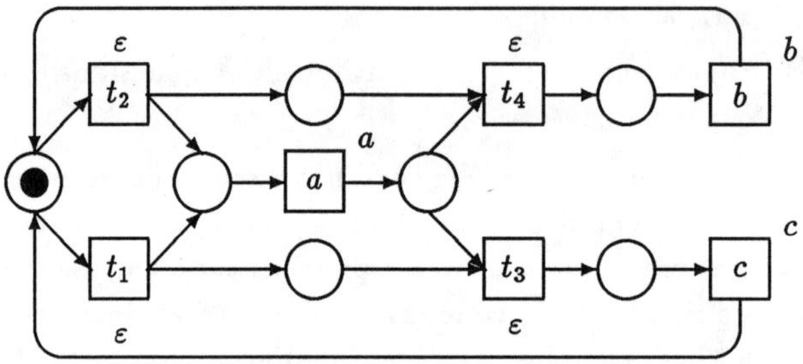

$CN^{st}(P_2) =$

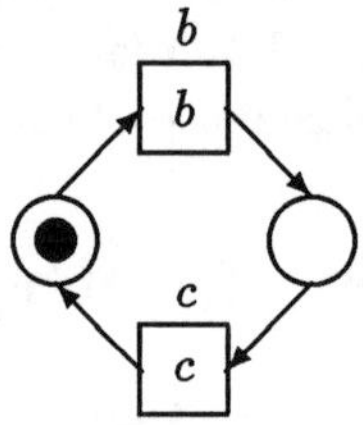

$CN^{st}(Pr_{12}) =$

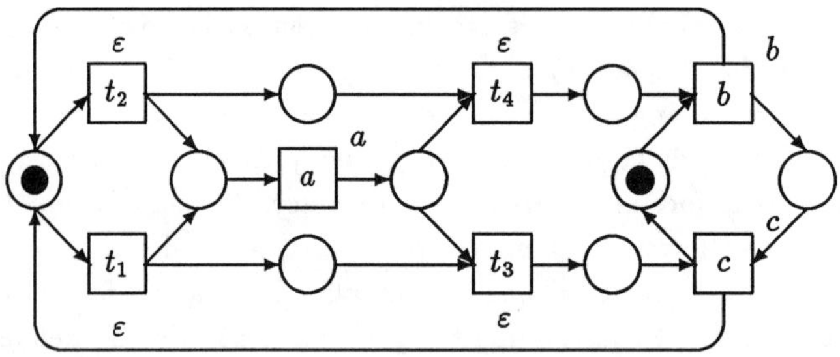

Figure 2.35. Path programs and nets with silent transitions.

$\underline{path}$ a $\underline{end}$ or $P = \underline{path}\ a_1;\ldots;a_m\ \underline{end}$ for some $m \geq 2$, where $a, a_1, \ldots, a_m$ are events.

A path program $Pr = P_1 \ldots P_n$ is a **syntactically conflict-free path program** (abbreviated SCF-path program) if every P_i, $i = 1, \ldots, n$, is conflict-free. Clearly every SCF-path program is conflict free in the sense that:

$$(\forall \underline{x} \in VFS)(\underline{xa} \in VFS \wedge \underline{xb} \in VFS \wedge (a,b) \notin ind) \Rightarrow a = b,$$

but not vice versa. For instance:

$$Pr_{13} = \begin{cases} P(1): & \underline{path}\ a, (b; c)\ \underline{end} \\ P(2): & \underline{path}\ a, (c; b)\ \underline{end} \end{cases}$$

is conflict free in the above sense, but it is not syntactically conflict-free.

Thus a SCF-path program is just a composition of cyclic, conflict-free sequential components, and Petri nets corresponding to SCF-path programs are nets with only unbranched places. It turns out that only in this case can we formulate a kind of "static" criterion for adequacy, although even in this case this criterion is not fully "static" (or "syntactic").

Let $P = \underline{path}\ a_1;\ldots;a_m\ \underline{end}$ be a conflict-free path. In this case $Cyc(P) = \{a_1 \ldots a_m\}$, so we can define $cyc(P) = a_1 \ldots a_m$ and we have $Cyc(P) = \{cyc(P)\}$. In the sequel, we shall write cyc_i rather than $cyc(P_i)$.

For every sequence of positive integers $r_1, \ldots, r_n$, let $hcf(r_1, \ldots, r_n)$ denote the greatest common divisor of $r_1, \ldots, r_n$.

The main result of this subsection can be formulated as follows:

Theorem 2.20 (Shields' Adequacy Criterion)

For every SCF-path program $Pr = P_1 \ldots P_n$ the following are equivalent:

(a) P is adequate

(b) $(\exists x \in Ev^*)(\exists r_1 \geq 1, \ldots, r_n \geq 1) hcf(r_1, \ldots, r_n) = 1 \wedge (\forall i = 1, \ldots, n) h_i(x) = (cyc_i)^{r_i}$.

Furthermore, let $x \in Ev^*$ be that of (b), then: $VFS(Pr) = Pref(\{\underline{x}\}^*)$.

Proof. (a) $\Rightarrow$ (b): Consider P_i. Let $R_i \subseteq FS_i \times FS_i$ be an equivalence relation defined by:

$$xR_iy :\Leftrightarrow length(x)\ mod\ length(cyc_i) = length(y)\ mod\ length(cyc_i).$$

From the definition of conflict-free path it follows that

$$(1)\ (\forall x, y \in FS_i)(\forall a, b \in Ev)\ xaR_iyb \Rightarrow (a = b \wedge xR_iy)$$

Define $R \subseteq Vev^* \times Vev^*$ as follows:

$$\underline{x}R\underline{y} :\Leftrightarrow (\forall i = 1,\ldots,n)\ pr_i(\underline{x})R_i pr_i(\underline{y}).$$

We observe that R is an equivalence relation, and R has only a finite number of equivalence classes (because for every i, $length(cyc_i)$ is finite). We also notice that:

(2) $(\forall x, y \in VFS)(\forall a \in Ev)\ \underline{xa}R\underline{ya} \Rightarrow \underline{x}R\underline{y}$.

To show (2) we observe that $\underline{xa}R\underline{ya}$ implies $h_i(x)h_i(a)R_ih_i(y)h_i(a)$ for $i = 1,\ldots,n$. Hence, from (1), it follows that $h_i(x)R_ih_i(y)$ for $i = 1,\ldots,n$.
Thus $\underline{x}R\underline{y}$.

Let $V = \{\underline{z} \in Vev^* | (\forall a \in Ev)\ \#_a(\underline{z}) \geq 1\}$. We will now show that:

(3) $(\exists \underline{x} \in VFS)(\exists \underline{z} \in V)\ \underline{x}R\underline{x}\,\underline{z}$.

To show (3) we first prove that:

(4) $(\forall \underline{w} \in VFS)(\exists \underline{u} \in V)\ \underline{w}\,\underline{u} \in VFS$.

Let $Ev = \{a_1,\ldots,a_k\}$. Since Pr is adequate we have:
$(\exists y_1)\ \underline{w}\,\underline{y}_1\,\underline{a}_1 \in VFS$, $(\exists y_2)\ \underline{w}\,\underline{y}_1\,\underline{a}_1\,\underline{y}_2\,\underline{a}_2 \in VFS,\ldots,$
$(\exists y_k)\underline{w}\,\underline{y}_1\,\underline{a}_1\,\underline{y}_2\,\underline{a}_2 \cdots\ \underline{y}_k\,\underline{a}_k \in VFS$.
Of course, $\underline{u} = \underline{y}_1\,\underline{a}_1\,\underline{y}_2\,\underline{a}_2 \cdots\ \underline{y}_k\,\underline{a}_k \in V$. Hence (4) holds.
Let $m = card(V/R)$. From (4) it follows that:

$$(\exists \underline{z}_1,\ldots,\underline{z}_{m+1} \in V)(\underline{z}_1 \in VFS \wedge \underline{z}_1\underline{z}_2 \in VFS \wedge \ldots \wedge \underline{z}_1\underline{z}_2 \cdots \underline{z}_{m+1} \in VFS)$$

There are $r < p \leq m+1$ such that $\underline{z}_1 \ldots \underline{z}_r R \underline{z}_1 \ldots \underline{z}_p$. Hence (3) holds.
Let $Z = \{(\underline{u}, \underline{u}\,\underline{t}) \in VFS \times VFS | \underline{t} \in V \wedge \underline{u}R\underline{u}\,\underline{t}\}$, and let $l_{(\underline{u},\underline{u}\,\underline{t})} = length(\underline{u})$, for all $(\underline{u}, \underline{u}\,\underline{t}) \in Z$. We will show that:

(5) $(\exists \underline{z} \in V)(\underline{\varepsilon}, \underline{z}) \in Z$.

We first observe that, by (3), $Z \neq \emptyset$. Hence there is $(\underline{x}, \underline{x}\,\underline{z}) \in Z$, such that $l_{(\underline{x},\underline{x}\,\underline{z})} \leq l_{(\underline{u},\underline{u}\,\underline{t})}$ for all $(\underline{u}, \underline{u}\,\underline{t}) \in Z$. Suppose $\underline{x} \neq \underline{\varepsilon}$, i.e. $\underline{x} = \underline{u}\,\underline{a}$ for some $a \in Ev$. Consider $i \in i(a)$. We have $pr_i(\underline{x}) = pr_i(\underline{u})a$. From $pr_i(\underline{x}\,\underline{z}) \neq \varepsilon$ and (1) it follows that $pr_i(\underline{x}\,\underline{z}) = y_ia$ for some y_i. Hence $\underline{x}\,\underline{z} = \underline{w}\,\underline{a}$ for some $\underline{w} \in VFS$. We have $\underline{u}\,\underline{a}R\underline{w}\,\underline{a}$. Hence, by (2), $\underline{u}R\underline{w}$. We also have: $\underline{x}\,\underline{z} = \underline{u}\,\underline{a}\,\underline{v}\,\underline{a}$ with $\underline{w} = \underline{u}\,\underline{a}\,\underline{v}$, $\underline{z} = \underline{v}\,\underline{a}$, $\underline{u} \in Pref(\underline{w})$ and for all $b \in Ev$, $\#_b(\underline{a}\,\underline{v}) = \#_b(\underline{v}\,\underline{a}) \geq 1$. Hence $(\underline{u}, \underline{w}) \in Z$, contradicting the minimality of $l_{(\underline{x},\underline{x}\,\underline{z})}$. Consequently (5) is satisfied. Let $(\underline{\varepsilon}, \underline{z}) \in T$, and let $\underline{z} = (z_1,\ldots,z_n)$. From $z_i \neq \varepsilon$ and the definition of R_i it follows that there are $q_1 \geq 1,\ldots,q_n \geq 1$ such that $z_i = (cyc_i)^{q_i}$ for $i = 1,\ldots,n$. Let q be the greatest common divisor of $q_1,\ldots,q_n$. Define $r_i = \frac{q_i}{q}$

for $i = 1, \ldots, n$. For every i, let $x_i = (cyc_i)^{r_i}$. Clearly, we have $x_i \in Pref(z_i)$ for all i. Furthermore, if $a \in Ev_i \cap Ev_j$, then $\#_a(x_i) = \#_a(x_j)$, which follows from:

$$\#_a(x_i) = r_i \cdot \#_a(cyc_i) = \frac{1}{q}(q_i \cdot \#_a(cyc_i)) = \frac{1}{q}(\#_a(\underline{z})).$$

Hence, by Lemma 2.3 (1), $(x_1, \ldots, x_n) \in Vev^*$. If we now take any $x \in Ev^+$ such that $\underline{x} = (x_1, \ldots, x_n)$ and $r_1, \ldots, r_n$ as above, then (b) holds.

Before showing (b) $\Rightarrow$ (a), we prove the second part of this theorem. Let $\underline{y} = (y_1, \ldots, y_n) \in VFS$. From the definition of conflict-free path it follows that for every i there is $q_i \geq 1$ such that $y_i \in Pref((cyc_i)^{q_i})$. Let $q = 1 + max\{q_1, \ldots, q_n\}$ and let $\underline{z} = (\underline{x})^q$. We have $\underline{z} \in \{\underline{x}\}^*$ and $\underline{y} \in Pref(\underline{z})$. Hence $VFS \subseteq Pref(\{\underline{x}\}^*)$.

Clearly, $\underline{x}^0 = \underline{\varepsilon} \in VFS$, hence $\{\underline{x}\}^* \subseteq VFS$. Thus by Lemma 2.6 (1), $Pref(\{\underline{x}\}^*) \subseteq VFS$. Hence $VFS = Pref(\{\underline{x}\}^*)$.

To show (b) $\Rightarrow$ (a) we take $\underline{y} \in VFS$ and $a \in Ev$. From the second part of the theorem it follows that $(\exists z \in Ev^*)\ \underline{y}\,\underline{z} = \underline{x}^r$, for some $r \geq 0$. Moreover, $(\exists x_1, x_2 \in Ev^*)\ \underline{x} = \underline{x}_1\,\underline{a}\,\underline{x}_2$. Note that $\{\underline{a}\} \in enabled(\underline{y}\,\underline{z}\,\underline{x}_1)$, hence Pr is adequate. □

It turns out that for connected SCF-path programs, adequacy and deadlock-freeness are equivalent properties.

Lemma 2.15 *Let Pr be a connected SCF-path program. Then:*
Pr is adequate $\Leftrightarrow$ Pr is deadlock-free. □

An easy proof is left to the reader. We shall prove this assertion later for a slightly more general case (see Chap. 2.4.4 and Theorem 2.29).

To illustrate Theorem 2.20 let us consider the following example.

Example 2.4.1 Let $Pr_{14} = P_1P_2P_3$, where:

$$Pr_{14} = \begin{cases} P(1): & \underline{path}\ a; b; a; a; b\ \underline{end} \\ P(2): & \underline{path}\ c; a; c\ \underline{end} \\ P(3): & \underline{path}\ c; c; b; c\ \underline{end} \end{cases}$$

Here we have $cyc_1 = abaab, cyc_2 = cac, cyc_3 = ccbc$, and $x = cacbcaccabc$ satisfies: $h_1(x) = abaab = (cyc_1)^1$, $h_2(x) = caccaccac = (cyc_2)^3$, $h_3(x) = ccbcccbc = (cyc_3)^2$, and $hcf(1,3,2) = 1$. Furthermore, one may easily check that $VFS(Pr_{14}) = Pref(\{cacbcaccabc\}^*)$. ■

2.4.2 Checking Adequacy of SCF-Path Programs

Theorem 2.20 creates a basis for finding an algorithm for checking adequacy of SCF-path programs. First observe that if a path program Pr is not connected, then it is adequate if and only if all its connected components are adequate, so we can restrict our attention to connected path programs. So assume that $n \geq 2$ and $Pr = P_1 \ldots P_n$ is a *connected* SCF-path program. Theorem 2.20 states, among other things, that whenever an event a occurs in P_i and P_j, then:

$$r_i \cdot \#_a(cyc_i) = r_j \cdot \#_a(cyc_j).$$

In other words, we may say that there is a *set of integer equations*:

$$(\forall i, j = 1, \ldots, n)(\forall a \in Ev_i \cap Ev_j) t_i \cdot \#_a(cyc_i) = t_j \cdot \#_a(cyc_j) \qquad (2.1)$$

where the t_i are *variables* for $i = 1, \ldots, n$, and *if* Pr is adequate then there is a *solution* $r_1, \ldots, r_n$ of (2.1) such that each $r_i \geq 1$, r_i is an integer, and $hcf(r_1, \ldots, r_n) = 1$. Note that the above set of equations does not depend on the events which appear in one path only, suggesting the following lemma.

Lemma 2.16 *Let $Pr' = P'_1 \ldots P'_n$ be a path program such that P'_i is obtained from P_i by removing all events a satisfying $card(i(a)) = 1$, together with superfluous semicolons, for $i = 1, \ldots, n$. Then:*
Pr is adequate $\Leftrightarrow$ Pr' is adequate.

(Note that Pr' is well-defined since $n \geq 2$ and Pr is connected)
Proof. Let $A = \{a \in Ev | card(i(a)) = 1\}$, $B = Ev \setminus A$.
($\Rightarrow$) Let x be a sequence from Theorem 2.20 (b), and let $x' = h_B(x)$. Then:

$$h_i(h_B(x)) = h_B(h_i(x)) = h_B((cyc_i)^{r_i}) = (h_B(cyc_i))^{r_i} = (cyc(P'_i))^{r_i},$$

for each i, thus Pr' is adequate by Theorem 2.20.
($\Leftarrow$) From Theorem 2.20 we have a sequence $r_1, \ldots, r_n$ for Pr'. Consider $(x_1, \ldots, x_n)$ where $x_i = (cyc_i)^{r_i}$. From Theorem 2.20 for Pr' we have $(h_B(x_1), \ldots, h_B(x_n)) \in VFS(Pr') \subseteq Vev^*$. Hence, by Lemma 2.3 (2) $(x_1, \ldots, x_n) \in Vev^*$. Let $\underline{x} = (x_1, \ldots, x_n)$, then $x, r_1, \ldots, r_n$ satisfy Theorem 2.20 (b) for Pr, and Pr is adequate. □

Thus, we may assume that our Pr is of the form of Pr' from Lemma 2.16. Of course, the existence of a solution to the set of equations (2.1) *does not* mean that Pr is adequate. Consider the example Pr_4 (Sect. 2.2) which is of the form:

$$Pr_4 = \begin{cases} P(1): & \underline{path}\ a; b\ \underline{end} \\ P(2): & \underline{path}\ b; a\ \underline{end} \end{cases}$$

Here the set of equations (2.1) is the following:

$$t_1 \cdot \#_a(cyc_1) = t_2 \cdot \#_a(cyc_2) \quad (2.2)$$
$$t_1 \cdot \#_b(cyc_1) = t_2 \cdot \#_b(cyc_2) \quad (2.3)$$

where $\#_a(cyc_1) = \#_a(cyc_2) = \#_b(cyc_1) = \#_b(cyc_2) = 1$. Clearly Pr_4 is not adequate, but $t_1 = t_2 = 1$ are solutions satisfying $hcf(t_1, t_2) = 1$.
The equation (2.1) *does not* depend on the particular shape of the P_i's, *it only depends* on the numbers of occurrences of common events. However *if* Pr is adequate, then finding $r_1, \ldots, r_n$ from Theorem 2.20 (b) may be reduced to solving the set of equations (2.1). If Pr is not adequate, the set of equations (2.1) may have no non-trivial solution, as the example below shows:

$$Pr_{15} = \begin{cases} P(1): & \underline{path}\ a; b; a\ \underline{end} \\ P(2): & \underline{path}\ a; b\ \underline{end} \end{cases}$$

Pr_{15} deadlocks after $\underline{aba}$, the set of equations (2.1) has the form:

$$t_1 \cdot 2 = t_2 \cdot 1 \quad (2.4)$$
$$t_1 \cdot 1 = t_2 \cdot 1 \quad (2.5)$$

and the only solution is $t_1 = t_2 = 0$. Furthermore, if the set of equations (2.1) has a non-trivial solution, it has an infinite number of solutions, since if $t_1, \ldots, t_n$ satisfy the set of equations (2.1) then for every $k \geq 1, k \cdot t_1, \ldots, k \cdot t_n$ also is a solution of (2.1).
Let us analyze the structure of (2.1) in detail. Assume that Pr is adequate, i.e., (2.1) has a non-trivial solution, and furthermore Pr is connected and for each $a \in Ev$, there are i, j such that $i \neq j$ and $a \in Ev_i \cap Ev_j$. We show that in such a case, we can transform (2.1) into an equivalent but simpler and much smaller set of equations.

Lemma 2.17 *Let $n_1, \ldots, n_k, m_1, \ldots, m_k$ be positive integers such that the following set of equations has a strictly positive ($\neq 0$) integer solution:*

$t \cdot n_1 = t' \cdot m_1$

$\ldots$

$t \cdot n_k = t' \cdot m_k$

where t, t' are variables. Then, if s, s' are positive integers such that $s \cdot n_1 = s' \cdot m_1$, then for every $i = 2, \ldots, k$, $s \cdot n_i = s' \cdot m_i$.

Proof. Let t, t' be positive integer solutions of the above set of equations, and let $i \in \{2, \ldots, k\}$. Thus: $t \cdot n_1 = t' \cdot m_1$ and $t \cdot n_i = t' \cdot m_i$, hence

$$m_i = \frac{m_1 \cdot n_i}{n_1},\ i.e.\ s \cdot n_i = \frac{s' \cdot m_1}{n_1} \cdot n_i = s' \cdot \frac{m_1 \cdot n_i}{n_1} = s' \cdot m_i.$$

□

The result above enables us to reduce the set of equations (2.1). For every $i, j = 1, \ldots, n$, if $Ev_i \cap Ev_j \neq \emptyset$ and $i \neq j$, we choose arbitrarily one element $a_{ij} \in Ev_i \cap Ev_j$, and then we define the following set of equations:

$$t_i \cdot \#_{a_{ij}}(cyc_i) = t_j \cdot \#_{a_{ij}}(cyc_j) \tag{2.6}$$

for all i, j such that $i \neq j$ and $Ev_i \cap Ev_j \neq \emptyset$.
From Lemma 2.17 it follows that if equation (2.1) has solutions then equation (2.1) and equation (2.6) have exactly the same sets of solutions.
Let $G_{\mathcal{E}} = (I, \mathcal{E})$ be a **connected undirected graph** with the set of vertices $I = \{1, 2, \ldots, n\}$ and the set of edges $\mathcal{E}$ (see Appendix C). In the sequel we shall usually identify $\mathcal{E}$ with $G_{\mathcal{E}}$.
For every $\{i, j\} \in \mathcal{E}$, let $k_{i,j}, k_{j,i}$ be positive integers such that the following set of equations (with variables $t_1, \ldots, t_n$) have positive integer solutions:

$$(\forall \{i, j\} \in \mathcal{E}) t_i \cdot k_{i,j} = t_j \cdot k_{j,i} \tag{2.7}$$

We shall denote the above set of equations by $EQ_{\mathcal{E}}$. Clearly, $EQ_{\mathcal{E}}$ is nothing else but a bit more abstract representation of equation (2.6). The lemma below shows the dependencies between different solutions of $EQ_{\mathcal{E}}$.

Lemma 2.18 *Let $r_1, \ldots, r_n$, and $s_1, \ldots, s_n$ be positive solutions of $EQ_{\mathcal{E}}$. Then:* $(\forall i, j \in I) \frac{r_i}{r_j} = \frac{s_i}{s_j}$

Proof. Let $i, j \in I$ and let $l_1, \ldots, l_s$ be a path of vertices in $G_{\mathcal{E}}$ connecting i with j (see definitions C.2 and C.3 in Appendix C).

$$t_{l_1} \cdot k_{l_1,l_2} = t_{l_2} \cdot k_{l_2,l_1}$$

$$\ldots$$

$$t_{l_{s-1}} \cdot k_{l_{s-1},l_s} = t_{l_s} \cdot k_{l_s,l_{s-1}}$$

where $t_{l_1} = t_i, t_{l_s} = t_j$. Hence:

$$\frac{r_i}{r_j} = \frac{k_{l_2,l_1} \cdot k_{l_3,l_2} \cdot \ldots \cdot k_{l_s,l_{s-1}}}{k_{l_1,l_2} \cdot k_{l_2,l_3} \cdot \ldots \cdot k_{l_{s-1},l_s}} = \frac{s_i}{s_j}.$$

□

In many cases we can replace $EQ_{\mathcal{E}}$ by a smaller equivalent set of equations $EQ_{\mathcal{E}'}$.

Lemma 2.19 *Let $\mathcal{E}' \subseteq \mathcal{E}$, $G_{\mathcal{E}'} = (I, \mathcal{E}')$ be a connected graph, and let $EQ_{\mathcal{E}'}$ be a subset of $EQ_{\mathcal{E}}$ (i.e. $k_{i,j}$ and $k_{j,i}$ are identical for $EQ'_{\mathcal{E}}$ and $EQ_{\mathcal{E}}$). Then:*

$s_1, \ldots, s_n$ is a positive solution for $EQ_{\mathcal{E}} \Leftrightarrow s_1, \ldots, s_n$ is a positive solution for $EQ_{\mathcal{E}'}$.

Proof. ($\Rightarrow$) Obvious, since $EQ_{\mathcal{E}'}$ is a subset of $EQ_{\mathcal{E}}$.
($\Leftarrow$) Let $\{i,j\} \in \mathcal{E} \setminus \mathcal{E}'$. We have to prove that $s_i \cdot k_{i,j} = s_j \cdot k_{j,i}$. Let $r_1, \ldots, r_n$ be any non-trivial solution of $EQ_{\mathcal{E}}$. By the definition of $\mathcal{E}$ there is at least one. Thus we have: $r_i \cdot k_{i,j} = r_j \cdot k_{j,i}$, i.e., $\frac{r_i}{r_j} = \frac{k_{j,i}}{k_{i,j}}$. Since $\mathcal{E}' \subseteq \mathcal{E}$, $r_1, \ldots, r_n$ is also a solution of $\mathcal{E}'$, so by Lemma 2.18, $\frac{r_i}{r_j} = \frac{s_i}{s_j}$. Hence: $\frac{s_i}{s_j} = \frac{r_i}{r_j} = \frac{k_{j,i}}{k_{i,j}}$, i.e., $s_i \cdot k_{i,j} = s_j \cdot k_{j,i}$. □

We next show that there is a *unique* positive solution for $EQ_{\mathcal{E}}$: $r_1, \ldots, r_n$ such that $hcf(r_1, \ldots, r_n) = 1$.

Lemma 2.20 *Let $s_1, \ldots, s_n$ and $p_1, \ldots, p_n$ be positive solutions for $EQ_{\mathcal{E}}$, $s = hcf(s_1, \ldots, s_n)$, $p = hcf(p_1, \ldots, p_n)$, and $r_i = \frac{s_i}{s}$, $r_i' = \frac{p_i}{p}$ for $i = 1, \ldots, n$. Then: $r_1, \ldots, r_n$ is a positive solution for $EQ_{\mathcal{E}}$ and $r_i = r_i'$ for $i = 1, \ldots, n$.*

Proof. Clearly $r_1, \ldots, r_n$ and $r_1', \ldots, r_n'$ are solutions and $hcf(r_1, \ldots, r_n) = hcf(r_1', \ldots, r_n') = 1$. From Lemma 2.18 we have: $(\exists q)(\forall i = 1, \ldots, n)\frac{r_i}{r_i'} = q$. Without loss of generality we can assume $q \geq 1$. If $q = 1$ then $r_i = r_i'$ for $i = 1, \ldots, n$, so assume $q > 1$. Then $q = \frac{k}{m}$, where $m \neq 1$, k, m are integers, $hcf(k, m) = 1$, and $k > m$. This means $r_i \cdot m = r_i' \cdot k$ for $i = 1, \ldots, n$. By $hcf(k, m) = 1$, k must be a divisor of r_i, for $i = 1, \ldots, n$. Hence $hcf(r_1, \ldots, r_n) \geq k > m \geq 1$, a contradiction, so $q = 1$. □

Now we are ready to formulate an algorithm for finding the unique positive solution $r_1, \ldots, r_n$ of $EQ_{\mathcal{E}}$ satisfying $hcf(r_1, \ldots, r_n) = 1$.
Let $\mathcal{E}_0$ be any *minimal* subset of $\mathcal{E}$ such that $G_{\mathcal{E}_0} = (I, \mathcal{E}_0)$ is a connected graph.

In graph theory every such $\mathcal{E}_0$ is called an *undirected spanning tree* for $\mathcal{E}$. Algorithms for finding spanning trees and their analysis can be found for instance in [AHU74]. See also Appendix C.

Let us choose $i_0 \in I$, and let $\mathcal{E}_0(i_0)$ be a *directed tree* derived from $\mathcal{E}_0$ by setting i_0 as a root. From graph theory we know (see Appendix C) that an undirected tree $\mathcal{E}_0$ and a node i_0 of $\mathcal{E}_0$ define the directed tree $\mathcal{E}_0(i_0)$ in an *unambiguous way*. Moreover, $\mathcal{E}_0(i_0)$ as a directed graph can be interpreted as a binary relation, so we can write that $\mathcal{E}_0(i_0) \subseteq I \times I$.
From Lemma 2.19 we have that the solutions for $EQ_{\mathcal{E}_0}$ and $EQ_{\mathcal{E}}$ are exactly the same. The theorem below gives us an effective way of finding the desired solution.

Theorem 2.21 *Let $s_1, \ldots, s_n$ be numbers defined recursively on the basis of the structure of $\mathcal{E}_0(i_0)$:*
$s_{i_0} = \prod_{(i,j) \in \mathcal{E}_0(i_0)} k_{i,j}$
$s_j = s_i \cdot \frac{k_{i,j}}{k_{j,i}}$ *for every $(i,j) \in \mathcal{E}_0(i_0)$.*

Let $s = hcf(s_1, \ldots, s_n)$*, and let* $r_1 = \frac{s_1}{s}, \ldots, r_n = \frac{s_n}{s}$.
Then $r_1, \ldots, r_n$ *is the positive unique solution of* $EQ_{\mathcal{E}}$ *such that* $hcf(r_1, \ldots, r_n) = 1$.

Proof. Since $Dom(\mathcal{E}_0(i_0)) = I = \{1, \ldots, n\}$, and $\mathcal{E}_0(i_0)$ is a directed tree, all s_i's are defined, and one can easily check by inspection that $s_1, \ldots, s_n$ is a solution for $EQ_{\mathcal{E}_0}$, so by Lemma 2.19 it is also a solution for $EQ_{\mathcal{E}}$. From Lemma 2.20 we have that $r_1, \ldots, r_n$ is thus the unique positive solution for $EQ_{\mathcal{E}}$ with $hcf(r_1, \ldots, r_n) = 1$. □

Since the set of Equations (2.6) is a particular case of $EQ_{\mathcal{E}}$ (where $k_{i,j} = \#_{a_{ij}}(cyc_i), k_{j,i} = \#_{a_{ij}}(cyc_j)$), then the above procedure can be applied to it, and since Equations (2.6) are equivalent to Equations (2.1) we can say that $r_1, \ldots, r_n$ from Theorem 2.20 is the only solution of Equations (2.1), if any, satisfying $hcf(r_1, \ldots, r_n) = 1$.
Recall that the results obtained above are true only under the assumption that Pr is adequate. Of course, we may always define the set $\mathcal{E}$ for a given path program Pr, and find numbers $r_1, \ldots, r_n$ according to the Theorem 2.21, but if Pr is not adequate, $r_1, \ldots, r_n$ might not be a solution for Equations (2.1). Furthermore, even if $r_1, \ldots, r_n$ are the unique solution for Equations (2.1) satisfying $hcf(r_1, \ldots, r_n) = 1$, their existence *does not guarantee* the adequacy of Pr. But given $r_1, \ldots, r_n$ defined as above, we can calculate the length of x from Theorem 2.20 (b) and perform a simulation of at most that length, in order to find out whether or not the given SCF-path program is adequate.
Thus, the algorithm for checking adequacy and finding $r_1, \ldots, r_n$ from Theorem 2.20 (b) is the following:

Algorithm 2.1 *Let* $Pr = P_1 \ldots P_n$ *be a connected SCF-path program.*

Step 1. *Delete from* Pr *all events belonging to one single path only together with superfluous semicolons. Denote this new SCF-path program by* $Pr' = P'_1, \ldots, P'_n$*, and let* $Ev'_i = Ev(P'_i)$*, and* cyc'_i *be a string such that* $Cyc(P'_i) = \{cyc'_i\}$*, for* $i = 1, \ldots, n$.

Step 2. *Define the undirected graph* $\mathcal{E}$ *as follows:* $\{i,j\} \in \mathcal{E} \Leftrightarrow Ev'_i \cap Ev'_j \neq \emptyset$ *and* $i \neq j$.

Step 3. *Label every edge* $\{i,j\} \in \mathcal{E}$ *by an arbitrary* $a_{ij} \in Ev'_i \cap Ev'_j$.

Step 4. *Choose an arbitrary* $i_0 \in \{1, \ldots, n\}$ *and define a directed spanning tree* $\mathcal{E}_0(i_0)$.

Step 5. *Define* $s_1, \ldots, s_n$ *recursively on the basis of* $\mathcal{E}_0(i_0)$ *in the following way:*

$$s_{i_0} = \prod_{(i,j) \in \mathcal{E}_0(i_0)} \#_{a_{i,j}}(cyc'_j)$$

$$s_j = s_i \cdot \frac{\#_{a_{i,j}}(cyc'_i)}{\#_{a_{i,j}}(cyc'_j)} \text{ for every } (i,j) \in \mathcal{E}_0(i_0).$$

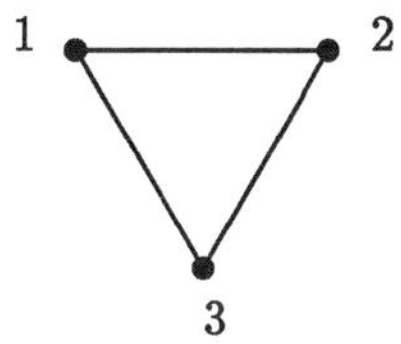

Figure 2.36. Graph $\mathcal{E}$ (Ex 2.4.2)

Step 6. *Set $s = hcf(s_1, \ldots, s_n)$, and let $r_i = \frac{s_i}{s}$ for $i = 1, \ldots, n$.*

Step 7. *Simulate x by the Pascal-like program below. We may simulate x in either Pr or Pr', the program below simulates in Pr, but after replacing cyc_i by cyc'_i, and Ev_i by Ev'_i, it will simulate in Pr'.*

$x := \varepsilon$;
deadlocked := *false*;
<u>*for*</u> $i := 1$ <u>*to*</u> n <u>*do*</u> $d_i := (cyc_i)^{r_i}$;
<u>*while*</u> $(\exists i) d_i \neq \varepsilon \wedge \neg deadlocked$ <u>*do*</u> <u>*begin*</u>
<u>*if*</u> $(\exists a)(\forall i) a \in Ev_i \Rightarrow ((\exists d'_i) d_i = a d'_i)$
<u>*then*</u> $x := xa; d_i := d'_i$ *for all i such that $a \in Ev_i$*
<u>*else*</u> *deadlocked* := *true*
<u>*end*</u> ∎

The program from Step 7 constructs either a deadlocking sequence or a sequence x from Theorem 2.20. Now we illustrate the above algorithm by proving adequacy of an SCF-path program.

Example 2.4.2 Consider:

$$Pr_{16} = \begin{cases} P(1): & \underline{path}\ a; b; a; d; a; b; d\ \underline{end} \\ P(2): & \underline{path}\ c; a; c\ \underline{end} \\ P(3): & \underline{path}\ c; c; b; c; d\ \underline{end} \end{cases}$$

<u>Step 1.</u> Since $a \in Ev_1 \cap Ev_2, b, d \in Ev_1 \cap Ev_3, c \in Ev_2 \cap Ev_3$, then $Pr' = Pr_{16}$.
<u>Step 2.</u> The graph $\mathcal{E}$ is depicted in Fig. 2.36.
<u>Step 3.</u> The graph $\mathcal{E}$ with labelled edges is depicted in Fig. 2.37. We chose $b \in Ev_1 \cap Ev_3 = \{b, d\}$.
<u>Step 4.</u> Let $i_0 = 1$, so the spanning tree $\mathcal{E}_0(1)$ is as in Fig. 2.38.
<u>Step 5.</u>

$$s_1 = \#_a(cyc_2) \cdot \#_b(cyc_3) = 1 \cdot 1 = 1$$

$$s_2 = s_1 \cdot \#_a(cyc_1)/\#_a(cyc_2) = 1 \cdot 3/1 = 3$$

$$s_3 = s_1 \cdot \#_b(cyc_1)/\#_b(cyc_3) = 1 \cdot 2/1 = 2$$

<u>Step 6.</u> $hcf(1,2,3) = 1$, so $r_1 = 1, r_2 = 3, r_3 = 2$.
<u>Step 7.</u> Note that at the beginning: $d_1 = abadabd, d_2 = caccaccac, d_3 =$

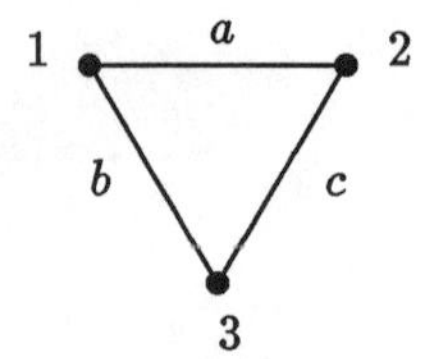

Figure 2.37. Labeled graph $\mathcal{E}$ (Ex 2.4.2)

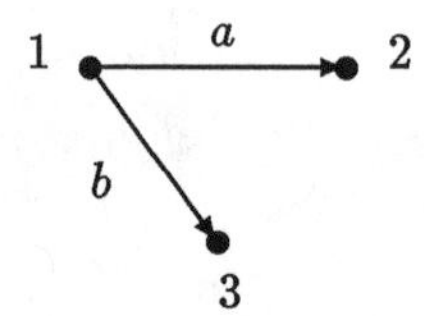

Figure 2.38. Spanning tree $\mathcal{E}_0(1)$ (Ex 2.4.2)

ccbcdccbcd. One may verify that the following non-deadlocking sequence x may be produced by the program from Step 7 of Algorithm 2.1: $x = cacbcadccabcd$ with $length(x) = 13$. Thus Pr_{16} is adequate. ■

Unfortunately, the computational overhead of Step 7 of Algorithm 2.1 can be quite considerable as we show below.

Theorem 2.22 *For any sequence of positive integers $r_1, \ldots, r_n$ with $hcf(r_1, \ldots, r_n) = 1$ an adequate and connected SCF-path program can be constructed such that the condition (b) of Theorem 2.20 is satisfied with these numbers.*

Proof. Let $Ev = \{a_{ij} | i < j \wedge i, j \in \{1, \ldots, n\}\}$, and, for $i = 1, \ldots, n$, let $Ev_i = \{a_{kj} | a_{kj} \in Ev \wedge (k = i \vee j = i)\}$. Note that $Ev_i \cap Ev_j = \{a_{ij}\}$, for every $i, j = 1, \ldots, n$, and $i \neq j$. Let $\preceq$ be any arbitrary total order in $Ev \times Ev$. Define $Eo \subseteq Ev \times \{1, 2, \ldots\}$ as: $Eo = \{(a_{ij}, r) | a_{ij} \in Ev \wedge 1 \leq r \leq r_i \cdot r_j\}$. Now we introduce the relation $\leq$ on E_0 as follows:

$$(a_{ij}, r) \leq (a_{kl}, r') :\Leftrightarrow (a_{ij} = a_{kl} \wedge r = r') \vee (\frac{r}{r_i \cdot r_j} < \frac{r'}{r_k \cdot r_l}) \vee$$

$$(\frac{r}{r_i \cdot r_j} = \frac{r'}{r_k \cdot r_l} \wedge a_{ij} \preceq a_{kl}).$$

We left to the reader to check that $\leq$ is a total order in $Eo \times Eo$, and since $\leq$ is a total order, Eo might be represented as:

$$(a_{i_1 j_1}, r'_1) \leq (a_{i_2 j_2}, r'_2) \leq \ldots \leq (a_{i_m j_m}, r'_m).$$

Define $x = a_{i_1 j_1} a_{i_2 j_2} \dots a_{i_m j_m}$. Clearly each a_{ij} from Ev occurs in x $r_i \cdot r_j$ times. Moreover, one can easily prove that for every $i = 1, \dots, n$, there is $x_i \in Ev_i^*$ such that $h_i(x) = (x_i)^{r_i}$. Now we may define $P_i = \underline{path}\ b_1; \dots; b_{k_i}\ \underline{end}$ where $x_i = b_1 \dots b_{k_i}$. Clearly $Pr = P_1 \dots P_n$ satisfies Theorem 2.20 (b) with $r_1, \dots, r_n$. □

For example, if $r_1 = 1, r_2 = 3, r_3 = 2$, we may set $a_{12} = a, a_{13} = b, a_{23} = c$ and $a \prec b \prec c$, then:
$Eo = (c,1) < (a,1) < (c,2) < (b,1) < (c,3) < (a,2) < (c,4) < (c,5) < (a,3) < (b,2) < (c,6)$,
$x = cacbcaccabc$, $h_1(x) = abaab$, i.e. $x_1 = abaab$, $h_2(x) = caccaccac = (cac)^3$, i.e. $x_2 = cac$, $h_3(x) = ccbcccbc = (ccbc)^2$, i.e. $x_3 = ccbc$, thus $P_1 = \underline{path}\ a; b; a; a; b\ \underline{end}$, $P_2 = \underline{path}\ c; a; c\ \underline{end}$, $P_3 = \underline{path}\ c; c; b; c\ \underline{end}$, and $Pr = P_1 P_2 P_3 = Pr_{14}$ from Section 2.4.1.
We can now calculate the computational overhead for an SCF-path program constructed from the proof of Theorem 2.22. From this proof it follows that

$$length(cyc_i) = \sum_{j \neq i} r_j$$

hence we have for the average length q of a path:

$$q = \frac{1}{n} \sum_{i=1}^{n} \sum_{j \neq i}^{n} r_i = \frac{n-1}{n} \sum_{j=1}^{n} r_j \approx \sum_{j=1}^{n} r_j = r \cdot n$$

where r is the average of the r_i's. For the length of x, on the other hand, we obtain:

$$length(x) = \sum_{i<j} r_i \cdot r_j = \frac{1}{2} (\sum_{i=1}^{n} r_i)^2 - \frac{1}{2} \sum_{i=1}^{n} r_i^2.$$

Consequently, $length(x) \approx \frac{1}{2} q^2 - \frac{1}{2} r \cdot q$.
But since for each element of x the entire set of simple paths has to be checked in the production of x, the overhead of the program from Step 1 calculates to:

$$\frac{1}{2} n \cdot (q^2 - r \cdot q) \approx \frac{1}{2} n \cdot q^2 \approx \frac{1}{2} r^2 \cdot n^3.$$

This is cubic in n and quadratic in r, while as a consequence of Theorem 2.22, n and r are independent quantities. Thus, in general, it might be worthwhile to try to decrease the number of simple paths in an SCF-path program we want to analyze.
If we are interested in checking adequacy of Pr only, and we are not interested in particular values of $r_1, \dots, r_n$, we can sometimes reduce checking the adequacy of $Pr = P_1 \dots P_n$ to checking the adequacy of $P_{i_1} \dots P_{i_k}$ where $k < n$ and $i_1, \dots, i_k \in \{1, \dots, n\}$. This reduction is based on the following two lemmas.

Lemma 2.21 *Let $Pr = P_1P_2$ be an SCF-path program, $Ev_1 \subseteq Ev_2$, $a \in Ev_1$, $s_i = \#_a(cyc_i)$, $i = 1,2$, and $r_1 = s_2/hcf(s_1,s_2)$, $r_2 = s_1/hcf(s_1,s_2)$. Then: Pr is adequate $\Leftrightarrow$ $(cyc_1)^{r_1} = h_1((cyc_2)^{r_2})$.*

Proof. ($\Rightarrow$) Since Pr is adequate, from the theory developed so far in this section it follows that there is $x \in Ev^*$ such that: $(cyc_1)^{p_1} = h_1(x)$, $(cyc_2)^{p_2} = h_2(x)$, where $hcf(p_1,p_2) = 1$, $p_1,p_2 \geq 1$, and p_1,p_2 form a positive solution of (t,t' are variables): $t \cdot s_1 = t' \cdot s_2$.

Since r_1, r_2 is also a positive solution $t \cdot s_1 = t' \cdot s_2$, by Lemma 2.20, $p_1 = r_1$ and $p_2 = r_2$. This and $h_2(x) = x$ yields $(cyc_1)^{r_1} = h_1((cyc_2)^{r_2})$.

($\Leftarrow$) It suffices to take $x = (cyc_2)^{r_2}$ and then apply Theorem 2.20. □

Lemma 2.22 *Let $Pr = P_1 \ldots P_n$ be an SCF-path program, and let $j,k \in \{1,\ldots,n\}$, $j \neq k$, be such that $Ev_j \subseteq Ev_k$. Then the following are equivalent:*

(1) Pr is adequate

(2) $Pr' = P_1 \ldots P_{j-1}P_{j+1} \ldots P_n$ and $Pr'' = P_kP_j$ are adequate.

Proof. ($\Rightarrow$): Let $x, r_1, \ldots, r_n$ satisfy Theorem 2.20 (b) for Pr. One can easily see that:

(a) $x, r_1/r, \ldots, r_{j-1}/r, r_{j+1}/r, \ldots, r_n/r$, where $r = hcf(r_1, \ldots, r_{j-1}, r_{j+1}, \ldots, r_n)$, satisfy Theorem 2.20 (b) for Pr'.

(b) $h_k(x), r_k/q, r_j/q$, where $q = hcf(r_k, r_j)$, satisfy Theorem 2.20 (b) for Pr''.

($\Leftarrow$) Let $x, r_1, \ldots, r_{j-1}, r_{j+1}, \ldots, r_n$ satisfy Theorem 2.20 (b) for Pr', and let y, p_1, p_2 satisfy Theorem 2.20 (b) for Pr''. Let $r = p_1/hcf(r_k, p_1)$ and $p = r_k/hcf(r_k, p_1)$. Define $r'_i = r_i \cdot r$ for $i = 1, \ldots, j-1, j+1, \ldots, n$, and $r'_j = p_2 \cdot p$. One can easily see that $x^r, r'_1, \ldots, r'_{j-1}, r'_j, r'_{j+1}, \ldots, r'_n$ satisfy Theorem 2.20 (b) for Pr. □

The above two lemmas enable us to make the algorithm for checking adequacy more "static", in the sense that Step 7 can sometimes be applied to a path program with less paths than n.

Algorithm 2.2 *Let $Pr = P_1 \ldots P_n$ be a connected SCF-path program.*

Step 1. *Delete from Pr all events belonging to one single path only together with superfluous semicolons. Denote this new SCF-path program by $Pr' = P'_1 \ldots P'_n$, and next* rename *Pr' back to Pr, i.e., let $P_i := P'_i$ for $i = 1, \ldots, n$, and $Pr := P_1 \ldots P_n$.*

Step 2. *Define the undirected graph $\mathcal{E}$ as follows:*

$$\{i,j\} \in \mathcal{E} \Leftrightarrow Ev_i \cap Ev_j \neq \emptyset \wedge i \neq j.$$

Step 3. *Label every edge $\{i,j\} \in \mathcal{E}$ by the set*

$$label_{i,j} = Ev_i \cap Ev_j.$$

Step 4. *Let I_0 be the set of all vertices of $\mathcal{E}$ all of whose adjacent edges carry the same label, i.e.*

$$I_0 = \{i | (\forall j,k)(i,j) \in \mathcal{E} \wedge (i,k) \in \mathcal{E} \Rightarrow label_{i,j} = label_{i,k}\}.$$

From Step 1 we have that for every $i \in I_0$, $\{i,j\} \in \mathcal{E} \Rightarrow Ev_i \subseteq Ev_j$ so we can use Lemma 2.21 and 2.22. If $I_0 = \emptyset$ then goto Step 7.

Step 5. *For every $i \in I_0$, and every j such that $\{i,j\} \in \mathcal{E}$, verify the adequacy of P_iP_j using Lemma 2.21. If there is $\{i_0, j_0\} \in \mathcal{E}$ such that $i_0 \in I_0$ and $P_{i_0}P_{j_0}$ is not adequate, then Pr is not adequate either, and we stop the procedure. If for all $i \in I_0$ and $\{i,j\} \in \mathcal{E}$, P_iP_j is adequate, goto next step.*

Step 6. *First set $\mathcal{E} := \mathcal{E} \setminus \{\{i,j\} | i \in I_0\}$, and next set $P := P_{i_1} \dots P_{i_k}$ where $\{i_1, \dots, i_k\} = \{i | \{i,j\} \in \mathcal{E}\}$. In other words $\{i_1, \dots, i_k\}$ is the set of vertices of the new $\mathcal{E}$. Next go to Step 4.*

Step 7. *Label every edge $\{i,j\} \in \mathcal{E}$ by an arbitrary $a_{ij} \in label_{i,j}$.*

Step 8. *Choose an arbitrary $i_0 \in \{i_1, \dots, i_k\}$ and define a directed spanning tree $\mathcal{E}_0(i_0)$.*

Step 9. *Define $s_{i_1}, \dots, s_{i_k}$ recursively on the basis of $\mathcal{E}_0(i_0)$ in the following way:*

$$s_{i_0} = \prod_{(i,j) \in \mathcal{E}_0(i_0)} \#_{a_{ij}}(cyc_j)$$

$$s_j = s_i \cdot \frac{\#_{a_{ij}}(cyc_i)}{\#_{a_{ij}}(cyc_j)} \text{ for every } (i,j) \in \mathcal{E}_0(i_0).$$

Step 10. *Set $s = hcf(s_{i_1}, \dots, s_{i_k})$ and let $r_i = s_i / s$ for $i = 1, \dots, n$.*

Step 11. *Simulate x for $P_{i_1}, \dots, P_{i_k}$ by the following (Pascal-like) program (we can assume that $i_1 < i_2 < \dots < i_k$):*

$x := \varepsilon;$
$deadlocked := false;$
<u>for</u> $i := i_1$ <u>to</u> i_k <u>do</u> $d_i := (cyc_i)^{r_i};$
<u>while</u> $(\exists i \in \{i_1, \dots, i_k\}) d_i \neq \varepsilon \wedge \neq \neg deadlocked$ <u>do</u> <u>begin</u>
<u>if</u> $(\exists a)(\forall i \in \{i_1, \dots, i_k\}) a \in Ev_i \Rightarrow ((\exists d_i') d_i = a d_i')$
<u>then</u> $x := xa; d_i := d_i'$ *for all* $i \in \{i_1, \dots, i_k\}$ *such that* $a \in Ev_i$
<u>else</u> $deadlocked := true$
<u>end</u>

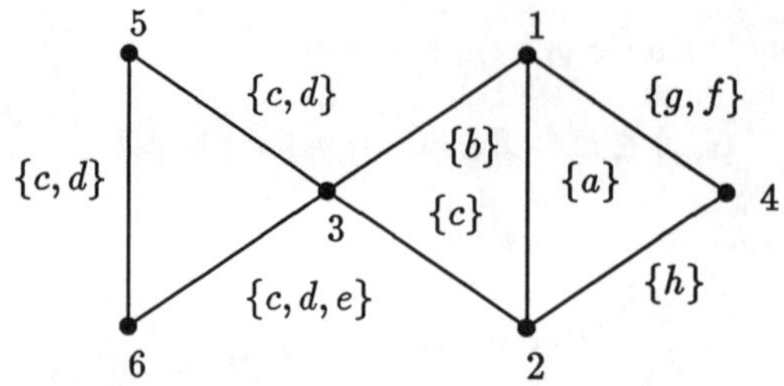

Figure 2.39. Labeled graph $\mathcal{E}$ (Step 3 Ex 2.4.3)

If the simulation is successful, Pr from the beginning of this algorithm is adequate, in the opposite case it is not. ■
We shall illustrate the above algorithm with the following example.

Example 2.4.3 Let $Pr_{17} = P_1 \ldots P_6$ where:

$$Pr_{17} = \begin{cases} P(1): & \underline{path}\ g; a; f; b; a; g; f; a; b\ \underline{end} \\ P(2): & \underline{path}\ h; c; a; c; h\ \underline{end} \\ P(3): & \underline{path}\ e; c; d; e; c; b; d; e; c; d\ \underline{end} \\ P(4): & \underline{path}\ h; g; h; h; f\ \underline{end} \\ P(5): & \underline{path}\ c; f'; f'; d; f'\ \underline{end} \\ P(6): & \underline{path}\ e; c; d; e; c; d\ \underline{end} \end{cases}$$

Step 1. f' appears in P_5 only, hence the new P_5 is of the form:

$$P_5 : \underline{path}\ c; d\ \underline{end}$$

Steps 2 and 3. The labelled undirected graph $\mathcal{E}$ is shown in Fig. 2.39.
Step 4. (First time). $I_0 = \{5\}$, note that $Ev_5 = \{c,d\} \subseteq Ev_3 = \{b,c,d,e\}$, and $Ev_5 = \{c,d\} \subseteq Ev_6 = \{c,d,e\}$.
Step 5. (First time). Consider P_5P_6 and take $c \in Ev_5$. We have $\#_c(cyc_5) = 1, \#_c(cyc_6) = 2$, hence $r_1 = 2, r_2 = 1$, and $(cyc_5)^{r_1} = (cyc_5)^2 = cdcd$, $(cyc_6)^{r_2} = cyc_6 = ecdecd$, i.e. $h_{Ev_5}(ecdecd) = cdcd$, thus P_5P_6 is adequate.
Consider P_5P_3 and take $d \in Ev_5$. Thus: $\#_d(cyc_5) = 1$, $\#_d(cyc_3) = 3$, i.e. $r_1 = 3, r_2 = 1$, $(cyc_5)^{r_1} = (cyc_5)^3 = cdcdcd$, $(cyc_3)^{r_2} = cyc_3 = ecdecbdecd$, hence $h_{Ev_5}(ecdecbdecd) = cdcdcd$, so P_5P_3 is adequate.
Step 6. (First time). The new $\mathcal{E}$ is defined as:

$$\mathcal{E} := \mathcal{E} \setminus \{\{i,j\} | i \in \{5\}\} = \mathcal{E} \setminus \{\{5,3\},\{6,5\}\}$$

and it is depicted in Fig. 2.40.
The new Pr is $P_1P_2P_3P_4P_6$.
Step 4. (Second time). $I_0 = \{6\}$, note that $Ev_6 = \{c,d,e\} \subseteq Ev_3$.
Step 5. (Second time). Consider P_6P_3, and take $e \in Ev_6$. Thus $\#_e(cyc_6) = 2$,

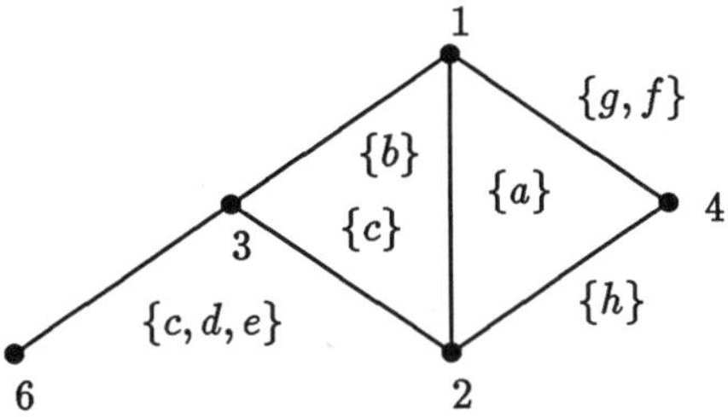

Figure 2.40. Labeled $\mathcal{E}$ (Step 6 (1st time) Ex 2.4.3)

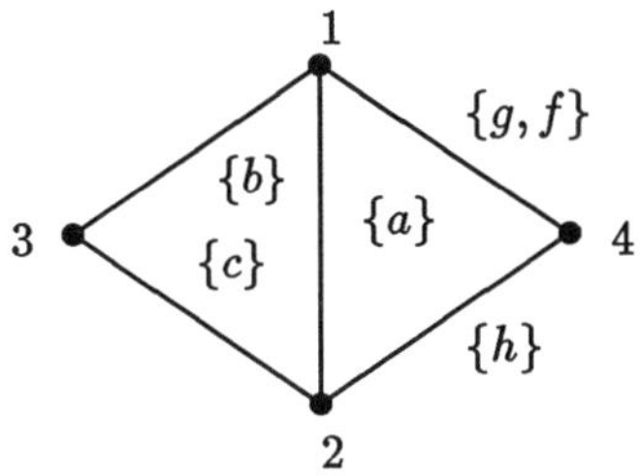

Figure 2.41. Labeled $\mathcal{E}$ (Step 6 (2nd time) Ex 2.4.3)

$\#_e(cyc_3) = 3$, i.e. $r_1 = 3, r_2 = 2$, $(cyc_6)^{r_1} = (cyc_6)^3 = ecdecdecdecdecdecd$, $(cyc_3)^{r_2} = (cyc_3)^2 = ecdecbdecdecdecbdecd$, and $h_{Ev_6}(ecdecbdecdecdecbdecd) = ecdecdecdecdecdecd$, so P_6P_3 is adequate.

Step 6. (Second time). The new $\mathcal{E}$ is defined as $\mathcal{E} := \mathcal{E} \setminus \{\{3,6\}\}$, and is shown in Fig. 2.41.

The new Pr is $P_1P_2P_3P_4$.

Step 4. (Third time). Now $I_0 = \emptyset$, so the next step is Step 7.

Step 7. After new labelling, the graph $\mathcal{E}$ is depicted in Fig. 2.42.

Step 8. Let us take $i_0 = 3$. The directed tree $\mathcal{E}_0(i_0)$ is depicted in Fig. 2.43.

Step 9.

$$s_3 = \#_b(cyc_1) \cdot \#_c(cyc_2) \cdot \#_g(cyc_4) = 2 \cdot 2 \cdot 1 = 4$$

$$s_1 = s_3 \cdot \#_b(cyc_3)/\#_b(cyc_1) = 4 \cdot 1/2 = 2$$

$$s_2 = s_3 \cdot \#_c(cyc_3)/\#_c(cyc_2) = 4 \cdot 3/2 = 6$$

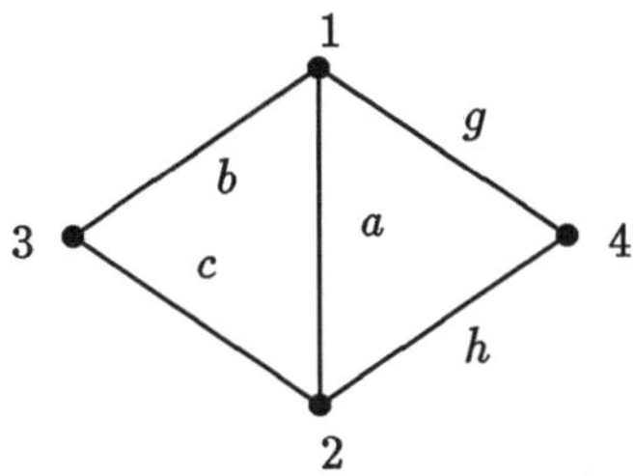

Figure 2.42. Labeled $\mathcal{E}$ (Step 7 Ex 2.4.3)

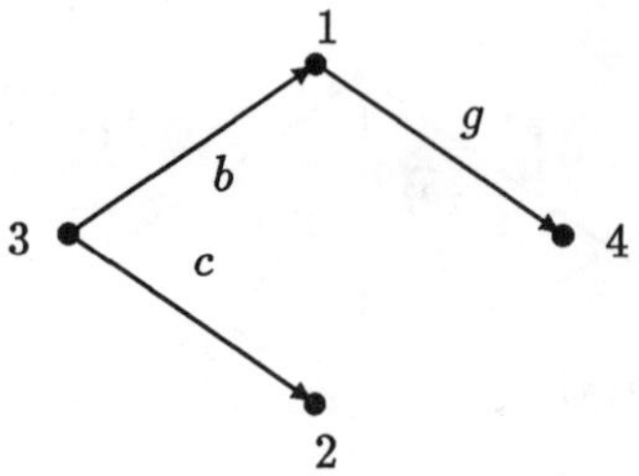

Figure 2.43. $\mathcal{E}_0(3)$ (Ex 2.4.3)

$$s_4 = s_1 \cdot \#_g(cyc_1)/\#_g(cyc_4) = 2 \cdot 2/1 = 4$$

Step 10.

$$s = hcf(s_1, s_2, s_3, s_4) = 2, r_1 = 1, r_2 = 3, r_3 = 2, r_4 = 2.$$

Step 11. The initial d_i's are:

$$d_1 = gafbagfab$$
$$d_2 = hcachhcachhcach$$
$$d_3 = ecdecbdecdecdecbdecd$$
$$d_4 = hghhfhghhf$$

One may verify that the following non-deadlocking sequence x may be produced by the program from Step 11 of Algorithm 2.2:

$$x = hcgachbfcachghcabchf \text{ } with \text{ } length(x) = 21.$$

2.4.3 Adequacy-Preserving Transformations

We shall now discuss adequacy-preserving transformation rules for COSY path programs. We shall distinguish two types of such transformations. The first type is called **program insertions** and can be briefly explained as follows. Given a path program $Qr = Q_1 \ldots Q_m$ and event $\omega \in Ev(Qr)$ we can, provided that some simple conditions are met, "replace" ω by a (possibly complex) path program $Pr = P_1 \ldots P_n$, $n = card(i(w)) \leq m$, which results in the new path program denoted $Pr \longrightarrow Qr$. Thus, in other words, Qr can be thought of as an abstraction of $Pr \longrightarrow Qr$. Without loss of generality we can assume that $\omega \in Ev(Q_i)$ for $i = 1, \ldots, n$, and $\omega \notin Ev(Q_i)$, for $i = n+1, \ldots, m$. Furthermore, assume that $P_i = \underline{path}\ seq_i\ \underline{end}$ for $i = 1, \ldots, n$. Now we can make precise the meaning of "replacing" ω by Pr, namely,

$$Pr \xrightarrow{\omega} Qr = Q'_1 \ldots Q'_n Q_{n+1} \ldots Q_m,$$

and every Q'_i is derived from Q_i by substituting (seq_i) for every occurrence of ω in Q_i. Notice that Q'_i is always a path.

Example 2.4.4 Let

$$Pr = \begin{cases} P(1): & \underline{path}\ \omega_1;\omega_2\ \underline{end} \\ P(2): & \underline{path}\ \omega_1;\omega_3\ \underline{end} \end{cases}$$

$$Qr = \begin{cases} Q(1): & \underline{path}\ \omega, a; b; \omega, a\ \underline{end} \\ Q(2): & \underline{path}\ a; b; \omega\ \underline{end} \\ Q(3): & \underline{path}\ c; a, d\ \underline{end} \end{cases}$$

After inserting Pr into Qr we obtain:

$$Pr \xrightarrow{\omega} Qr = \begin{cases} Q'(1): & \underline{path}\ (\omega_1;\omega_2), a; b; (\omega_1;\omega_2), a\ \underline{end} \\ Q'(2): & \underline{path}\ a; b; (\omega_1;\omega_3)\ \underline{end} \\ Q'(3): & \underline{path}\ c; a, d\ \underline{end} \end{cases}$$

■

In this chapter we shall determine the conditions under which an insertion is adequacy-preserving, i.e., when the adequacy of Qr implies the adequacy of $Pr \xrightarrow{\omega} Qr$, and vice versa.

The second type of adequacy-preserving transformation is **deleting** events, which can briefly be explained as follows. Let $Pr = P_1 \dots P_n$ be a path program, and let $a \in Ev(Pr)$. Define Pr/a to be the path program obtained from Pr by deleting all a's and resulting superfluous operators. Under which condition can we say that Pr is adequate if and only if Pr/a is adequate? How Pr/a can be defined formally? When is Pr/a defined correctly? We shall try to answer these types of questions in this chapter.

Example 2.4.5 Let:

$$Pr = \begin{cases} P(1): & \underline{path}\ (b; c; a), (c; b)\ \underline{end} \\ P(2): & \underline{path}\ c; b\ \underline{end} \end{cases}$$

Then:

$$Pr/a = \begin{cases} P(1): \underline{path}\ (b; c), (c; b)\ \underline{end} \\ P(2): \underline{path}\ c; b\ \underline{end} \end{cases}$$

$$Pr/b = \begin{cases} P(1): & \underline{path}\ (c; a), c\ \underline{end} \\ P(2): & \underline{path}\ c\ \underline{end} \end{cases}$$

$$Pr/c = \begin{cases} P(1): & \underline{path}\ (b; a), b\ \underline{end} \\ P(2): & \underline{path}\ b\ \underline{end} \end{cases}$$

■

Insertions Preserving Adequacy

Let EV and EV_ω be two disjoint non-empty sets of events, *fixed* for the rest of this section, such that EV_ω contains a distinguished symbol ω. We shall use symbols $\omega_1, \omega_2, \ldots$, to denote the elements of EV, and $a, b, c, \ldots$, to denote the elements of $EV_\omega \setminus \{\omega\}$. Also, we *fix* an integer $n \geq 2$.
Let Γ denote the set of all path programs $Pr = P_1 \ldots P_n$ such that $Ev(Pr) \subseteq EV$, and let Ω denote the set of all path programs $Qr = Q_1 \ldots Q_m$ such that $m \geq n$, $Ev(Qr) \subseteq EV_\omega$ and $\omega \in Ev(Q_i) \Leftrightarrow i \leq n$, for all $i \leq m$.
We can now formally define the *program insertion* discussed above. Let $Pr = P_1 \ldots P_n \in \Gamma$, $P_i = \underline{path}\ seq_i\ \underline{end}$ for $i = 1, \ldots, n$, $Qr = Q_1 \ldots Q_m \in \Omega$. We shall use $Pr \xrightarrow{\omega} Qr$ to denote the path program resulting from the *insertion of Pr into Qr*, which is defined in the following way:

(1) $Pr \xrightarrow{\omega} Qr = Q'_1 \ldots Q'_n Q_{n+1} \ldots Q_m$ (or $Pr \xrightarrow{\omega} Qr = Q'_1 \ldots Q'_n$ if $m = n$).

(2) for $i = 1, \ldots, n$, the path Q'_i is derived from Q_i by substituting "(seq_i)" for every occurrence of ω in Q_i.

In the sequel, if it does not lead to confusion we shall write rather $Pr \longrightarrow Qr$ instead of $Pr \xrightarrow{\omega} Qr$.
The Example 2.4.4 from the beginning of Sect. 2.4.3 is an illustration of the above definition. Of course in general an insertion may, or may not, preserve adequacy as the example below shows.

Example 2.4.6 [5]
Let $n = 2$, and let $Pr_1 \in \Gamma$ and $Qr_1 \in \Omega$ be the following path programs:

$$Pr_1 = \begin{cases} P(1): & \underline{path}\ \omega_1; \omega_2\ \underline{end} \\ P(2): & \underline{path}\ \omega_1; \omega_3\ \underline{end} \end{cases}$$

$$Qr_1 = \begin{cases} Q(1): & \underline{path}\ \omega, a; b; \omega, a\ \underline{end} \\ Q(2): & \underline{path}\ a; b; \omega, a\ \underline{end} \end{cases}$$

After inserting Pr_1 into Qr_1 we obtain:

$$Pr_1 \longrightarrow Qr_1 = \begin{cases} Q(1): & \underline{path}\ (\omega_1; \omega_2), a; b; (\omega_1; \omega_2), a\ \underline{end} \\ Q(2): & \underline{path}\ a; b; (\omega_1; \omega_3), a\ \underline{end} \end{cases}$$

Note that Qr_1 and $Pr_1 \longrightarrow Qr_1$ are both adequate since $VFS(Qr_1) = Pref(\{\underline{ab\omega}, \underline{aba}\}^*)$ and $VFS(Pr_1 \longrightarrow Qr_1) = Pref(\{\underline{ab\omega_1\omega_2\omega_3}, \underline{aba}\}^*)$.

[5] To preserve pattern $Pr_i \longrightarrow Qr_i$ the numbering of example programs starts with Pr_1 in this section.

The situation changes if we take:

$$Pr_2 = \begin{cases} P(1): & \underline{path}\ \omega_2; \omega_1\ \underline{end} \\ P(2): & \underline{path}\ \omega_3; \omega_1\ \underline{end} \end{cases}$$

Now we have

$$Pr_2 \longrightarrow Qr_1 = \begin{cases} Q(1): & \underline{path}\ (\omega_2;\omega_1), a; b; (\omega_2;\omega_1), a\ \underline{end} \\ Q(2): & \underline{path}\ a; b; (\omega_3;\omega_1), a\ \underline{end} \end{cases}$$

and $Pr_2 \longrightarrow Qr_1$ deadlocks after $\underline{\omega_2}$. ■

We shall now present a result which provides a means of deriving the adequacy of $Pr \longrightarrow Qr$ from the adequacy of Qr. Before that we have to introduce the concept of **vector insertion** which is a "dynamic" counterpart of the "static" program insertion.

Let $Pr = P_1 \dots P_n \in \Gamma$ and $Qr = Q_1 \dots Q_m \in \Omega$. Let $insert_{Pr,Qr}$ denote the set of all sequences $< \underline{v}_1, \dots, \underline{v}_k, \underline{t} >$ such that: $\underline{t} \in VFS(Qr)$, $k = \#_\omega(t) \geq 0$, $\underline{v}_i \in Vcyc(Pr)$ for $i = 1, \dots, k$. If $< \underline{v}_1, \dots, \underline{v}_k, \underline{t} > \in insert_{Pr,Qr}$ then $< \underline{v}_1, \dots, \underline{v}_k > \longrightarrow \underline{t}$ denotes the element of $Vev(Pr \longrightarrow Qr)^*$ defined by:

$$< \underline{v}_1, \dots, \underline{v}_k > \longrightarrow \underline{t} = \underline{t_1 v_1} \dots \underline{t_k v_k t_{k+1}}\ where\ \underline{t_1} \dots \underline{t_{k+1}} \in Vev(Qr)^*$$

is any vector sequence satisfying:

$$\underline{t} = \underline{t_1 \omega t_2} \dots \underline{\omega t_k \omega t_{k+1}}.$$

Attention: in the above definition, the vector $\underline{v_i}$ in $< \underline{v}_1, \dots, \underline{v}_k > \longrightarrow t$ belongs to $Vev(Pr)^*$ and has n coordinated, while $\underline{v_i}$ in $\underline{t_1}\ \underline{v_1} \dots \underline{t_k}\ \underline{v_k}\ \underline{t_{k+1}}$ belong to $Vev(Qr)^*$ and has m coordinates.

It is not difficult to see that $< \underline{v}_1, \dots, \underline{v}_k > \longrightarrow \underline{t}$ is a well defined notion since it does not depend on a particular choice of t_i's. Let

$$Insert(Pr, Qr) = \{< \underline{v}_1, \dots, \underline{v}_k > \longrightarrow \underline{t} \mid < \underline{v}_1, \dots, \underline{v}_k, \underline{t} > \in insert_{Pr,Qr}\}.$$

It is also easy to prove that $Insert(Pr, Qr) \subseteq VFS(Pr \longrightarrow Qr)$. We left this proof to the reader. (Hint. Note that $\underline{x} \in Insert(Pr, Qr) \Rightarrow pr_i(\underline{x}) \in FS(Q_i')$ if $i \leq n$, or $pr_i(\underline{x}) \in FS(Q_i)$ if $i \geq n$).

Theorem 2.23 *If $Pr \in \Gamma$ and $Qr \in \Omega$ satisfy:*
(1) $VFS(Pr \longrightarrow Qr) = Pref(Insert(Pr, Qr))$ and
(2) $Ev(Vcyc(Pr)) = Ev(Pr)$,
then: Qr is adequate $\Rightarrow$ $Pr \longrightarrow Qr$ is adequate.

Proof. Let $\underline{x} \in VFS(Pr \longrightarrow Qr)$ and $a \in Ev(Pr \longrightarrow Qr)$. It is sufficient to show that there is $\underline{xv} \in VFS(Pr \longrightarrow Qr)$ such that $a \in Ev(\underline{v})$. By (1) we may assume that $\underline{x} \in Insert(Pr, Qr)$. Let $\underline{x} =< \underline{v}_1, \ldots, \underline{v}_k > \longrightarrow \underline{t}$. Since Qr is adequate, there is $\underline{tz} \in VFS(Qr)$ such that $Ev(\underline{z}) = Ev(Qr)$. We can represent $\underline{z}$ as $\underline{z} = \underline{z_1\omega \ldots z_l\omega z_{l+1}}$, where $l \geq 1$, $z_i \in Ev(Qr)^*$, $\omega \notin Ev(\underline{z_1 z_2 \ldots z_{l+1}})$, and then consider two cases.
Case 1: $a \in Ev(Pr)$. Then, by (2), there is $\underline{w} \in Vcyc(Pr)$ with $a \in Ev(\underline{w})$, and we can take $\underline{v} = \underline{z_1 w}$.
Case 2: $a \in Ev(Qr)$. Then we choose any $\underline{w} \in Vcyc(Pr) \neq \emptyset$ and take $\underline{v} = \underline{z_1 w \ldots z_l w z_{l+1}}$. □

Note that if we dropped (1) or (2) then the thesis would not be true. To show that one cannot drop (1) we take:

$$Pr_3 = \begin{cases} P(1): & \underline{path}\ \omega_1\ \underline{end} \\ P(2): & \underline{path}\ \omega_1 *\ \underline{end} \end{cases}$$

$$Qr_3 = \begin{cases} Q(1): & \underline{path}\ \omega; a\ \underline{end} \\ Q(2): & \underline{path}\ \omega; b; a\ \underline{end} \end{cases}$$

$$Pr_3 \longrightarrow Qr_3 = \begin{cases} Q'(1): & \underline{path}\ \omega_1; a\ \underline{end} \\ Q'(2): & \underline{path}\ \omega_1 *; b; a\ \underline{end} \end{cases}$$

Clearly, $Ev(Vcyc(Pr_3)) = \{\omega_1\} = Ev(Pr_3)$ and Qr_3 is adequate. On the other hand, $Pr_3 \longrightarrow Qr_3$ is not deadlock-free, it may deadlock for instance after $\underline{b}$. To show that one cannot drop (2) we can take any Pr such that $VFS(Pr) = \{\underline{\varepsilon}\}$.
The implication in Theorem 2.23 cannot be reversed either. For example we can take:

$$Pr_4 = \begin{cases} P(1): & \underline{path}\ \omega_1 *\ \underline{end} \\ P(2): & \underline{path}\ \omega_1 *\ \underline{end} \end{cases}$$

$$Qr_4 = \begin{cases} Q(1): & \underline{path}\ \omega; (\omega; a), (c; a; b)\ \underline{end} \\ Q(2): & \underline{path}\ \omega; (\omega; b), (c; a; b)\ \underline{end} \end{cases}$$

We have $VFS(Pr_4 \longrightarrow Qr_4) = Pref((\{\underline{\omega_1}\}^*\{\underline{cab}\})^*), VFS(Qr_4) = Pref(\{\underline{\omega cab}\}^* \cup \{\underline{\omega cab}\}^*\{\underline{\omega\omega}\}), Vcyc(Pr_4) = VFS(Pr_4) = \{\underline{\omega_1}\}^*$. Thus $VFS(Pr_4 \longrightarrow Qr_4) = Pref(Insert(Pr_4, Qr_4))$,$VFS(Pr_4) = Pref(Vcyc(Pr_4))$, $Ev(Vcyc(Pr_4)) = Ev(Pr_4)$, and Pr_4 is adequate. On the other hand, $Pr_4 \longrightarrow Qr_4$ is an adequate program, whereas Qr_4 is not.
To reverse the implication in Theorem 2.23 we need an additional auxiliary notion. Let $Pr \in \Gamma$, $Qr \in \Omega$. For every $\underline{x} \in Insert(Pr, Qr)$, let

$$image_{Qr}(\underline{x}) = \{\underline{t} | (\exists \underline{v}_1, \ldots, \underline{v}_k \in Vcyc(Pr)) < \underline{v}_1, \ldots, \underline{v}_k > \longrightarrow \underline{t} = \underline{x}\}.$$

Note that it may happen that $\underline{t}, \underline{u} \in image_{Qr}(\underline{x})$ and $\#_\omega(\underline{t}) \neq \#_\omega(\underline{u})$. For example, for Pr_4 and Qr_4, $image_{Qr_4}(\underline{\omega_1}) = \{\underline{\omega}, \underline{\omega\omega}\}$, since $< \underline{\omega_1} > \longrightarrow \underline{\omega} = \underline{\omega_1}$ and $< \underline{\omega_1}, \underline{\varepsilon} > \longrightarrow \underline{\omega\omega} = \underline{\omega_1}$ and no other ones.

Theorem 2.24 *If $Pr \in \Gamma, Qr \in \Omega$ satisfy:*
(1) $VFS(Pr \longrightarrow Qr) = Pref(Insert(Pr, Qr))$
(2) $(\forall \underline{x}, \underline{z} \in Insert(Pr, Qr))\underline{x} \in Pref(\underline{z}) \Rightarrow$
$image_{Qr}(\underline{x}) \subseteq Pref(image_{Qr}(\underline{z}))$
then: $Pr \longrightarrow Qr$ *is adequate* $\Rightarrow$ Qr *is adequate.*

Proof. Let $\underline{t} \in VFS(Qr)$ and $a \in Ev(Qr)$. We shall find $\underline{u} \in VFS(Qr)$ such that $\underline{t} \in Pref(\underline{u})$ and $\#_a(\underline{t}) < \#_a(\underline{u})$ which clearly implies adequacy of Qr. We have to consider two cases.
Case 1: $a = \omega$. Let $Pr = P_1 \ldots P_n, Qr = Q_1 \ldots Q_m$. Note that the theorem is true if $Ev(Q_i) = \{\omega\}$ for all $i \leq n$. Thus, without loss of generality, we may assume that there is $b \in Ev(Q_1)$ where $b \neq \omega$. If $Vcyc(Pr) = \emptyset$ then $Insert(Pr, Qr) = \{t \in VFS(Qr) | \#_\omega(t) = 0\}$, so by (1), $Pr \longrightarrow Qr$ is not adequate. Hence $Vcyc(Pr) \neq \emptyset$. Let $\underline{w} \in Vcyc(Pr)$. We have $<\underbrace{\underline{w}, \ldots, \underline{w}}_{k-times}, \underline{t}> \in insert_{Pr,Qr}$ where $\#_\omega(\underline{t}) = k$. Let c be any event of $Ev(P_1)$, and let $\underline{y} = <\underbrace{\underline{w}, \ldots, \underline{w}}_{k-times}> \longrightarrow \underline{t}$. By the adequacy of $Pr \longrightarrow Qr$ and (1), there are $\underline{t}_1, \ldots, \underline{t}_{k+1}, \underline{v}_1, \ldots, \underline{v}_{k+1} \in Vev(Pr \longrightarrow Qr)^*$. such that:

$$\underline{x} = \underline{yt_1bv_1c \ldots t_kbv_kct_{k+1}bv_{k+1}c} \in Insert(Pr, Qr).$$

Now, from $b \in Ev(Q_1)$ and $c \in Ev(P_1)$ it follows that for all $\underline{v} \in image_{Qr}(\underline{x}), \#_\omega(\underline{v}) \geq k+1$. We have $\underline{y} \in Pref(\underline{x})$ and $\underline{t} \in image_{Qr}(\underline{y})$. Hence, by (2), there is a $\underline{u} \in image_{Qr}(\underline{x})$ such that $\underline{t} \in Pref(\underline{u})$. On the other hand, $\#_\omega(\underline{t}) = k < \#_\omega(\underline{u})$, which completes the analysis of this case.

Case 2: $a \neq \omega$. Let $\underline{y}$ be defined as in Case 1, i.e., $\underline{w} \in Vcyc(Pr), \#_\omega(\underline{t}) = k$, $\underline{y} = <\underbrace{\underline{w}, \ldots, \underline{w}}_{k-times}> \longrightarrow \underline{t}$. By the adequacy of $Pr \longrightarrow Qr$ and (1), there is $\underline{x} = \underline{yz} \in Insert(Pr, Qr)$ such that $a \in Ev(\underline{z})$. We have $\underline{y} \in Pref(\underline{x})$ and $\underline{t} \in image_{Qr}(\underline{y})$. Hence, again by (2), there is $\underline{u} \in image_{Qr}(\underline{x})$ such that $\underline{t} \in Pref(\underline{u})$. Clearly, $\#_a(\underline{t}) < \#_a(\underline{u})$, which completes the proof. □

The path programs Pr_4 and Qr_4 demonstrate that the condition (2) above cannot be dropped. For Pr_4 and Qr_4 we have $\underline{x} = \underline{\omega_1 cab\omega_1\omega_1} \in Insert(Pr_4, Qr_4)$, $\underline{z} = \underline{\omega_1 cab\omega_1\omega_1 cab} \in Insert(Pr_4, Qr_4)$, and $image_{Qr}(\underline{x}) = \{\underline{\omega cab\omega}, \underline{\omega cab\omega\omega}\}$, while $image_{Qr}(\underline{z}) = \{\underline{\omega cab\omega cab}\}$, so $\underline{\omega cab\omega\omega} \in image_{Qr}(\underline{x}) \setminus Pref(image_{Qr}(\underline{z}))$. The next example shows that we cannot delete (1) either. Let

$$Pr_5 = \begin{cases} P(1): & \underline{path}\ \omega_1 *\ \underline{end} \\ P(2): & \underline{path}\ \omega_1\ \underline{end} \end{cases}$$

$$Qr_5 = \begin{cases} Q(1): & \underline{path}\ \omega; a; b; \omega\ \underline{end} \\ Q(2): & \underline{path}\ a; b; \omega\ \underline{end} \end{cases}$$

We have $insert_{Pr_5,Qr_5} = \{<\underline{\varepsilon}>\}$, so that (2) is satisfied. On the other hand, $Pr_5 \longrightarrow Qr_5$ is adequate (since $VFS(Pr_5 \longrightarrow Qr_5) = Pref(\{\underline{ab\omega_1}\}^*)$), whereas Qr_5 is not (since $VFS(Qr_5) = \{\underline{\varepsilon}\}$).
By combining Theorem 2.23 and Theorem 2.24 we obtain a sufficient condition for an insertion to be adequacy-preserving.

Corollary 2.14 (Conditions for Adequacy Preserving Insertion)
If $Pr \in \Gamma$ and $Qr \in \Omega$ satisfy:
(1) $VFS(Pr \longrightarrow Qr) = Pref(Insert(Pr, Qr))$
(2) $Ev(Vcyc(Pr)) = Ev(Pr)$
(3) $(\forall \underline{x}, \underline{z} \in Insert(Pr, Qr))\underline{x} \in Pref(\underline{z}) \Rightarrow$
$image_{Qr}(\underline{x}) \subseteq Pref(image_{Qr}(\underline{z}))$
then: Qr *is adequate* $\Leftrightarrow$ $Pr \longrightarrow Qr$ *is adequate.* □

Although of theoretical relevance, the above result is rather unattractive as a means of verification. The conditions of the above corollary are, firstly, likely to be very difficult to check, and, secondly, (1) and (3) directly refer to dynamic properties of Qr. The second property is especially restrictive since, for a typical verification process, there will be a sequence of insertions, $Q_{i+1} = P_i \longrightarrow Q_i$, with the "size" of the Q_i's constantly growing which makes analysis even more problematic. We shall try to overcome these difficulties below, although, as it will turn out, very little can be done in the general case.
Let us now define the concept of a **smooth insertion.**
Let $Qr = Q_1 \dots Q_m \in \Theta \subseteq \Omega$ where Θ is any subset of Ω . We shall use $Smooth(Qr)$ to denote the set of all path programs $Pr \in \Gamma$ such that:

$$Qr \text{ is adequate } \Leftrightarrow Pr \longrightarrow Qr \text{ is adequate.}$$

Furthermore, for every non-empty $\Theta \subseteq \Omega$, let

$$Smooth(\Theta) = \{Pr \in \Gamma | (\forall Qr \in \Theta) Pr \in Smooth(Qr)\}.$$

Having introduced the notion of a smooth insertion we can investigate the properties of path programs belonging to $Smooth(\Theta)$, for some sets $\Theta \subseteq \Omega$. We shall start with the "simplest" case where $\Theta = \Omega$, but first we analyze two examples of non-smooth insertions.
Consider the following three path programs:

$$Pr_6 = \begin{cases} P(1): & \underline{path}\ \omega_1; \omega_1, \omega_2\ \underline{end} \\ P(2): & \underline{path}\ \omega_1; \omega_1\ \underline{end} \end{cases}$$

$$Qr_6 = \begin{cases} Q(1): & \underline{path}\ \omega; a; b\ \underline{end} \\ Q(2): & \underline{path}\ \omega; b\ \underline{end} \end{cases}$$

$$Pr_6 \longrightarrow Qr_6 = \begin{cases} Q'(1): & \underline{path}\ (\omega_1; \omega_1, \omega_2); a; b\ \underline{end} \\ Q'(2): & \underline{path}\ (\omega_1; \omega_1); b\ \underline{end} \end{cases}$$

As one can see, $Pr_6 \notin Smooth(Qr_6)$. For example, $\underline{x} = \underline{\omega_1\omega_2 a}$ belongs to $VFS(Pr_6 \longrightarrow Qr_6)$ and leads to a deadlock while Qr_6 and Pr_6 are adequate. A reason for this is that there is a vector firing sequence $\underline{w} = \underline{\omega_1\omega_2} = (\omega_1\omega_2, \omega_1) \in VFS(Pr_6)$ (a "projection" of $\underline{x} = \underline{\omega_1\omega_2 a}$ onto Pr_6) such that:

(i) P_1 has completed one full cycle ($\omega_1\omega_2 \in Cyc(P_1)$);

(ii) P_2 has not completed the cycle it began ($\omega_1 \notin Cyc(P_2)$);

(iii) to complete the cycle in P_2 it is necessary for P_2 to interact with P_1.

Thus, if we insert Pr_6 into a path program, such as Qr_6, which can make (iii) impossible to accomplish, then the cycle in P_2 will never be completed. The above observation leads us to the informal formulation of the following requirement, which is to be satisfied by every path program $Pr \in Smooth(\Omega)$:

> "if $\underline{w} \in VFS(Pr)$ then it is possible to complete cycles in all of the paths of Pr without executing any event in those paths which have already completed their cycles."

This requirement corresponds to the point (2) of the formal definition of Γ_0 which is give later on.
The second example illustrates a different kind of non-smooth insertion. Let

$$Pr_7 = \begin{cases} P(1): & \underline{path}\ \omega_1; (\omega_2, (\omega_1; \omega_2))\ \underline{end} \\ P(2): & \underline{path}\ \omega_1\ \underline{end} \end{cases}$$

$$Qr_7 = \begin{cases} Q(1): & \underline{path}\ \omega; (a; b), (b; c; a)\ \underline{end} \\ Q(2): & \underline{path}\ \omega; (\omega; a; c), (b; c; a)\ \underline{end} \end{cases}$$

$$Pr_7 \longrightarrow Qr_7 = \begin{cases} Q'(1): & \underline{path}\ (\omega_1; (\omega_2, (\omega_1; \omega_2))); (a; b), (b; c; a)\ \underline{end} \\ Q'(2): & \underline{path}\ (\omega_1); ((\omega_1); a; c), (b; c; a)\ \underline{end} \end{cases}$$

We first notice that Pr_7 satisfies the requirement formulated above since $VFS(Pr_7) = \{\underline{\omega_1\omega_2}, \underline{\omega_1\omega_1\omega_2}\}^*\{\underline{\omega_1\omega_1}, \underline{\omega_1}, \underline{\varepsilon}\} = Pref(Vcyc(Pr_7)^*)$. As before, we have $Pr_7 \notin Smooth(Qr_7)$. $VFS(Qr_7) = Pref(\{\underline{\omega bca}\}^*)$, whereas, $\underline{\omega_1\omega_1\omega_2 a}$ belongs to $VFS(Pr_7 \longrightarrow Qr_7)$ and leads to a deadlock. An intuitive reason for this is that having executed $\underline{\omega_1\omega_1\omega_2 a}$ in $Pr_7 \longrightarrow Qr_7$ one may say that P_1 has performed one complete cycle ($\omega_1\omega_1\omega_2 \in Cyc(P_1)$), which corresponds to the execution of ω in Q_1, whereas P_2 has performed two complete cycles ($\omega_1\omega_1 \in Cyc(P_2)Cyc(P_2)$) which corresponds to the execution of $\omega\omega$ in Q_2. Thus, the execution of $\underline{\omega_1\omega_1\omega_2 a}$ in $Pr_7 \longrightarrow Qr_7$ induces the "execution" of "impossible" vector $(\omega a, \omega\omega a)$ in Qr_7. All path programs $Pr \in Smooth(\Omega)$ will, therefore, have to satisfy an additional (informal) requirement, corresponding to (3) of the formal definition of Γ_0 given later, which essentially says that:

> "for any history of $Pr = P_1 \dots P_n$ the number of (complete or incomplete) cycles performed by P_i's is always the same for all $i = 1, \dots, n$."

To express formally the last requirement, we have to introduce a new notion, called a **cyclic decomposition.** Let P be a single path and $x \in FS(P)$.
A sequence $\sigma = < u_1, \dots, u_k >, k \geq 1, u_i \in Ev(P)^*$, is said to be a **cyclic decomposition** of x if and only if:

(1) $x = u_1 \dots u_k$

(2) $u_1, \dots, u_{k-1} \in Cyc(P)$

(3) $u_k \in Cyc(P) \vee u_k \in Pref(Cyc(P)) \setminus \{\varepsilon\}$.

The set of all cyclic decompositions of $x \in FS(P)$ will be denoted by $cd_P(x)$. Let $<>$ denote the empty sequence, and assume that $<> \in cd_p(\varepsilon)$. If $card(cd_p(x)) = 1$ then we let $cd_p(x)$ denote its unique member. Notice that if $\varepsilon \in Cyc(P)$ then $cd_p(x)$ is an infinite set for all $x \in FS(P)$; and that $<>$ and $< \varepsilon >$ are two different cyclic decompositions, namely $<>$ is the empty cyclic decomposition, whereas $< \varepsilon >$ is a cyclic decomposition which comprises one empty cycle.
We can now define the class of path programs satisfying the two requirements mentioned above.
Let Γ_0 denote the set of all path programs $Pr = P_1 \dots P_n \in \Gamma$ such that:

(1) $Ev(Vcyc(Pr)) = Ev(Pr)$

(2) $(\forall \underline{x} \in Pref(Vcyc(Pr)))(\exists \underline{y} \in Vev(Pr)^*)\underline{xy} \in Vcyc(Pr) \wedge$
$$(\forall i = 1, \dots, n)(pr_i(\underline{y}) = \underline{\varepsilon} \Leftrightarrow pr_i(x) \in Cyc(P_i))$$

(3) $(\forall \underline{x} \in VFS(Pr))(\forall \sigma, \mu \in cd_{P_1}(pr_1(\underline{x})) \cup \dots \cup cd_{P_n}(pr_n(\underline{x})))$
$$\sigma = < u_1, \dots, u_k > \wedge \mu = < v_1, \dots, v_l > \Rightarrow k = l.$$

It is not difficult to see that the table in Fig. 2.44 is correct, where

$$Pr_8 = \begin{cases} P(1): & \underline{path}\ (\omega_1; \omega_2), \omega_3\ \underline{end} \\ P(2): & \underline{path}\ (\omega_2; \omega_1), \omega_3\ \underline{end} \end{cases}$$

$$Pr_9 = \begin{cases} P(1): & \underline{path}\ \omega_1; \omega_2 *\ \underline{end} \\ P(2): & \underline{path}\ \omega_1; \omega_2; \omega_2\ \underline{end} \end{cases}$$

Hence the conditions (1)–(3) are independent.

It turns out that Γ_0 contains exactly those path programs which can be smoothly inserted into **all** path programs of Ω.

Theorem 2.25 (Characterization of Smooth Insertions)
$\Gamma_0 = Smooth(\Omega)$. □

	Pr_1	Pr_2	Pr_3	Pr_4	Pr_5	Pr_6	Pr_7	Pr_8	Pr_9
(1) of Γ_0	+	+	+	+	+	+	+		+
(2) of Γ_0	+	+					+	+	
(3) of Γ_0	+							+	+

Figure 2.44. A "+" means that condition (i) for program Pr_j holds.

The proof is in Appendix D.
Unfortunately, even a brief look at the conditions (1)–(3) of Γ_0 gives the impression that the set Γ_0 is a restricted collection of path programs, since each $Pr \in \Gamma_0$ has a very regular,"strongly cyclic", behaviour. None the less, the next result seems to be rather surprising. For there are relatively "small" sets $\Omega_0 \subseteq \Omega$ and $\Omega_1 \subseteq \Omega$ such that $Smooth(\Omega_0) = Smooth(\Omega_1) = \Gamma_0$ and, furthermore, the path programs of Ω_0 and Ω_1 admit only very restrictive forms of conflicts involving event ω. Boldly speaking, any reasonable $\Theta \subseteq \Omega$ with path programs which allow conflicts involving ω should include Ω_0 and/or Ω_1.
The set Ω_0 is defined as follows.
Let $a_1, \ldots, a_n, a, b, c$ be distinct elements of $Ev_\omega \setminus \{\omega\}$, and let $\xi, \psi_1, \phi_1, \psi_2, \phi_2, \ldots$ be single paths defined by:

$$\xi = \underline{path}\ \omega, a\ \underline{end}$$

$$\psi_1 = \underline{path}\ b; (\omega; b), a; c; \omega, a\ \underline{end}$$

$$\phi_1 = \underline{path}\ a; c; \omega, a\ \underline{end}$$

and for $k \geq 2$:

$$\psi_k = \underline{path}\ \underbrace{\omega, a; \ldots; \omega, a}_{k-1\ times}; b; (\omega; b), a; c\ \underline{end}$$

$$\phi_k = \underline{path}\ \overbrace{\omega, a; \ldots; \omega, a}^{k-1\ times}; a; c\ \underline{end}$$

Let Ω_0 be the least subset of Ω such that:

(1) The path program:

$$\underline{path}\ \omega; a_1; a\ \underline{end}$$

$$\dots$$

$$\underline{path}\ \omega; a_n; a\ \underline{end}$$

belongs to Ω_0

(2) For every $k \geq 1$ any path program that can be constructed from ψ_k, ϕ_k and $n-2$ copies of ξ, belongs to Ω_0.

The set Ω_1 is defined in the same way as Ω_0. The only difference is in the choice of the paths $\xi, \psi_1, \phi_1, \psi_2, \phi_2, \dots$ which are defined in the following way:

$$\xi = \underline{path}\ \omega\ \underline{end}$$

$$\psi_1 = \underline{path}\ b; (\omega; b), a; c; \omega\ \underline{end}$$

$$\phi_1 = \underline{path}\ a; c; \omega\ \underline{end}$$

and for $k \geq 2$:

$$\psi_k = \underline{path}\ \underbrace{\omega; \dots; \omega}_{k-1\ times}; b; (\omega; b), a; c\ \underline{end}$$

$$\phi_k = \underline{path}\ \overbrace{\omega; \dots; \omega}^{k-1\ times}; a; c\ \underline{end}$$

It turns out that a path program Pr can be smoothly inserted into any path program $Qr \in \Omega$ if and only if it can be smoothly inserted into any path program of Ω_0 or Ω_1.

Theorem 2.26 (Limits of Adequacy Preserving Insertions)
$\Gamma_0 = Smooth(\Omega_0) = Smooth(\Omega_1)$. □

The proofs of Theorem 2.25 and Theorem 2.26 are long and rather complex, so we have relegated them to Appendix D.
So far we have demonstrated that for any $\Theta \subseteq \Omega$ such that $\Omega_0 \subseteq \Theta$ or $\Omega_1 \subseteq \Theta$, we have $Smooth(\Theta) = \Gamma_0 = Smooth(\Omega)$. Also, the path programs of Ω_0 and Ω_1 admit only a restrictive form of conflicts involving ω. A natural question which may now arise is whether the situation changes if we assume that the programs in Θ do not admit conflicts involving ω.

Let $\Omega_{\omega CF}$ denote the set of all path programs $Qr = Q_1, \dots, Q_m \in \Omega$ such that for all i: $(x\omega \in FS_i \wedge xb \in FS_i) \Rightarrow \omega = b$. Clearly $\Omega_{\omega CF} \subseteq \Omega$, so $Smooth(\Omega) \subseteq Smooth(\Omega_{\omega CF})$. We show that the inclusion is a proper one, i.e., $Smooth(\Omega) \neq Smooth(\Omega_{\omega CF})$.

Theorem 2.27 $Smooth(\Omega) \subseteq Smooth(\Omega_{\omega CF}) \wedge Smooth(\Omega) \neq Smooth(\Omega_{\omega CF})$.

Proof. We shall show that the following path program belongs to $Smooth(\Omega_{\omega CF}) \setminus Smooth(\Omega)$:

$$Pr_{10} = \begin{cases} P(1): & \underline{path}\ \omega_1; \omega_2\ \underline{end} \\ P(2): & \underline{path}\ \omega_2\ \underline{end} \\ & \dots \\ P(n): & \underline{path}\ \omega_2\ \underline{end} \end{cases}$$

where $\omega_1 \neq \omega_2$. We first observe that $Pr_{10} \notin \Gamma_0$ since $\omega_1 \in VFS(Pr_{10})$ does not satisfy (3) of Γ_0. Since, by Theorem 2.25, $\Gamma_0 = Smooth(\Omega)$, then $Pr_{10} \notin Smooth(\Omega)$. We shall now show that $Pr_{10} \in Smooth(\Omega_{\omega CF})$. Let $Qr = Q_1 \dots Q_m$ be any program in Ω (not necessarily in $\Omega_{\omega CF}$), and let $Pr = Pr_{10} \longrightarrow Qr = P'_1 \dots P'_m$. For every $x \in Ev(Qr)^*$, let $f_1(x)$ be the string obtained from x by replacing each ω by $\omega_1\omega_2$ and let $f_2(x)$ be the string obtained from x by replacing each ω by ω_2.
From the definition of Pr_{10} above it follows that

(1) $FS(P'_1) = \{f_1(x) \mid x \in FS(Q_1)\} \cup \{f_1(x)\omega_1 \mid x\omega \in FS(Q_1)\}$

(2) $FS(P'_i) = \{f_2(x) \mid x \in FS(Q_i)\}, i = 2, \dots, m$.

We first show the following fact:

(3) if $\underline{x} \in VFS(Pr)$ then there are $\underline{x}_0, \dots, \underline{x}_r \in Vev(Qr)^*$ such that $r \geq \emptyset$ and one of the following holds:

(a) $\underline{x} = \underline{x_0\omega_1\omega_2x_1} \dots \underline{\omega_1\omega_2x_r}$

(b) $\underline{x} = \underline{x_0\omega_1\omega_2x_1} \dots \underline{\omega_1\omega_2x_r\omega_1}$

The proof of (3) is by induction on the length of $\underline{x}$. Clearly, if $\underline{x} = \underline{\varepsilon}$ then (3) holds. Suppose that $\underline{y} = \underline{xa} \in VFS(Pr)$, and that (3) holds for $\underline{x}$. From (1) it follows that if $\underline{x}$ satisfies (3a) then $a \neq \omega_2$, and, if $\underline{x}$ satisfies (3b) then $a \neq \omega_1$. If $\underline{x}$ satisfies (3a) and $a \in Ev(Qr) \cup \{\omega_1\}$ then (3) holds for $\underline{y}$. Similarly, y satisfies (3) if $\underline{x}$ satisfies (3b) and $a = \omega_2$. Finally, if $\underline{x}$ satisfies (3b) and $a \in Ev(Qr)$ then, by (1), $a \notin Ev(Q_1)$. Hence $\underline{\omega_1 a} = \underline{a\omega_1}$. Thus $\underline{y}$ satisfies (3b). Consequently, (3) holds for $\underline{y} = \underline{xa}$. We further observe that $\underline{z} = \underline{x_0\omega x_1} \dots \underline{\omega x_r} \in VFS(Qr)$. Indeed, from (1) and (2) it follows that $pr_i(\underline{z}) \in FS(Q_i)$ for $i = 1, 2, \dots, m$. From (1), (2) and $\underline{z} \in VFS(Qr)$ it follows that

(4) $\underline{x} \in VFS(Pr) \wedge \#_{\omega_1}(\underline{z}) = \#_{\omega_2}(\underline{z}) \Rightarrow \underline{x} \in Insert(Pr_{10}, Qr)$.

From (1), (2) and $\varepsilon \notin Cyc(P_1) \cup \dots \cup Cyc(P_n)$, it follows that

(5) $\underline{x}, \underline{z} \in Insert(Pr_{10}, Qr) \wedge \underline{x} \in Pref(\underline{z}) \Rightarrow image_{Qr}(\underline{x}) \subseteq image_{Qr}(\underline{z})$.

Suppose now that the following is satisfied:

(6) $(\exists \underline{x} \in VFS(Qr))pr_1(\underline{x})\omega \in FS(Q_1) \wedge ((\forall \underline{xz} \in VFS(Qr))\ \#_\omega(\underline{z}) = 0)$.

Then, clearly, Qr is not adequate. Suppose that Pr is adequate, and let

$$\underline{w} = <\underbrace{\omega_1\omega_2, \ldots, \omega_1\omega_2}_{k-times}> \longrightarrow \underline{x}$$

where $k = \#_\omega(\underline{x})$. Clearly, $\underline{x} \in image_{Qr}(\underline{w})$.
From (1), $pr_1(\underline{x})\omega \in FS(Q_1)$ and $\omega_1 \notin Ev(P_2) \cup \ldots \cup Ev(P_m)$, it follows that $\underline{w\omega_1} \in VFS(Pr)$. From the adequacy of Pr and (1) it follows that there is $\underline{v}$ such that $\underline{u} = \underline{w\omega_1 v} \in VFS(Pr)$ and $\#_{\omega_1}(\underline{u}) = \#_{\omega_2}(\underline{u}) = k+1$. By (4), $\underline{u} \in Insert(Pr_{10}, Qr)$. Let $\underline{t} \in image_{Qr}(\underline{u})$. From (1), $\varepsilon \notin Cyc(P_1)$ and $\#_{\omega_2}(\underline{u}) = k+1$, it follows that $\#_\omega(\underline{t}) = k+1$. Hence, because $\underline{x} \in image_{Qr}(\underline{w})$ and (5) holds, we obtain:

$$\underline{t} \in VFS(Qr) \wedge \underline{x} \in Pref(\underline{t}) \wedge \#_\omega(\underline{t}) = k+1 > \#_\omega(\underline{x}).$$

This, however, contradicts (6). Hence we have shown that $Pr_{10} \longrightarrow Qr$ is a smooth insertion if (6) holds. To complete the proof we thus need to show that $Pr_{10} \longrightarrow Qr$ is smooth if (6) does not hold. It is only now that we **do** assume that $Qr \in \Omega_{\omega CF}$. What we essentially need to prove is that $VFS(Pr) \subseteq Pref(Insert(Pr_{10}, Qr))$. Suppose that $\underline{x} \in VFS(Pr)$. If $\underline{x}$ satisfies (3a) then, by (4), $\underline{x} \in Insert(Pr_{10}, Qr)$. Suppose that $\underline{x}$ satisfies (3b), i.e., $\underline{x} = \underline{x_0\omega_1\omega_2x_1 \ldots \omega_1\omega_2x_r\omega_1}$, and let $\underline{y} = \underline{x_0\omega_1\omega_2x_1 \ldots \omega_1\omega_2x_r}$. Clearly, $\underline{y}$ satisfies (3a) and thus, by (4), $\underline{y} \in Insert(Pr_{10}, Qr)$. Let $\underline{z} \in image_{Qr}(\underline{y})$. By (1) and $\varepsilon \notin Cyc(P_1)$, we have $pr_1(\underline{z})\omega \in FS(Q_1)$.
Thus, since (6) does not hold, there is $\underline{v}$ such that $\underline{zv} \in VFS(Qr)$ and $\#_\omega(\underline{zv}) = \#_\omega(\underline{z}) + 1$. Let

$$\underline{u} = <\underbrace{\omega_1\omega_2, \ldots, \omega_1\omega_2}_{(k+1)-times}> \longrightarrow \underline{zv}.$$

Clearly, $pr_i(\underline{x}) = pr_i(\underline{y}) \in Pref(pr_i(\underline{u}))$ for $i = 2, \ldots, m$. Furthermore, $pr_1(\underline{x}) = pr_1(\underline{y})\omega_1$ and, by (1) and $\varepsilon \notin Cyc(P_1)$, $pr_1(\underline{y}) \in Pref(pr_1(\underline{u})) \setminus \{pr_1(\underline{u})\}$. Hence, by the definition of $\Omega_{\omega CF}$, $pr_1(\underline{y})\omega_1 \in Pref(pr_1(\underline{u}))$. Thus $pr_1(\underline{x}) \in Pref(pr_1(\underline{u}))$. Consequently, $\underline{x} \in Pref(\underline{u})$ and $\underline{u} \in Insert(Pr_{10}, Qr)$. Thus $\underline{x} \in Pref(Insert(Pr_{10}, Qr))$. We have shown $VFS(Pr) \subseteq Pref(Insert(Pr_{10}, Qr))$ which means that Corollary 2.14(1) holds. Moreover, Corollary 2.14(2) is trivially satisfied, and Corollary 2.14(3) follows from (5). Hence we can apply Corollary 2.14 to conclude that: Qr is adequate $\Leftrightarrow$ $Pr = Pr_{10} \longrightarrow Qr$ is adequate, which completes the proof. □

To obtain some characterization of programs that can be smoothly inserted into path programs without conflicts involving ω, we shall define two sets, $\Gamma_1 \subseteq \Gamma$ and $\Omega_3 \subseteq \Omega_{\omega CF}$.

Let Γ_1 denote the set of all path programs $Pr = P_1 \dots P_n \in \Gamma$ such that:

(1) $Ev(Vcyc(Pr)) = Ev(Pr)$

(2) $(\forall k \geq 1)(\forall \underline{x} \in Pref(Vcyc(Pr)^k))(\exists \underline{y} \in Vev(P)^*)$:
$$\underline{xy} \in Vcyc(Pr)^k \wedge (pr_i(\underline{y}) = \varepsilon \Leftrightarrow pr_i(\underline{x}) \in Cyc(P_i)^k,\ for\ i \leq n)$$

(3) $(\forall \underline{x} \in VFS(Pr))(i \neq j \wedge pr_i(\underline{x}) \in Cyc(P_i)^k \wedge pr_j(\underline{x}) \in Cyc(P_j)^m) \Rightarrow k = m$.

We observe that (1)–(3) of Γ_1 are quite similar to (1)–(3) of Γ_0. More precisely, (1) of Γ_1 is just (1) of Γ_0; (2) of Γ_1 is stronger than (2) of Γ_0; and (3) of Γ_1 is weaker than (3) of Γ_0. Also, we note that $\Gamma_0 \neq \Gamma_1$. For Example $Pr_{10} \in \Gamma_1 \setminus \Gamma_0$, where Pr_{10} is defined in the proof of Theorem 2.27. We shall show in the sequel that $\Gamma_0 \subset \Gamma_1$.

Let $a_1, \dots, a_n, a, b, c, d$ be distinct elements of $EV_\omega \setminus \{\omega\}$, and let $\xi, \psi_1, \psi_2, \dots$ and $\phi_{k,m}$ $(for\ k > m \geq 0)$ be paths defined by:

$$\xi = \underline{path}\ \omega\ \underline{end}$$

and for $k \geq 1$,

$$\psi_k = \underline{path}\ \underbrace{\omega; \dots; \omega}_{k\ times}; b, (c; c); d; c\ \underline{end}$$

$$\phi_{k,0} = \underline{path}\ a, (c; d); \underbrace{\omega; \dots; \omega}_{k\ times}; d; c\ \underline{end}$$

and for $k > m > 0$,

$$\phi_{k,m} = \underline{path}\ \underbrace{\omega; \dots; \omega}_{m\ times}; a, (c; d); \underbrace{\omega; \dots; \omega}_{k-m\ times}; d; c\ \underline{end}$$

Let Ω_3 be the least subset of Ω such that:

(1) For every $k \geq 1$, the following path program $R_k \in \Omega_3$:

$$R_k = \begin{cases} R(1): & \underline{path}\ \underbrace{\omega; \dots; \omega}_{k\ times}; a_1; a\ \underline{end} \\ & \dots \\ R(n): & \underline{path}\ \underbrace{\omega; \dots; \omega}_{k\ times}; a_n; a\ \underline{end} \end{cases}$$

(2) For every $k > m \geq 0$ any path program that can be formed from: $\psi_k, \phi_{k,m}$ and $n-2$ copies of ξ, belongs to Ω_3 (we assume $n \geq 2$ in this chapter).

Clearly $\Omega_3 \subseteq \Omega_{\omega CF}$, hence $Smooth(\Omega_{\omega CF}) \subseteq Smooth(\Omega_3)$.

Theorem 2.28
$Smooth(\Omega_{\omega CF}) \subseteq Smooth(\Omega_3) \subseteq \Gamma_1$. □

The proof of Theorem 2.28 is in Appendix D. Furthermore, we suspect that both inclusions in Theorem 2.28 can be replaced by equalities. As an immediate consequence of Theorem 2.26, Theorem 2.27 and the fact that $Pr_{10} \in \Gamma_1 \setminus \Gamma_0$ we obtain the corollary:

Corollary 2.15
$\Gamma_0 \subseteq \Gamma_1 \wedge \Gamma_0 \neq \Gamma_1$ □

Deletions Preserving Adequacy

Even a perfunctory look at communication between paths shows that we cannot expect very general results about deletions preserving adequacy. The two path programs below show that we may have Pr adequate although Pr/a may deadlock, and vice versa, Pr/a may be adequate and Pr may deadlock. Let[6]

$$Pr_{18} = \begin{cases} P(1): & \underline{path}\ a; b\ \underline{end} \\ P(2): & \underline{path}\ b; a\ \underline{end} \end{cases}$$

Clearly $VFS(Pr_{18}) = \{\underline{\varepsilon}\}$, while $Pr_{18}/a = \underline{path}\ b\ \underline{end}\ \underline{path}\ b\ \underline{end}$ and $VFS(Pr_{18}/a) = \{\underline{b}\}^*$, i.e. Pr_{18}/a is adequate. Let

$$Pr_{19} = \begin{cases} P(1): & \underline{path}\ (b; a; c; d), (c; a; c)\ \underline{end} \\ P(2): & \underline{path}\ a; c; d\ \underline{end} \end{cases}$$

Here we have:

$$Pr_{19}/a = \begin{cases} P(1): & \underline{path}\ (b; c; d), (c; c)\ \underline{end} \\ P(2): & \underline{path}\ c; d\ \underline{end} \end{cases}$$

hence Pr_{19}/a deadlocks after $\underline{c}$, while $VFS(Pr_{19}) = Pref(\{\underline{bacd}\}^*)$, i.e. Pr_{19} is adequate. The precise definition of Pr/a has not been given yet, and the intuitive meaning described at the beginning of this section is ambiguous. For instance, if $Pr = \underline{path}\ a, b; c\ \underline{end}$, is Pr/b equal to $\underline{path}\ a; c\ \underline{end}$ or to $\underline{path}\ a, c\ \underline{end}$? On the other hand the deletions preserving adequacy are highly desirable from the methodological viewpoint. They are the simplest possible transformations, and they seem to have a great "handling" power. Such a situation suggests a "custom-tailored" approach. We do not have general results but we may try to obtain a number of "custom-tailored" results which are valid for some special classes of path programs. These special classes will either conform to some simple but frequently used and rather general patterns, or, be generalizations of

[6] In this section we return to the old numbering of path programs.

some specifications of real problems. In this sub-section we shall formulate four results concerning some rather simple general patterns of path programs, and in the next subsection we shall present one more result, "custom-tailored" to a particular class of path programs.
We start with three results showing when and how Pr/a can be constructed and used.

Lemma 2.23 *Let $Pr = P_1 \dots P_n$ and $a \in Ev$ satisfy the following:*

(1) $i(a) = \{j\}$, $1 \le j \le n$

(2) There are $c_1, \dots, c_k \in Ev_j \setminus \{a\}$ and constants $\alpha > 0$, $\beta \ge 0$ such that $(\forall x \in FS_j)\ \#_a(x) \ge \alpha \cdot (\#_{c_1}(x) + \dots + \#_{c_k}(x)) - \beta$.

Then: Pr is adequate $\Leftrightarrow$ Pr/a is adequate,
where $Pr/a = P_1 \dots P_{j-1} P_j' P_{j+1} \dots P_n$ and P_j' is any path such that $Cyc(P_j') = h_{Ev\setminus\{a\}}(Cyc(P_j))$.

Proof. ($\Rightarrow$) Let $\underline{x} = (x_1, \dots, x_n) \in VFS(Pr/a)$. From $Cyc(P_j') = h_{Ev\setminus\{a\}}(Cyc(P_j))$ it follows that there is $x_j' \in FS(P_j)$ such that $h_{Ev\setminus\{a\}}(x_j') = x_j \in FS(P_j')$. Consider $(x_1, \dots, x_j', \dots, x_n)$. Since $i(a) = \{j\}$, by Lemma 2.3 (2), we have $(x_1, \dots, x_j', \dots, x_n) \in Vev(Pr)^*$. Hence $\underline{y} = (x_1, \dots, x_j', \dots, x_n) \in VFS(Pr)$ and $\underline{x} = h_{Ev\setminus\{a\}}(\underline{y})$. From the adequacy of Pr it follows that there is $\underline{z}$ such that $\underline{y}\,\underline{z} \in VFS(Pr)$ and $\#_b(\underline{z}) \ge 1$ for all $b \in Ev$. Clearly, $\underline{x}\,\underline{w} \in VFS(Pr/a)$ where $\underline{w} = h_{Ev\setminus\{a\}}(\underline{z})$. Hence Pr/a is adequate. Note that we have not used the condition (2) yet.
($\Leftarrow$) Take $\underline{x} \in VFS(Pr)$, and let $\underline{y} = h_{Ev\setminus\{a\}}(\underline{x})$. By $i(a) = \{j\}$ and $Cyc(P_j') = h_{Ev\setminus\{a\}}(Cyc(P_j))$, we have $\underline{y} \in VFS(Pr/a)$. Hence, since Pr/a is adequate, for every $l \ge 1$ there exist $\underline{z}'$ such that $\underline{y}\,\underline{z}' \in VFS(Pr/a)$ and $\#_b(z') \ge l$ for all $b \in Ev(Pr/a)$.

In particular, there exist $\underline{z}$ such that $\underline{y}\,\underline{z} \in VFS(Pr/a)$, $\#_b(z) \ge 1$ for all $b \in Ev(Pr/a)$, and $\#_{c_1}(\underline{z}) + \dots + \#_{c_k}(\underline{z}) \ge \frac{1}{\alpha}(\#_a(\underline{x}) + \beta + 1)$. Again, one can find $\underline{w}$ such that $\underline{z} = h_{Ev\setminus\{a\}}(\underline{w})$ and $\underline{x}\,\underline{w} \in VFS(Pr)$ (see the construction from the proof of "$\Rightarrow$"). Thus:

$$\#_b(\underline{x}\,\underline{w}) = \#_b(\underline{x}\,\underline{z}) \ge \#_b(\underline{x}) + 1$$

for all $b \in Ev(Pr) \setminus \{a\}$, and

$$\#_a(\underline{x}\,\underline{w}) \ge \alpha \cdot (\#_{c_1}(\underline{x}\,\underline{w}) + \dots + \#_{c_k}(\underline{x}\,\underline{w})) - \beta \ge \alpha \cdot (\#_{c_1}(\underline{w}) + \dots + \#_{c_k}(\underline{w})) - \beta =$$

$$= \alpha \cdot (\#_{c_1}(\underline{z}) + \dots + \#_{c_k}(\underline{z})) - \beta \ge \alpha \cdot (\frac{1}{\alpha}(\#_a(\underline{x}) + \beta + 1)) - \beta = \#_a(\underline{x}) + 1.$$

Hence Pr is adequate. □

As an example of paths P_j satisfying the condition (2) of the above lemma, we can take: $\underline{path}\ (a;c_1),(a;c_2)\ \underline{end}$, where $\alpha = 1$, $\beta = 0$, and $\underline{path}\ (c_1;a),(c_2;a)\ \underline{end}$, where $\alpha = 1$, $\beta = 1$. The path program Pr_{18} shows that we can not easily drop the condition (1). The following path program:

$$Pr_{20} = \begin{cases} P(1): & \underline{path}\ (c;b;a),(b;c)\ \underline{end} \\ P(2): & \underline{path}\ b;c\ \underline{end} \end{cases}$$

shows that the condition (2) cannot be dropped either. In this case $VFS(Pr_{20}) = \{\underline{b}\,\underline{c}\}^*\{\underline{b},\ \underline{\varepsilon}\}$, so Pr_{20} is not adequate, and the condition (2) is not satisfied either. As Pr_{20}/a we may take

$$Pr_{20}/a = \begin{cases} P(1): & \underline{path}\ (c;b),(b;c)\ \underline{end} \\ P(2): & \underline{path}\ b;c\ \underline{end} \end{cases}$$

where $VFS(Pr_{20}) = \{\underline{b}\,\underline{c}\}^*\{\underline{b},\ \underline{\varepsilon}\}$, so it is adequate.

Lemma 2.24 *Let $Pr = P_1 \ldots P_n$ and $a, b \in Ev$ satisfy:*

(1) $i(a) \subseteq i(b)$

(2) for every $j \in i(a)$: $b \notin FS_j$ and

$$(\forall x \in FS_j)(\forall c \in Ev_j)\ (xcb \in FS_j \Rightarrow c = a) \wedge (xac \in FS_j \Rightarrow c = b).$$

Then: Pr is adequate $\Leftrightarrow$ Pr/a is adequate, where $Pr/a = P_1' \ldots P_n'$, and $P_j' = P_j$ if $j \notin i(a)$ and P_j' is any path such that $Cyc(P_j') = h_{Ev\setminus\{a\}}(Cyc(P_j))$ for $j \in i(a)$.

Proof. From (2) it follows immediately (since $b \notin FS_j$ for $j \in i(a)$):
(2a) for every $j \in i(a)$: $Cyc(P_j) = h_{b\longrightarrow ab}(h_{Ev\setminus\{a\}}(Cyc(P_j)))$, where $h_{b\longrightarrow ab} : Ev^* \longrightarrow Ev^*$ is a homomorphism defined as a natural extension of:

$$h_{b\longrightarrow ab}(c) = \begin{cases} c & c \neq b \\ ab & c = b \end{cases}$$

We extend standardly this homomorphism to $h_{b\longrightarrow ab} : Vev^* \longrightarrow Vev^*$.
As a direct consequence of (2a), we have:
(2b) for every $\underline{x} \in VFS(Pr/a)$, $h_{b\longrightarrow ab}(\underline{x}) \in VFS(Pr)$.
And, from the definition of Pr/a:
(2c) for every $\underline{x} \in VFS(Pr)$, $h_{Ev\setminus\{a\}}(\underline{x}) \in VFS(Pr/a)$.
($\Rightarrow$) Let $\underline{x} \in VFS(Pr/a)$. Define $\underline{y} = h_{b\longrightarrow ab}(\underline{x})$. Clearly $\underline{x} = h_{Ev\setminus\{a\}}(\underline{y})$, and, by (2b), $\underline{y} \in VFS(Pr)$. Since Pr is adequate, there is $\underline{z}$ such that $\underline{y}\,\underline{z} \in VFS(Pr)$ and $\#_c(\underline{z}) \geq 1$ for all $c \in Ev(Pr)$. Let $\underline{w} = h_{Ev\setminus\{a\}}(\underline{z})$. Since $\underline{x}\,\underline{w} = h_{Ev\setminus\{a\}}(\underline{y}\,\underline{z})$, from (2c) we have $\underline{x}\,\underline{w} \in VFS(Pr/a)$, and clearly $\#_c(\underline{w}) \geq 1$ for all $c \in Ev(Pr/a) = Ev \setminus \{a\}$. Hence Pr/a is adequate.

($\Leftarrow$) Let $\underline{x} \in VFS(Pr)$. We have to consider two cases: (1) $\#_a(x) = \#_b(x)$ and (2) $\#_a(x) = \#_b(x) + 1$.
Case 1. We have $\#_a(x) = \#_b(x)$. Define $\underline{y} = h_{Ev\setminus\{a\}}(\underline{x})$. by (2a) we have $\underline{x} = h_{b\longrightarrow ab}(\underline{y})$, and by (2c), $\underline{y} \in VFS(Pr/a)$. From adequacy of Pr/a it follows that there is $\underline{z}$ such that $\underline{y}\,\underline{z} \in VFS(Pr/a)$ and $\#_c(\underline{z}) \geq 1$ for all $c \in Ev(Pr/a) = Ev \setminus \{a\}$. Define $\underline{w} = h_{b\longrightarrow ab}(\underline{z})$. By (2b) we have $\underline{x}\,\underline{w} = h_{b\longrightarrow ab}(\underline{y}\,\underline{z}) \in VFS(Pr)$. Note that for all $c \in Ev(Pr)$, $\#_c(\underline{z}) \geq 1$, so Pr is adequate.
Case 2. Now $\#_a(x) = \#_b(x) + 1$. Define $\underline{y} = h_{Ev\setminus\{a\}}(\underline{x})$. From (2) it follows that $\underline{x} = \underline{x}'\,\underline{a}$, and from (2a) we have $\underline{x}' = h_{b\longrightarrow ab}(\underline{y})$. Define $\underline{x}'' = \underline{x}\,\underline{b}$. We claim that $\underline{x}'' \in VFS(Pr)$. Suppose that $\underline{x}\,\underline{b} \notin VFS(Pr)$. From (2) it follows that for all $j \in i(a)$, $pr_j(\underline{x})b \in FS(P_j)$, so there must be $i_0 \in i(b) \setminus i(a)$ such that $pr_i(\underline{x})b \notin FS(P_{i_0})$. Since $h_{Ev\setminus\{a\}}(\underline{x}\,\underline{b}) = \underline{y}\,\underline{b}$ and $P'_{i_0} = P_{i_0}$, we have $pr_{i_0}(\underline{y})b \notin FS(P'_{i_0})$. On the other hand, from (2) we have that for all $j \in i(a)$: $pr_i(\underline{y})b \in FS(P'_j)$ and $pr_i(\underline{y})c \in FS(P'_j) \Rightarrow c = b$. Hence Pr/a deadlocks after $\underline{y}$, a contradiction since Pr/a is adequate. Thus, $\underline{x}'' = \underline{x}\,\underline{b} \in VFS(Pr)$. Since $\#_a(\underline{x}'') = \#_b(\underline{x}'')$ we may now apply Case 1. □

Lemma 2.25 *Let $Pr = P_1 \ldots P_n$ satisfy the following:*

(1) $i(a) = \{j\}, 1 \leq j \leq n$.

(2) $P_j = \underline{path}\ a, (sequence; b)\ \underline{end}$

(3) both a and b occur in P_j only once,

where sequence denote any string of terminal symbols derived from the nonterminal "sequence" of the Basic COSY grammar.
Then: Pr is adequate $\Leftrightarrow$ Pr/a is adequate,
where $Pr/a = P_1 \ldots P_{j-1} P'_j P_{j+1} \ldots P_n$ and $P'_j = \underline{path}\ sequence; b\ \underline{end}$.

Proof.
($\Rightarrow$) Let $\underline{x} \in VFS(Pr/a)$. Note that also $\underline{x} \in VFS(Pr)$. From adequacy of Pr it follows that there is $\underline{z}$ such that $\underline{x}\,\underline{z} \in VFS(Pr)$ and $\#_c(z) \geq 1$ for all $c \in Ev(Pr)$. Let $\underline{y} = h_{Ev\setminus\{a\}}(\underline{z})$. Note that $\underline{x}\,\underline{y} \in VFS(Pr/a)$ and $\#_c(\underline{y}) \geq 1$ for all $c \in Ev(Pr/a)$, so Pr/a is adequate.
($\Leftarrow$) Let $\underline{x} \in VFS(Pr)$. Define $\underline{y} = h_{Ev\setminus\{a\}}(\underline{x})$. Clearly $\underline{y} \in VFS(Pr/a)$. Since Pr/a is adequate there is $\underline{z}$ such that $\#_c(\underline{z}) \geq 1$ for all $c \in Ev(Pr/a)$ and $\underline{y}\,\underline{z}\,\underline{b} \in VFS(Pr/a)$. Note that $\underline{x}\,\underline{z}\,\underline{b}\,\underline{a} \in VFS(Pr)$, and $\#_c(\underline{z}\,\underline{b}\,\underline{a}) \geq 1$ for all $c \in Ev(Pr)$, so Pr is adequate. □

The fourth result will concern the case when we erase a whole single path of a special form. Let $A = \{a_1, \ldots, a_m\}$ be a set of events, $Pr = P_1 \ldots P_n$ be a path program, and let P_A be the following path:

$$P_A = \underline{path}\ a_1, \ldots, a_m\ \underline{end}.$$

Define $Pr_A = P_1 \dots P_n P_A$.

Lemma 2.26 *Pr is adequate if and only if Pr_A is adequate.*

Proof. Let $(\underline{x})_{Pr}$ denote an element of $Vev(Pr)^*$, and $(\underline{x})_{Pr_A}$ denote an element of $Vev(Pr_A)^*$. Note that $(\underline{x})_{Pr_A} \in VFS(Pr_A) \Leftrightarrow (\underline{h_{Ev}(x)})_{Pr} \in VFS(Pr)$, where $Ev = Ev(Pr)$.
($\Rightarrow$) Let $(\underline{x})_{Pr_A} \in VFS(Pr_A)$. If $a \in A \setminus Ev(Pr)$ then $(\underline{xa})_{Pr_A} \in VFS(Pr_A)$. Let $a \in Ev$. Consider $x' = h_{Ev}(x)$. Clearly, there is $(\underline{y})_{Pr}$ such that $(\underline{x'ya})_{Pr} \in VFS(Pr)$. But $h_{Ev}(xya) = x'ya$, i.e., $(\underline{xya})_{Pr_A} \in VFS(Pr_A)$.
($\Leftarrow$) Let $(\underline{x})_{Pr} \in VFS(Pr)$, $a \in Ev(Pr)$. Clearly $(\underline{x})_{Pr_A} \in VFS(Pr_A)$, and there is $(\underline{y})_{Pr_A}$ such that $(\underline{xya})_{Pr_A} \in VFS(Pr_A)$. Let $y' = h_{Ev}(y)$. Note that $(\underline{xy'a})_{Pr} \in VFS(Pr)$. □

Although the assumptions in the last two lemmas are rather strong and very particular, the results turn out to be very usefull as we show in the next subsection. Moreover, they show the kind of results we can get for deletions preserving adequacy. Such a situation is quite frequent in Mathematics. For example, in the Integral Calculus, we have very few effective results concerning general integrals without restrictions, while there is a vast number of extremely useful results about various particular cases, as you can see in any Encyclopedia of Mathematics.

Applications of Adequacy-Preserving Transformations

In this subsection we shall deal with specifications of three simplified real non-sequential systems, and we shall prove that all three specifications are adequate. In the process of proving adequacy we shall use both adequacy- preserving transformations and the results about adequate SCF-path programs.
The first system we shall analyze is a system consisting of a linereader and lineprinter communicating accross a ring buffer. We also provide the system with an interrupt handler. What this does is to register an error message arriving from somewhere, halts the whole system and then restarts it again. We might picture the situation as one in which both the registering of an error and the restarting of the system results from an intervention from the system's environment, say a human operator. The path program Pr_{LL} which specifies this system is presented in Fig. 2.45, where event names stand for their meanings.

The part of Pr_{LL} called Pr_B describes the ring buffer with n frames. Let us consider the buffer, i.e., Pr_B, first. We suppose that it has n frames and that we are not concerned with the actual content of the buffer at any time. Given this, the only operations associated with it are *deposits* and *removes* involving the various frames. Assuming that all the frames are empty initially, the desired pattern of behaviour of an individual frame must be a sequence of alternating

$$\left.\begin{array}{ll} \left.\begin{array}{ll} P(1): & \underline{path}\ deposit_1; remove_1\ \underline{end} \\ & \dots \\ P(n): & \underline{path}\ deposit_n; remove_n\ \underline{end} \\ P(n+1): & \underline{path}\ deposit_1; \dots; deposit_n\ \underline{end} \\ P(n+2): & \underline{path}\ remove_1; \dots; remove_n\ \underline{end} \end{array}\right\} Pr_B \\ P(n+3): \ \underline{path}\ readline; deposit_1, \dots, deposit_n\ \underline{end} \\ P(n+4): \ \underline{path}\ remove_1, \dots, remove_n; printline\ \underline{end} \end{array}\right\} Pr^1_{LL}$$

$$\begin{array}{ll} P(n+5): & \underline{path}\ deposit_1, (getmessage; stopsystem; startsystem)\ \underline{end} \\ & \dots \\ P(2n+4): & \underline{path}\ deposit_n, (getmessage; stopsystem; startsystem)\ \underline{end} \\ P(2n+5): & \underline{path}\ remove_1, (getmessage; stopsystem; startsystem)\ \underline{end} \\ & \dots \\ P(3n+4): & \underline{path}\ remove_n, (getmessage; stopsystem; startsystem)\ \underline{end} \\ P(3n+5): & \underline{path}\ readline, (getmessage; stopsystem; startsystem)\ \underline{end} \\ P(3n+6): & \underline{path}\ printline, (getmessage; stopsystem; startsystem)\ \underline{end} \end{array}$$

Figure 2.45. Pr_{LL} Reader and printer with n-frame ring buffer and interrupt handler

deposits and removes beginning with a deposit. This constraint is expressed by the paths $P_1, \dots, P_n$ for frames $1, \dots, n$, respectively. The paths P_{n+1} and P_{n+2} impose a ring discipline on deposits and removes. The path P_{n+3} expresses the constraint that after performing readline, the message (line) read must be deposited in one of the frames, and the path P_{n+4} says that after removing a message (line) from any frame, the message must be printed by executing printline. The paths P_{n+5} to P_{3n+6} express the fact that the system may be interrupted and restarted after any event. Clearly Pr_B is an SCF-path program, and one may easily check (or use the algorithm from Sect. 2.4.2) that

$$\underline{deposit_1} \dots \underline{deposit_n remove_1} \dots \underline{remove_m} =$$

$$(deposit_1 remove_1, \dots, deposit_n remove_n,$$

$$deposit_1 \dots deposit_n, remove_1 \dots remove_n) \in VFS(Pr_B),$$

so by Theorem 2.20, Pr_B is adequate.
Let *Removes*, *Deposits* be the following sets of events:

$$Removes = \{remove_1, \dots, remove_n\},$$

$$Deposits = \{deposit_1, \dots, deposit_n\}.$$

By Lemma 2.26, the path program

$$Pr^0_{LL} = P_1 \dots P_n P_{n+1} P_{n+2} P_{Removes} P_{Deposits}$$

is adequate (recall $P_{Removes} = \underline{path}\ remove_1, \ldots, remove_n\ \underline{end}$ and similarly for $P_{Deposits}$). Consider $Pr^1_{LL} = P_1 \ldots P_{n+2}P_{n+3}P_{n+4}$ (see Fig. 2.45). Note that P_{n+3} satisfies the condition (2) of Lemma 2.23 with $a = readline$, $c_i = deposit_i$, $i = 1, \ldots, n$, $\alpha = 1, \beta = 0$, and P_{n+4} satisfies the condition (2) of Lemma 2.23 with $a = printline$, $c_i = remove_i$, $i = 1, \ldots, n$, $\alpha = 1, \beta = 1$.

Furthermore, in Pr^1_{LL}, *readline* occurs only in P_{n+3} and *printline* occurs only in P_{n+4}. Thus all the conditions of Lemma 2.23 are satisfied. Finally, note that $P_{n+3}/readline = P_{Deposits}$ and $P_{n+4}/printline = P_{Removes}$. Thus, from Lemma 2.23 we have that Pr^1_{LL} is adequate if and only if Pr^0_{LL} is adequate, that is to say, Pr^1_{LL} is adequate.

Define:

$$\begin{aligned} P'_{n+5} &= \underline{path}\ deposit_1, \omega\ \underline{end} \\ &\ldots \\ P'_{2n+4} &= \underline{path}\ deposit_n, \omega\ \underline{end} \\ P'_{2n+5} &= \underline{path}\ remove_1, \omega\ \underline{end} \\ &\ldots \\ P'_{3n+4} &= \underline{path}\ remove_n, \omega\ \underline{end} \\ P'_{3n+5} &= \underline{path}\ readline, \omega\ \underline{end} \\ P'_{3n+6} &= \underline{path}\ printline, \omega\ \underline{end} \end{aligned}$$

Obviously, $P'_{n+5} = P_{\{deposit_1, \omega\}}, \ldots, P'_{3n+6} = P_{\{printline, \omega\}}$, thus by Lemma 2.26, we have that $Pr^2_{LL} = P_1 \ldots P_{n+4}P'_{n+5} \ldots P'_{3n+6}$ is adequate. Let Pr' be the following path program:

$Pr' =$

$$\left.\begin{array}{l} \underline{path}\ getmessage; stopsystem; startsystem\ \underline{end} \\ \ldots \\ \underline{path}\ getmessage; stopsystem; startsystem\ \underline{end} \end{array}\right\} (2n+2) times$$

Note that $Pr' \in \Gamma_0$, and $Pr' \longrightarrow Pr^2_{LL} = Pr_{LL}$, hence, by Theorem 2.25, Pr_{LL} is adequate.

Our second example is a specification of a mechanism for controlling access to a critical section by a collection of k processes. A typical process trying to use the system will first request use of the critical section, and then, after the request has been granted, it will enter a pipeline queue. After it has been transferred out of kth frame of the queue, it may then use the critical section, but on completion it must send a signal to say that the next process in the queue may be transferred out of the kth frame. Requests are serviced in cyclic order. The mechanism tries each process one at a time in turn and if the currently considered process is not making a request to use the critical section, skips to the next process. The path program Pr_{CS} specifying this system is listed in Fig. 2.46, where, as ususal, event names stand for their meanings.

We shall now prove that Pr_{CS} is adequate. First observe that Pr_{CS} satisfies the conditions of Lemma 2.23 with $a = signal$, $c_1 = transfer_{k+1}$, $\alpha = 1$, $\beta = 0$,

$$
Pr_{CS} =
$$

$$
\begin{array}{ll}
P(1): & \underline{path}\ transfer_1; transfer_2\ \underline{end} \\
 & \ldots \\
P(k): & \underline{path}\ transfer_k; transfer_{k+1}\ \underline{end} \\
P(k+1): & \underline{path}\ transfer_{k+1}; signal\ \underline{end}
\end{array}
$$

$$
Pr^0_{CS} \left\{ \begin{array}{ll}
P(k+2): & \underline{path}\ skip_1, (request_1; granted_1)\ \underline{end} \\
 & \ldots \\
P(2k+1): & \underline{path}\ skip_k, (request_k; granted_k)\ \underline{end} \\
P(2k+2): & \underline{path}\ skip_1, (granted_1; transfer_1); \\
 & \ldots; skip_k, (granted_k; transfer_1)\ \underline{end}
\end{array} \right.
$$

Figure 2.46. Critical section for k processes

so Pr_{CS} is adequate if and only if $Pr_{CS}/signal = P_1 \ldots P_k P_{k+1}/signal \ldots P_{2k+2}$ is adequate. But $P_{k+1}/signal = P_{\{transfer_{k+1}\}}$, so by Lemma 2.26, $Pr_{CS}/signal$ is adequate if and only if $Pr^k_{CS} = P_1 \ldots P_k P_{k+2} \ldots P_{2k+2}$ is adequate. Now, Pr^k_{CS} satisfies the conditions of Lemma 2.23 with $a = transfer_{k+1}$, $c_1 = transfer_k$, $\alpha = 1$, $\beta = 0$, and $P_k/transfer_{k+1} = P_{\{transfer_k\}}$, so by Lemmas 2.23 and 2.26 we obtain that Pr^k_{CS} is adequate if and only if $Pr^{k-1}_{CS} = P_1 \ldots P_{k-1} P_{k+2} \ldots P_{2k+2}$ is adequate. We may repeat the same reasoning until we get $Pr^1_{CS} = P_1 P_{k+2} \ldots P_{2k+2}$. Thus, Pr_{CS} is adequate if and only if Pr^1_{CS} is adequate. Since now Pr^1_{CS} satisfies the conditions of Lemma 2.23 with $a = transfer_2$, $c_1 = transfer_1$, $\alpha = 1$, $\beta = 0$, and $P_1/transfer_2 = P_{\{transfer_1\}}$, using Lemmas 2.23 and 2.26, we may say that Pr_{CS} is adequate if and only if $Pr^0_{CS} = P_{k+2} \ldots P_{2k+2}$ is adequate.

Note that for every $i = 1, \ldots, k$: the conditions of Lemma 2.24 are satisfied with $a = request_i$, $b = granted_i$. Thus, from Lemma 2.24 it follows that $Pr^0_{CS} = P_{k+2} \ldots P_{2k+2}$ is adequate if and only if $Pr'_{CS} = P'_{k+2} \ldots P'_{2k+1} P_{2k+2}$ is adequate, where $P'_{k+1+i} = \underline{path}\ skip_i, granted_i\ \underline{end}$ for $i = 1, \ldots, k$. Using Lemma 2.26 we may say that Pr'_{CS} is adequate if and only if P_{2k+2} is adequate. But P_{2k+2} is a single path, so it is obviously adequate. Thus, both Pr_{CS} and all intermediate path programs $Pr^k_{CS}, \ldots, Pr^1_{CS}, Pr^0_{CS}$ and Pr'_{CS}, are adequate.

The third system we shall analyze in this section is a specification of a mechanism regulating the use of a critical section by m classes of processes having distinct priorities. With each class is associated a k-frame ring buffer, which a process from the class must enter in order to eventually gain access to the critical section. The mechanism looks at each buffer in turn. If a process is waiting in the next frame due to be emptied, the mechanism allows that process to proceed to the critical section, otherwise the mechanism skips to the next buffer. The path program $Pr^{k,m}_{CS}$ which specifies this system is presented in Fig. 2.47, where as usual, event names stand for their meanings. The particular case $Pr^{3,2}_{CS}$, where

$$\left.\begin{array}{l}\underline{path}\ deposit_{11}; remove_{11}\ \underline{end}\\ \ldots\\ \underline{path}\ deposit_{k1}; remove_{k1}\ \underline{end}\\ \underline{path}\ deposit_{11};\ldots;deposit_{k1}\ \underline{end}\\ \underline{path}\ remove_{11};\ldots;remove_{k1}\ \underline{end}\end{array}\right\} Pr'_1$$

$\ldots$

$$\left.\begin{array}{l}\underline{path}\ deposit_{1m}; remove_{1m}\ \underline{end}\\ \ldots\\ \underline{path}\ deposit_{km}; remove_{km}\ \underline{end}\\ \underline{path}\ deposit_{1m};\ldots;deposit_{km}\ \underline{end}\\ \underline{path}\ remove_{1m};\ldots;remove_{km}\ \underline{end}\end{array}\right\} Pr'_m$$

$$\begin{array}{ll} Q(1): & \underline{path}\ skip_1, (remove_{11},\ldots,remove_{k1}; criticalsection_1)\ \underline{end}\\ & \ldots\\ Q(m) & \underline{path}\ skip_m, (remove_{1m},\ldots,remove_{km}; criticalsection_m)\ \underline{end}\\ R: & \underline{path}\ (skip_1; (skip_2; (\ldots(skip_{m-1}; skip_m, criticalsection_m),\\ & criticalsection_{m-1})\ldots), criticalsection_2), criticalsection_1\ \underline{end}\end{array}$$

Figure 2.47. $Pr^{k,m}_{CS}$ Critical section for m classes of processes with distinct priorities

$k = 3$ and $m = 2$ is presented in Fig. 2.48.

Proof that $Pr^{k,m}_{CS}$ is adequate requires a "custom-tailored" lemma.

Lemma 2.27 *Let $k, m \geq 1$, and for $i = 1, \ldots, m$, let $Pr_i = Pr'_i Q_i$ be a path program such that:*

Pr'_i *is a path program*

$Q_i = \underline{path}\ s_i, (r_{1i}, \ldots, r_{ki}; c_i)\ \underline{end}$

$Ev(Pr_i) \cap Ev(Pr_j) = \emptyset$ *if* $i \neq j$

Let $Pr = Pr'_1 \ldots Pr'_m Q_1 \ldots Q_m R$,
where $R = \underline{path}\ (s_1; (s_2; (\ldots(s_{m-1}; s_m, c_m), c_{m-1})\ldots), c_2), c_1\ \underline{end}$
Assume also that for all $i = 1, \ldots, m$, s_i and c_i occur in Q_i and R only. Then: Pr is adequate $\Leftrightarrow (\forall i = 1, \ldots, m) Pr_i$ is adequate.

Proof. ($\Rightarrow$) Let $\underline{x} \in VFS(Pr_j)$, some $j \leq m$. Define y as x if $j = 1$, or as a string built from x by replacing every occurrence of s_j by a sequence $s_1 \ldots s_j$ and every occurrence of c_j by $s_1 \ldots s_{j-1} c_j$ if $j > 1$. Note that $\underline{y} \in VFS(Pr)$. From the adequacy of Pr it follows that there is $\underline{z}$ such that $\underline{y}\,\underline{z} \in VFS(Pr)$ and $\#_b(\underline{z}) \geq 1$ for all $b \in Ev(Pr)$. Define $\underline{w} = h_{Ev(Pr_i)}(\underline{z})$ and note that (as $Ev(Pr_i) \cap Ev(Pr_j) = \emptyset$ if $i \neq j$) $\underline{x}\,\underline{w} \in VFS(Pr_i)$, and clearly $\#_b(\underline{w}) \geq 1$ for all $b \in Ev(Pr_i)$, so Pr_i is adequate.

($\Leftarrow$) Let $\underline{x} \in VFS(Pr)$, $a \in Ev(Pr)$. There must be j such that $a \in Ev(Pr_j)$ and $a \notin Ev(Pr_i)$ if $i \neq j$. Consider $h_{Ev(Q_i)}(\underline{x})$ for $i = 1,\dots,m$. Note that either $h_{Ev(Q_i)}(\underline{x})s_i \in FS(Q_i)$ and $h_{Ev(Q_i)}(x)r_{pi} \in FS(Q_i), p = 1,\dots,k$, or $h_{Ev(Q_i)}(x)c_i \in FS(Q_i)$. Let $i_1 < \dots < i_t$ be the sequence of all numbers such that $h_{Ev(Q_{i_r})}(x)c_{i_r} \in FS(Q_{i_r})$, $r = 1,\dots,t$, $t \geq 0$. Suppose that $t \geq 1$.

Consider now $h_{Ev(R)}(x)$. Not that there is l such that $h_{Ev(R)}(x)b \in FS(R) \Rightarrow b = s_l$ or $b = c_l$.

Define the string v as follows:

$$v = s_l s_{l+1} \dots s_{i_1-1}c_{i_1} \text{ if } l < i_1,$$

$$v = c_l \text{ if } l = i_r,\ r \in \{1,\dots,t\},$$

$$v = s_l s_{l+1} \dots s_m s_1 s_2 \dots s_{i_1-1}c_{i_1} \text{ if } i_t < l,$$

$$v = s_l s_{l+1} \dots s_{i_q-1}c_{i_q} \text{ if } i_1 < \dots < i_{q-1} < l < i_q < \dots < i_t.$$

Define strings $v_{i_r} = s_1 s_2 \dots s_{i_r-1}c_{i_r}$ for $r = 1,\dots,t$ and let $\underline{x}' = \underline{x}\,\underline{v}\,\underline{v}_{i_1} \dots \underline{v}_{i_t}$. If $t = 0$ then $\underline{x}' = \underline{x}\,\underline{s}_l\,\underline{s}_{l+1} \dots \underline{s}_m$. Note that after a history $\underline{x}'$ the paths $Q_1 \dots Q_n$, and R are ath their "initial states". Let $\underline{y} = h_{Ev(Pr_j)}(\underline{x})$. Clearly $\underline{y} \in VFS(Pr_j)$, so let $\underline{z}$ be such that $\underline{y}\,\underline{z}\,\underline{a} \in VFS(Pr_j)$. Define w as z if $j = 1$, or a string built from z by replacing every occurrence of s_j by a sequence $s_1 \dots s_j$ and every occurrence of c_j by $s_1 \dots s_{j-1}c_j$ if $j > 1$. We now have $\underline{x}\,\underline{v}\,\underline{v}_{i_1} \dots \underline{v}_{i_t}\,\underline{w}\,\underline{a} \in VFS(Pr)$ if $t > 0$, or $\underline{x}\,\underline{s}_l \dots \underline{s}_m\,\underline{w}\,\underline{a} \in VFS(Pr)$ if $t = 0$, so Pr is adequate. □

From Lemma 2.27 it immediately follows that $Pr_{CS}^{k,m}$ is adequate if and only if $Pr_i = Pr_i'Q_i$ are adequate for all $i = 1,\dots,m$. Define $Q_i^1 =$ <u>path</u> $remove_{1i},\dots,remove_{ki}; criticalsection_i$ <u>end</u>, and $Pr_i^1 = Pr_i'Q_i^1$ for $i = 1,\dots,m$. Every Pr_i satisfies the conditions of Lemma 2.25 with $a = skip_i$, $b = criticalsection_i$, so Pr_i is adequate if and only if Pr_i^1 is adequate. The conditions of Lemma 2.23 are satisfied by every Pr_i^1 with $a = criticalsection_i$, $\alpha = 1, \beta = 0$, hence Pr_i^1 is adequate if and only if Pr_i^2 is adequate, where $Pr_i^2 = Pr_i'P_{\{remove_{1i},\dots,remove_{ki}\}}$. We can now apply Lemma 2.26 to Pr_i^2, which means that Pr_i is adequate if and only if Pr_i' is adequate. On the other hand, each Pr_i' describes a ring buffer, and is isomorphic with Pr_B from Fig. 2.45. As we have already proven Pr_B is adequate, hence each Pr_i' is adequate also, and consequently $Pr_{CS}^{k,m}$ is adequate.

The above three examples should serve to explain what we mean by "custom tailored" approach. The results of Sect. 2.4.3 show that strong general transformations are hardly to be expected, so we find the "custom tailored" approach the only practical solution to the adequacy problem.

2.4.4 Free-Choice Path Programs

In the general case, adequacy implies deadlock-freeness but not necessarily vice versa. However, for some subclasses of path programs deadlock-freeness and

$$\left.\begin{array}{l}
\underline{path}\ deposit_{11}; remove_{11}\ \underline{end} \\
\underline{path}\ deposit_{21}; remove_{21}\ \underline{end} \\
\underline{path}\ deposit_{31}; remove_{31}\ \underline{end} \\
\underline{path}\ deposit_{11}; deposit_{21}; deposit_{31}\ \underline{end} \\
\underline{path}\ remove_{11}; remove_{21}; remove_{31}\ \underline{end}
\end{array}\right\} Pr'_1$$

$$\left.\begin{array}{l}
\underline{path}\ deposit_{12}; remove_{12}\ \underline{end} \\
\underline{path}\ deposit_{22}; remove_{22}\ \underline{end} \\
\underline{path}\ deposit_{32}; remove_{32}\ \underline{end} \\
\underline{path}\ deposit_{12}; deposit_{22}; deposit_{32}\ \underline{end} \\
\underline{path}\ remove_{12}; remove_{22}; remove_{32}\ \underline{end}
\end{array}\right\} Pr'_2$$

$$\begin{array}{ll}
Q(1): & \underline{path}\ skip_1, (remove_{11}, remove_{21}, remove_{31}; criticalsection_1)\ \underline{end} \\
Q(2) & \underline{path}\ skip_2, (remove_{12}, remove_{22}, remove_{32}; criticalsection_2)\ \underline{end} \\
R: & \underline{path}\ (skip_1; skip_2, criticalsection_2), criticalsection_1\ \underline{end}
\end{array}$$

Figure 2.48. $Pr^{3,2}_{CS}$ Critical section for 2 classes of processes with distinct priorities and 3-frame buffers

adequacy are equivalent. We claimed such a property for SCF-path programs. In this section we define a class of path programs much wider than SCF-path programs, and we shall show that for path programs belonging to this class, adequacy and deadlock-freeness are equivalent properties. In this class, every local choice, i.e., a choice in a particular single path, will imply a global choice with respect to the whole path program. We shall call this class: **free-choice path programs.**

Let $Pr = P_1 \dots P_n$.

A path program Pr is said to be a **free-choice path program** (abbreviated FC-path program) if and only if the following property is satisfied:

$$(\forall \underline{x} \in VFS)(\forall a, b \in Ev)((\exists i)pr_i(\underline{x})a \in FS_i \wedge pr_i(\underline{x})b \in FS_i) \Rightarrow$$
$$(\underline{xa} \in VFS \Leftrightarrow \underline{xb} \in VFS).$$

Corollary 2.16 *Every SCF-path program is an FC-path program.*

Proof. If Pr is a SCF-path program then we have:
$(\forall \underline{x} \in VFS)(\forall a, b \in Ev)(\forall i)(pr_i(\underline{x})a \in FS_i \wedge pr_i(\underline{x})b \in FS_i) \Rightarrow a = b$. Which clearly implies the formula from the definition of FC-path programs. □

We shall illustrate the above definition by presenting examples of path programs not being free-choice. The simplest example of a non FC-path program is probably the following:

$$Pr_{21} = \left\{ \begin{array}{ll} P(1): & \underline{path}\ a; b\ \underline{end} \\ P(2): & \underline{path}\ a, b\ \underline{end} \end{array} \right.$$

We have: $pr_2(\underline{\varepsilon})a \in FS_2$, $pr_2(\underline{\varepsilon})b \in FS_2$ and $\underline{a} \in VFS$, while $\underline{b} \notin VFS = \{\underline{ab}\}^*\{\underline{a}, \underline{\varepsilon}\}$, i.e., there is a local conflict between a and b in P_2 but there is no conflict between $\underline{a}$ and $\underline{b}$ at all in the whole path program Pr_{21}. Let us consider next:

$$Pr_{22} = \begin{cases} P(1): & \underline{path}\ d; a, b\ \underline{end} \\ P(2): & \underline{path}\ b, c; d\ \underline{end} \end{cases}$$

In this case, $pr_2(\underline{\varepsilon})b \in FS_2$, $pr_2(\underline{\varepsilon})c \in FS_2$, $\underline{c} \in VFS = \{\underline{cda}\}^*\{\underline{cd}, \underline{c}, \underline{\varepsilon}\}$ but $\underline{b} \notin VFS$. Note that here b never occurs in VFS, furthermore neither Pr_{21} nor Pr_{22} contains global conflict at all.
Let Pr_{23} be as follows:

$$Pr_{23} = \begin{cases} P(1): & \underline{path}\ c; a, b\ \underline{end} \\ P(2): & \underline{path}\ a, (c; a, b)\ \underline{end} \end{cases}$$

In this case $VFS = \{\underline{ca}, \underline{cb}\}^*\{\underline{c}, \underline{\varepsilon}\}$, so we have global conflicts after every $\underline{xc}$ from VFS. Pr_{23} is not free-choice either, since for instance $pr_2(\underline{\varepsilon})a \in FS_2$, $pr_2(\underline{\varepsilon})c \in FS_2$, $\underline{c} \in VFS$ but $\underline{a} \notin VFS$.

The SCF-path programs do not involve local and global conflicts (other than those resulting from concurrency) at all. The next simple class of FC-path programs may be defined as follows: if a and b involve a local conflict, they belong to one single path only. Formally, this leads to the following lemma.

Lemma 2.28 *Let $Pr = P_1 \ldots P_n$ be a path program satisfying the following property:*
$(\forall i = 1, \ldots, n)(\forall a, b \in Ev_i)(\exists x \in Ev_i^*)(xa \in FS_i \wedge xb \in FS_i \wedge a \neq b) \Rightarrow i(a) = i(b) = \{i\}$.
Then Pr is a FC-path program.

Proof. Let $\underline{x}, i, a, b$ be those from the left-hand side of the implication from the definition of FC-path programs. From the condition of this lemma, it follows that a and b belong to Ev_i only, but this means that $\underline{xa} \in VFS$ and $\underline{xb} \in VFS$, i.e., Pr is free-choice. □

Now we shall formulate the basic property of FC-path programs.

Theorem 2.29 (Adequacy of Free-Choice path programs)
Let Pr be a connected FC-path program. Then the following are equivalent:
(1) Pr is adequate
(2) Pr is deadlock-free. □

The proof of Theorem 2.29 is quite complex and will be presented in Appendix F. Lemma 2.15 is an immediate consequence of Theorem 2.29 and Corollary 2.16.

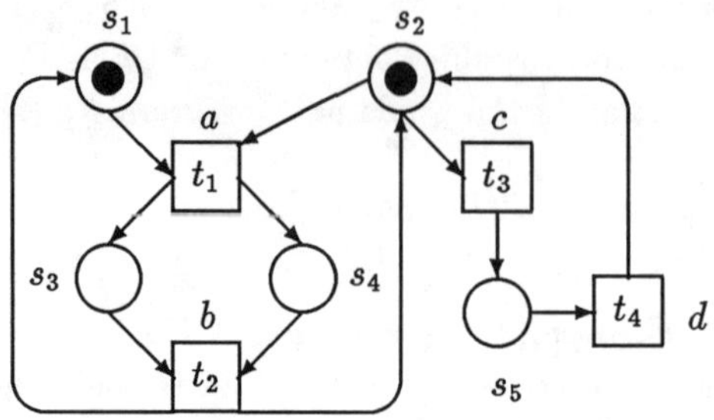

Figure 2.49. Labelled smd-net for FC-path Pr_{23}

It is worth while to discuss the relationship between FC-path programs and free-choice Petri nets (see Sect. 2.3.1). The free-choice property in net theory is a structural, that is to say "graph theoretical" property of nets. A net is free choice if and only if whenever two transitions share an input place then it is, for both of them, their only input place. This has a dynamic consequence, namely, in any marking, one of them is enabled (i.e., may fire) if and only if the other one is enabled. The free choice property for path programs is essentially the latter dynamic free-choice property. If two events are in local conflict then whenever one is enabled the other is also enabled.
Not all nets corresponding to free-choice path programs are free-choice nets. The following FC-path prograsm Pr_{24} defines, under the net semantics of Sect. 2.3, a labelled net depicted in Fig. 2.49 which is not free-choice:

$$Pr_{24} = \begin{cases} P(1): & \underline{path}\ a; b\ \underline{end} \\ P(2): & \underline{path}\ (a;b),(c;d)\ \underline{end} \end{cases}$$

This net is not free-choice since $(s_2, t_1) \in F$, $s_2^\bullet = \{t_1, t_3\} \neq \{t_1\}$, and ${}^\bullet t_1 = \{s_1, s_2\} \neq \{s_2\}$.

2.5 Execution Semantics for COSY

In this section we develop a formal framework for specifying and analyzing execution semantics for basic COSY path programs. There are at least three different kinds of execution semantics which have been widely investigated for a number of models of non-sequential computation. The first is the **interleaving** semantics that represents all possible computations using **sequences of event occurrences.** In this approach, the meaning of the path program Pr_2 from Sect. 2.1 (recall $Pr_2 = \underline{path}\ a; c\ \underline{end}\ \underline{path}\ b; c\ \underline{end}$) would be defined as the following set of **sequences** of events of Pr_2:

$$\{\varepsilon, a, b, ab, ba, abc, bac, abca, abcb, \ldots\} = Pref(\{abc, bac\}^*).$$

Another approach is that of **full** semantics which may be intuitively expressed as "execute as possible (but not necessarily with maximal parallelism)". In other words, **full semantics allows all possible execution steps to be actually executed.** The vector firing sequence semantics described in Sect. 2.2 represents exactly the full semantics in the above sense, i.e., the meaning of Pr_2 is defined by $VFS(Pr_2)$ which is equal to:

$$Pref(\{\underline{abc}\}^*),\ where\ \underline{a} = (a,\varepsilon), \underline{b} = (\varepsilon, b), \underline{c} = (c,c).$$

A distinguished variant of the full semantics is the **maximally concurrent** semantics. Here, all computation steps are **maximal in the sense that no additional independent events can be added.** For the path program Pr_2, the set of all computations defined by the maximally concurrent semantics is the following:

$$\{\underline{\varepsilon}, \underline{ab}, \underline{abc}, \underline{abcab}, \underline{abcabc}, \ldots\} = \{\underline{abc}\}^*\{\underline{\varepsilon}, \underline{ab}\}.$$

Furthermore, we can use the interleaving approach, but **replace events by vector events**, and as the result we get the so called **vector event interleaving semantics**, which for the path program Pr_2 produces the following set of sequences of vector events:

$$\underline{\varepsilon}, \underline{a}, \underline{b}, \underline{ab}, \underline{ba}, \underline{abc}, \ldots$$

but since $\underline{ab} = \underline{ba} = (a,b)$, this set is exactly the set $\{\underline{\varepsilon}, \underline{a}, \underline{b}, \underline{abc}, \ldots\} = VFS(Pr_2)$.

In this section we shall introduce the notion of a **strategy** that is intended to represent an **arbitrary execution semantics**, not only those mentioned above, and will be defined as a kind of acyclic binary relation on vector sequences representing the system's histories. In fact, we have already defined two strategies, namely the relation CON, in Sect. 2.2.3, which represents the full semantics, and the relation SEQ, in the same section, which represents the vector event interleaving semantics. Fig. 2.50 shows the relations SEQ, CON and MAX, where MAX represents the maximally concurrent semantics and is not defined formally as yet, for the path program Pr_2.

A formal definition of strategies allows us to speak about some relevant aspect of sytem behaviours. We shall define the notion of behavioural equivalence of two strategies and show that such an equivalence preserves a number of behavioural properties including those of deadlock-freeness and adequacy. We shall also introduce the notion of an optimal strategy that intuitively represents the most efficient execution semantics, and we then show that for each strategy one can find an optimal and behaviourally equivalent strategy. Furthermore, we shall analyze the sufficient and necessary conditions for the equivalence of the full strategy and the maximally concurrent strategy. To illustrate the concepts mentioned above let us consider the path program Pr_2 and the relations

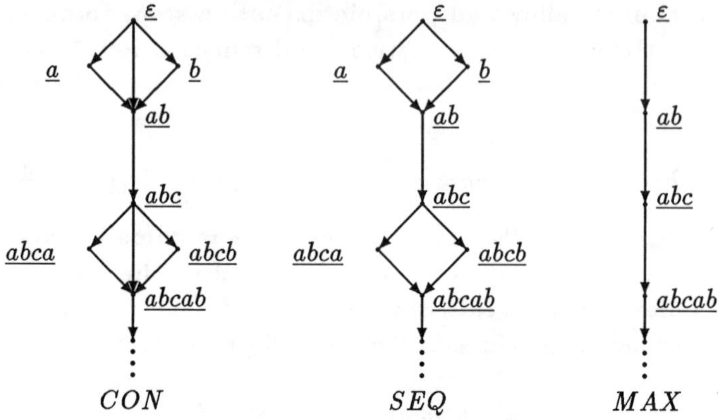

Figure 2.50. SEQ, CON and MAX for Pr_2

SEQ, CON and MAX from Fig. 2.50. Clearly, SEQ and CON can be regarded as behaviourally equivalent since they define, starting from $\underline{\varepsilon}$, the same set of vector firing sequences, namely, $\{\underline{abc}\}^*\{\underline{\varepsilon}, \underline{a}, \underline{b}, \underline{ab}\}$ which is equal to $VFS(Pr_2)$. The set of vector firing sequences defined by MAX is equal to $\{\underline{abc}\}^*\{\underline{\varepsilon}, \underline{ab}\}$, so it differs from $VFS(Pr_2)$, but it is all we can obtain from MAX according to our intuition. It seems to be intuitively correct that for every (possibly concurrent) history of a given system, each subhistory of this history is also a history of the system. In formal terms this means that if a vector sequence $\underline{x}$ represents a history of the system, then each $\underline{y} \in Pref(\underline{x})$ also represents a history of this system. This means that the set of histories actually defined by MAX is not $\{\underline{abc}\}^*\{\underline{\varepsilon}, \underline{ab}\}$ but rather $Pref(\{\underline{abc}\}^*\{\underline{\varepsilon}, \underline{ab}\}) = Pref(VFS(Pr_2)) = VFS(Pr_2)$. Thus one can say that in this case SEQ, CON and MAX are equivalent, and it is rather obvious that MAX is an optimal strategy, i.e., the most efficient strategy equivalent to CON. As the next example consider the folowing path program:

$$Pr_{25} = \begin{cases} P(1): & \underline{path}\ a, c; d\ \underline{end} \\ P(2): & \underline{path}\ b; c, d\ \underline{end} \end{cases}$$

Here at the beginning events a and b are enabled; since they are independent either a or b may occur, or both of them may occur simultaneously. In the latter case, after the simultaneous occurrence of a and b, only event d is enabled; after its occurrence the system is again in the initial state. Note that in this way, if a and b always occur simultaneously, the event c will never be enabled. On the other hand, if it is possible for event b to occur alone, event c becomes enabled. Thus, for Pr_{25} the behaviour defined by CON is not the same as that defined by MAX even together with the operator $Pref$. The CON and MAX for Pr_{25}

are presented in Fig. 2.51.

Figure 2.51 presents also the third relation, denoted by OPT. One may check by inspection that OPT has two desirable properties. First, it is equivalent to CON in the sense that OPT under the operator $Pref$ produces the same set of vector sequences as CON, and second, OPT is maximally concurrent in the collection of strategies equivalent to CON. Moreover, if $\underline{x} = \underline{A_1} \ldots \underline{A_k}$ where $A_i \in Ind(Pr_{25})$ and $\underline{\varepsilon} OPT \underline{A_1} OPT \underline{A_1 A_2} \ldots OPT \underline{A_1} \ldots \underline{A_k}$ then $< A_1, \ldots, A_k >$ is the left-normal form of $\underline{x}$. We show in the sequel that this property holds for every optimal strategy.

2.5.1 Execution Strategies

In this subsection we shall introduce execution strategies representing possible execution semantics of the path program $Pr = P_1 \ldots P_n$. A **strategy** will be defined as a kind of acyclic binary relation on VFS satisfying three conditions. The first two conditions can be informally formulated as follows:

- Every history (vector) the strategy involves can be reached from the initial history, i.e., from $\underline{\varepsilon}$.
- All pairs of histories (vectors) the strategy relates may differ only by a set of independent events, i.e., an element of Ind.

The above two conditions seem to be rather obvious and we hope they do not require further comments. The third condition deals with the relationship between the strategy and the prefix operator. The kind of situation we want to avoid is illustrated in Fig. 2.52, where we have a relation on vector sequences W satisfying the two conditions above and a non-empty set of vector sequences $V \subseteq Dom(W)$ (see Appendix A for the definition of Dom) such that:

(1) V is closed under W, i.e., $W(V) \subseteq V$;

(2) V can be fully derived through the prefix operator from other histories of W, i.e., $V \subseteq Pref(Dom(W) \setminus V)$.

Note that in such a case, on the one hand V is a set of subhistories of $Dom(W) \setminus V$, and on the other hand, there is no W-connection between any element of V and any element of $Dom(W) \setminus V$.

The third condition can be regarded as a **weaker** version of the rule stating that "if W is a strategy then there must be a W-connection from subhistories to histories". The "weaker" stands for the fact that in some cases we allow the situation that $\underline{y} \in Pref(\underline{x})$ and $\underline{x} \notin W^*(\underline{y})$.

Consider the relation OPT from Fig. 2.51. We have $\underline{b} \in Pref(\underline{ab})$, and $\underline{ab} \notin OPT^*(\underline{b})$, but in this case only an occurrence of b alone enables c, while an occurrence of a and b in one step does not, and the W-continuation after b, i.e.,

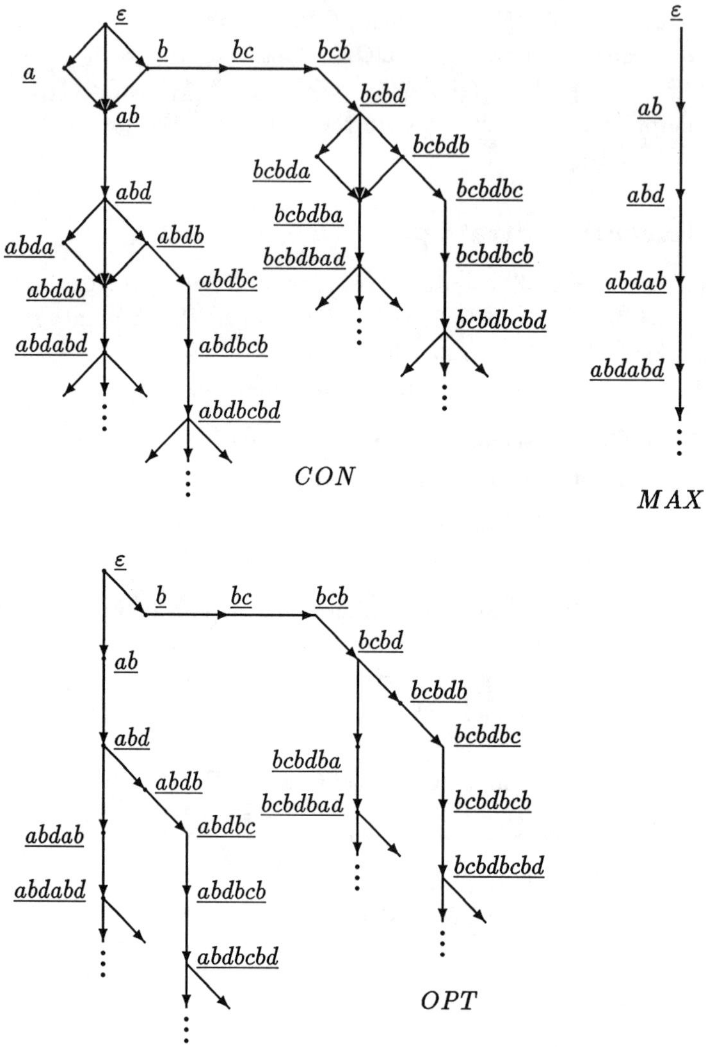

Figure 2.51. CON, MAX and OPT for Pr_{25}

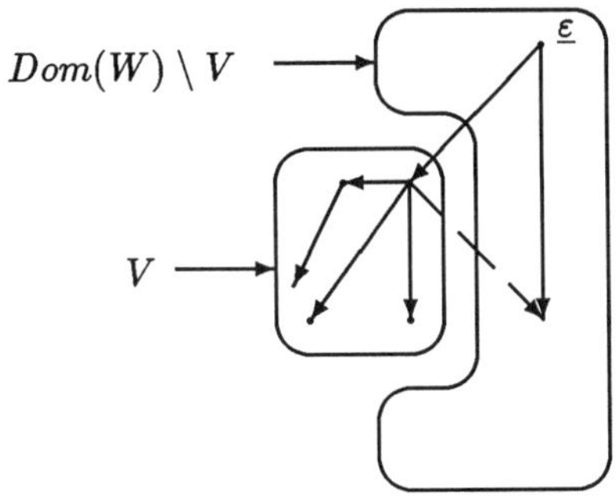

Figure 2.52. ($\longrightarrow$ represents W^*; - - -> represents prefix relation)

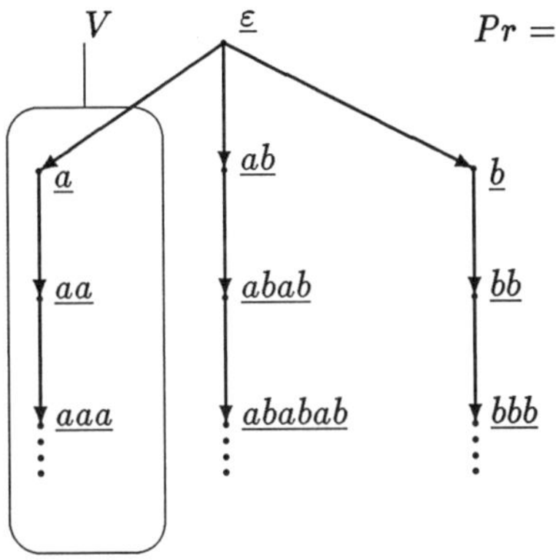

$$Pr = \begin{cases} \underline{path}\ a\ \underline{end} \\ \underline{path}\ b\ \underline{end} \end{cases}$$

Figure 2.53. Satisfies strategy definition (1-2) but not (3)

$W^*(\underline{b})$, cannot be obtained from the W-continuation after $\underline{ab}$, i.e., from $W^*(\underline{ab})$, through the prefix operation. Thus $\underline{b}$ is a "distinctive" subhistory of $\underline{ab}$, and for such subhistories we relax the constraint about the W-connection. The third condition for a relation to be a **strategy** can therefore be formulated as follows:

- the strategy does not involve the situation from Fig. 2.52.

A simple example of a relation which satisfies the first two conditions but does not satisfy the third condition is presented in Fig. 2.53.

A non-empty relation $W \in VFS \times VFS$ is said to be a **strategy** if and only if:

(1) $Dom(W) = W^*(\underline{\varepsilon})$

(2) $(\forall \underline{x} \in Dom(W))\ W(\underline{x}) \subseteq \{\underline{xA} | A \in Ind\}$

(3) $(\forall V \subseteq Dom(W))(V \neq \emptyset \wedge W(V) \subseteq V) \Rightarrow V \setminus Pref(Dom(W) \setminus V)) \neq \emptyset$.

The condition (3) says exactly that the situation from Fig. 2.52 cannot happen. Let $STRG$ denote the family of all strategies over VFS. We first prove a simple lemma which allows us to show that some well motivated relations, including CON and SEQ, are strategies in the sense of the above definition.

Lemma 2.29 *Let $W \subseteq VFS \times VFS$ satisfy (1) and (2) of the strategy definition and let:*
$(\forall \underline{x}, \underline{y} \in Dom(W))\underline{y} \in Pref(\underline{x}) \Rightarrow \underline{x} \in W^*(\underline{y})$.
Then W is a strategy.

Proof. Assume that there is $V \subseteq Dom(W)$ such that $V \neq \emptyset$, $W(V) \subseteq V$ and $V \setminus Pref(Dom(W) \setminus V) = \emptyset$, i.e., $V \subseteq Pref(Dom(W) \setminus V)$. Let $\underline{y} \in V$ and $\underline{x} \in Dom(W) \setminus V$ such that $\underline{y} \in Pref(\underline{x})$. Since $W(V) \subseteq V$ we have $\underline{x} \notin W^*(\underline{y})$ which produces a contradiction. □

The condition from Lemma 2.29 can be interpreted as "there is always a W-connection from a subhistory to a history".

Theorem 2.30 *CON and SEQ are strategies.*

Proof. It is easy to check that both CON and SEQ satisfy the conditions of Lemma 2.29. □

For every strategy $W \subseteq VFS \times VFS$, and every $\underline{x} \in Dom(W)$, let $ev_W(\underline{x})$ denote the set of events which may potentially occur after $\underline{x}$, i.e.,

$$ev_W(\underline{x}) = \{a \in Ev | (\exists \underline{y} \in Vev^*)(\underline{x}, \underline{xya}) \in W^*\}.$$

We can also extend the concept of "enabled" events to an arbitrary strategy W, not necessarily CON, namely, for every startegy $W \in VFS \times VFS$ and every $\underline{x} \in Dom(W)$, let *enabled*$(\underline{x})$ denote the set of all sets of independent events that may occur immediately after $\underline{x}$, i.e.,

$$enabled_W(\underline{x}) = \{A \in Ind | (\underline{x}, \underline{xA}) \in W\}.$$

Of course, $enabled_{CON}(\underline{x}) = enabled(\underline{x})$ for every $\underline{x} \in VFS$.
For every strategy $W \subseteq VFS \times VFS$ and every $\underline{x} \in Dom(W)$, let $maxenabled_W(\underline{x})$ denote the set of those elements of $enabled_W(\underline{x})$ which are maximal with respect to set inclusion ordering, i.e.,

$$maxenabled_W(\underline{x}) = \{A \in enabled_W(\underline{x}) | (B \in enabled_W(\underline{x}) \wedge A \subseteq B) \Rightarrow A = B\}.$$

Finally, for every strategy $W \subseteq VFS \times VFS$, let $Scope_W = Pref(Dom(W))$. The set $Scope_W$ will be called **the scope of** W.

We now can formally define the relation MAX, discussed at the beginning of this section, which represents the maximally concurrent execution semantics.

Recall that the maximally concurrent semantics is one all of whose execution steps are maximal, meaning that they cannot be extended by adding any further independent events.
Let $MAX \subseteq VFS \times VFS$ be the following relation:

$$\underline{x}MAX\underline{y} :\Leftrightarrow (\exists A \in maxenabled_{CON}(\underline{x}))\underline{y} = \underline{x}\underline{A}.$$

We shall call MAX the **maximal concurrent reachability in one step.**

Corollary 2.17
$(\forall \underline{x} \in Dom(MAX))enabled_{MAX}(\underline{x}) = maxenabled_{CON}(\underline{x}) = maxenabled_{MAX}(\underline{x}).$
□

We shall show that MAX is indeed a strategy, but first we show the relationship between MAX and the **left-normal form** considered in Sect. 2.2.

Theorem 2.31 *(Below if $i = 1$ then $\underline{A_1} \dots \underline{A_{i-1}}$ denotes $\underline{\varepsilon}$.)*

(1) If $\underline{x} = \underline{A_1} \dots \underline{A_k} \in VFS \setminus \{\underline{\varepsilon}\}$ is such that

(a) $A_i \in Ind$ for $i = 1, \dots, k$
(b) $A_i \in enabled_{MAX}(\underline{A_1} \dots \underline{A_{i-1}})$ for $i \geq 2$

then: $\underline{x} \doteq \underline{A_1} \dots \underline{A_k}$.

(2) if $\underline{x} \in MAX^+(\underline{\varepsilon})$ and $\underline{x} \doteq \underline{A_1} \dots \underline{A_k}$ then $A_i \in enabled_{MAX}(\underline{A_1} \dots \underline{A_{i-1}})$ for $i = 1, \dots, k$

(3) if $\underline{x} \in MAX^+(\underline{\varepsilon})$, $\underline{x}CON^*\underline{y}$ and $\underline{y} \doteq \underline{A_1} \dots \underline{A_m}$ then $\underline{x} \doteq \underline{A_1} \dots \underline{A_k}$ for some $1 \leq k \leq m$.

(4) if $\underline{\varepsilon}CON^+\underline{x}CON^*\underline{y}$, $\underline{x} \doteq \underline{A_1} \dots \underline{A_k}$, $\underline{y} \doteq \underline{B_1} \dots \underline{B_m}$ and $A_k \in enabled_{MAX}(\underline{A_1} \dots \underline{A_{k-1}})$ then $k \leq m$ and $A_i = B_i$ for $i = 1, \dots, k$.

Proof.

(1) Suppose that for some $j \geq 2$ and $a \in A_j$ we have $(a,b) \in ind$, for all $b \in A_{j-1}$. Then $A_{j-1} \cup \{a\} \in enabled(A_1 \dots A_{j-2})$, $A_{j-1} \subset A_{j-1} \cup \{a\}$ and $A_{j-1} \neq A_{j-1} \cup \{a\}$, contradicting $A_{j-1} \in enabled_{MAX}(A_1 \dots A_{j-2})$.

(2) Since $\underline{x} \in MAX^+(\underline{\varepsilon})$ we have: $\underline{\varepsilon}MAX\underline{B_1}MAX \dots MAX\underline{B_1} \dots \underline{B_m} = \underline{x}$, where $B_i \in enabled_{MAX}(\underline{B_1} \dots \underline{B_{i-1}})$ for $i = 1, \dots, m$. From (1) it follows that $\underline{x} \doteq \underline{B_1} \dots \underline{B_m}$. Hence, by Theorem 2.1, $k = m$ and $A_i = B_i$ for $i = 1, \dots, k$.

(3) Follows immediately from (4).

(4) Let $\underline{y} = \underline{x}\,\underline{z}$, $\underline{z} \neq \underline{\varepsilon}$, and $z \neq \varepsilon$, and $\underline{z} \doteq \underline{C_1} \ldots \underline{C_j}$. Since $A_k \in enabled_{MAX}(\underline{A_1} \ldots \underline{A_{k-1}})$, we have $(\forall c \in C_1)(\exists a \in A_k)\ (a,c) \notin ind$. Thus $\underline{x}\,\underline{z} \doteq \underline{A_1} \ldots \underline{A_k}\underline{C_1} \ldots \underline{C_j}$. □

We can now prove that MAX is a strategy.

Theorem 2.32 *MAX is a strategy, i.e., $MAX \in STRG$.*

Proof. We shall show that MAX satisfies the conditions of Lemma 2.29. It is easy to check that (1) and (2) of the definition of strategy are satisfied. Let $\underline{x}, \underline{y} \in Dom(MAX) \setminus \{\underline{\varepsilon}\} \subseteq VFS$, $\underline{y} \in Pref(\underline{x})$ and let $\underline{x} \doteq \underline{A_1} \ldots \underline{A_m}$, $\underline{y} \doteq \underline{B_1} \ldots \underline{B_k}$. From Theorem 2.31 (4) we have that $k \leq m$ and $B_i = A_i$ for $i = 1, \ldots, k$. From Theorem 2.31 (2) we have that $A_{k+1} \in enabled_{MAX}(\underline{B_1} \ldots \underline{B_k}) = enabled_{MAX}(\underline{y})$, and $A_j \in enabled_{MAX}(\underline{B_1} \ldots \underline{B_k}\underline{A_{k+1}} \ldots \underline{A_{j-1}})$ for $j = k+2, \ldots, m$, which means that $\underline{y} MAX^{m-k} \underline{x}$, so the conditions of Lemma 2.29 are satisfied. □

So far we have introduced a framework for discussing various aspects of the general execution model. We shall now attempt to characterize formally behaviourally equivalent execution semantics. Suppose that W is a strategy. It generates the set of histories $W^*(\underline{\varepsilon}) = Dom(W)$. In addition, as we have argued at the beginning of this section, using the prefix operator, we can derive the whole set $Scope_W$ and reason about histories it comprises. Now, if another strategy, say U, satisfies $U^*(\underline{\varepsilon}) \subseteq Scope_W$ we may say that anything that U describes is also described by W, and thus U is **characterized by** W, leading us to the following definition.

Two strategies U and W are said to be **behaviourally equivalent** if and only if $Scope_U = Scope_W$. We shall denote this by $U \approx W$.

Obviously, there is no ultimate notion of behavioural equivalence, since each particular formalization reflects to some extent a subjective point of view of its author(s) and the kind of applications it has been introduced for. We therefore do not claim that $\approx$ is the only choice, however, we think that any reasonable equivalence should imply the scope consistency rule. The justification of our definition will be provided by demonstrating that behavioural equivalence in the above sense implies equivalence with respect to some important behavioural properties of execution semantics.

We shall now introduce a set of properties and a satisfaction relation for which it will follow that if $U \approx W$, then both U and W enjoy the same subsets of properties in that set. The properties will be expressed as formulas of a simple language LAN_φ. The alphabet of LAN_φ consists of symbols $\neg, \wedge, (,)$, and φ_E for every non-empty $E \subseteq Ev$.

The language LAN_φ is the least set such that:

(1) for every non-empty set of events $E \subseteq Ev$, φ_E belongs to LAN_φ.

(2) if $f, g \in LAN_\varphi$ then $(\neg f)$ and $(f \wedge g)$ belong to LAN_φ.

The elements of LAN_φ will be called **formulas of LAN_φ**.
The satisfaction relation $\models$ (written in the infix manner) is a binary relation in $STRG \times LAN_\varphi$. It is defined as the least relation such that:

(1) for every non-empty set of events $E \subseteq Ev$ and every strategy $W \in STRG$,

$W \models \varphi_E :\Leftrightarrow (\forall \underline{x} \in Dom(W)) ev_W(\underline{x}) \cap E \neq \emptyset$

(2) for all $f, g \in LAN_\varphi$ and every $W \in STRT$,

$W \models (\neg f) :\Leftrightarrow$ *not* $W \models f$

$W \models (f \wedge g) :\Leftrightarrow W \models f$ *and* $W \models g$.

We can see that each formula in LAN_φ expresses a liveness-like property. In particular, we can express in this way both deadlock-freeness and adequacy. Standardly, we can extend the concepts of deadlock-freeness and adequacy for the general strategy in the following way.
For every strategy $W \in STRT, Pr = P_1 \ldots P_n$ is **W-deadlock-free** if and only if for every $\underline{x} \in W^*(\underline{\varepsilon}) : enabled_W(\underline{x}) \neq \emptyset$.
For every strategy $W \in STRT, Pr = P_1 \ldots P_n$ is **W-adequate** if and only if for every $\underline{x} \in W^*(\underline{\varepsilon})$ and every $\underline{a} \in Ev$ there is $\underline{y} \in W^*(\underline{x})$ and $A \in enabled_W(\underline{y})$ such that $a \in A$.

Corollary 2.18
(1) Pr is W-deadlock-free if and only if $W \models \varphi_{Ev}$.
(2) Pr is W-adequate if and only if $W \models \varphi_{\{a_1\}} \wedge \ldots \wedge \varphi_{\{a_m\}}$, where $Ev = \{a_1, \ldots, a_m\}$. □

Of course adequacy is equivalent to CON-adequacy and deadlock-freeness is equivalent to CON-deadlock-freeness, therefore our two formulas φ_{Ev} and $\varphi_{\{a_1\}} \wedge \ldots \wedge \varphi_{\{a_m\}}$ provide a natural generalization of these important properties of path programs.
In the language LAN_φ we may of course express not only W-deadlock-freeness and W-adequacy, but also a number of other properties. For example, $W \models \varphi_{\{a\}} \wedge \varphi_{\{b,c\}}$ can be read "after all histories in W we can execute event a and at least one of events b or c". We shall not discuss here the question of the expressive power of the whole language LAN_φ.
It is now possible to formulate the main result about the bahavioural equivalence.

Theorem 2.33 (Equivalence of Strategies)
If $U \approx W$ then for every $f \in LAN_\varphi$: $W \models f \Leftrightarrow U \models f$.

Proof. It suffices to show that if $W \models \varphi_E$ then $U \models \varphi_E$ for every non-empty $E \subseteq Ev$. Suppose that $W \models \varphi_E$ and $U \models \neg\varphi_E$. Let $\underline{x} \in Dom(U) = U^*(\underline{\varepsilon})$ and $E \cap ev_U(\underline{x}) = \emptyset$. Define $V = U^*(\underline{x})$. Clearly, $U(V) \subseteq V$. Hence, by (3) of the definition of a strategy, there is $\underline{y} \in V$ such that $\underline{y} \notin Pref(Dom(U) \setminus V)$. By $Scope_W = Scope_U$ there is $\underline{u} \in Vev^*$ such that $\underline{yu} \in Dom(W) = W^*(\underline{\varepsilon})$. Thus, by $W \models \varphi_E$ there is $\underline{w} \in Vev^*$ such that $\underline{yuwa} \in Dom(W)$ for some $a \in E$. Now, again by $Scope_W = Scope_U$, there is $\underline{z} \in Vev^*$ such that $\underline{v} = \underline{yuwaz} \in Dom(U)$. Since $E \cap ev_U(\underline{v}) = \emptyset$ and $a \in E$, then $\underline{v} \notin V$, contradicting $\underline{y} \notin Pref(Dom(U) \setminus V)$. This completes the proof. □

The reciprocal of the theorem is clearly false. To prove this we can take the path program consisting of one path $\underline{path}\ a, b\ \underline{end}$ only, and the following two strategies: $W = \{(\underline{\varepsilon}, \underline{a})\}$ and $U = \{(\underline{\varepsilon}, \underline{b})\}$. Clearly $Scope_W \neq Scope_U$. On the other hand, for every non-empty $E \subseteq Ev = \{a, b\}$ we have $W \models \neg\varphi_E$ and $U \models \neg\varphi_E$, hence W and U satisfy exactly the same formulas LAN_φ.

Another aspect of the non-sequential behaviour we intend to deal with has to do with the efficiency of a strategy (execution semantics). More precisely, we ask whether a given equivalence class of $\approx$ contains a strategy that could be regarded as the most efficient (or optimal). Before attempting to answer this question we have to make precise the notion of an optimal strategy. Intuitively, an optimal strategy should at least avoid redundant execution paths which is a property embodied in the definition below.

A **minimal semi-strategy** is any non-empty relation $W \subseteq VFS \times VFS$ satisfying (1) and (2) of the strategy definition, and such that for every proper subrelation U of W (i.e., $U \subseteq W$ and $U \neq W$) we have $Pref(U^*(\underline{\varepsilon})) \neq Pref(W^*(\underline{\varepsilon}))$.

Lemma 2.30 *Every minimal semi-strategy is a strategy.*

Proof. Let W be a minimal semi-strategy that does not satisfy (3) of the definition of strategy. There is a non-empty set $V \subseteq Dom(W)$ such that $W(V) \subseteq V$ and $V \subseteq Pref(Dom(W) \setminus V)$. Let $U = \{(\underline{x}, \underline{y}) \in W | \underline{x} \notin V \wedge \underline{y} \notin V\}$. Clearly, $U \subseteq W$ and $U \neq W$. On the other hand, $Pref(U^*(\underline{\varepsilon})) = Pref(Dom(W) \setminus V) = Pref(Dom(W)) = Pref(W^*(\underline{\varepsilon}))$, a contradiction. □

We say that a strategy is **minimal** if it is a minimal semi-strategy. In other words, a strategy is minimal when it cannot be reduced without losing its expressive power.

Although restricting to only minimal strategies dramatically reduces the number of strategies that might be regarded as optimal, we still may have to consider more than one strategy. Let us take for example:

$$Pr_{26} = \begin{cases} P(1): & \underline{path}\ a; d\ \underline{end} \\ P(2): & \underline{path}\ b\ \underline{end} \\ P(3): & \underline{path}\ d, c\ \underline{end} \end{cases}$$

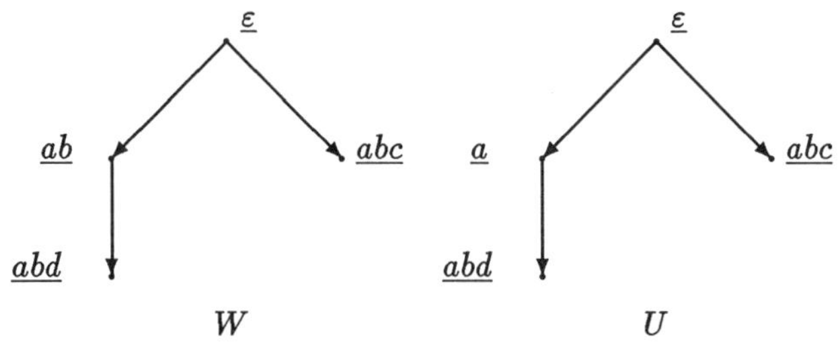

Figure 2.54. Two minimal strategies for Pr_{26}

and two minimal strategies:

$$W = \{(\underline{\varepsilon}, \underline{ab}), (\underline{ab}, \underline{abd}), (\underline{\varepsilon}, \underline{abc})\}$$

and

$$U = \{(\underline{\varepsilon}, \underline{a}), (\underline{a}, \underline{abd}), (\underline{\varepsilon}, \underline{abc})\}$$

which have the same scopes. Consider the graphs of W and U shown in Fig. 2.54.

Though the graphs are isomorphic and thus indistinguishable in graph-theoretic terms, we can find a rather subtle property which designates W as the more efficient strategy of the two. For suppose that we are about to decide whether a leads to deadlock. Taking the left graph we get the answer after one step regardless which branch is chosen. On the other hand, the right graph gives the answer after one step (the right branch) or after two steps (the left branch). To turn the above observation into a formal definition we need some auxiliary notions.

Lemma 2.31 *If W is a minimal strategy, then the graph $(Dom(W), W)$ is a directed tree rooted at $\underline{\varepsilon}$.*

Proof. Clearly $(\forall \underline{x} \in Dom(W))(\underline{x}, \underline{\varepsilon}) \notin W$, so (2) of the directed tree definition is satisfied (see Appendix C). Let $\underline{x}, \underline{x}', \underline{x}'' \in Dom(W)$ and $(\underline{x}', \underline{x}) \in W, (\underline{x}'', \underline{x}) \in W$. We need to prove $\underline{x}' = \underline{x}''$. Assume that $\underline{x}' \neq \underline{x}''$. Define $U = W \setminus \{(\underline{x}'', \underline{x})\}$. Clearly $U \subseteq W$ and $U \neq W$, but on the other hand $U^*(\underline{\varepsilon}) = W^*(\underline{\varepsilon})$, and consequently $Pref(U^*(\underline{\varepsilon})) = Pref(W^*(\underline{\varepsilon}))$, hence W is not minimal. □

Let W be a minimal strategy. We can introduce the notion of the **immediate predecessor of** $\underline{x} \in Dom(W) \setminus \{\varepsilon\}$ in W as being the element $\underline{y} \in Dom(W)$ satisfying $(\underline{y}, \underline{x}) \in W$. We shall denote the immediate predecessor of $\underline{x}$ in W by $pred_W(\underline{x})$, i.e., for every $\underline{x} \in Dom(W) \setminus \{\underline{\varepsilon}\}$, $(pred_W(\underline{x}), \underline{x}) \in W$. Because W is a tree then $pred_W(\underline{x})$ is defined unambiguously. Also, in an obvious way, we can introduce the notion of the **distance**, $dist_W(\underline{x})$, between the root $\underline{\varepsilon}$ and a vertex $\underline{x} \in Dom(W)$ in the graph $(Dom(W), W)$, namely, $dist_W(\underline{\varepsilon}) = 0$ and

$dist_W(\underline{x}) = 1 + dist_W(pred_W(\underline{x}))$ for $\underline{x} \in Dom(W) \setminus \{\underline{\varepsilon}\}$. Again, the definition of $dist_W(\underline{x})$ is correct because W is a directed tree. Now we can introduce a kind of efficiency measure for minimal strategies.

For every minimal strategy W, let $\delta_W : Scope_W \longrightarrow \{0,1,\ldots\}$ be a mapping such that:

$\delta_W(\underline{\varepsilon}) = 0$, and for every $\underline{x} \in Scope_W \setminus \{\underline{\varepsilon}\}$,

$\delta_W(\underline{x}) = max\{dist_W(\underline{y}) | \underline{y} \in Dom(W) \wedge \underline{x} \in Pref(\underline{y}) \wedge \underline{x} \notin Pref(pred_W(\underline{y}))\}$.

Intuitively, regardless which execution path involving $\underline{x}$ is taken, after **at most** $\delta_W(\underline{x})$ steps we can reach $\underline{x}$ or a history from which $\underline{x}$ can be derived as a subhistory. Consider the strategy U from Fig. 2.54 and $\underline{ab} \in Scope_U$. The only $\underline{y}$'s from $Dom(U) = \{\underline{\varepsilon}, \underline{a}, \underline{abc}, \underline{abd}\}$ satisfying $\underline{ab} \in Pref(\underline{y})$ and $\underline{ab} \notin Pref(pred_W(\underline{y}))$ are $\underline{abd}$ and $\underline{abc}$. Since $dist_U(\underline{abd}) = 2$ and $dist_U(\underline{abc}) = 1$, we have $\delta_U(\underline{ab}) = 2$. If we take W from Fig. 2.54 and the same $\underline{ab} \in Scope_W = Scope_U$, we find that the only $\underline{y}$'s from $Dom(W) = \{\underline{\varepsilon}, \underline{ab}, \underline{abc}, \underline{abd}\}$ satisfying $\underline{ab} \in Pref(\underline{y})$ and $\underline{ab} \notin Pref(pred_W(\underline{y}))$ are $\underline{ab}$ itself and $\underline{abc}$. Since $dist_W(\underline{ab}) = dist_W(\underline{abc}) = 1$, then $\delta_W(\underline{ab}) = 1$, and we obtain $\delta_W(\underline{ab}) < \delta_U(\underline{ab})$. For two minimal strategies U and W satisfying $U \approx W$ we shall write $\delta_W \leq \delta_U$ if and only if $(\forall \underline{x} \in Scope_U = Scope_W)\delta_W(\underline{x}) \leq \delta_U(\underline{x})$, and $\delta_W < \delta_U$ if and only if $\delta_W \leq \delta_U$ and there is $\underline{x} \in Scope_U = Scope_W$ such that $\delta_W(\underline{x}) < \delta_U(\underline{x})$. The reader may check that $\delta_W < \delta_U$ for the strategies of Fig. 2.54.

Let W and U be two minimal strategies satisfying $W \approx U$. Strategy W is said to be **more efficient** than strategy U if and only if $\delta_W < \delta_U$.

The optimal strategy can now be defined as being more efficient than every other behaviourally equivalent minimal strategy, i.e., a minimal startegy W is said to be **optimal** if and only if:

$(W \approx U$ and U is minimal$) \Rightarrow \delta_W < \delta_U$ or $W = U$.

Having introduced the notion of an optimal strategy we must answer the basic question, namely, whether such a strategy always exists (i.e., whether each equivalence class of $\approx$ contains an optimal startegy). The answer turns out to be positive and we shall now construct such a strategy.

Suppose that μ is an arbitrary equivalence class of $\approx$ and let $Scope_\mu$ denote the common scope of the strategies of μ.

In the following definition we describe how to construct the unique optimal strategy that belongs to μ. In this definition we shall use the concept of the left-normal form discussed in Sect. 2.2.

First we introduce the length of a left-normal form as a function $\kappa : Vev^* \Leftrightarrow \{0,1,2,\ldots\}$ defined by $\kappa(\underline{\varepsilon}) = 0$ and $\kappa(\underline{x}) = m$ if $\underline{x} \doteq \underline{A_1} \ldots \underline{A_m}$. Theorem 2.1 guarantees that the above definition of κ is correct.

For every $\mu \in STRT/\approx$ let $OPT_\mu \subseteq CON$ be a relation defined in the following three steps:

(Step 1) Let $T_\mu \subseteq CON \cap Scope_\mu \times Scope_\mu$ be a relation such that:

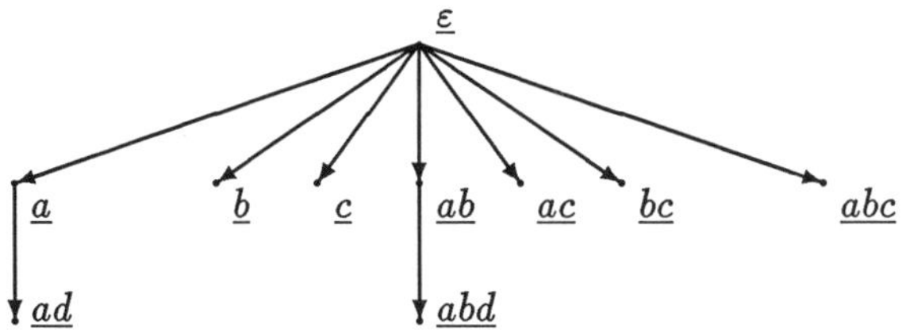

Figure 2.55. Representation of T_μ

$\underline{x}T_\mu\underline{y} \Leftrightarrow (\underline{x} = \underline{\varepsilon} \wedge (\exists A \in Ind)\underline{y} = \underline{A} \in Scope_\mu) \vee$
$(\underline{x}, \underline{y} \in Scope_\mu \setminus \{\underline{\varepsilon}\} \wedge (\exists A_1, \ldots, A_m, A \in Ind)\underline{x} \doteq \underline{A_1} \ldots \underline{A_m}, \underline{y} \doteq \underline{A_1} \ldots \underline{A_m}\underline{A})$.

Note that usually T_μ is not a strategy, although it always satisfies (1) and (2) of the definition of strategy.

(Step 2) Let $R_\mu \subseteq Scope_\mu \setminus \{\underline{\varepsilon}\}$ be the set of all vector sequences satisfying:

$$\underline{x} \in R_\mu :\Leftrightarrow (\underline{x} \in Scope_\mu \setminus \{\underline{\varepsilon}\} \wedge (\forall a \in Ev)\underline{xa} \in Scope_\mu \Rightarrow \kappa(\underline{xa}) > \kappa(\underline{x})).$$

(Step 3) Define OPT_μ to be the least relation $W \subseteq T_\mu$ such that: $R_\mu \subseteq W^*(\underline{\varepsilon})$.

The relation OPT_μ is well defined since the graph of T_μ is a tree, see Lemma 2.32 below. To illustrate the process of the construction of OPT_μ we again take the path program Pr_{25} and the equivalence class μ containing the strategies U and W shown in Fig. 2.54. We obtain the following:

- $Scope_\mu = Scope_W = Scope_U = \{\underline{\varepsilon}, \underline{a}, \underline{b}, \underline{c}, \underline{ab}, \underline{ac}, \underline{bc}, \underline{ad}, \underline{abc}, \underline{abd}\}$,
  $\underline{a} \doteq \underline{\{a\}}, \underline{b} \doteq \underline{\{b\}}, \underline{c} \doteq \underline{\{c\}}$,
  $\underline{ab} \doteq \underline{\{a,b\}}, \underline{ac} \doteq \underline{\{a,c\}}, \underline{bc} \doteq \underline{\{b,c\}}, \underline{ad} \doteq \underline{\{a\}\{d\}}$,
  $\underline{abc} \doteq \underline{\{a,b,c\}}, \underline{abd} \doteq \underline{\{a,b\}\{d\}}$.
- T_μ is presented in Fig. 2.55.
- $R_\mu = \{\underline{abc}, \underline{abd}\}$
- $OPT_\mu = \{(\underline{\varepsilon}, \underline{ab}), (\underline{ab}, \underline{abd}), (\underline{\varepsilon}, \underline{abc})\}$.

Note that OPT_μ here is the relation W of Fig. 2.54.

We prove that OPT_μ is always an optimal strategy providing the positive answer to the question we have posed earlier. Before proving this we need to formulate an auxiliary lemma about T_μ.

Lemma 2.32
(1) The graph $(Dom(T_\mu), T_\mu)$ is a directed tree rooted at $\underline{\varepsilon}$.
(2) OPT_μ is well defined.

Proof. (1) The graph $(Dom(T_\mu), T_\mu)$ is a directed tree since each $\underline{x} \in Scope_\mu \setminus \{\underline{\varepsilon}\}$ has exactly one left-normal form, and thus exactly one immediate predecessor in T_μ.
(2) First observe that $Dom(T_\mu) = Scope_\mu$, hence $R_\mu \subseteq Dom(T_\mu)$. Since T_μ is a directed tree rooted at $\underline{\varepsilon}$, for every $\underline{x} \in Dom(T_\mu)$ there is exactly *one* sequence $\underline{x}_0, \ldots, \underline{x}_m$ of elements of $Dom(T_\mu)$ such that $\underline{x}_0 = \underline{\varepsilon}, \underline{x}_m = \underline{x}$ and $(\underline{x}_{i-1}, \underline{x}_i) \in T_\mu$ for $i = 1, \ldots, m$. Thus, for every $\underline{x} \in Dom(T_\mu)$ we may define $T_\mu^{\underline{x}} \subseteq T_\mu$, where $T_\mu^{\underline{x}} = \{(\underline{x}_0, \underline{x}_1), (\underline{x}_1, \underline{x}_2), \ldots, (\underline{x}_{m-1}, \underline{x}_m)\}$. Define $U = \bigcup_{\underline{x} \in R_\mu} T_\mu^{\underline{x}}$. Note that $\underline{\varepsilon} \in Dom(U), U \subseteq T_\mu$ and $R_\mu \subseteq U^*(\underline{\varepsilon})$. On the other hand, if $W \subseteq T_\mu$ and $R_\mu \subseteq W^*(\underline{\varepsilon})$ then for every $\underline{x} \in R_\mu$, W must satisfy: $T_\mu^{\underline{x}} \subseteq W$, which means that $U \subseteq W$. Hence $OPT_\mu = U$ is a correctly defined unique relation. □

Theorem 2.34 (Existence of Optimal Strategy)
For every equivalence class $\mu \in STRT/\approx$, the relation OPT_μ is the optimal strategy that belongs to μ.

Proof. From Theorem 2.2 we have the following two facts.
Fact 1: Suppose that $\underline{x} \in VFS$, $\underline{x} \in Pref(\underline{y})$, $\underline{y} \in Scope_\mu$, $\underline{x} \doteq \underline{A}_1 \ldots \underline{A}_k$ and $\underline{y} \doteq \underline{B}_1 \ldots \underline{B}_m$. Then $k \leq m$ and $(\forall i \leq k) A_i \subseteq B_i$.
Fact 2: If we additionally assume that $\underline{x} \in R_\mu$ then, by proceeding similarly as in the proof of Theorem 2.31 (4), we have: $(\forall i \leq k) A_i = B_i$.

The third fact we will use is the following.
Fact 3: If $U \subseteq T_\mu$ and $\underline{x} \in R_\mu \setminus U^*(\underline{\varepsilon})$ then $\underline{x} \notin Pref(U^*(\underline{\varepsilon}))$.
To prove this fact assume that $\underline{x} \in Pref(U^*(\underline{\varepsilon}))$. Let $\underline{x} \doteq \underline{A}_1 \ldots \underline{A}_k$, $\underline{y} \in U^*(\underline{\varepsilon})$ and $\underline{x} \in Pref(\underline{y})$. From Fact 2 it follows that $\underline{y} \doteq \underline{A}_1 \ldots \underline{A}_k \underline{A}_{k+1} \ldots \underline{A}_m$ where $m > k$. Since by Lemma 2.32 (1) T_μ is a tree rooted at $\underline{\varepsilon}$, and $Dom(T_\mu) = Scope_\mu = T_\mu^*(\underline{\varepsilon})$, then: $\underline{x} \doteq \underline{A}_1 \ldots \underline{A}_k \in T_\mu^*(\underline{\varepsilon})$, and $\underline{y} \doteq \underline{A}_1 \ldots \underline{A}_k \underline{A}_{k+1} \ldots \underline{A}_m \in U^*(\underline{\varepsilon}) \subseteq T_\mu^*(\underline{\varepsilon})$, which means that $\underline{x} \in (T_\mu^{-1})^*(\underline{y})$. But since T_μ is a directed tree rooted at $\underline{\varepsilon}$, then $U \subseteq T_\mu$ and $\underline{y} \in U^*(\underline{\varepsilon})$ implies that $(T_\mu^{-1})^*(\underline{y}) = (U^{-1})^*(\underline{y})$, i.e., $\underline{x} \in U^*(\underline{\varepsilon})$, which yields a contradiction and completes the proof of Fact 3.

We can now start to prove that OPT_μ is an optimal strategy. We first prove that OPT_μ is a minimal semi-strategy. Clearly, OPT_μ satisfies (1) and (2) of the definition of strategy. Let $U \subset OPT_\mu$ and $U \neq OPT_\mu$. By Lemma 2.32 (1) and $R_\mu \neq \emptyset$, the graph of OPT_μ is also a directed tree. Moreover, OPT_μ is the least subrelation W of T_μ such that $R_\mu \subseteq W^*(\underline{\varepsilon})$. Thus there is $\underline{x} \in R_\mu$ such that $\underline{x} \notin U^*(\underline{\varepsilon})$. Consequently, by Fact 3, we obtain $Pref(U^*(\underline{\varepsilon})) \neq Pref(OPT_\mu^*(\underline{\varepsilon}))$. Hence OPT_μ is a minimal semi-startegy. Thus, by Lemma 2.30, OPT_μ is a strategy.

Suppose now that $\underline{x} \in Scope_\mu \setminus \{\underline{\varepsilon}\}$. Let $Z_{\underline{x}} = \{\underline{y} \in Scope_\mu | \kappa(\underline{x}) = \kappa(\underline{y}) \wedge \underline{x} \in Pref(\underline{y})\}$. Clearly $Z_{\underline{x}}$ is a non-empty finite set. Consider $Z_{\underline{x}} \cap R_\mu$, and assume $Z_{\underline{x}} \cap R_\mu = \emptyset$. Thus, $(\forall \underline{y} \in Z_{\underline{x}})(\exists a \in Ev)\underline{ya} \in Scope_\mu \wedge \kappa(\underline{ya}) = \kappa(\underline{y})$. But this implies that $Z_{\underline{x}}$ is an infinite set, a contradiction, hence $Z_{\underline{x}} \cap R_\mu \neq \emptyset$. Let $\underline{z} \in Z_{\underline{x}} \cap R_\mu$. Then $\underline{x} \in Pref(\underline{z})$ and $\underline{z} \in R_\mu$, i.e., $\underline{x} \in Pref(R_\mu)$. Consequently, $Scope_\mu \subseteq Pref(R_\mu) \subseteq Scope_{OPT_\mu}$, i.e., $Scope_{OPT_\mu} = Scope_\mu$ which means that $OPT_\mu \in \mu$.

What is more, by Lemma 2.32 (2) and Fact 3, OPT_μ is the only minimal strategy included in T_μ which has the scope $Scope_\mu$. Therefore, to prove that OPT_μ is optimal it suffices to show that if $W \in \mu$ is a minimal strategy such that $W \setminus T_\mu \neq \emptyset$ then OPT_μ is more efficient than W. From Fact 1 it follows that $\delta_{OPT_\mu}(\underline{x}) = \kappa(\underline{x}) \leq \delta_W(\underline{x})$ for every $\underline{x} \in Scope_\mu \setminus \{\underline{\varepsilon}\}$. Hence it is enough to demonstrate that $\kappa(\underline{x}) < \delta_W(\underline{x})$ for at least one $\underline{x} \in Scope_\mu$. From $W \setminus T_\mu \neq \emptyset$ it follows that there is a sequence of vector sequences $\underline{x}_0, \underline{x}_1, \ldots, \underline{x}_k$ such that: $k \geq 2$ and $\underline{x}_0 = \underline{\varepsilon}$, $(\underline{x}_{i-1}, \underline{x}_i) \in W \cap T_\mu$ for $i = 1, \ldots, k-1$ and $(\underline{x}_{k-1}, \underline{x}_k) \in W \setminus T_\mu$. Let $A_i \in Ind$ satisfy $\underline{x}_i = \underline{x}_{i-1}\underline{A_i}$ for all $i = 1, \ldots, k$. Clearly $\underline{x}_{k-1} \doteq \underline{A}_1 \ldots \underline{A}_{k-1}$ since $\underline{x}_{k-1} \in T_\mu^*(\underline{\varepsilon})$. Now, since $(\underline{x}_{k-1}, \underline{x}_k) \notin T_\mu$, there is $a \in A_k$ such that $(\forall b \in A_{k-1})(a, b) \in ind$. Consider $\underline{y} = \underline{x}_{k-1}\underline{a}$. Clearly $\underline{y} \in Pref(\underline{x}_k)$, i.e., $\underline{y} \in Scope_\mu$. From Fact 1, we have that $\kappa(\underline{y}) \leq k-1$. Consequently, $\kappa(\underline{y}) < k \leq \delta_W(\underline{y})$, since $dist_W(\underline{x}_k) = k$, $\underline{y} \in Pref(\underline{x}_k)$ and $\underline{y} \notin Pref(pred_W(\underline{x}_k)) = Pref(\underline{x}_{k-1})$, which completes the proof. □

As an immediate consequence of Theorem 2.34 we obtain:

Corollary 2.19 *For every strategy W there is the optimal strategy U such that $W \approx U$.* □

Referring to the example from Fig. 2.51, OPT is the optimal strategy such that $OPT \approx CON$.

The result below shows that OPT_μ can be thought of as a natural generalization of the concurrent semantics.

Theorem 2.35 (Optimal and Maximally Concurrent Strategies) *If $MAX \in \mu$ then $OPT_\mu = MAX$.*

Proof. If $MAX \in \mu$ then $R_\mu = MAX^*(\underline{\varepsilon}) \setminus \{\underline{\varepsilon}\}$, thus clearly MAX is the least relation W satisfying $W \subseteq T_\mu \wedge R_\mu \subseteq W^*(\underline{\varepsilon})$. □

Although the above theory holds for any equivalence class of $\approx$, the most interesting case seems to be when $CON \in \mu$. Unfortunately, even in this case we have no "local" characterization of OPT_μ, i.e., we cannot express OPT_μ without referring to the whole set $Scope_\mu$. On the other hand, the strategy MAX has such a "local" characterization, it can be defined (and simulated) step by step using the $maxenabled_{CON}$. Therefore in the next subsection we shall ask the question what conditions have to be satisfied in order to guarantee $CON \approx MAX$.

2.5.2 Maximally Concurrent Strategies

In practice, when we analyze the dynamic properties of formal specifications (in any formalism) of real non-sequential systems, we are quite often forced to analyze something like the "reachability graph" of the system. This is usually a very long and complex task, often impossible without the assistance of a kind of Computer-Aided-Design Tool, and even though we have such tools implemented, the task may still be difficult. The reachability graphs defined by maximally concurrent behaviours are much smaller than those defined by the full behaviour, according to practical experience of the authors, by a factor of more than ten. This means that the systems for which the maximally concurrent behaviour determines the full behaviour can be much more easily analyzed, particularly when we cannot use deep theoretical results because of the generality and/or complexity of the system to be analyzed. In the COSY formalism the equivalence between the full semantics and maximally concurrent semantics can be understood as the equivalence $CON \approx MAX$, which, as the example Pr_{25} and Fig. 2.51 show, is not true in general. In this subsection we shall find sufficient and necessary conditions for such equivalence.
From the previous subsection we get the following characterization of the equivalence $CON \approx MAX$.

Corollary 2.20 $CON \approx MAX \Leftrightarrow OPT_\mu = MAX$ *where* $\mu = [CON]_\approx$. □

Unfortunately finding OPT_μ is not an easy task since, as we pointed out in the previous subsection, there is no "local" characterization of OPT_μ, while we have such characterizations for both MAX and CON. In fact, if there were a "local" characterization of OPT_μ, we would be much less interested in the properties of MAX. Thus, we shall try to formulate the conditions for the equivalence $CON \approx MAX$ without any explicit reference to OPT_μ. It turns out that it is easier to formulate the conditions for $CON \not\approx MAX$ than for $CON \approx MAX$. Furthermore, note that we have *always* $Scope_{MAX} \subseteq Scope_{CON}$ (compare the definitions of MAX and CON).

Theorem 2.36 (Concurrent and Maximally Concurrent Strategies I)
The following properties are equivalent:
(1) $CON \not\approx MAX$
(2) $(\exists \underline{x} \in VFS)\underline{x} \stackrel{\bullet}{=} \underline{A}_1 \ldots \underline{A}_k \wedge k \geq 2 \wedge A_{k-1} \notin maxenabled_{CON}(\underline{A}_1 \ldots \underline{A}_{k-2}) \wedge A_k \in maxenabled_{CON}(\underline{A}_1 \ldots \underline{A}_{k-1})$, *where* $\underline{A}_1 \ldots \underline{A}_{k-2} = \underline{\varepsilon}$ *if* $k = 2$.

Proof. ($2 \Rightarrow 1$): Suppose $Scope_{CON} = Scope_{MAX}$ and $\underline{x}$ is as in (2). Then $\underline{x}CON^*\underline{y}$ for some $\underline{y} \in MAX^*(\underline{\varepsilon})$. Let $\underline{y} \stackrel{\bullet}{=} \underline{B}_1 \ldots \underline{B}_m$. From Theorem 2.31 (4), we have $m \geq k$ and $A_i = B_i$ for $i = 1, \ldots, k$. On the other hand, by Theorem 2.31 (2), we have $B_{k-1} \in maxenabled_{CON}(\underline{B}_1 \ldots \underline{B}_{k-2})$, a contradiction

with $A_{k-1} \notin maxenabled_{CON}(\underline{A_1} \dots \underline{A_{k-1}})$.
$(1 \Rightarrow 2)$: Let $k \geq 1$ be a number such that:

$$Z = \{\underline{x} | \underline{x} \in Scope_{CON} \setminus Scope_{MAX} \wedge \kappa(\underline{x}) = k\} \neq \emptyset.$$

Clearly, $(\forall \underline{y} \in Z) length(\underline{y}) \leq card(Ev)^k$. Hence there is $\underline{x}_0 \in Z$ such that $(\forall z \in Z)$ $length(\underline{z}) \leq length(\underline{x}_0)$. Let $\underline{x}_0 \doteq \underline{A_1} \dots \underline{A_k}$ and suppose that $A_k \notin maxenabled_{CON}(\underline{A_1} \dots \underline{A_{k-1}})$. Thus, for some $a \in Ev$, we have $(\forall b \in A_k)(a,b) \in ind$ and $\underline{y} = \underline{A_1} \dots \underline{A_k}\underline{a} \in Scope_{CON}$. Since $\underline{y} \in Z$ (see the proof of Theorem 2.2 Cases 1 and 2) and $length(\underline{x}_0) < length(\underline{y})$, we get a contradiction so that, necessarily, $A_k \in maxenabled_{CON}(\underline{A_1} \dots \underline{A_{k-1}})$. Because $\underline{x}_0 \notin Scope_{MAX}$, we have $A_i \notin maxenabled_{CON}(\underline{A_1} \dots \underline{A_{i-1}})$ for some $1 \leq i < k$. Thus, $k \geq 2$ and there exists a $j < k$ such that $A_j \notin maxenabled_{CON}(\underline{A_1} \dots \underline{A_{j-1}})$ and $A_{j+1} \in maxenabled_{CON}(\underline{A_1} \dots \underline{A_j})$. Hence $\underline{x} = \underline{A_1} \dots \underline{A_{j+1}}$ satisfies (2). □

In the example Pr_{25}, for instance, $\underline{x} = \underline{bc}$, $A_1 = \{b\}$, $A_2 = \{c\}$ satisfy the conditions of Theorem 2.36 (2). We have: $\underline{bc} \doteq \underline{\{b\}\{c\}}$, $A_1 = \{b\} \notin maxenabled_{CON}(\underline{\varepsilon}) = \{\{a,b\}\}$, and $A_2 \in maxenabled_{CON}(\underline{\{b\}}) = maxenabled_{CON}(\underline{b}) = \{\{a\},\{c\}\}$. The above theorem characterizes the inequivalence $CON \not\approx MAX$ using the properties of some $\underline{x} \in VFS = CON^*(\underline{\varepsilon})$. We shall now formulate a theorem which characterizes $CON \not\approx MAX$ by an element of $MAX^*(\underline{\varepsilon})$.

Theorem 2.37 (Concurrent and Maximally Concurrent Strategies II)
The following properties are equivalent:
(1) $CON \not\approx MAX$
(2) there are $\underline{x} \in MAX^*(\underline{\varepsilon})$, $\underline{xA_1} \dots \underline{A_k} \in Scope_{MAX}$ *and* $A_1, \dots, A_k \in Ind$, *where* $k \geq 2$, *such that*

(a) $\underline{y} = \underline{xA_1} \dots \underline{A_k} \in VFS \wedge \underline{x}\,\underline{A_1} \dots \underline{A_{k-1}} \in Scope_{MAX}$

(b) $(\forall a \in Ev)\underline{ya} \in VFS \Rightarrow (\exists b \in A_{k-1} \cup A_k)(a,b) \notin ind$

(c) $< A_1, \dots, A_k >$ *is the left-normal form of* $\underline{z} = \underline{A_1} \dots \underline{A_k}$,
i.e., $\underline{z} \doteq \underline{A_1} \dots \underline{A_k}$.

(d) $(\exists b \in Ev)\underline{xb} \in VFS \wedge (\forall c \in A_1 \cup \dots \cup A_{k-1})(b,c) \in ind.$ □

As one can see the formulation is in this case much more complicated. The proof is also quite complicated and is presented in Appendix E. In the case of Pr_{25} and Fig. 2.51, for example $\underline{x} = \underline{abd}$, $A_1 = \{b\}$ and $A_2 = \{c\}$ satisfy the condition (2) of Theorem 2.37. Here we have: $\underline{y} = \underline{abdbc} \in VFS$ (Condition (2)(a)); $enabled_{CON}(\underline{y}) = \{\{b\}\}$, thus, $\underline{ya'} \in VFS \wedge \underline{a'} \in Ev \Rightarrow a' = b$, $A_1 \cup A_2 = \{b,c\}$ and for instance $(b,c) \notin ind$ (Condition (2)(b)); $\underline{A_1}\underline{A_2} = \underline{bc} \doteq \underline{\{b\}\{c\}} = \underline{A_1}\underline{A_2}$

(Condition (2)(c)), $\underline{xa} = \underline{abda} \in VFS$, $A_1 = \{b\}$ and $(b, a) \in ind$ (Condition (2)(d)).

Theorems 2.36 and 2.37 give us both the necessary and sufficient conditions for the equivalence of the full semantics and the maximally concurrent semantics, but the criteria they provide are of dynamic nature, i.e., they depend on the existence of some type of history, and they are generally rather difficult to verify. For application purposes we are often satisfied by sufficient conditions for the equivalence $CON \approx MAX$ (i.e., necessary conditions for $CON \not\approx MAX$), but we would prefer the conditions to be relatively easy to check. In particular, it would be useful to have some structural criteria, i.e., depending on the static description of the system, instead of the dynamic ones like those of Theorems 2.36 and 2.37. The following theorem gives a basis for defining such static criteria.

Theorem 2.38 *If $CON \not\approx MAX$, then there are $\underline{x} \in Scope_{MAX}$, $a, b, c \in Ev$ satisfying:*
(1) $\underline{xab} \in VFS, \underline{xac} \in VFS$
(2) $a \neq b, (a, b) \notin ind, (b, c) \notin ind, (a, c) \in ind$. □

The proof of Theorem 2.38 is based on the proof of Theorem 2.37 and can be found in Appendix E. For the path program Pr_{25} and Fig. 2.51, $\underline{\varepsilon}, b, c, a$ (in this order) satisfy the right-hand side of the implication in Theorem 2.38. The reverse of Theorem 2.38 is not true. To prove this, let us consider the following example:

$$Pr_{27} = \begin{cases} P(1): & \underline{path}\ a; b; d\ \underline{end} \\ P(2): & \underline{path}\ c, (b; d)\ \underline{end} \\ P(3): & \underline{path}\ c, e; d\ \underline{end} \end{cases}$$

In this case we have:
$MAX^*(\underline{\varepsilon}) = \{\underline{acbd}, \underline{aebd}\}^*$, $VFS = \{\underline{acbd}, \underline{aebd}\}^*\{\underline{\varepsilon}, \underline{a}, \underline{ab}, \underline{aeb}, \underline{c}, \underline{ac}, \underline{acb}, \underline{e}, \underline{ae}\}$, so $Scope_{CON} = VFS = Scope_{MAX} = Pref(MAX^*(\underline{\varepsilon}))$. On the other hand, $\underline{\varepsilon} \in VFS$, $\underline{\varepsilon ab} \in VFS$, $\underline{\varepsilon ac} \in VFS$, $(a, b) \notin ind$, $(b, c) \notin ind$, $(a, c) \in ind$, thus $\underline{\varepsilon}, a, b, c$ satisfy the right-hand side of the implication in Theorem 2.38.
From Theorem 2.38 it follows that if $CON \approx MAX$, then Ev consists of at least three elements. The example Pr_{25} has $Ev(Pr_{25}) = \{a, b, c, d\}$ and it is not the "smallest" one. Here is probably the simplest one:

$$Pr_{28} = \begin{cases} P(1): & \underline{path}\ a, b\ \underline{end} \\ P(2): & \underline{path}\ c, b\ \underline{end} \end{cases}$$

Note that here we have, $\underline{cb} \in Scope_{CON} \setminus Scope_{MAX}$.
Let us analyze the right-hand side of the implication from Theorem 2.38. We have $\underline{xab} \in VFS$ and $(a, b) \notin ind$, which means that there is j such that $x_j ab = h_j(x)ab \in FS_j$. We have also that $\underline{xab} \in VFS$, $\underline{xac} \in VFS$ and $(b, c) \notin$

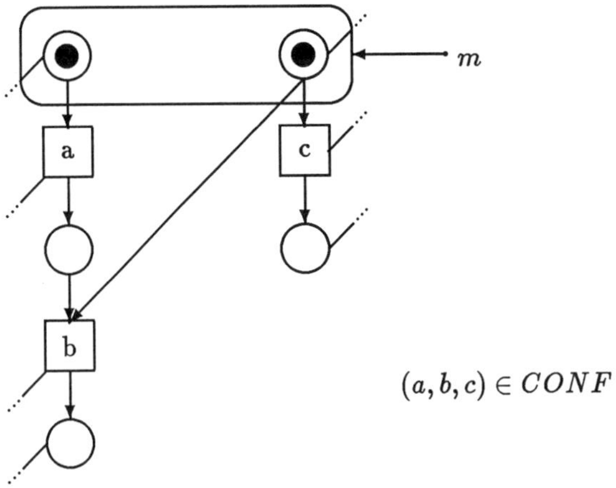

Figure 2.56. "Confusion". Dotted lines denote context, m is a marking reachable from initial marking

ind which implies the existence of i such that $y_i b = h_i(xa)b \in FS_i$ and $y_i c = h_i(xa)c \in FS_i$. In terms of Petri nets associated with the path program $Pr = P_1 \ldots P_n$, such a situation is called **confusion** and is depicted in Fig. 2.56.
The above leads us to the following definitions. Let $CONF \subseteq Ev \times Ev \times Ev$ be the following relation:

$$(a,b,c) \in CONF :\Leftrightarrow ((\exists \underline{x} \in VFS)\underline{xab} \in VFS \wedge \underline{xac} \in VFS) \wedge$$

$$a \neq b \wedge (a,b) \notin ind \wedge (b,c) \notin ind \wedge (a,c) \in ind.$$

The name $CONF$ stands for "confusion".
A path program $Pr = P_1 \ldots P_n$ is said to be **confusion-free** if and only if $CONF = \emptyset$.
As an immediate consequence of Theorem 2.38 we obtain the following result.

Corollary 2.21 *If Pr is confusion-free then $CON \approx MAX$.* □

The reciprocal of Corollary 2.21 is not true since the reciprocal of Theorem 2.38 is not true either, as the path program Pr_{27} shows.
The particular shape of the relation $CONF$ gives rise to a whole family of sufficient conditions for the equivalence $CON \approx MAX$.

Define the following relations on $Ev \times Ev$:

- $(a,b) \in pre :\Leftrightarrow a \neq b \wedge (\exists i)(\exists x \in FS_i)xab \in FS_i$
- $(a,b) \in \underline{pre} :\Leftrightarrow a \neq b \wedge (a,b) \notin ind \wedge (\exists \underline{x} \in VFS)\underline{xab} \in VFS$

- $(a,b) \in pre_1 :\Leftrightarrow a \neq b \wedge (a,b) \notin ind$
  $\wedge (\forall i \in i(a) \cap i(b))(\exists x \in FS_i) xab \in FS_i$
- $(a,b) \in exc :\Leftrightarrow a \neq b \wedge (\exists i)(\exists x \in FS_i) xa \in FS_i \wedge xb \in FS_i$
- $(a,b) \in \underline{exc} :\Leftrightarrow a \neq b \wedge (a,b) \notin ind \wedge (\exists \underline{x} \in VFS) \underline{xa} \in VFS \wedge \underline{xb} \in VFS$
- $(a,b) \in exc_1 :\Leftrightarrow a \neq b \wedge (a,b) \notin ind$
  $\wedge (\forall i \in i(a) \cap i(b))(\exists x \in FS_i) xa \in FS_i \wedge xb \in FS_i$
- $(a,b) \in con :\Leftrightarrow (a,b) \in ind \wedge (\exists \underline{x} \in VFS) \underline{xab} \in VFS.$

Corollary 2.22

(1) $\underline{pre} \subseteq pre_1 \subseteq pre.$

(2) $\underline{exc} \subseteq exc_1 \subseteq exc.$

(3) $con \subseteq ind.$ □

In terms of Petri net associated with the path program $Pr = P_1 \ldots P_n$, the relations $pre, \underline{pre}, exc, \underline{exc}$, and con capture the situations shown in Fig. 2.57. Note that there are cases where $(a,b) \in pre \cap exc$, as for instance for the path program Pr_{28}. The important thing about these relations is that pre, pre_1, exc, exc_1 and ind can be derived by analyzing $Pr = P_1 \ldots P_n$ on the *syntax level*, i.e., by analyzing $P_1, \ldots, P_n$ treated as strings of symbols, or by analyzing the graph of the Petri net associated with Pr.
Let $PDT_0, PDT_1, PDT_2, PDT_3 \subseteq Ev \times Ev \times Ev$ be the following relations:

- $(a,b,c) \in PDT_0 :\Leftrightarrow (a,b) \in pre \wedge (b,c) \in exc \wedge (a,c) \in ind$
- $(a,b,c) \in PDT_1 :\Leftrightarrow (a,b) \in pre \wedge (b,c) \in exc \wedge (a,c) \in con$
- $(a,b,c) \in PDT_2 :\Leftrightarrow (a,b) \in \underline{pre} \wedge (b,c) \in \underline{exc} \wedge (a,c) \in con$
- $(a,b,c) \in PDT_3 :\Leftrightarrow (a,b) \in pre_1 \wedge (b,c) \in exc_1 \wedge (a,c) \in ind$

The name PDT is an abbreviation for Potentially Dangerous Triple.

Corollary 2.23

$CONF \subseteq PDT_2 \subseteq PDT_i \subseteq PDT_0$, *for* $i = 1, 3$. □

As an immediate consequence of Corollary 2.22 and Theorem 2.38 we obtain the following useful result.

Theorem 2.39

$PDT_0 = \emptyset \Rightarrow PDT_i = \emptyset \Rightarrow PDT_2 = \emptyset \Rightarrow$
$CONF = \emptyset \Rightarrow CON \approx MAX$, *for* $i = 1, 3$. □

Note that the verification of the emptiness of PDT_0 and PDT_3 can be done on the *syntax level only*. In general, no implication from Theorem 2.39 can be replaced by equivalence, although it may happen that for some subclass of path programs we have some of the equivalences hold.
The next (and last) family of sufficient conditions for the equivalence of the full semantics and the maximally concurrent semantics is based on the observation that some special forms of the relations exc, exc_1 and $\underline{exc}$ guarantee the emptiness of suitable PDT's.

Theorem 2.40
(1) $((\forall a,b \in Ev)(a,b) \in exc \Rightarrow i(a) = i(b)) \Rightarrow PDT_0 = \emptyset.$
(2) $((\forall a,b \in Ev)(a,b) \in exc_1 \Rightarrow i(a) = i(b)) \Rightarrow PDT_3 = \emptyset.$
(3) $((\forall a,b \in Ev)(a,b) \in \underline{exc} \Rightarrow i(a) = i(b)) \Rightarrow PDT_2 = \emptyset.$

Proof. (1) Suppose $(\forall a,b \in Ev)(a,b) \in exc \Rightarrow i(a) = i(b)$ and $(a,b,c) \in PDT_0$. This means that $(a,b) \in pre, (b,c) \in exc, (a,c) \in ind$. But $(a,b) \in pre \Rightarrow i(a) \cap i(b) \neq \emptyset$, $(a,c) \in ind \Rightarrow i(a) \cap i(c) = \emptyset$. On the other hand, $(b,c) \in pre \Rightarrow i(b) = i(c)$, a contradiction. For (2) and (3), we proceed similarly. □

The full relationships among the various sufficient conditions for the equivalence $CON \approx MAX$ is presented in Fig. 2.58.

2.6 Semantics of COSY with Priorities

In non-sequential systems, **priority** denotes the order of preference in which events in conflict obtain service. It may be based upon the nature of the service being requested or the importance of events. Consider a simple system which copies values from a file x to a file y and which has an interrupt handler. The interrupt handler registers an error message from somewhere, halts the whole system and then restarts it again. The system might be specified as:

$$Pr_{29} = \begin{cases} P(1): & \underline{path}\ readx; writey\ \underline{end} \\ P(2): & \underline{path}\ readx, (getmessage; stopsystem; startsystem)\ \underline{end} \\ P(3): & \underline{path}\ writey, (getmessage; stopsystem; startsystem)\ \underline{end} \end{cases}$$

However the above specification has some serious disadvantages. We would like our system to have the following properties:

(a) *getmessage* should be able to occur concurrently with *readx* and concurrently with *writey*, so the conflict should be rather between *readx* and *stopsystem*, and between *writey* and *stopsystem*,

(b) the sequence "*getmessage*; *stopsystem*; *startsystem*" should be atomic, i.e. incapable of being interrupted,

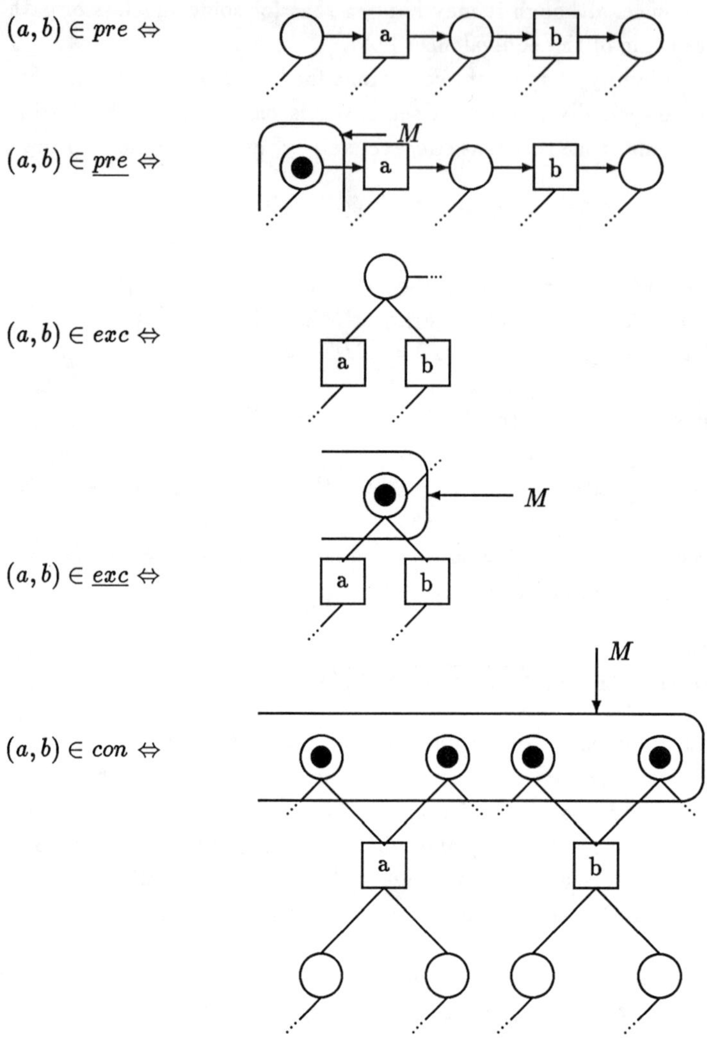

Figure 2.57. Interpretation of $pre, \underline{pre}, exc, \underline{exc}$ and con in terms of Petri nets. Dotted lines denote context and M denotes a marking reachable from the initial marking

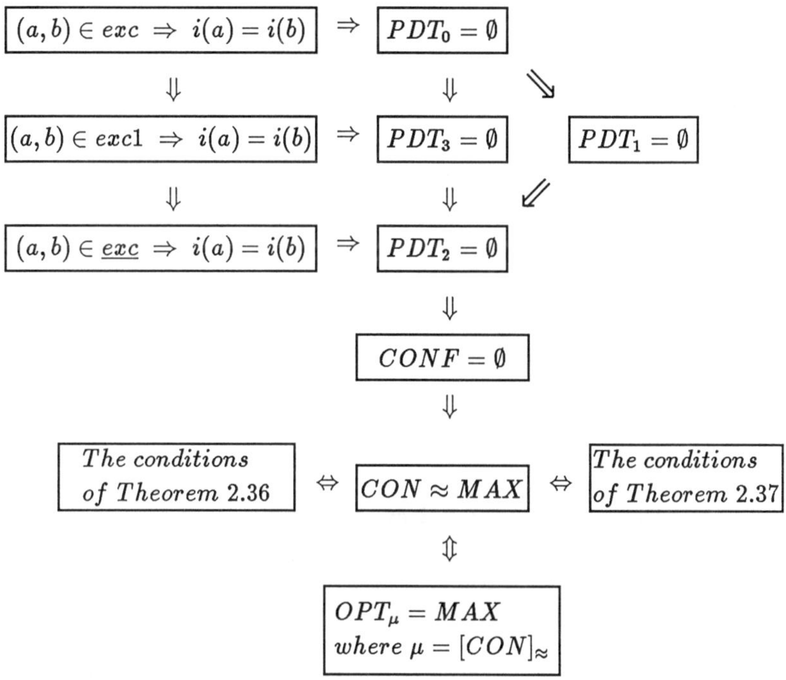

Figure 2.58. Conditions for the equivalence $CON \approx MAX$

(c) the interrupt handler should have priority over other system events.

Pr_{29} satisfies neither (a) nor (c), it satisfies only (b). The path program Pr_{30} below satisfies (a) but neither (b) nor (c), and it is rather incorrect since after *getmessage* it allows *readx*.

$$Pr_{30} = \begin{cases} P(1): & \underline{path}\ readx; writey\ \underline{end} \\ P(2): & \underline{path}\ readx, (stopsystem; startsystem)\ \underline{end} \\ P(3): & \underline{path}\ writey, (stopsystem; startsystem)\ \underline{end} \\ P(4): & \underline{path}\ getmessage; stopsystem; startsystem\ \underline{end} \end{cases}$$

Note that in fact it is *impossible* to find a COSY specification satisfying both (a) and (b). The reason is simple: if for instance *readx* and *getmessage* can occur concurrently after some $\underline{x} \in VFS(Pr_{30})$, then clearly $\{readx, getmessage\} \subseteq enabled(\underline{x})$, which means that $readx \in enabled(\underline{x.getmessage})$, i.e. *readx* is enabled after *getmessage*. This follows from the standard interpretation of partial orders representing behaviours: no order, i.e. concurrency, means that the events concerned may occur simultaneously or in any order one after the other.
Condition (c) is clearly beyond the descriptive power of the basic COSY notation considered up to this point. Thus, to meet the requirements (a-c) we need to extend the notation.
Let "$<$" (or "$>$") denote a new separator between events. This operator may be used in a path whenever a comma is used, and $a > b$ means that *in case of conflict between events a and b, this conflict is always resolved in favour of a.* We may use this construct in Pr_{30} to obtain the following solution:

$$Pr_{31} = \begin{cases} P(1): & \underline{path}\ readx; writey\ \underline{end} \\ P(2): & \underline{path}\ readx < (stopsystem; startsystem)\ \underline{end} \\ P(3): & \underline{path}\ writey < (stopsystem; startsystem)\ \underline{end} \\ P(4): & \underline{path}\ getmessage; stopsystem; startsystem\ \underline{end} \end{cases}$$

One may check, using the intuitive meaning of "$<$", that Pr_{31} satisfies (a-c), i.e. *getmessage* can be performed concurrently with *readx* or *writey*, the sequence "*getmessage*; *stopsystem*; *startsystem*" can be regarded as atomic, and *stopsystem* has priority over *readx* and *writey*. Nevertheless, the formal meaning of Pr_{31} is not clear. We *cannot use vector sequences* because of the following problem. The sequence *readx.getmessage* is allowed and the simultaneous occurrence of *readx* and *getmessage* is also allowed, so $\underline{readx.getmessage}$ is a possible behaviour. On the other hand, after *getmessage* alone, *stopsystem* must occur since the conflict between *readx* and *stopsystem* must be resolved in favour of *stopsystem*. Thus, $\underline{getmessage.readx}$ is not a permitted behaviour. The problem is that $\underline{readx.getmessage} = \underline{getmessage.readx} = (readx, readx, \varepsilon, getmessage)$! To analyze this situation in detail, let us consider an even simpler system, namely a system which can only read from file x and be interrupted. Such a system can be specified by:

$$Pr_{32} = \begin{cases} P(1): & \underline{path}\ readx < stopandrestart\ \underline{end} \\ P(2): & \underline{path}\ getmessage; stopandrestart\ \underline{end} \end{cases}$$

which represents the essence of the problematic part of Pr_{31}.
In this case, using the *intuitive* meaning of "<", at the outset events *getmessage* and *readx* may occur, either both simultaneously or only one of them. After an occurrence of *getmessage* alone, only the event *stopandrestart* may occur, because there is a conflict between *readx* and *stopandrestart* which has to be resolved in favour of *stopandrestart*. Also, after simultaneous occurrence of *readx* and *getmessage* only *stopandrestart* may occur, but after an occurrence of *readx* alone, *getmessage* and *readx* may again occur simultaneously or one of them. Thus the sequence *readx.getmessage* and the simultaneous occurrence of *readx* and *getmessage* are possible behaviours of Pr_{32}, but the sequence *getmessage.readx* is not. This is a *desired behaviour*, however it *cannot be modelled by vector sequences* (since $\underline{readx.getmessage} = \underline{getmessage.readx} = (readx, getmessage)$!) and it *cannot be interpreted as a partial order* (if a history were a partial causal order, a possibility of simultaneous occurences of *readx* and *getmessage* would imply that both sequences *readx.getmessage* and *getmessage.readx* are possible).
If we forget about priority, the Petri net corresponding to Pr_{32} represents the situation which is called **confusion** in net theory, and which we already met in the analysis of the relationship between the full and maximally concurrent execution semantics (see Fig. 2.56 with $a = getmessage, b = stopandrestart, c = readx$). However structures like confusion are *not the only cause of "badness" in systems with priorities*. Let us consider the following example:

$$Pr_{33} = \begin{cases} P(1): & \underline{path}\ a, d; e\ \underline{end} \\ P(2): & \underline{path}\ b, d; e\ \underline{end} \\ P(3): & \underline{path}\ c < d; e\ \underline{end} \end{cases}$$

One may easily check that here the sequence abc is a permitted behaviour, but cab is not since c cannot occur as the first event (because of "$c < d$"). Since $\underline{abc} = \underline{cab} = (a, b, c)$, vector sequences cannot be used to model this behaviour. What about partial orders? The vector sequences represent only some kinds of partial orders, so the impossibility of representing system behaviours by vector sequences might not necessarily imply that the same holds for partial orders. So, what kind of partial order would correspond to the history consisting of events a, b, c, each ocurring only once (so we may identify a, b, c with their occurrences $a^{(1)}, b^{(1)}, c^{(1)}$) such that a and b may occur simultaneously, followed by an occurrence of c, so there should be no order between a and b. But a may also occur alone first, and next b and c in one step, or first b alone and next a and c in one step. This means there should be no order between b and c, and between a and c as well! The only three element partial order which satifies these requirements is a degenerated partial order consisting of one maximal antichain $\{a, b, c\}$ only! On the other hand, the simultaneous ocurrence of a, b and c is not allowed because of the priority constraint $d > c$, so the antichain $\{a, b, c\}$ cannot possibly be an adequate description of this history, a discrepancy.

Essentially the problem discussed above might be solved in two ways. Either one disallows priorities when the partial orders cannot be used, and searches for a characterization of stuctures that permit priorities without problems within the partial order semantics. Or, one searches for other semantics. We choose the second solution here since we feel that the first solution is more an adaptation of the reality to the theory than a real solution. General priority schemes are popular in software practice, partial orders create only a model (although truly general) of non-sequential behaviours, so we prefer to state that partial orders simply do not capture all aspects of reality, and we shall use another model.

Another problem, of a different nature, is that although the operator "$>$" works well in many particular examples, it causes many problems for obtaining general results. First and probably the least important problem is that it does not cope with a conflict induced by the Kleene star followed by semicolon, as in "$(a; b)*; c$". Second, the meaning of patterns like the ones shown below is not entirely clear:

$$Pr_{34} = \begin{cases} P(1): & \underline{path} \ \ldots; a > b; \ldots \ \underline{end} \\ P(2): & \underline{path} \ \ldots; b < a; \ldots \ \underline{end} \\ P(3): & \underline{path} \ \ldots; a, b; \ldots \ \underline{end} \end{cases}$$

or "$a* > c$", "$a > b > c$","$(a, b, c) > (a, d)$", "$a > b, a, b$", and many others, should they be allowed. However, the most difficult problem is that it allows priority to be defined **dynamically** as in $\underline{path}\ a > b; b > a\ \underline{end}$. Although a dynamic priority is a very attractive and powerful tool, e.g. it might help to specify a fairness-like property, the formal theory that supports such a notion has yet to be developed.

To avoid all these problems, we shall use a separate specification of the priority rule and write it before the first path. For instance, Pr_{31}, Pr_{32}, and Pr_{33} would look as follows in the new notation:

$$Pr_{35} = \begin{cases} & \underline{priority}\ readx < stopsystem\ \underline{end} \\ P(1): & \underline{path}\ readx; writey\ \underline{end} \\ P(2): & \underline{path}\ readx, (stopsystem; startsystem)\ \underline{end} \\ P(3): & \underline{path}\ writey, (stopsystem; startsystem)\ \underline{end} \\ P(4): & \underline{path}\ getmessage; stopsystem; startsystem\ \underline{end} \end{cases}$$

$$Pr_{36} = \begin{cases} & \underline{priority}\ readx < stopandrestart\ \underline{end} \\ P(1): & \underline{path}\ readx, stopandrestart\ \underline{end} \\ P(2): & \underline{path}\ getmessage; stopandrestart\ \underline{end} \end{cases}$$

$$Pr_{37} = \begin{cases} & \underline{priority}\ d > c\ \underline{end} \\ P(1): & \underline{path}\ a, d; e\ \underline{end} \\ P(2): & \underline{path}\ b, d; e\ \underline{end} \\ P(3): & \underline{path}\ c, d; e\ \underline{end} \end{cases}$$

This is clearly a kind of **statically** defined priority and the pattern

$$\underline{priority}\ \ldots, a > b, \ldots\ \underline{end}$$

means that **every** conflict between a and b must be resolved in favour of a. Moreover, $a > b$ makes sense only if there is at least one single path involving a local conflict between a and b. We shall also show, that dynamic priorities like $\underline{path}\ a > b; b > a\ \underline{end}$ can be properly simulated by this static approach.

2.6.1 Observations and Multiple Sequences

A partial order representing a history is nothing but a causality relation among event occurrences. No order between two event occurrences means that they are regarded as causaly independent, and such a model does not require the introduction of a global clock. However we can reason about behaviour in another way.

Assume that our history is observable, and a possible observer can *only* detect either **sequential** or **simultaneous** occurrences of events, at any one point in time. Under such an assumption every observation can be adequately represented as a sequence of sets of events, for instance the observation: "first simultaneously a, b and c; next d alone; and next a and e simultaneously", can be adequately represented as a sequence:

$$\{a, b, c\}\{d\}\{a, e\}$$

where $\{a, b, c\}$ denotes a simultaneous occurrence of a, b and c, $\{d\}$ denotes an occurrence of d alone, and $\{a, e\}$ denotes a simultaneous occurrence of a and e. Every such sequence will be called a **multiple sequence**, and every multiple sequence represents a special partial order, for instance $\{a, b, c\}\{d\}\{a, e\}$ represents the order shown in Fig. 2.59.

We leave the definition of the formal relationship between multiple sequences and partial orders to the reader. Partial orders generated by multiple sequences can also be interpreted as observations. These orders have some special properties, namely *the complement of the partial ordering relation* (see Appendix A) *is always an equivalence relation*, with maximal antichains as equivalence classes. Furthermore, the maximal antichains correspond directly to sets of events in

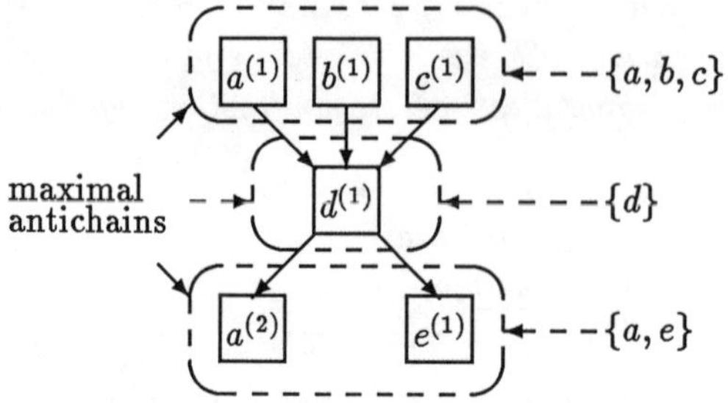

Figure 2.59. Partial order for multiple sequence $\{a, b, c\}\{d\}\{a, e\}$

multiple sequences (see Fig. 2.59). This is a simple consequence of the fact that simultaneity is assumed to be a transitive[7] relation (concurrency clearly is not). So, what is the role of partial orders representing causality relations in the approach based on multiple sequences? From the observer's viewpoint, such a partial order represents an *invariant*, i.e. the maximal order which is satisfied by every observation. In the approach we have used so far in this book, an *invariant represents a history and describes unambiguously all possible observations* (and vice versa).

For instance, a history represented by the partial (causal) order from Fig. 2.60, which may be generated by the specification:

$$Pr_{38} = \begin{cases} P(1): & \underline{path}\ a; c\ \underline{end} \\ P(2): & \underline{path}\ b\ \underline{end} \end{cases}$$

is equivalent to the following set of observations:

$$\{a\}\{b\}\{c\}, \{a\}\{c\}\{b\}, \{b\}\{a\}\{c\}, \{a, b\}\{c\}, \textit{ and } \{a\}\{b, c\}.$$

Partial orders corresponding to these observations are depicted in Fig. 2.60.

For systems with priorities some observations are not allowed, although they can satisfy a given invariant, so, in this case an invariant usually describes more observations than are really possible.

On the other hand, we may use *observations* for representing non-sequential behaviours directly. Fig. 2.61 depicts the behaviour of Pr_{36} in terms of observations (multiple sequences), and Fig. 2.62 does the same for Pr_{37}, and these behaviours seem to be exactly as we would intuitively expect.

Unfortunately the approach described above has some disadvantages. The first disadvantage is that the number of possible observations is usually large, and they seem to have less structure than the rather regularly structured vector sequences. Furthermore, the assumption that an observer can only detect *either*

[7] This assumption is widely accepted, for another approach see [JK91a].

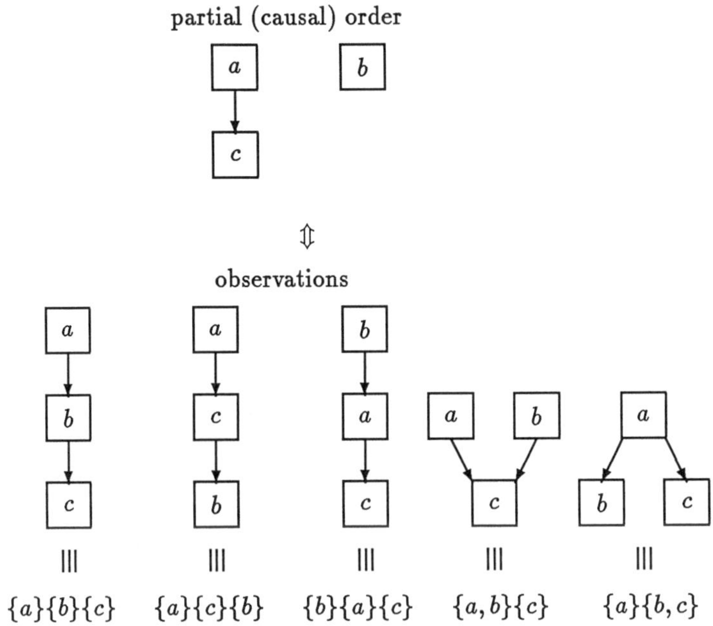

Figure 2.60. Partial (causal) order and its observations

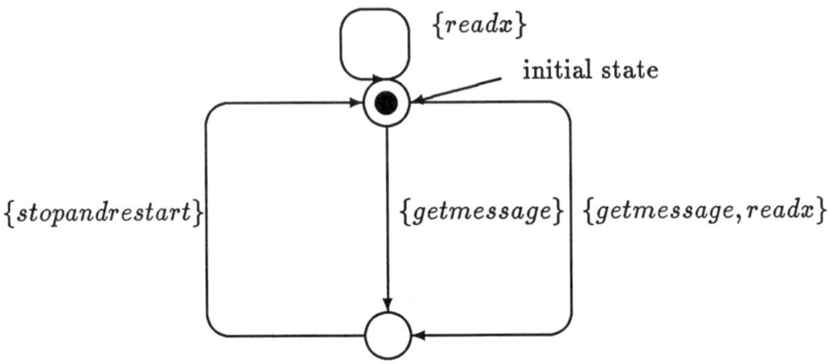

Figure 2.61. Observation behaviour of Pr_{36}

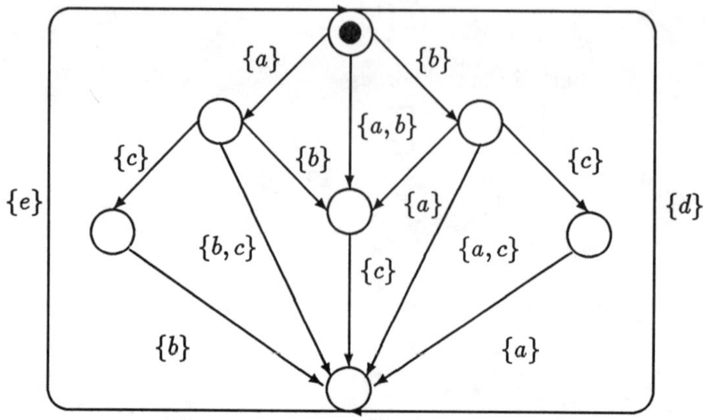

Figure 2.62. Observation behaviour of Pr_{37}

sequence *or* simultaneity disallows some observations. For instance, in the case of the partial order from Fig. 2.60, it disallows the observation: "simultaneously b and the sequence ac", i.e. something like $\{b, ac\}$, although the partial (causal) order does describe such a possibility very well. Giving more power to the observers will result in a much more complex formal theory, but probably will not increase the results significantly. In the approach we take observers can detect only sequence or simultaneity, and one has to remember interpretational and formal limits resulting from this assumption. In some cases, the precise interpretation of "simultaneous occurrences" is not obvious and might be ambiguous. All those problems disappear for partial (causal) orders, so in many respects the partial order approach we have used in previous sections is better, *if applicable.*

2.6.2 MFS Semantics for Priority COSY

Let $Pr = P_1 \ldots P_n$ be a path program (without priority), and let $exc \subseteq Ev \times Ev$ be the relation defined at the end of Sect. 2.5. Recall that:

$$(a,b) \in exc :\Leftrightarrow a \neq b \wedge (\exists i)(\exists x \in FS_i) xa \in FS_i \wedge xb \in FS_i.$$

We shall call the relation exc the **local conflict** relation. Let $<\subseteq Ev \times Ev$ be a relation satisfying for all $a, b \in Ev$:

(1) $a < b \Rightarrow \neg(b < a)$

(2) $a < b \Rightarrow (a,b) \in exc$

The relation $<$ will be called a **priority** (we do *not* assume that $<$ is transitive). By a **priority path program** we then mean a sequence :

$$Pr =< P_1 \ldots P_n$$

where:

$$<= \underline{priority}\ a_1 < b_1, \ldots, a_k < b_k\ \underline{end}$$

is the definition of a priority relation $<$, $<$ is empty if $k = 0$, and $Pr' = P_1 \ldots P_n$ is an ordinary path program.
We shall assume that Pr inherits Ev, ind and Ind from Pr', i.e. $Ev(Pr) = Ev(Pr'), ind(Pr) = ind(Pr'), Ind(Pr) = Ind(Pr')$. We shall later omit Pr and write simply Ev, ind, and Ind.
Let $Pr =< P_1 \ldots P_n$ be a priority path program fixed for the rest of this section. To distinguish between the empty string (denoted by ε) and the empty multiple sequence, the latter is denoted by λ. Let Ind^* denote the family of all multiple sequences over the alphabet Ind including λ, and let, for $i = 1, \ldots, n$:
$\hat{h}_i : Ind^* \longrightarrow Ev^*$ be the following homomorphism:

(i) $\hat{h}_i(\lambda) = \varepsilon$

(ii) $(\forall A \in Ind)\hat{h}_i(A) = \begin{cases} \varepsilon & if \quad A \cap Ev_i = \emptyset \\ a & if \quad A \cap Ev_i = \{a\} \end{cases}$

The correctness of the definition of $\hat{h}_i$ follows from the fact that:

$$(\forall i = 1, \ldots, n)(\forall A \in Ind) card(A \cap Ev_i) \leq 1.$$

Since $\hat{h}_i$ is a homomorphism, $\hat{h}_i(A_1 \ldots A_k) = \hat{h}_i(A_1) \ldots \hat{h}_i(A_k)$ for $A_1, \ldots, A_k \in Ind$. The simple relationship between $\hat{h}_i$ and h_i, $i = 1, \ldots, n$, is characterized by the following corollary.

Corollary 2.24 $(\forall A \in Ind)\hat{h}_i(A) = h_i(\underline{A})$. □

A set $A \in Ind$ is said to be **potentially enabled** (i.e. without taking into account priority constraints) at $t \in Ind^*$ if and only if:

$$(\forall i = 1, \ldots, n)\hat{h}_i(tA) \in FS_i.$$

Let $mpenabled(t)$ denote the family of all sets (multiples) potentially enabled at t.
A set $A \in Ind$ is said to be **enabled** at $t \in Ind^*$ if and only if:

(1) A is potentially enabled at t (i.e. $A \in mpenabled(t)$),

(2) $(\forall a \in A)(\forall B \in mpenabled(t))(\forall b \in B)\neg(a < b)$.

The second condition says that there is no potentially enabled event which has a greater priority than an element of A. The family of all sets (multiples) enabled at

t will be denoted by $menabled(t)$ (or $menabled_{Pr}(t)$ if the name Pr is necessary). For example in the case of Pr_{36} we have:

$$\begin{aligned} &mpenabled(\lambda) = menabled(\lambda) \\ &= \{\{readx\}, \{getmessage\}, \{getmessage, readx\}\}, \\ &mpenabled(\{readx\}) = menabled(\{readx\}) = menabled(\lambda), \\ &mpenabled(\{getmessage\}) = \{\{readx\}, \{stopandrestart\}\}, \\ &menabled(\{getmessage\}) = \{\{stopandrestart\}\}, \\ &\text{so } mpenabled(\{getmessage\}) \neq menabled(\{getmessage\}), \\ &mpenabled(\{getmessage, readx\}) = menabled(\{getmessage, readx\}) \\ &= \{\{stopandrestart\}\}, \text{ and so on.} \end{aligned}$$

Let $R_< \subseteq Ind^* \times Ind^*$ be the following relation:

$$(\forall t, r \in Ind^*) t R_< r :\Leftrightarrow (\exists A \in menabled(t)) r = tA.$$

The relation $R_<$ will be called the **reachability in one step**, and may also be denoted by $R_<^{Pr}$.

The set of **possible observations**, the **multiple firing sequences** of Pr, denoted by MFS (or $MFS(Pr)$ if needed), is defined by:

$$MFS = \{t | \lambda R_<^* t\}.$$

For the path program Pr_{36} from the beginning of this section, we have:
$MFS(Pr_{36}) =$
$Pref((\{\{readx\}\}^* \{\{getmessage, readx\}, \{getmessage\}\} \{\{stopandrestart\}\})^*)$,
and for instance
$\{readx\}\{readx\}\{getmessage, readx\}\{stopandrestart\}\{getmessage\}$
$\in MFS(Pr_{36})$. To improve readability in the sequel, we shall write a regular expression over the alphabet Ind rather than the explicit description of a multiple language. In such a notation
$MFS(Pr_{36}) =$
$Pref((\{readx\}^*(\{getmessage, readx\} \cup \{getmessage\})\{stopandrestart\})^*)$.
Let $\sim$ be the following equivalence relation on MFS:

$$(\forall t, r \in MFS) t \sim r :\Leftrightarrow (\forall i = 1, \ldots, n) \hat{h}_i(t) = \hat{h}_i(r).$$

Let $[t]_\sim$ be the equivalence class of $\sim$ containing $t \in MFS$. The interpretation of the relation $\sim$ is the following. Suppose we have n local observers and the i-th observer, $i = 1, \ldots, n$, can only observe events belonging to Ev_i. Formally, for every $t \in MFS$, the i-th observer observes only $\hat{h}_i(t)$. If $t \sim r$, then t and r are different observations of the same history, and $[t]_\sim$ is the set of all observations of a history. Note that in general $[t]_\sim$ *cannot* be constructed from the $\hat{h}_i(t)$'s alone! For instance in the case of Pr_{36}:

$$[\{readx\}\{getmessage\}]_\sim = [\{readx, getmessage\}]_\sim =$$

$$\{\{readx\}\{getmessage\}, \{readx, getmessage\}\},$$

and

$$\hat{h}_1(\{readx\}\{getmessage\}) = \hat{h}_1(\{readx, getmessage\}) = readx,$$

$$\hat{h}_2(\{readx\}\{getmessage\}) = \hat{h}_2(\{readx, getmessage\}) = getmessage.$$

Note that we also have:

$$\hat{h}_1(\{getmessage\}\{readx\}) = readx,$$

$$\hat{h}_2(\{getmessage\}\{readx\}) = getmessage,$$

although $\{getmessage\}\{readx\} \notin MFS(Pr_{36})$.
The multiple firing sequence semantics may also be applied to ordinary COSY path programs without priorities, simply by assuming that the priority relation is empty. In such a case, the set MFS can be described without the use of the relation $R_<$, using only $FS_1, \ldots, FS_n$, that is, more or less in the style of the definition of VFS.

Theorem 2.41
If $<= \emptyset$ *then* $MFS = \{t \in Ind^* | (\forall i = 1, \ldots, n)\ \hat{h}_i(t) \in FS_i\}$.

Proof. Let us denote $Q = \{t \in Ind^* | (\forall i = 1, \ldots, n) \hat{h}_i(t) \in FS_i\}$. Clearly $\lambda \in MFS \cap Q$. Let $t \neq \lambda$ and $t \in MFS$. Assume that $t = t'A$, where $t' \in MFS \cap Q$. Hence $A \in menabled(t)$, since $<= \emptyset$, $\hat{h}_i(t'A) \in FS_i$ for $i = 1, \ldots, n$. Thus $MFS \subseteq Q$. Let $t \neq \lambda$ and $t \in Q$. Assume that $t = A_1 \ldots A_k$. Because each $A_i \in Ind$, we have $\hat{h}_i(A_1 \ldots A_j) \in FS_i$ for every $j = 1, \ldots, k, i = 1, \ldots, n$. But this means that $\lambda R_< A_1 R_< A_1 A_2 R_< \ldots R_< A_1 \ldots A_{k-1} R_< t$, i.e. $t \in MFS$. □
One immediate consequence of Theorem 2.41 is that if $<= \emptyset$ then for every $t \in MFS$ we can write $[t]_\sim = \{r \in Ind^* | \hat{h}_i(t) = \hat{h}_i(r), i = 1, \ldots, n\}$, i.e. $[t]_\sim$ can be constructed from the $\hat{h}_i(t)$'s alone (in the general case we only have $[t]_\sim = \{r \in MFS | \hat{h}_i(t) = \hat{h}_i(r), i = 1, \ldots, n\} \subseteq \{r \in Ind^* | \hat{h}_i(t) = \hat{h}_i(r), i = 1, \ldots, n\}$.)
If $<\neq \emptyset$, the behaviour of a path program *might not adequately be represented by interleavings*. Consider the following example:

$$Pr_{39} = \begin{cases} & \underline{priority}\ b > c, d > a\ \underline{end} \\ P(1): & \underline{path}\ (a; b, e), d\ \underline{end} \\ P(2): & \underline{path}\ (c; d, e), b\ \underline{end} \end{cases}$$

The multiple firing sequences generated by Pr_{39} are represented in Fig. 2.63, and one may easily check that

$$MFS(Pr_{39}) = Pref((\{a\}\{b\} \cup \{c\}\{d\} \cup \{a, c\}\{e\})^*).$$

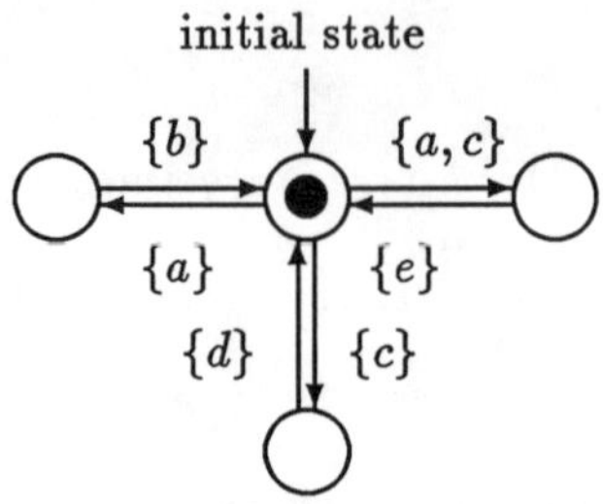

Figure 2.63. Observation behaviour of Pr_{39}

Note that in order for e to occur *a and c must have occurred simultaneously just before that.* Thus, *under interleaving semantics, the event e will never occur, whilst it may occur if the simultaneous occurrence of a and c is allowed.*
The above example suggests that it might be useful to distinguish a class of priority path programs whose behaviours can be simulated by interleavings.
We say that $Pr =< P_1 \ldots P_n$ is **sequentializable** if and only if:

$$(\forall t \in MFS \setminus \{\lambda\})(\exists a_1, \ldots, a_k \in Ev)\{a_1\} \ldots \{a_k\} \sim t.$$

Clearly Pr_{36} and Pr_{33} are sequentializable, but Pr_{39} is not. The theorem below provides a sufficient condition for sequentializability.

Theorem 2.42 (Sequentializability Criterion)
If $(\forall t \in MFS)(\forall A \in menabled(t))(B \subseteq A \wedge \emptyset \neq B \neq A) \Rightarrow A \setminus B \in menabled(tB)$, *then* Pr *is sequentializable.*

Proof. Let $t = A_1 \ldots A_k$, where $A_i \in Ind$ for $i = 1, \ldots, k$, and let $A_1 = \{a_1, \ldots, a_m\}, m \geq 2$. This means that $A_1 \in menabled(\lambda)$, $\{a_1\} \in menabled(\lambda)$, so $\{a_1\} \in MFS$. From the condition of the theorem, since $\{a_1\} \subset A_1$, we have $\{a_2, \ldots, a_m\} \in menabled(\{a_1\})$, in particular $\{a_2\} \in menabled(\{a_1\})$ and $\{a_1\}\{a_2\} \in MFS$. Repeating this reasoning for all a_i, we obtain $\{a_1\} \ldots \{a_m\} \in MFS$. Since $A_1 \in Ind$ then $A_1 \sim \{a_1\} \ldots \{a_m\}$. Repeating the whole reasoning for $A_2, \ldots, A_k$ we get the implication. □

The condition of Theorem 2.42 is not a necessary condition. The path program Pr_{36} is sequentializable, but $\{readx, getmessage\} \in menabled(\lambda)$, while $\{readx\} \notin menabled(\{getmessage\})$.
The multiple firing sequence model of behaviour allows us to speak formally of dynamic properties of a system specified by a priority path program $Pr =< P_1 \ldots P_n$.
We say that $Pr =< P_1 \ldots P_n$ is **deadlock-free** if and only if:

$$(\forall t \in MFS) menabled(t) \neq \emptyset.$$

We say that $Pr =< P_1 \ldots P_n$ is **adequate** if and only if:
$(\forall t \in MFS)(\forall a \in Ev)(\exists r \in R^*_<(t))(\exists A \in Ind) a \in A \wedge A \in menabled(r)$.
Using the obvious fact that: $A \in menabled(t) \Leftrightarrow (\forall B \subseteq A) B \in menabled(t)$, we may define adequacy in a somewhat shorter way, namely, we say that $Pr =< P_1 \ldots P_n$ is adequate if and only if:

$$(\forall t \in MFS)(\forall a \in Ev)(\exists r \in R^*_<(t))\{a\} \in menabled(r).$$

In fact we should prove that for $<= \emptyset$ the above definitions of deadlock-freeness and adequacy coincide with those from Sect. 2.2.3. This is true, but we prove it later for a somewhat more general case. .

Corollary 2.25
(1) Pr is deadlock-free $\Leftrightarrow (\forall t \in MFS)(\exists A \in Ind) tA \in MFS$.
(2) Pr is adequate $\Leftrightarrow (\forall t \in MFS)(\forall a \in Ev)(\exists r \in Ind^*) tr\{a\} \in MFS$. □

2.6.3 Starvation Problem

Consider a system consisting of a single resource involving read and write operations and a set of "reader" processes and "writer" processes which repeatedly use the operations to read from and to write to the resource, respectively. It is required that any number of readers may be concurrently using the resource, but each writer must have exclusive use of it. Also, no writer may jointly use the resource with a reader. Furthermore, no reader should be kept waiting unless a writer is using the resource. The problem specified here is frequently called "the first reader-writer problem". To simplify the problem assume that the number of readers is two and the number of writers is also two. As we shall observe, this assumption does not change the generality of our discussion, yet makes it much simpler. The following path program may be regarded as a possible specification of the "two readers- two writers problem":

$$Pr_{RW1} = \begin{cases} P(1): & \underline{path}\ (readbegin_1; readend_1), write_1, write_2\ \underline{end} \\ P(2): & \underline{path}\ (readbegin_2; readend_2), write_1, write_2\ \underline{end} \\ P(3): & \underline{path}\ readrequest_1; readbegin_1; readend_1\ \underline{end} \\ P(4): & \underline{path}\ readrequest_2; readbegin_2; readend_2\ \underline{end} \\ P(5): & \underline{path}\ writerequest_1; write_1\ \underline{end} \\ P(6): & \underline{path}\ writerequest_2; write_2\ \underline{end} \end{cases}$$

Since every writer has exclusive access to the resource, we can use "$write_i$" instead of the corresponding "$(writebegin_i; writeend_i)$" without any substantial loss of generality. It is easily seen that Pr_{RW1} is adequate. One can prove it formally by reasoning similar to that of Sect. 2.4.3. However, since the conflicts between readers and writers can be resolved arbitrarily, the following is a possible behaviour of Pr_{RW2}:

(a) $readrequest_1, readrequest_2, writerequest_1, writerequest_2$ simultaneously or in any order,

(b) $readbegin_1, readbegin_2$ simultaneously or in any order, i.e. the conflict is resolved in favour of the readers,

(c) $readend_1, readend_2$ simultaneously or in any order,

(d) $readrequest_1, readrequest_2$ simultaneously or in any order.

We may now cycle through (b),(c),(d) indefinitely. The events $write_1, write_2$ never occur. In this case, we would say that the writers are *starving* because of *unfair conflict resolution*.
In general, *starvation occurs when some part of a system is constantly blocked from progressing*. We should point out that a system may involve starvation and yet be adequate; indeed this is the case for Pr_{RW1}.
Suppose we would like to exclude the possibility of starvation of writers by readers. We can require additionally that once a writer is ready to write, he performs his write event as soon as possible. This problem is usually called *the second reader-writer problem*, and a straightforward solution to it seems to be the following:

$$Pr_{RW2} = \begin{cases} & \underline{priority}\ write_1 > readbegin_1, write_1 > readbegin_2, \\ & \quad write_2 > readbegin_1, write_2 > readbegin_2\ \underline{end} \\ P(1): & \underline{path}\ (readbegin_1; readend_1), write_1, write_2\ \underline{end} \\ P(2): & \underline{path}\ (readbegin_2; readend_2), write_1, write_2\ \underline{end} \\ P(3): & \underline{path}\ readrequest_1; readbegin_1; readend_1\ \underline{end} \\ P(4): & \underline{path}\ readrequest_2; readbegin_2; readend_2\ \underline{end} \\ P(5): & \underline{path}\ writerequest_1; write_1\ \underline{end} \\ P(6): & \underline{path}\ writerequest_2; write_2\ \underline{end} \end{cases}$$

Unfortunately, readers can still starve writers. However in this case starvation is not a result of an unfair conflict resolution. Consider the following possible system activity:

(a) $readrequest_1$ followed immediately by $readbegin_1$,

(b) $writerequest_1, writerequest_2$ simultaneously or in any order,

(c) $readrequest_2$,

(d) $readend_1, readbegin_2$ simultaneously,

(e) $readrequest_1$,

(f) $readend_2, readbegin_1$ simultaneously.

We may now cycle through (c),(d),(e),(f) indefinitely. The writers are starving and yet there is never any conflict within the system in the sense that at no time

$Pr'_{RW2} =$

$\underline{priority}\ (write_1, write_2) > (readbegin_1, readbegin_2),$
$(readend_1, readend_2) > (write_1, write_2)\underline{end}$

$P(1): \underline{path}\ readbegin_1, readend_1, write_1, write_2\ \underline{end}$
$P(2): \underline{path}\ readbegin_2, readend_2, write_1, write_2\ \underline{end}$
$P(3): \underline{path}\ readrequest_1; readbegin_1; readend_1\ \underline{end}$
$P(4): \underline{path}\ readrequest_2; readbegin_2; readend_2\ \underline{end}$
$P(5): \underline{path}\ writerequest_1; write_1\ \underline{end}$
$P(6): \underline{path}\ writerequest_2; write_2\ \underline{end}$

Figure 2.64. The program Pr'_{RW2}

is it possible for any writer to become active. Here, the reader processes are "conspiring" to ensure that at no time the resource is free, and this "conspiracy" can be successful despite the fact that writers have higher priority. To prevent such conspiracies we have to assure that no $readbegin_i$ may be performed between $writerequest_j$ and $write_j$, for $i, j = 1, 2$. Note that just replacing the first two paths at Pr_{RW2} by

$$\underline{path}\ readbegin_1, write_1, write_2\ \underline{end}$$
$$\underline{path}\ readbegin_2, write_1, write_2\ \underline{end}$$

does not work properly since it allows for example $write_i$ to be performed *between* $readbegin_j$ and $readend_j$. To simplify the specification of large priority relations we shall write

$$(a_1, \ldots, a_k) > (b_1, \ldots, b_l)$$

to denote

$$(\forall i = 1, \ldots, k)(\forall j = 1, \ldots, l) a_i > b_j.$$

For example, using this notation we shall write:

$$\underline{priority}\ (a, b, c) > (d, e)\ \underline{end}$$

instead of:

$$\underline{priority}\ a > d, a > e, b > d, b > e, c > d, c > e\ \underline{end}.$$

The new specification of a solution to the second reader-writer problem is Pr'_{RW2} in Fig. 2.64.

One may show that Pr'_{RW2} guarantees no starvation of writers by readers. The reason is that now the performance of $writerequest_i$ immediately makes $write_i$ potentially enabled. The priority $write_i > readbegin_j$ guarantees that no $readbegin_j$ can be performed before $write_i$, and the priority $readend_j > write_i$ enforces all enabled $readend_j$ to occur before $write_i$.

$Pr''_{RW2} =$

$\underline{\textit{priority}}\ (write_1, write_2) > (readbegin_1, readbegin_2, skip)$
$(readend_1, readend_2) > (write_1, write_2)\underline{\textit{end}}$

$P(1):\ \underline{\textit{path}}\ readbegin_1, readend_1, write_1, write_2\ \underline{\textit{end}}$
$P(2):\ \underline{\textit{path}}\ readbegin_2, readend_2, write_1, write_2\ \underline{\textit{end}}$
$P(3):\ \underline{\textit{path}}\ readrequest_1; readbegin_1; readend_1\ \underline{\textit{end}}$
$P(4):\ \underline{\textit{path}}\ readrequest_2; readbegin_2; readend_2\ \underline{\textit{end}}$
$P(5):\ \underline{\textit{path}}\ writerequest_1; write_1\ \underline{\textit{end}}$
$P(6):\ \underline{\textit{path}}\ writerequest_2; write_2\ \underline{\textit{end}}$
$P(7):\ \underline{\textit{path}}\ write_1, skip; write_2, skip\ \underline{\textit{end}}$

Figure 2.65. The program Pr''_{RW2}

However, the prevention of starvation of a writer by other writers is still not attained by this solution. We can eliminate such a possibility by adding constraints on the order in which writers may use the resource. This can be done with the aid of "skip" events. Consider the specification Pr''_{RW2} in Fig. 2.65. Suppose $write_1$ has just been activated. If $writerequest_2$ has just been activated, then $write_2$ is ready to write, and the events $write_2$ and $skip$ are in conflict. The presence of priority determines that $write_2$ is performed. If, on the other hand, there has been no $writerequest_2$, then $write_2$ and $skip$ are not in conflict and indeed $skip$ may occur, after which $write_1$ may be allowed, and so on. This clearly prevents the starvation of writers by writers due to unfair conflict resolution. However, a "suicide" of one writer and a "suicide" of $skip$ still cause starvation of the second writer. Consider the following system behaviour:

(a) $writerequest_1$ followed immediately by $write_1$,

(b) $writerequest_1$,

(c) $readrequest_1$ followed immediately by $readbegin_1$ and next by $readend_1$.

We may now cycle through (c) indefinitely. Both $writerequest_2$ and $skip$ are enabled at any time after (a), however, there is no way to force any of them to become active. The first writer performed a $writerequest_1$, but it is starving because of "cooperative suicide" of the other writer and the "skip" event. Furthermore, note that $skip$ is permanently potentially enabled, for instance $\{skip\}\{skip\}\ldots\{skip\}$ is a possible behaviour, as well as $\{readrequest_1\}\{readbegin_1, skip\}\{skip, readend_1\}$, etc. All these seem to be rather undesirable properties. Intuitively $skip$ should be enabled only in case of need and in this case it should be forced to occur. This can be achieved at a cost of a slightly more complex solution, and a possible specification of this solution is shown in Fig. 2.66.

$Pr'''_{RW2} =$

$\underline{priority}\ (write_1, write_2) > (readbegin_1, readbegin_2, skip_1, skip_2),$
$(readend_1, readend_2) > (write_1, write_2, skip_1, skip_2),$
$(skip_1, skip_2) > (readbegin_1, readbegin_2)\ \underline{end}$

$P(1):\ \underline{path}\ readbegin_1, readend_1, write_1, write_2, skip_1, skip_2\ \underline{end}$
$P(2):\ \underline{path}\ readbegin_2, readend_2, write_1, write_2, skip_1, skip_2\ \underline{end}$
$P(3):\ \underline{path}\ readrequest_1; readbegin_1; readend_1\ \underline{end}$
$P(4):\ \underline{path}\ readrequest_2; readbegin_2; readend_2\ \underline{end}$
$P(5):\ \underline{path}\ writerequest_1; skip_1^*; write_1\ \underline{end}$
$P(6):\ \underline{path}\ writerequest_2; skip_2^*; write_2\ \underline{end}$
$P(7):\ \underline{path}\ write_1, skip_2; write_2, skip_1\ \underline{end}$

Figure 2.66. Solution of second reader-writer problem starvation-free for writers

Now the situation is different. The paths P_1, P_2 guarantee that $skip_i$'s cannot be performed simultaneously with $readbegin_j$'s or $readend_j$'s, the priority $readend_j > skip_i$ gurantees that $skip_i$'s cannot occur between $readbegin_j$ and $readend_j$, P_5, P_6 locate occurences of $skip_i$ precisely between $writerequest_i$ and $write_i$. Arguing almost identically as for Pr''_{RW2} we can show that the solution prevents starvation of writers by writers due to unfair conflict resolution. To show that starvation of writers by readers is impossible we can use the same arguments as for Pr'_{RW2}.

We now show that starvation by "cooperative suicide" is also disallowed. Consider the case of $write_1$. An event $write_1$ is starved by a "cooperative suicide" of $writerequest_2$ and $skip_1$ if both $writerequest_2$ and $skip_1$ are permanently enabled, but their performance is not necessary to keep the program running. Suppose that $writerequest_2$ and $skip_1$ are enabled.

By the structure of P_7 we have that if $skip_1$ is enabled, then $write_1$ is not. But now we have the priority $skip_1 > readbegin_i, i = 1,2$, and the paths P_3, P_4 are just conflict-free cycles, so after a few steps the only enabled events are $writerequest_2$ and $skip_1$. Performing $skip_1$ immediately enables $write_1$, while the performance of $writerequest_2$ enables $write_2$ which now, because of the priority $write_2 > (readbegin_i, skip_i), i = 1,2$, is the only enabled event. The performance of $write_2$ also immediately enables $write_1$.

Identical reasoning can be applied for $write_2$, so starvation due to a "cooperative suicide" is also impossible. Thus, we may conclude that Pr'''_{RW2} is starvation-free for writers.

Of course, our reasoning is informal as a formal definition of starvation-freeness has not been given yet, and as the above examples show, starvation-freeness is neither easy to define (formally), nor to achieve.

It is useful to point out that starvation is not always an undesirable property. Let us consider the specifications Pr_{35} and Pr_{36}. A constant stream of error messages will cause conflicts always to be resolved in favour of *stopsystem* for Pr_{35}, or *stopandrestart* for Pr_{36}, thereby starving the rest of the system. However, this could be the case only if the system was constantly malfunctioning in some respect, in which case starvation would be a desirable property.

2.6.4 Infinite MFS and Formal Definition of Starvation

Let $P = \underline{path}\ seq\ \underline{end}$ be a single path.
The set of **infinite behaviours**, or **infinite firing sequences** of P, denoted by IFS (or $IFS(P)$ if needed), is defined as:

$$IFS = \{a_1a_2\ldots a_i\ldots|(\forall i \geq 1)a_i \in Ev(P) \wedge a_1\ldots a_i \in FS\}.$$

Let $Pr =< P_1\ldots P_n$ be a priority path program. The set of **infinite multiple firing sequences** of Pr, denoted by $IMFS$ (or $IMFS(Pr)$ if needed), is defined by:

$$IMFS = \{A_1A_2\ldots A_i\ldots|(\forall i \geq 1)A_i \in Ind \wedge A_1\ldots A_i \in MFS\}.$$

The infinite multiple firing sequence model permits us to define formally the starvation-freeness property for $Pr =< P_1\ldots P_n$, however the formal definition of this property requires the introduction of some new auxiliary notions.
The first of these is the concept of a **simple request system**. Let $a, b \in Ev$. We say that $Pr =< P_1\ldots P_n$ is a **simple request system for** a, and b is the **request of** a **in** Pr, if and only if there is $i, 1 \leq i \leq n$, such that:

(1) $i(b) = \{i\} \wedge ((\forall c \in Ev_i)(b,c) \notin exc)$,

(2) $(\forall x \in \hat{h}_i(MFS))h_{\{a,b\}}(x) \in \{ba\}^*\{b,\varepsilon\}$.

If Pr is a request system for a, and b is the request of a, we shall write $req(a)$ instead of b.
The condition (1) of the definition of a request system says that $b = req(a)$, the request of a, occurs in one path only and it is not involved in any conflict. The condition (2) says that the path containing $b = req(a)$ has a special form, namely, if x is a finite firing sequence of this path, generated by a finite behaviour of Pr, and if we erase from x all event occurrences except those of a and b, the result, if non-empty string, is either $baba\ldots aba$, or $baba\ldots ab$, i.e. $req(a).a.req(a).a.\ldots.req(a).a$, or $req(a).a.req(a).\ldots.req(a)$.
For instance the COSY priority path programs

$$Pr_{RW1}, Pr_{RW2}, Pr'_{RW2}, Pr''_{RW2}, Pr'''_{RW2}$$

are all request systems for: $write_1$,$write_2$,$readbegin_1$,$readbegin_2$ with $writerequest_i = req(write_i)$, and $readrequest_i = req(readbegin_i), i = 1,2$.

The priority path program Pr_{35} is a request system for *stopsystem* with $getmessage = req(stopsystem)$, and Pr_{36} is a request system for *stopandrestart* with $getmessage = req(stopandrestart)$.
The next auxiliary definition is the function called **number of possible enablings**, denoted *npe*, and defined as follows:

$$npe : Ev \times (MFS \cup IMFS) \longrightarrow \{0, 1, \ldots\} \cup \{\infty\}$$

and:

(1) $(\forall b \in Ev) npe(b, \lambda) = \begin{cases} 0 & \text{if } \{b\} \notin menabled(\lambda) \\ 1 & \text{if } \{b\} \in menabled(\lambda) \end{cases}$

(2) $(\forall b \in Ev)(\forall tA \in MFS)$

$$npe(b, tA) = \begin{cases} npe(b,t) & \text{if } \{b\} \notin menabled(tA) \\ npe(b,t) + 1 & \text{if } \{b\} \in menabled(tA) \end{cases}$$

(3) $(\forall b \in Ev)(\forall k \in \{0, 1, \ldots\})(\forall A_1A_2A_3 \ldots \in IMFS)$

$npe(b, A_1A_2A_3 \ldots) = k :\Leftrightarrow (\exists j)(\forall i \geq j) npe(b, A_1 \ldots A_i) = k,$

(4) $(\forall b \in Ev)(\forall A_1A_2A_3 \ldots \in IMFS) npe(b, A_1A_2A_3 \ldots) = \infty :\Leftrightarrow (\forall i)(\exists j) npe(b, A_1 \ldots A_j) = i.$

where $A, A_i \in Ind$, $t \in Ind^*$. The function $npe(b,t)$ describes how many times b was enabled during t.
Now we can formally define the starvation-freeness property for request systems. Let $Pr =< P_1 \ldots P_n$ be a **request system** for $a \in Ev$. We shall say that Pr is **starvation-free** for a if and only if:

(1) Pr is deadlock-free.

(2) $(\forall t \in IMFS) npe(req(a), t) = \infty$.

(3) $(\forall A_1A_2A_3 \ldots \in IMFS)(\forall i)(req(a) \in A_i \Rightarrow (\exists k \geq 1) a \in A_{i+k}),$

where $A_1, A_2, \ldots \in Ind$.
The first condition is obvious. A deadlock implies the starvation of the whole system. If Pr is deadlock-free then its infinite behaviour defines unambiguously its finite behaviour, since in such a case MFS can be defined as the set of all finite prefixes of elements of $IMFS$. The condition (2) says that the number of possible enablings of $req(a)$ is infinite for any infinite multiple firing sequence, so $req(a)$ may be disabled only temporarily. Since $req(a)$ is not involved in any conflict (see the definition of a request system), if Pr is deadlock-free, nothing can prevent $req(a)$ from being active an infinite number of times. The last condition (3) says that any request for a is served after a certain finite amount of time. Hence, the above definition seems to meet the intuitive requirements of starvation-freeness. Unfortunately, unlike in the case of adequacy, we have no

$Pr_{RW2}^{n,m} =$

$\underline{priority}\ (write_1, \ldots, write_m) >$
$(readbegin_1, \ldots, readbegin_n, skip_1, \ldots, skip_m),$
$(readend_1, \ldots, readend_n) > (write_1, \ldots, write_m, skip_1, \ldots, skip_m),$
$(skip_1, \ldots, skip_m) > (readbegin_1, \ldots, readbegin_n)\ \underline{end}$

$P(1):$ $\underline{path}\ readbegin_1, readend_1, write_1, \ldots, write_m,$
$skip_1, \ldots, skip_m\ \underline{end}$

$\ldots$

$P(n):$ $\underline{path}\ readbegin_n, readend_n, write_1, \ldots, write_m,$
$skip_1, \ldots, skip_m\ \underline{end}$

$P(n+1):$ $\underline{path}\ readrequest_1; readbegin_1; readend_1\ \underline{end}$

$\ldots$

$P(n+n):$ $\underline{path}\ readrequest_n; readbegin_n; readend_n\ \underline{end}$

$P(2n+1):$ $\underline{path}\ writerequest_1; skip_1^*; write_1\ \underline{end}$

$\ldots$

$P(2n+m):$ $\underline{path}\ writerequest_m; skip_m^*; write_m\ \underline{end}$

$P(2n+m+1):$ $\underline{path}\ write_1, skip_2, skip_3, \ldots, skip_m; write_2, skip_1, skip_3, \ldots, skip_m;$
$\ldots; write_i, skip_1, \ldots, skip_{i-1}, skip_{i+1}, \ldots, skip_m;$
$\ldots; write_m, skip_1, \ldots, skip_{m-1}\ \underline{end}$

Figure 2.67. Solution of second reader-writer problem starvation-free for writers ($m, n \geq 2$)

formal results concerning the starvation property, so in every case we have to use the definition. To show how this definition can be used, we prove formally that the specification of the second reader-writer problem for n readers and m writers, shown in Fig. 2.67, is starvation-free for all m writers.

First observe that since $readrequest_i$'s and $writerequest_i$'s occur each in one path only, cases where none of them is enabled occurs infrequently, and to prove deadlock-freeness it is sufficient to show that these particular cases do not deadlock the system either.

Suppose that $t \in MFS(Pr_{RW2}^{n,m})$ and $\{a\} \notin menabled(t)$ for $a \in \{readrequest_1, \ldots, readrequest_n, writerequest_1, \ldots, writerequest_m\}$ (we omit $Pr_{RW2}^{n,m}$ later).

This means that for every $i = 1, \ldots, n$, either $\hat{h}_{n+i}(t) = x_i readrequest_i$ or $\hat{h}_{n+i}(t) = x_i readbegin_i$, where $x_i \in FS_{n+i}$. If $\hat{h}_{n+i}(t) = x_i readbegin_i$, then $\{readend_i\} \in menabled(t)$, so t does not deadlock. Suppose that for every $i = 1, \ldots, n$, $\hat{h}_{n+i}(t) = x_i readrequest_i$. Note that in such a case $\{readbegin_1, \ldots, readbegin_n, write_k, skip_1, \ldots, skip_{k-1}, skip_{k+1}, \ldots, skip_m\} \in mpenabled(t)$, where k satisfies $1 \leq k \leq m$ and depends on $\hat{h}_{2n+m+1}(t)$ in the following way. If $\hat{h}_{2n+m+1}(t) \in (Cyc_{2n+m+1})^*$ then $k = 1$, otherwise, if $\hat{h}_{2n+m+1}(t) = xa_1a_2 \ldots a_j$ where $x \in (Cyc_{2n+m+1})^*$, $a_1, \ldots, a_j \in Ev_{2n+m+1}$, and $1 \leq j \leq m-1$, then $k = j+1$. Since $write_k > (readbegin_1, \ldots, readbegin_n, skip_1, \ldots, skip_m)$, we have that $\{write_k\} \in menabled(t)$, i.e. t does not deadlock in this case either. Thus $Pr^{n,m}_{RW2}$ is deadlock-free.

We now prove that for $Pr^{n,m}_{RW2}$ we have the implication (3) $\Rightarrow$ (2), where (2) and (3) are those from the definition of starvation-freeness. First note that since $\{writerequest_i\} \in menabled(\lambda)$, for every $t \in IMFS$, $npe(writerequest_i, t) \geq 1$. Suppose that $t = A_1A_2A_3 \ldots \in IMFS$ and $npe(writerequest_i, t) < \infty$. This means that there is j such that $\{writerequest_i\} \in menabled(A_1 \ldots A_j)$ and $(\forall l \geq 1)\{writerequest_i\} \notin menabled(A_1 \ldots A_{j+l})$. But $writerequest_i$ belongs to Ev_{2n+i} only, and $P_{2n+i} =$ <u>path</u> $writerequest_i; skip_i^*; write_i$ <u>end</u>, and $writerequest_i$ is not involved in the priority constraints, so $writerequest_i \notin A_{j+1} \Rightarrow writerequest_i \in menabled(A_1 \ldots A_{j+1})$, which means that $writerequest_i \in A_{j+1}$. Since $writerequest_i = req(write_i)$, from (3) it follows that there is $k \geq 1$ such that $write_i \in A_{j+k+1}$, which implies that $\{writerequest_i\} \in menabled(A_1 \ldots A_{j+k+1})$, a contradiction. Thus (3) $\Rightarrow$ (2), and we need to prove only (3).

Suppose that there is $t = A_1A_2A_3 \ldots \in IMFS$ such that $writerequest_i \in A_{j_1}$ and for all $j > j_1$, $write_i \notin A_j$. The fact that $writerequest_i \in A_{j_1}$ and $write_i \notin A_{j_1}$ implies that either $write_i$ or $skip_i$ is potentially enabled at every $A_1 \ldots A_j, j \geq j_1$, i.e., $\{write_i\} \in mpenabled(A_1 \ldots A_j) \vee \{skip_i\} \in mpenabled(A_1 \ldots A_j)$, for $j \geq j_1$. Because of the priority constraint $(skip_i, write_i) > readbegin_l, l = 1, \ldots n$, no $readbegin_l$ is ever enabled after A_{j_1}, i.e., $\{readbegin_l\} \notin menabled(A_1 \ldots A_j)$ for $j \geq j_1$, $l = 1, \ldots, n$. This means that $readbegin_l \notin A_j$ for $j > j_1, l = 1, \ldots n$. Hence, due to the structure of the paths $P_{n+1}, \ldots, P_{2n}$, each $readend_l$ and $readrequest_l$, $l = 1, \ldots, n$, might be performed after A_{j_1} *at most* once. Thus, there is $j_2 \geq j_1$ such that $(\forall j \geq j_2) A_j \subseteq \bigcup_{i=1}^{m} \{writrequest_l, skip_l, write_l\}$.

Since t is infinite, there is at least one event occurring an infinite number of times in t, and clearly it has to be either $write_k$ or $skip_k$ or $writerequest_k$ for some k. Suppose that it is $writerequest_k$. But every occurrence of $writerequest_k$ except the first must be preceded by an occurrence of $write_k$, so $write_k$ must also occur an infinite number of times. In order to perform $write_k$ twice we must perform one full cycle of P_{2n+m+1} which means that in particular the conflict "$write_i, skip_1, \ldots, skip_{i-1}, skip_{i+1}, \ldots, skip_m$" must be resolved an infinite number of times for some i. Thus there is A_{j_3}, $j_3 > j_2$ such that $A_{j_3} \cap \{write_i, skip_1, \ldots, skip_{i-1}, skip_{i+1}, \ldots, skip_m\} \neq \emptyset$ and $\hat{h}_{2n+m+1}(A_1 \ldots A_{j_3-1})a \in FS(P_{2n+m+1}) \Rightarrow a \in \{write_i, skip_1, \ldots, skip_{i-1}, skip_{i+1}, \ldots, skip_i\}$.

This means that $skip_i \notin mpenabled(A_1 \ldots A_{j_3-1})$, hence (because either $write_i$ or $skip_i$ is potentially enabled at any $A_1 \ldots A_j, j \geq j_1$) we have $\{write_i\} \in mpenabled(A_1 \ldots A_{j_3-1})$. This and the priority constraint $write_i > skip_j$, imply that $\{skip_j\} \notin menabled(A_1 \ldots A_{j_3-1})$, $j = 1, \ldots, m$. Hence $A_{j_3} \cap \{write_i, skip_1, \ldots, skip_{i-1}, skip_{i+1}, \ldots, skip_m\} = \{write_i\}$, i.e., $write_i \in A_{j_3}$, a contradiction as $j_3 > j_1$ and $write_i \notin A_j$ for all $j > j_1$.

Thus $write_k$ cannot occur infinitely often in t for any k, so the only event left is $skip_k$. But the occurrence of $skip_k$ more than m times requires more than one full cycle of P_{3n+m+1}, hence in particular the conflict "$write_i, skip_1, \ldots, skip_{i-1}, skip_{i+1}, \ldots, skip_m$" must be resolved an infinite number of times, and we can repeat the same arguments as for the case $write_k$. Thus the negation of the assumption (3) at the definition of starvation-freeness leads to the conclusion that there is no event occurring an infinite number of times in t, i.e. t is not an infinite multiple sequence, a contradiction. Hence (3) must hold for $Pr^{n,m}_{RW2}$ which means that $Pr^{n,m}_{RW2}$ is starvation-free for writers.

2.6.5 Simulation of Dynamic Priorities

As we already mentioned, dynamic priority is a very attractive tool and specifications like $\underline{path}\ a > b; b > a\ \underline{end}$ would make it easier, or at least conceptually more natural, to specify starvation-free systems. For instance, adding to Pr'_{RW2} a path like $\underline{path}\ write_1 > write_2; write_2 > write_1\ \underline{end}$ solves completely the problem of starvation of writers by writers, and this solution is clearly simpler than Pr'''_{RW2}. Unfortunately, we have no formal theory which allows us to cope with this kind of dynamic priority. Fortunately, however, we can quite easily simulate specifications like $\underline{path}\ a > b; b > a\ \underline{end}$ using static priorities, defined in Sect. 2.6.2, and so called "ghost" events.
The crux of the method is that *reflexivity is not a property required of our priority relation*. Consider the following fragment of a system specification:

$$\underline{priority}\ gh_b > a, a > gh_a, gh_a > b, b > gh_b, \ldots\ \underline{end}$$

$\ldots$

$$\underline{path}\ a, gh_a, b; b, gh_b, a\ \underline{end}$$

$\ldots$

and assume additionally that in this specification a potential enabling of gh_a always implies a conflict between gh_a and a, and a potential enabling of gh_b always implies a conflict between gh_b and b. This additional assumption might be (but not necessarily) achieved for instance by the requirement that in the rest of the system, a occurs only in the context "a, gh_a", or in the context "a, b", and similarly for b. Note that the above pattern is just equivalent to $\underline{path}\ a > b; b > a\ \underline{end}$, where the first conflict, if it occurs, must be resolved in favour of a, and the second must be resolved in favour of b. The "ghost" events gh_a, gh_b are *never* enabled, their role is to "move" priority through a path $\underline{path}\ a, gh_a, b; b, gh_b, a\ \underline{end}$. Using this technique for the second reader-writer problem with two readers and two writers results in the priority path path program of Fig. 2.68.

We leave the reader to convince himself that Pr^{iv}_{RW2} is indeed starvation-free for writers. The arguments are very similar to those used for Pr'''_{RW2}.

$Pr^{iv}_{RW2} =$

$\underline{priority}\ (write_1, write_2) > (readbegin_1, readbegin_2),$
$(readend_1, readend_2) > (write_1, write_2),$
$gh_2 > write_1, write_1 > gh_1, gh_1 > write_2 >, write_2 > gh_2\ \underline{end}$

$P(1):\ \underline{path}\ readbegin_1, readend_1, write_1, write_2\ \underline{end}$
$P(2):\ \underline{path}\ readbegin_2, readend_2, write_1, write_2\ \underline{end}$
$P(3):\ \underline{path}\ readrequest_1; readbegin_1; readend_1\ \underline{end}$
$P(4):\ \underline{path}\ readrequest_2; readbegin_2; readend_2\ \underline{end}$
$P(5):\ \underline{path}\ writerequest_1; write_1, gh_1\ \underline{end}$
$P(6):\ \underline{path}\ writerequest_2; write_2, gh_2\ \underline{end}$
$P(7):\ \underline{path}\ write_1, gh_1, write_2; write_2, gh_2, write_1\ \underline{end}$

Figure 2.68. New solution of second reader-writer problem

Next we give a formal definition of ghost events.
Let $Pr =< P_1 \dots P_n$ be a priority path program, and let $a \in Ev$. We shall say that $b \in Ev$ is a **ghost event** of a if and only if the following are satisfied:

(1) $(\forall t \in MFS)\{b\} \in mpenabled(t) \Rightarrow \{a\} \in mpenabled(t)$.

(2) $a > b$.

The set of all ghost events of a will be denoted by $gh(a)$. Directly from the definition we have the following result.

Corollary 2.26
$(\forall b \in gh(a))(\forall t \in MFS)\{b\} \notin menabled(t)$. □

This means that if Pr contains ghost events it is *not adequate*, however, this is a desired property in this case.
So far we discussed the simulation of dynamic priorities involving two events only. What about more complex structures involving more than two events? To illustrate the ghost event technique let us consider the case with two readers, and three writers. In this case we have to simulate a path like:

$$\underline{path}\ write_1 > write_2 > write_3; write_2 > write_3 > write_1;$$

$$write_3 > write_1 > write_2\ \underline{end}$$

where "$a > b > c$" in this path stands for "any conflict between a, b and c , or a and b only, or a and c only, must be resolved in favour of a, and any conflict between b and c only, must be resolved in favour of b." The solution requires six ghost events and is shown in Fig. 2.69.

The dynamic priority is simulated by the path P_8 and the priority relation, whose most important part (writers and ghost events) is depicted in Fig. 2.70.

$Pr^{2,3}_{RW2} =$

$\underline{priority}\ (write_1, write_2, write_3) > (readbegin_1, readbegin_2),$
$(readend_1, readend_2) > (write_1, write_2, write_3),$
$write_1 > gh^1_1, gh^1_1 > (write_2, write_3), write_2 > gh^1_2, gh^1_2 > write_3,$
$write_2 > gh^2_2, gh^2_2 > (write_3, write_1), write_3 > gh^2_3, gh^2_3 > write_1,$
$write_3 > gh^3_3, gh^3_3 > (write_1, write_2), write_1 > gh^3_1, gh^3_1 > write_2\ \underline{end}$

$P(1):$ $\underline{path}\ readbegin_1, readend_1, write_1, write_2, write_3\ \underline{end}$
$P(2):$ $\underline{path}\ readbegin_2, readend_2, write_1, write_2, write_3\ \underline{end}$
$P(3):$ $\underline{path}\ readrequest_1; readbegin_1; readend_1\ \underline{end}$
$P(4):$ $\underline{path}\ readrequest_2; readbegin_2; readend_2\ \underline{end}$
$P(5):$ $\underline{path}\ writerequest_1; write_1, gh^1_1, gh^3_1\ \underline{end}$
$P(6):$ $\underline{path}\ writerequest_2; write_2, gh^2_2, gh^1_2\ \underline{end}$
$P(7):$ $\underline{path}\ writerequest_3; write_3, gh^3_3, gh^2_3\ \underline{end}$
$P(8):$ $\underline{path}\ write_1, gh^1_1, write_2, gh^1_2, write_3; write_2, gh^2_2, write_3, gh^2_3, write_1;$
$write_3, gh^3_3, write_1, gh^3_1, write_2;\ \underline{end}$

Figure 2.69. Solution to second reader-writer problem

The part of the priority relation that is involved in the first conflict of P_8 is depicted in Fig. 2.71 a, the part involved in the second conflict is depicted in Fig. 2.71 b, and the part involved in the third conflict is depicted in Fig. 2.71 c.

In this case $gh(write_1) = \{gh^1_1, gh^3_1\}$, $gh(write_2) = \{gh^2_2, gh^1_2\}$, $gh(write_3) = \{gh^3_3, gh^2_3\}$. Every ghost event is involved in one conflict of P_8 only. The indices in gh^i_j stand for: i – the number of the conflict in P_8 that gh^i_j represents, j – the number of the writer whose ghost event gh^i_j is. The case of 2 readers and 3 writers can easily be extended to n readers and m writers. A solution, denoted $(Pr^{n,m}_{RW2})'$, is shown in Fig. 2.72. To simplify the specification of the priority relation of $(Pr^{n,m}_{RW2})'$ we will write the sequence:

$$a_1 > [b_1] > a_2 > [b_2] > \ldots > a_{m-1} > [b_{m-1}] > a_m$$

and take it to be equivalent to:

$$(\forall i = 1, \ldots, m-1)(a_i > b_i \wedge (\forall j = i+1, \ldots, m) b_i > a_j).$$

Finally, one may observe that $(Pr^{n,m}_{RW2})'$ is indeed a simple request system with $req(write_i) = writerequest_i$ for $i = 1, \ldots, m$, so starvation-freeness is formally defined in this case. The formal proof that $(Pr^{n,m}_{RW2})'$ is starvation free for writers is almost identical to the proof of the same property for $Pr^{n,m}_{RW2}$ in the previous section and we leave it as an exercise for the reader.

2.6.6 VFS Semantics for Priority COSY

The reader might be surprised by the above title since at the beginning of this section we argued that vector sequences cannot be used to reason about the

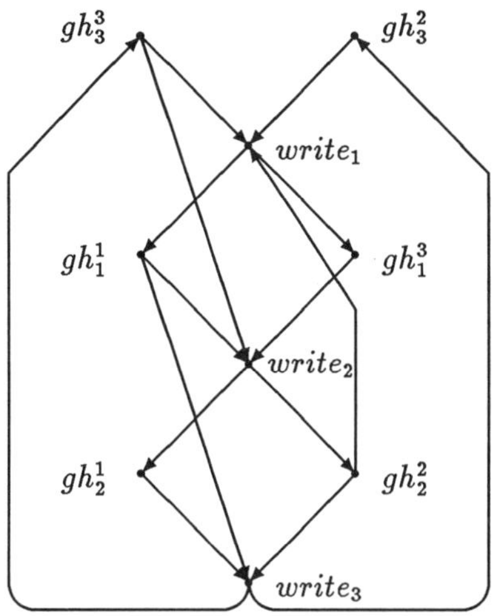

Figure 2.70. Priority relation of $Pr_{RW2}^{2,3}$ for writers and ghost events

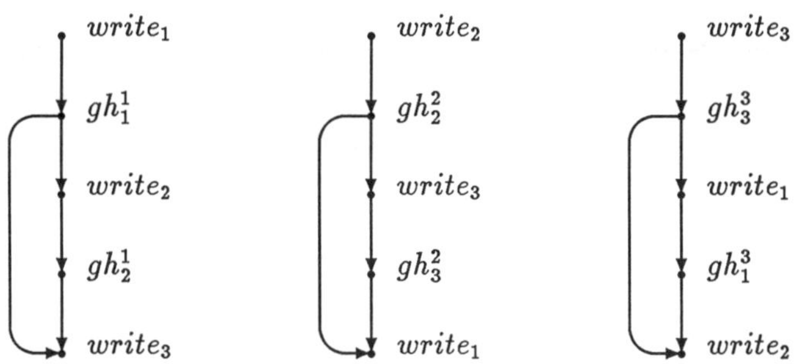

(a) The Part of a priority relation involved in the 1st conflict in P_{10}

(b) The Part of a priority relation involved in the 2nd conflict in P_{10}

(c) The Part of a priority relation involved in the 3rd conflict in P_{10}

Figure 2.71.

$$
\begin{array}{ll}
(Pr^{n,m}_{RW2})' = & \\
& \underline{priority}\ (write_1, \ldots, write_m) > (readbegin_1, \ldots, readbegin_n), \\
& (readend_1, \ldots, readend_n) > (write_1, \ldots, write_m), \\
& write_1 > [gh^1_1] > write_2 > [gh^1_2] > \ldots > write_{m-1} > [gh^1_{m-1}] > write_m, \\
& write_2 > [gh^2_2] > write_3 > [gh^2_3] > \ldots > write_{m-1} > \\
& [gh^2_{m-1}] > write_m > [gh^2_m] > write_1, \\
& \ldots \\
& write_i > [gh^i_i] > write_{i+1} > [gh^i_{i+1}] > \ldots > write_m > \\
& [gh^i_m] > write_1 > [gh^i_1] > \ldots > write_{i-2} > [gh^i_{i-2}] > write_{i-1}, \\
& \ldots \\
& write_m > [gh^m_m] > write_1 > [gh^m_1] > \ldots > write_{m-2} > \\
& [gh^m_{m-2}] > write_{m-1}\ \underline{end} \\
P(1): & \underline{path}\ readbegin_1, readend_1, write_1, \ldots, write_m\ \underline{end} \\
& \ldots \\
P(n): & \underline{path}\ readbegin_n, readend_n, write_1, \ldots, write_m\ \underline{end} \\
P(n+1): & \underline{path}\ readrequest_1; readbegin_1; readend_1\ \underline{end} \\
& \ldots \\
P(n+n): & \underline{path}\ readrequest_n; readbegin_n; readend_n\ \underline{end} \\
P(2n+1): & \underline{path}\ writerequest_1; write_1, gh^1_1, gh^3_1, \ldots, gh^m_1\ \underline{end} \\
P(2n+2): & \underline{path}\ writerequest_2; write_2, gh^1_2, gh^2_2, gh^4_2, \ldots, gh^m_2\ \underline{end} \\
& \ldots \\
P(2n+i): & \underline{path}\ writerequest_i; write_i, gh^1_i, gh^2_i, \ldots, gh^i_i, gh^{i+2}_i, \ldots, gh^m_i\ \underline{end} \\
& \ldots \\
P(2n+m-1): & \underline{path}\ writerequest_{m-1}; write_{m-1}, gh^1_{m-1}, gh^2_{m-1}, \ldots, gh^{m-1}_{m-1}\ \underline{end} \\
P(2n+m): & \underline{path}\ writerequest_m; write_m, gh^2_m, gh^3_m, \ldots, gh^m_m\ \underline{end} \\
P(2n+m+1): & \underline{path}\ write_1, gh^1_1, write_2, gh^1_2, \ldots, write_{m-1}, gh^1_{m-1}, write_m; \\
& write_2, gh^2_2, write_3, gh^2_3, \ldots, write_{m-1}, gh^2_{m-1}, write_m, gh^2_m, write_1; \\
& \ldots \\
& write_i, gh^i_i, write_{i+1}, gh^i_{i+1}, \ldots, write_m, gh^i_m, write_1, gh^i_1, \\
& \ldots, write_{i-2}, gh^i_{i-2}, write_{i-1}; \\
& \ldots \\
& write_m, gh^m_m, write_1, gh^m_1, \ldots, write_{m-2}, gh^m_{m-2}, write_{m-1}\ \underline{end}
\end{array}
$$

Figure 2.72. Solution to second reader-writer problem with ghost events

properties of priority systems. However, our remarks concerned vector sequences interpreted as system behaviours or partial orders. Another way of reasoning is to forget about any interpretation and start with a formal definition of vector sequences instead, and check what kinds of results we can get. The main motivation for such an approach is the following.

There are a number of results concerning vector sequences, many of which can be applied to priority systems. Usually priorities occur only in some part of the system. The rest of the system might be analyzed independently, and only after proving some properties of the priority-free part of a system might one analyze the system as a whole, using the already demonstrated properties. Clearly, such an approach becomes easier if we can use the same tools for each part of a system. We feel, in many cases this common tool would prove to be vector sequences.

To extend the vector firing sequence concept to deal with priority path programs we need only change the definition of $enabled(\underline{x})$ and to use the second equality of Lemma 2.7 as the definition of VFS.

Let $Pr =< P_1 \ldots P_n$ be a priority path program fixed for the rest of this section. Let $Pr' = P_1 \ldots P_n$. In the definition of multiple firing sequence semantics we assumed that Pr has inherited Ev, Ind and ind from Pr'. Now we additionally assume that Pr inherits Vev, (i.e. $Vev(Pr) = Vev(Pr')$), and mappings $h_1, \ldots, h_n$, $\_$, and $pr_1, \ldots, pr_n$, which are defined by the structure of Vev.

For every $\underline{x} \in Vev^*$, an event $a \in Ev$ is said to be **potentially enabled** at $\underline{x}$ if and only if:

$$(\forall i = 1, \ldots, n) a \in Ev_i \Rightarrow pr_i(\underline{x})a \in FS_i.$$

For every $\underline{x} \in Vev^*$, a set $A \in Ind$ is said to be **potentially enabled** at $\underline{x}$ if and only if every $a \in A$ is potentially enabled at $\underline{x}$.

The set of all potentially enabled sets of events at $\underline{x}$ will be denoted by $penabled(\underline{x})$.

We shall say that a set $A \in Ind$ is **enabled** at $\underline{x} \in Vev^*$ if and only if the following are satisfied:

(1) A is potentially enabled at $\underline{x}$ (i.e. $A \in penabled(\underline{x})$),

(2) $(\forall a \in A)(\forall B \in penabled(\underline{x}))(\forall b \in B)\neg(a < b)$.

The family of all sets enabled at $\underline{x}$ will be denoted by $enabled(\underline{x})$. Note that if $<= \emptyset$ then $enabled(\underline{x})$ as defined above is identical to $enabled(\underline{x})$ defined in Sect. 2.2.3.

Let $SEQ, CON \subseteq Vev^* \times Vev^*$ be the following relations (compare similar definitions in Sect. 2.2.3):

$$\underline{x} SEQ \underline{y} :\Leftrightarrow (\exists a \in Ev)\{a\} \in enabled(\underline{x}) \wedge \underline{y} = \underline{xa}$$

$$\underline{x} CON \underline{y} :\Leftrightarrow (\exists A \in enabled(\underline{x}))\underline{y} = \underline{aA}$$

We may now define the set of **vector firing sequences** VFS of $Pr = < P_1 \ldots P_n$ as

$$VFS = \{\underline{x} | \underline{\varepsilon} CON^* \underline{x}\}.$$

Clearly, if $<= \emptyset$ then the above definitions are the same as those of Sect. 2.2.3. Recall the program Pr_{36} from the beginning of Sect. 2.6:

$$Pr_{36} = \begin{cases} & \underline{priority}\ readx < stopandrestart\ \underline{end} \\ P(1): & \underline{path}\ readx, stopandrestart\ \underline{end} \\ P(2): & \underline{path}\ getmessage; stopandrestart\ \underline{end} \end{cases}$$

One can show that
$VFS(Pr_{36}) = Pref((\{\underline{readx}\}^*\{\underline{getmessage}.\underline{stopandrestart}\})^*)$.
Let Pr_{40} be the following ordinary path program:

$$Pr_{40} = \begin{cases} P(1): & \underline{path}\ readx, stopandrestart\ \underline{end} \\ P(2): & \underline{path}\ getmessage; stopandrestart\ \underline{end} \end{cases}$$

Note that $VFS(Pr_{36}) = VFS(Pr_{40})$!!!, which shows that vector firing sequences, (even without interpretation), could have only limited application. For instance they cannot express the fact that "$getmessage; stopandrestart$" is atomic and cannot be interrupted in Pr_{36}. However, both Pr_{36} and Pr_{40} are adequate, so as far as adequacy and deadlock-freeness are concerned, Pr_{36} and Pr_{40} can be regarded as equivalent.
The next example we will consider is Pr_{41} defined as follows:

$$Pr_{41} = \begin{cases} & \underline{priority}\ a > b\ \underline{end} \\ P(1): & \underline{path}\ a, d; c\ \underline{end} \\ P(2): & \underline{path}\ a, b; c\ \underline{end} \end{cases}$$

In this case $VFS(Pr_{41}) = ((\underline{db} \cup \underline{a})\underline{c})^*(\underline{db} \cup \underline{a} \cup \underline{d} \cup \underline{\varepsilon})$, where we use regular expressions over the alphabet $Vev(Pr_{41})$ to describe $VFS(Pr_{41})$ instead of sets of vectors explicitly. Since $\underline{b} \in Pref(\underline{db}) \subseteq Pref(VFS(Pr_{41}))$, $\underline{db} = \underline{bd}$, and $\underline{b} \notin VFS(Pr_{41})$, we have $VFS(Pr_{41}) \neq Pref(VFS(Pr_{41}))$, which means that in general *the equality* $VFS = Pref(VFS)$ *does not hold for priority path programs.* We now establish the relationship between vector firing sequences and multiple firing sequences for priority path programs.
Let $v : Ind^* \longrightarrow Vev^*$ be the following mapping:

$$(\forall t \in Ind^*) v(t) = (\hat{h}_1(t), \ldots, \hat{h}_n(t)).$$

The relationship between MFS and VFS is characterized by the following lemma.

the partial order defined by $[\{d\}\{b\}]_\sim$ $PO_{\underline{db}}$

Figure 2.73. Invariant of $[\{d\}\{b\}]_\sim$ and $PO_{\underline{db}}$ for Pr_{41}

Lemma 2.33
(1) $(\forall t \in MFS) menabled(t) = enabled(v(t))$.
(2) $VFS = \{v(t) | t \in MFS\}$, *i.e.* $VFS = v(MFS)$.
(3) $(\forall t, r \in MFS) t \sim r \Leftrightarrow v(t) = v(r)$.
(4) $(\forall \underline{x} \in VFS)(\forall t \in v^{-1}(\underline{x}) \cap MFS)[t]_\sim = v^{-1}(\underline{x}) \cap MFS$.

Proof. This easy and straightforward proof is left to the reader. □
The path programs Pr_{36} and Pr_{40} show that in general MFS cannot be derived from VFS, as $VFS(Pr_{36}) = VFS(Pr_{40})$ whilst $\{getmessage\}\{readx\} \in MFS(Pr_{40}) \setminus MFS(Pr_{36})$, i.e. $MFS(Pr_{40}) \neq MFS(Pr_{36})$. The example Pr_{41} shows that, despite Lemma 2.33 (3), for some $t \in MFS$ the partial order generated by $v(t)$, i.e. $PO_{v(t)}$, might differ from the invariant defined by the observations from $[t]_\sim$. Let us consider $t = \{d\}\{b\} \in MFS(Pr_{41})$. Clearly, $v(t) = \underline{db} = (d, b)$ and $[t]_\sim = \{\{d\}\{b\}\}$.

The partial order: $PO_{v(t)}$ and the invariant generated by $[t]_\sim$ are depicted in Fig. 2.73, and they are clearly different as may be seen. In general, the interpretation of vector firing sequences for priority path programs is not quite clear. All we can say is that there is a homomorphism v from multiple sequences into vector sequences, which preserves concatenation, and when we restrict the domain of this homomorphism to MFS, then the relation $\sim$ is nothing but the "natural equivalence relation" defined by v. This means that all the notions (implicitly or explicitly) based on the concept of concatenation *might* be defined in terms of vector firing sequences alone. We shall prove that this is true for such concepts as sequentializability, deadlock-freeness and adequacy.

Theorem 2.43
$Pr = < P_1 \dots P_n$ *is sequentializable* $\Leftrightarrow CON^* = SEQ^*$.

Proof. $(\Rightarrow)$ Clearly $SEQ \subseteq CON$, so $SEQ^* \subseteq CON^*$. Let $A = \{a_1, \dots, a_k\}$ and $\underline{x} CON \underline{y}$ such that $\underline{y} = \underline{x}\underline{A}$. Let $t \in v^{-1}(\underline{x}) \cap MFS$. From Lemma 2.33 (4) and sequentializability of Pr it follows that we may assume $t = \{b_1\}\dots\{b_m\}$. By Lemma 2.33 (1) we have $A \in menabled(t)$, so $tA \in MFS$. But Pr is sequentializable, thus there is a permutation $i_1, \dots, i_k$ of $1, \dots, k$ such that $tA \sim r$, where $r = \{b_1\}\dots\{b_m\}\{a_{i_1}\}\dots\{a_{i_m}\}$. From Lemma 2.33 (3) we have that $v(r) = \underline{y}$, and from Lemma 2.33 (1) that

$$\underline{x} SEQ \underline{x}\underline{a_{i_1}} SEQ \underline{x}\underline{a_{i_1}}\underline{a_{i_2}} SEQ \dots SEQ \underline{x}\underline{a_{i_1}}\underline{a_{i_2}}\dots\underline{a_{i_k}} = \underline{y}.$$

Thus $CON \subseteq SEQ^*$ which means $CON^* \subseteq SEQ^{**} = SEQ^*$. Thus $CON^* = SEQ^*$.

($\Leftarrow$) Let $t \in MFS$, and $t = A_1 \dots A_m$ where $A_i \in Ind$ for $i = 1, \dots, m$. Assume that $A_1 = \{a_1, \dots, a_k\}$. Thus $\underline{\varepsilon} CON \underline{A_1}$. Since $CON^* = SEQ^*$, there is a permutation $i_1, \dots, i_k$ of $1, \dots, k$ such that $\underline{\varepsilon} SEQ \underline{a_{i_1}} SEQ \underline{a_{i_1}}\underline{a_{i_2}} SEQ \dots SEQ \underline{a_{i_1}} \dots \underline{a_{i_k}} = \underline{A_1}$. Clearly, $\lambda \in MFS$, so by Lemma 2.33 (1) we have $\{a_{i_1}\} \in menabled(\lambda)$, which means that $\{a_{i_1}\} \in MFS$. Repeating this reasoning k times we get $\{a_{i_1}\}\dots\{a_{i_k}\} \in MFS$. Repeating the whole argument for $A_2, \dots, A_m$ we get the right-to-left implication. □

Theorem 2.44
$Pr =< P_1 \dots P_n$ is deadlock-free $\Leftrightarrow (\forall \underline{x} \in VFS) enabled(\underline{x}) \neq \emptyset$.

Proof. ($\Rightarrow$) Let $\underline{x} \in VFS$. By Lemma 2.33 (2) we have $(\exists t \in MFS) v(t) = \underline{x}$. From Lemma 2.33 (1) we have $menabled(t) = enabled(\underline{x})$, and since Pr is deadlock-free $menabled(t) \neq \emptyset$, hence $enabled(\underline{x}) \neq \emptyset$.

($\Leftarrow$) Let $t \in MFS$. Clearly $v(t) \in VFS$, and $enabled(v(t)) \neq \emptyset$. From Lemma 2.33 (1) we have that $menabled(t) \neq \emptyset$, so Pr is deadlock-free. □

Theorem 2.45
$Pr =< P_1 \dots P_n$ is adequate $\Leftrightarrow$
$(\forall \underline{x} \in VFS)(\forall a \in Ev)(\exists \underline{y} \in CON^*(\underline{x}))\{a\} \in enabled(\underline{y})$.

Proof. ($\Rightarrow$) Let $\underline{x} \in VFS, a \in Ev$. By Lemma 2.33 (2) we have $(\exists t \in MFS) v(t) = \underline{x}$. Since Pr is adequate, then $(\exists r \in R^*_<(t))\{a\} \in menabled(r)$. Let $r = tA_1 \dots A_m$, where $A_1 \dots A_m \in Ind$. From Lemma 2.33 (1) it follows that:

$$menabled(t) = enabled(\underline{x}), menabled(tA_1 \dots A_i) = enabled(\underline{x}\underline{A_1} \dots \underline{A_i})$$

for $i = 1, \dots, m$, and $menabled(r) = enabled(v(r))$. This means that $v(r) \in CON^*(\underline{x})$, and because of $menabled(r) = enabled(v(r))$, this completes the proof of the implication.

($\Leftarrow$) Let $t \in MFS, a \in Ev$. Clearly $v(t) \in VFS$, so we have $(\exists \underline{y} \in$

$Pr_{LL}^{new} =$

$\underline{priority}\ stopsystem > (deposit_1, \ldots, deposit_n, remove_1, \ldots, remove_n,$
$printline, readline)\ \underline{end}$

$$\left.\begin{array}{ll} P(1): & \underline{path}\ deposit_1; remove_1\ \underline{end} \\ & \ldots \\ P(n): & \underline{path}\ deposit_n; remove_n\ \underline{end} \\ P(n+1): & \underline{path}\ deposit_1; \ldots; deposit_n\ \underline{end} \\ P(n+2): & \underline{path}\ remove_1; \ldots; remove_n\ \underline{end} \\ P(n+3): & \underline{path}\ readline; deposit_1, \ldots, deposit_n\ \underline{end} \\ P(n+4): & \underline{path}\ remove_1, \ldots, remove_n; printline\ \underline{end} \end{array}\right\} Pr_{LL}^1$$

$$\begin{array}{ll} P(n+5): & \underline{path}\ deposit_1, (stopsystem; startsystem)\ \underline{end} \\ & \ldots \\ P(2n+4): & \underline{path}\ deposit_n, (stopsystem; startsystem)\ \underline{end} \\ P(2n+5): & \underline{path}\ remove_1, (stopsystem; startsystem)\ \underline{end} \\ & \ldots \\ P(3n+4): & \underline{path}\ remove_n, (stopsystem; startsystem)\ \underline{end} \\ P(3n+5): & \underline{path}\ readline, (stopsystem; startsystem)\ \underline{end} \\ P(3n+6): & \underline{path}\ printline, (stopsystem; startsystem)\ \underline{end} \\ P(3n+7): & \underline{path}\ getmessage; stopsystem; startsystem\ \underline{end} \end{array}$$

Figure 2.74. New linereader/lineprinter with n-frame ring buffer and interrupt handler

$CON^*(v(t)))\{a\} \in enabled(\underline{y})$. Suppose that $A_1, \ldots, A_k \in Ind$ are such that $v(t)CONv(t)\underline{A_1}CONv(t)\underline{A_1}\underline{A_2}CON \ldots CONv(t)\underline{A_1} \ldots \underline{A_k} = \underline{y}$. From Lemma 2.33 (1) it follows tha $tR_<tA_1R_< \ldots R_<tA_1 \ldots A_k$, i.e. $tA_1 \ldots A_k \in R_<^*(t)$. Clearly $v(tA_1 \ldots A_k) = \underline{y}$, and using Lemma 2.33 (1) again, $menabled(tA_1 \ldots A_k) = enabled(\underline{y})$, i.e. Pr is adequate. □

From Theorems 2.43, 2.44 and 2.45 it follows that if we are interested in proving sequentializability, deadlock-freeness and/or adequacy only, and we are not interested in atomicity of certain sequences of events for example, we may use vector firing sequences to prove these properties.

To show how we can use vector sequences for proving properties of priority path programs, we shall consider a new version of a linereader and lineprinter communicating via a ring buffer, a system which was analyzed in Sect. 2.4.3. This new version, called Pr_{LL}^{new} is presented in Fig. 2.74. The difference is that now we give priority to the interrupt handler. We shall prove that Pr_{LL}^{new} is adequate. In Sect. 2.4.3 we already proved that Pr_{LL}^1, a part of Pr_{LL}^{new} and Pr_{LL}, is adequate. Let $\underline{x} \in VFS(Pr_{LL}^{new})$ and $a \in Ev(Pr_{LL}^{new})$. We have to consider two cases.

Case 1.

$pr_{3n+7}(\underline{x}) \in \{getmessage.stopsytem.startsystem\}^*$, i.e. either the last event in $pr_{3n+7}(\underline{x})$ is $startsystem$ or $pr_{3n+7}(\underline{x}) = \varepsilon$. First note that in this case:

$\{getmessage\} \in enabled(\underline{x})$, $\{stopsystem\} \in enabled(\underline{x.getmessage})$, and $\{startsystem\} \in enabled(\underline{x.getmessage.stopsystem})$, hence, the condition for adequacy is satisfied for $a \in \{getmessage, stopsystem, startsystem\}$. Let $a \in \{deposit_1, \ldots, deposit_n, remove_1, \ldots, remove_n, readline, printline\}$. We need to distinguish the elements of $Vev(Pr_{LL}^{new})^*$ from the elements of $Vev(Pr_{LL}^1)^*$, so we shall write $\underline{x}$ for $\underline{x} \in Vev(Pr_{LL}^{new})^*$ and $(\underline{x})$ for $\underline{x} \in Vev(Pr_{LL}^1)^*$. In Sect. 2.4.3 we proved that Pr_{LL}^1 is adequate, which means that:

$$(\forall(\underline{z}) \in VFS(Pr_{LL}^1))(\exists(\underline{y}) \in Vev(Pr_{LL}^1)^*)\{a\} \in enabled_{Pr_{LL}^1}((\underline{zy})).$$

Let x' be a string obtained from x by removing all event occurrences which do not belong to $Ev(Pr_{LL}^1)$. Clearly, $(\underline{x}') \in VFS(Pr_{LL}^1)$. Hence, from the adequacy of Pr_{LL}^1 we have:

$$(\exists(\underline{y}') \in Vev(Pr_{LL}^1)^*)\{a\} \in enabled_{Pr_{LL}^1}((\underline{x}'\underline{y}')).$$

From a particular form of paths $P_{n+5}, \ldots, P_{3n+6}$ and the priority specification, it follows that $\underline{y}' \in Vev(Pr_{LL}^{new})^*$, $\underline{xy}' \in CON^*_{Pr_{LL}^{new}}(\underline{x})$, and $\{a\} \in enabled_{Pr_{LL}^{new}}(\underline{xy}')$. Hence the condition for adequacy is satisfied for every $a \in Ev(Pr_{LL}^{new})$.

Case 2.

$pr_{3n+7}(\underline{x}) \notin \{getmessage.stopsystem.startsystem\}^*$. This means that the last event in $pr_{3n+7}(\underline{x})$ is either *getmessage* or *stopsystem*. If this last event is *getmessage* then $enabled_{Pr_{LL}^{new}}(\underline{x}) = \{\{stopsystem\}\}$, if it is *stopsystem* then $enabled_{Pr_{LL}^{new}}(\underline{x}) = \{\{startsystem\}\}$, in either case $\underline{x}\,\underline{z} \in VFS(Pr_{LL}^{new})$ and $pr_{3n+7}(\underline{xz}) \in \{getmessage.stopsystem.startsystem\}^*$ where $z \in \{stopsystem.startsystem, startsystem\}$, so that the Case 2 can be reduced to Case 1. Thus Pr_{LL}^{new} is adequate.

The vector firing sequences formalism enabled us to prove the adequacy of Pr_{LL}^{new} using already proven results for a non-priority system, and we were able to use these results without translation into another formalism. However, we can not use vector firing sequences to prove, for instance, that the sequence "getmessage;stopsystem;startsystem" from path P_{3n+7} is atomic, i.e., it cannot be interrupted. To do this we have to use multiple sequences.

2.7 Bibliographic Notes

The *Basic COSY notation* was invented by Lauer and Campbell [LC75], and despite similarities **it should not be mistaken** for *Path Expressions* by Campbell and Haberman [CH74]. The Basic COSY notation was later refined by Lauer, Torrigiani and Shields in [LTS79]. The concept of vector firing sequences and vector firing sequence semantics follows from Mike Shields [Shi79a], and has been subsequently developed in [SL80], [Shi81], [Shi85], [JLKD86], [JK89]. Theorem 2.1 is also known in the theory of partially commutative monoids, and originates from [CF69] (see [Lal79] Chap. 11). The formulation and the proof of this theorem in the book follows from [JLKD86]. The relationships between partial orders and vector sequences were developed in [Shi85] and [Jan85a]. The latter used a style similar to that used in this book. Concepts similar to the concurrent product of regular expressions have been used in [Niv79] and [Bes87].

The basic ideas of the Petri Net semantics for Basic COSY presented in this book are due to Eike Best. In [LSB79] he presented an ambiguity preserving inductive construction for star-free path programs. This construction corresponds to $LN^{induc}(Pr)$ in our notation. In [Bes87] he extended this construction by allowing the Kleene star to occur in some special contexts, but he restricted the construction to uniquely named path programs. The construction from [Bes87] allows subexpressions of the form "$((sequence)*; b)$" where $b \in Ev$. In [Bes82] he suggested a construction which corresponds to our $LN^{msa}(Pr)$ and does not preserve ambiguity. To overcome the problems caused by ambiguity and the Kleene star occurring together we have used the concept of labelled uniquely named regular expressions. The idea of using uniquely named regular expressions instead of general regular expressions has its roots in the classical [MY60]. The first Petri net semantics for Basic COSY was presented by P. Lauer and R. Campbell in [LC75]. It was restricted to star-free path programs, preserved ambiguity, and it used the concept of Petri nets with silent transitions. In principle, the construction from [LC75] corresponds to our $CN^{st}(Pr)$. Recently this construction has been analysed and extended to the general case (Kleene star allowed) in [Pró90]. The process semantics of nets presented here is based on [BD87] and [Rei85]. The reader seeking more information about Petri nets is referred to [Rei85], [Pet81], [BRR86a] and [BRR86b].

Theorem 2.20 characterizing the conditions for adequacy of syntactically conflict-free (i.e. "semicolons only") path programs, as well as the concept of adequacy itself is due to Mike Shields [Shi79a]. The algorithm for checking adequacy of syntactically conflict- free path programs follows from the results obtained by Mike Shields, R. Devillers and Eike Best [Bes82]. The complexity analysis of this algorithm including Theorem 2.22 is due to E. Best [Bes82]. The adequacy preserving transformations of path programs were introduced by M. Shields in [Shi79a] and further developed in [LSB79] and [Kou88], [Kou]. The general results concerning adequacy preserving insertions, including the surprising Theorems 2.25 and 2.26 are due to Maciej Koutny [Kou88], [Kou]. Adequacy preserving deletions and the "custom tailored" approach is based on results of [Shi79a], [LSB79] and [LSC81]. The concept of free-choice path programs is due to M. Shields [Shi81].

The concept of strategy originates from R. Janicki and M. Koutny [JK87] and has been subsequently developed and refined in [JK89]. Optimal strategy and Theorems 2.33 and 2.34 follow from [JK89]. In [JK91b] the concept of optimal strategy (called optimal simulation) has been used to find an algorithm minimizing the reachability graph of a given concurrent system. The general concept of maximally concurrent execution semantics should probably be credited to [SM81]. The properties of maximally concurrent execution strategies were studied by R. Janicki, P.Lauer, M. Koutny and R. Devillers in [JLKD86] and [JLD85]. Theorems 2.36 and 2.37 follow from [JLKD86].

The first formal semantics for priority path programs is due to M. Shields [Shi81]. His priority relation was transitive, the semantics was defined in terms of vector firing sequences, and the approach was restricted to priority path programs generating behaviours which can be completely defined in terms of causal partial orders. The priority path program corresponding to Pr_{36} from Section 2.5 is also analysed in [Shi81], but it is rejected as an ill-defined specification, because its behaviour cannot properly be modelled in terms of causal partial orders (see also [Shi89]). For priority path programs with transitive priority relation, if their behaviour is completely definable in terms of causal partial orders, the semantics of [Shi81] is equivalent to our vector firing sequence semantics of priority path programs (Section 2.6.6). The paper [Shi81] is mainly concerned with a characterization of structures that permit priorities without problems within the partial order semantics. The approach to formal semantics for Basic COSY with priorities presented in this book originates from [Jan87]. The relationship between the semantics of [Shi81] and [Jan87] is discussed in detail in [Oku90]. We have to point out that Shields' viewpoint is not unique. [Rei87] and (implicitly) [Lam85] also suggest that the specification of priorities should be restricted to those cases in which priorities cannot mix with concurrency (as they do in path programs like Pr_{36}). Others, such as [BK] share our viewpoint and allow specifications like Pr_{36}. [BK] also provides analysis and some motivations underlying various approaches to priority systems. The starvation problem for COSY path programs with priorities has been formally analysed in [JL88], and semi-formally, without a formal theory of priorities in [LS78]. The concept of "ghost events" is due to F. Okulicka [Oku86]. The Reader-Writer problem analyzed originates from [CHP71]. The proofs of Theorems 2.4, 2.5, 2.13 and 2.20 presented in this chapter are based on M. Koutny's suggestions.

Chapter 3

High-level COSY Programs and System Design

3.1 High-level COSY Syntax and Semantics

Throughout Chap. 2 we have encountered two types of basic COSY programs. First, *uninterpreted* programs, which involve a (usually small) number of uninterpreted events to permit the illustration of notation, to support the definition of concepts arising in the context of concurrency theory, and to support the formulation and proof of important laws relating these concepts. Second, *interpreted* programs, which involve a (usually much larger) number of interpreted events which permit the application of the results of concurrency theory to develop verified solutions to problems arising in the area of concurrent system design and specification.

An uninterpreted basic COSY program involved (maximally 9) events which essentially only needed to have *distinct names* $a, b, c, \ldots$. Sometimes these "programs" also involved *variables* $\omega, \omega_1, \omega_2, \ldots$, to state transformations of programs into programs, but these variables are *auxiliary concepts* and not part of the basic notation. For events one sees no need to introduce any other notational feature than some set of distinct event names to choose from. The use of variables suggests that *parametrized* basic COSY programs are useful in the statement of relations between such programs. We shall find that such parametrization has been introduced in the system notation discussed later. In uninterpreted programs we have also used *ellipses* to indicate "arbitrary context" (see Pr_{34}) or "regular repetitions of some pattern" (see Pr_{10} in the proof of Theorem 2.27).

An interpreted basic COSY program generally involved many interpreted events, usually partitioned into subsets whose elements are distinguished by affixing *indices to the common (generic) event name*. To indicate that we were dealing with interpreted program we often wrote something like "event names stand for their meanings". Here we can see that appropriate notational support would be useful for expressing this partitioning explicitly, and thus to facilitate the comprehension of large systems in terms of its sub-sytems. Some interpreted programs also involved ellipses to indicate "regular repetitions of some pattern" (see Pr_{LL}). This second type of use of ellipses is highly context dependent and it is difficult to discover related ellipses occurring at different places in a program. Hence it would be very useful to have ways of expressing the meanings of such ellipses explicitly to facilitate comprehension of the overall structure of a complex specification.

Furthermore, an interpreted program usually also involves a large number (e.g. $3n+6$ many, where n is an integer parameter) of paths of similar structure, or which share operations from some of the partitions of the event space. Again, these interconnections among sub-systems of paths should bc made as explicit as possible to aid in the conceptualization of the system.

Finally, interpreted programs were shown to be correct under the assumption that the operations (events) mentioned in the paths are used by processes in a specific and regular way. Thus in the description of the problem specified by the program Pr_{CS}, see Fig.2.46 it is stated that:

> "A typical process trying to use the system will first request use of the critical section, and then, after the request has been granted, it will enter the pipeline queue. After it has been transferred out of the k-th frame of the queue, it may then use the critical section, but on completion it must send a signal to say that the next process in the queue may be transferred out of the k-th frame".

But the assumption made about "typical processes" is not expressed explicitly in the path program Pr_{CS}. To do so we will need either the process notation, compare Fig. 3.3, or the system notation with its ability to define composite operations (events). Both alternatives will be presented in the sequel. In the rest of this section we will explain the high level notations by means of the example programs in Chap. 2. We will start from the reader's understanding of these examples as developed in Chap. 2 and translate them into high-level notation. In this way the reader should obtain an intuitive understanding of this notation which will facilitate his understanding of the formal definition of this notation in the subsequent sections of this chapter.

Process Notation.

Recall that a **basic COSY path program** is a collection of paths in program and endprogram parentheses. A path is an expression, similar to a regular expression, built from event names, semicolons, commas, conventional parentheses and Kleene stars, enclosed by path and end parentheses.

Intuitively, each path specifies a sequential subsystem and expresses a constraint on the order of occurrences of events belonging to the subsystem. During any period of system activity, the events belonging to a single path may only be active or occur one at a time, and the sequence of event occurrences constituting a history of this activity must be either a concatenation of cyclic behaviours specified by the path, or a prefix of such a concatenation. From this it follows that each occurrence of an event of the system specified by a program is constrained by every path in which it is named, that is, the synchronization among paths is due to common events. This kind of synchronization is often called by "handshake" or by "rendez-vous". A **basic COSY program** is a basic COSY *path* program with processes which have the same syntax as paths except that *path* is replaced by *process*. Considered individually, the meaning of a process is the same as the meaning of the corresponding path, i.e. they define the same set of firing sequences. However, a program consisting only of processes and no paths is interpreted as consisting of a collection of *disjoint* finite state automata. Hence, when processes are composed, occurrences of shared events in distinct processes do not represent "handshake" synchronization as is the case for paths, but represent an "inclusive or" or a "concurrent or".

Thus an occurrence of event name a in two processes means that event a can occur concurrently "for the first process" and "for the second process", i.e. there are really two events of type a, one belonging to the first process and one belonging to the second process, which can, for instance, be made explicit by subscripting the event name by the process in which it occurs. Our formal treatment of processes will be based on this approach.

If a COSY program involves both paths and processes then the composition of paths and paths, and paths and processes is interpreted as "handshake synchronization", as in the case of path programs. This will all be made more precise in the next subsection.

For the present we will explain the meaning of Pr_{42} in terms of partial orders as we did for Pr_2 in Sect. 2.1. In fact, Pr_2 is a part of Pr_{42}, which additionally involves two processes:

$$Pr_{42} = \begin{cases} & \underline{program} \\ P(1): & \underline{path}\ a; c\ \underline{end} \\ P(2): & \underline{path}\ b; c\ \underline{end} \\ Q(1): & \underline{process}\ a; b\ \underline{end} \\ Q(2): & \underline{process}\ b; a\ \underline{end} \\ & \underline{endprogram} \end{cases}$$

We can take an initial segment of the partial order for Pr_2 given in Fig. 2.1 of Sect. 2.1 and add to it sequences consisting of indexed arrows, one for each process, to indicate how events become active for the process whose index is on the arrow. In this way Fig. 3.1 represents an initial segment of a possible decomposition of the partial order for the path part of Pr_{42} into total orders, and it does this in a way which corresponds to the process part of Pr_{42}. In Fig. 3.1, unbroken arrows correspond to the path part of the program, and the

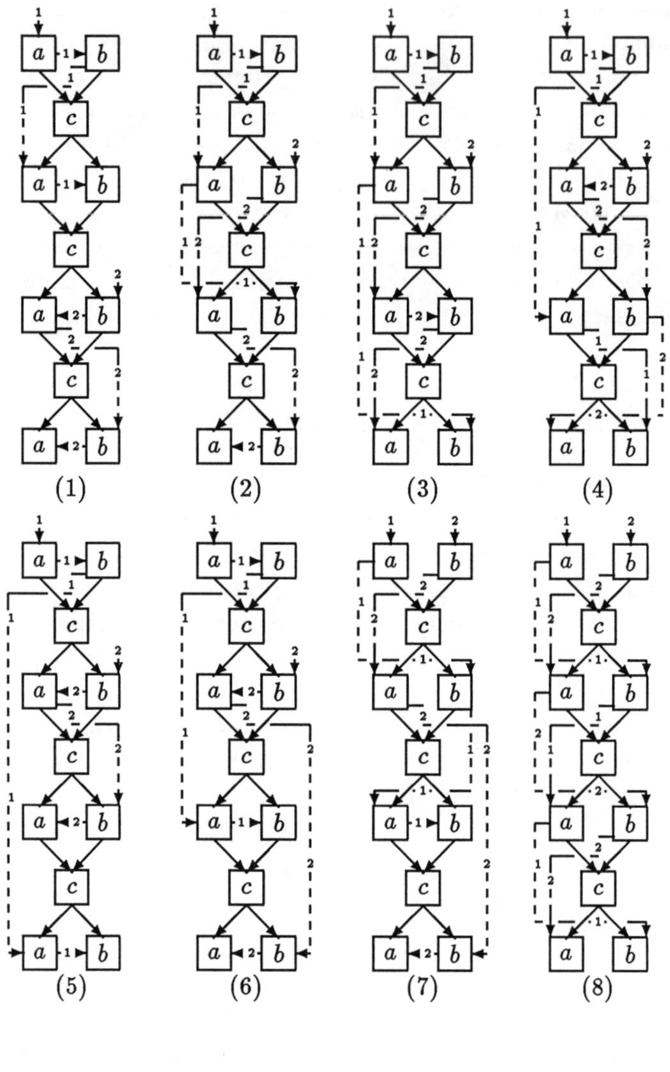

Figure 3.1. Partial order corresponding to the paths in Pr_{42} with possible decompositions into total orders corresponding to the processes in Pr_{42}

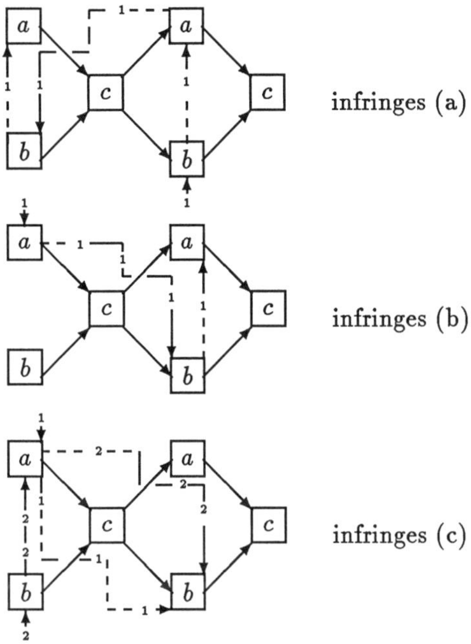

Figure 3.2. Graphs not satisfying (a), (b), or (c)

broken arrows involving integers show which event occurrences are part of which process.

The figure illustrates how processes progress by means of event occurrences. Thus process $Q(1)$ progresses by means of the sequence of event occurrences *abab* (the sequence of arrows –1–1–1–>) and process $Q(2)$ progresses by means of the sequence of event occurrences *baba* (the sequence of arrows –2–2–2–>). Processes can progress only by event occurrences in a way that does not conflict with the order specified by the paths. This fact is expressed in the figure by drawing the broken arrows on top of the graph of the partial order for Pr_2 in such a way that the following constraints are satisfied:

(a) the broken arrows should not introduce cycles,

(b) if some event mentioned in a path is also mentioned in a process then the event can only occur if it occurs for some process, and

(c) an event mentioned in a path may not occur for two distinct processes concurrently (coincidentally).

The graphs in Fig. 3.2 do not satisfy these constraints, and hence do not represent possible behaviours of program Pr_{42}.

$Pr_{CS,proc} =$

$$
\begin{array}{ll}
P(1): & \underline{path}\ transfer_1; transfer_2\ \underline{end} \\
 & \ldots \\
P(k): & \underline{path}\ transfer_k; transfer_{k+1}\ \underline{end} \\
P(k+1): & \underline{path}\ transfer_{k+1}; critical section; signal\ \underline{end} \\
Pr^0_{CS} \left\{ \begin{array}{ll} P(k+2): & \underline{path}\ skip_1, (request_1; granted_1)\ \underline{end} \\ & \ldots \\ P(2k+1): & \underline{path}\ skip_k, (request_k; granted_k)\ \underline{end} \\ P(2k+2): & \underline{path}\ skip_1, (granted_1; transfer_1); \\ & \ldots; skip_k, (granted_k; transfer_1)\ \underline{end} \end{array} \right. & \\
Q(1): & \underline{process}\ request_1; granted_1; transfer_1; \ldots; transfer_{k+1}; \\
 & critical section; signal\ \underline{end} \\
 & \ldots \\
Q(k): & \underline{process}\ request_k; granted_k; transfer_1; \ldots; transfer_{k+1}; \\
 & critical section; signal\ \underline{end}
\end{array}
$$

Figure 3.3. Critical section with k processes in process notation

Condition (c) indicates that in a program containing processes having an event name in common which occur in at least one path, there is implicit mutual exclusion of the processes from participating in the same occurrence of that event. For instance, in Pr_{42}, after a has occurred for the first process, the two processes are in competition over the next occurrence of b. Here we have another way in which a system may give rise to conflict. The conflict possibilities are "implicit", that is hidden away in the semantics of the programs, but they can be made explicit by means of the translation of programs (involving processes) into path programs which we will use in the formal definition of COSY processes.

With this intuitive understanding of the relationship of paths to COSY processes the reader should be able to understand how $Pr_{CS,proc}$ in Fig. 3.3, which is obtained from Pr_{CS} (in Section 2.4.3) by adding processes, renders explicit the content of the above quote concerning a typical process using the system specified by Pr_{CS}.

Similarly, it should not be too difficult for the reader to determine that $Pr^{k,m}_{CS,proc}$ (Fig. 3.4) has been obtained by adding processes to $Pr^{k,m}_{CS}$ (from Section 2.4.3) so as to make explicit the intent of the following quote about a typical process using the system specified by $Pr^{k,m}_{CS}$:

> "... a specification of a mechanism regulating the use of a critical section by m classes of processes having distinct priorities. With each class is associated a k-frame ring buffer, which a process from the class must enter in order to eventually gain access to the critical section."

Macro Notation

As discussed earlier, there are a number of reasons for introducing notation with which it is possible to express long basic COSY programs in a succinct and structured way. In this subsection we will illustrate such a notation by translating some of the example programs from Chap. 2, whose meaning the reader should already comprehend from the discussion in that chapter.

Collectivisors

If we consider the linereader/lineprinter with interrupt handler specification Pr_{LL} (Section 2.4.3) we see that it contains Pr^1_{LL}, the linereader/lineprinter, which in turn contains Pr_B, which specifies a ring buffer. First consider what we need to express the n-frame ring buffer specified by Pr_B:

$$\left.\begin{array}{ll} P(1): & \underline{path}\ deposit_1; remove_1\ \underline{end} \\ & \dots \\ P(n): & \underline{path}\ deposit_n; remove_n\ \underline{end} \\ P(n+1): & \underline{path}\ deposit_1;\dots;deposit_n\ \underline{end} \\ P(n+2): & \underline{path}\ remove_1;\dots;remove_n\ \underline{end} \end{array}\right\} Pr_B$$

For each of the buffer frames we need a *deposit* and a *remove* operation, and there are n frames altogether. So we need a collective operation name for each operation type from which to obtain specific operation names by indexing. In the macro notation such collective names and their index ranges are defined as follows:

$$\underline{array}\ deposit\ remove\ (n)\ \underline{endarray}$$

The above definition is called a *collectivisor*, it assumes that index bounds begin at 1 and it generates two *totally ordered sets* of indexed events without gaps:

$$deposit(1) \prec \dots \prec deposit(n), and$$

$$remove(1) \prec \dots \prec remove(n).$$

Once the indexed events have been defined by a collectivisor, they may be used by the rest of the macro notation constructions. In particular, the total order defined for each individual array will be used to expand so called distributors.

Before we consider the use of these collective names in paths and processes we will show how multidimensional collective names can be used to express the

$Pr^{k,m}_{CS,proc} =$

$$\left.\begin{array}{l}
\text{path}\ deposit_{11}; remove_{11}\ \text{end} \\
\dots \\
\text{path}\ deposit_{k1}; remove_{k1}\ \text{end} \\
\text{path}\ deposit_{11}; \dots ; deposit_{k1}\ \text{end} \\
\text{path}\ remove_{11}; \dots ; remove_{k1}\ \text{end}
\end{array}\right\} Pr_1$$

$\dots$

$$\left.\begin{array}{l}
\text{path}\ deposit_{1m}; remove_{1m}\ \text{end} \\
\dots \\
\text{path}\ deposit_{km}; remove_{km}\ \text{end} \\
\text{path}\ deposit_{1m}; \dots ; deposit_{km}\ \text{end} \\
\text{path}\ remove_{1m}; \dots ; remove_{km}\ \text{end}
\end{array}\right\} Pr_m$$

$Q(1):$ path $skip_1, (remove_{11}, \dots, remove_{k1}; criticalsection_1)$ end

$\dots$

$Q(m)$ path $skip_m, (remove_{1m}, \dots, remove_{km}; criticalsection_m)$ end

$R:$ path $(skip_1; (skip_2; (\dots (skip_{m-1}; skip_m, criticalsection_m),$
$criticalsection_{m-1}) \dots), criticalsection_2), criticalsection_1$ end

process $(deposit_{11}; remove_{11}), \dots, (deposit_{k1}; remove_{k1});$
$criticalsection_1$ end

$\dots k - many\ identical\ processes$

process $(deposit_{11}; remove_{11}), \dots, (deposit_{k1}; remove_{k1});$
$criticalsection_1$ end

$\dots$

process $(deposit_{1m}; remove_{1m}), \dots, (deposit_{km}; remove_{km});$
$criticalsection_m$ end

$\dots k - many\ identical\ processes$

process $(deposit_{1m}; remove_{1m}), \dots, (deposit_{km}; remove_{km});$
$criticalsection_m$ end

Figure 3.4. $Pr^{k,m}_{CS,proc}$ Critical section with m distinct priorities

operations needed for specifying the system of m ring buffers, each consisting of k frames, from specification $Pr_{CS}^{k,m}$ (Section 2.4.3.2) shown below:

$$\left.\begin{array}{l} \underline{path}\ deposit_{11}; remove_{11}\ \underline{end} \\ \ldots \\ \underline{path}\ deposit_{k1}; remove_{k1}\ \underline{end} \\ \underline{path}\ deposit_{11}; \ldots; deposit_{k1}\ \underline{end} \\ \underline{path}\ remove_{11}; \ldots; remove_{k1}\ \underline{end} \end{array}\right\} Pr_1$$

$$\ldots$$

$$\left.\begin{array}{l} \underline{path}\ deposit_{1m}; remove_{1m}\ \underline{end} \\ \ldots \\ \underline{path}\ deposit_{km}; remove_{km}\ \underline{end} \\ \underline{path}\ deposit_{1m}; \ldots; deposit_{km}\ \underline{end} \\ \underline{path}\ remove_{1m}; \ldots; remove_{km}\ \underline{end} \end{array}\right\} Pr_m$$

The $2(k \times m)$ frame operations can be defined by:

$$\underline{array}\ deposit\ remove\ (k, m)\ \underline{endarray}$$

Later we will see that index ranges can have lower bounds other than 1, and indices in the range, as well as bounds, can be generated by arbitrary arithmetic expressions, and even restricted by means of some predicates.

Body Replicators

Now that we have a way of defining collections of names distinguished by indices we will want to use these collective names to obtain more succinct notation for collections of paths and processes of identical structure but possibly with some operations differing in indices. Body replicators were introduced into the macro notation for this purpose, and we will see how their use will permit the elimination of a number of occurrences of ellipses in the example programs from Chap. 2. The general form of a *body replicator* is:

$$\#index : lower, upper, step[paths|processes|bodyreplicator]$$

and, intuitively speaking, this construct can be thought of as a "for loop" which concatenates successively generated copies of the text pattern between [and], where at each step, all occurrences of *index* in the text pattern are replaced by the current value of the index. For example, the n paths representing the individual buffer frames from Pr_{LL} can now be written as:

$$\begin{array}{rl} C(1): & \underline{array}\ deposit\ remove\ (n)\ \underline{endarray} \\ M(1): & \#i : 1, n, 1[\underline{path}\ deposit(i); remove(i)\ \underline{end}] \end{array}$$

Similarly, paths P_{n+5} through P_{3n+4} from Pr_{LL} can now be expressed succinctly as:

$$\begin{array}{l} C(1): \underline{array}\ deposit\ remove\ (n)\ \underline{endarray} \\ M(1): \\ \#i:1,n,1 \\ \left[\begin{array}{l} \underline{path}\ deposit(i),(getmessage; stopsystem; startsystem)\ \underline{end} \\ \underline{path}\ remove(i),(getmessage; stopsystem; startsystem)\ \underline{end} \end{array}\right] \end{array}$$

which shows the utility of having more than one path in a body replicator.

Before we can give the concise form of all of Pr_{LL} we need to introduce replicators which may occur inside of path and process expressions.

Sequence Replicators

Consider the two paths:

$$\begin{array}{l} \underline{path}\ deposit(1);\ldots;deposit(n)\ \underline{end} \\ \underline{path}\ remove(1);\ldots;remove(n)\ \underline{end} \end{array}$$

which specify the ring discipline deposits and removals from the frames of Pr_B must obey. To give a more concise formulation not involving ellipses we will need to use a so called *concatenator*, which concatenates consecutive copies of text patterns and has the form:

$$\#index : lower, upper, step[p(index)\ sep@]$$

where $p(index)$ denotes a text pattern (possibly involving occurrences of $index$), and sep is one of the separators ";" or ",". The @ strips the separator in front of it in the final copy produced from the pattern $p(index)$.

A concise formulation of the two paths from above then becomes:

$$\begin{array}{l} \underline{path}\ \#i:1,n,1[deposit(i)\ ;@]\ \underline{end} \\ \underline{path}\ \#i:1,n,1[remove(i)\ ;@]\ \underline{end} \end{array}$$

and the concise formulation of Pr_{LL} is depicted in Fig. 3.5:

The notation we have thus far explained also allows us to give a concise formulation of $Pr_{CS,proc}$ (Fig. 3.3), which is called $MPr_{CS,rep}$ and depicted in Fig. 3.6:

Distributors

There is a simple form of concatenator which appears so frequently, e.g. in MPr_{LL}, that we use a very convenient shorthand for it, called a distributor, which turns out to have very natural modelling properties as well. A later theorem asserts that a large class of concatenators may be expressed by distributors, but not all.

Consider the path:

$$\underline{path}\ \#i:1,n,1[deposit(i)\ ;@]\ \underline{end}$$

$$
\begin{array}{l}
MPr_{LL} = \\
\underline{program} \\
\quad \left.\begin{array}{l}
\left.\begin{array}{ll}
C(1): & \underline{array}\ deposit\ remove\ (n)\ \underline{endarray} \\
M(1): & \#i : 1, n, 1[\underline{path}\ deposit(i); remove(i)\ \underline{end}] \\
M(2): & \underline{path}\ \#i : 1, n, 1[deposit(i)\ ; @]\ \underline{end} \\
M(3): & \underline{path}\ \#i : 1, n, 1[remove(i)\ ; @]\ \underline{end}
\end{array}\right\} Pr_B \\
\begin{array}{ll}
M(4): & \underline{path}\ readline; \#i : 1, n, 1[deposit(i)\ , @]\ \underline{end} \\
M(5): & \underline{path}\ \#i : 1, n, 1[remove(i)\ , @]; printline\ \underline{end}
\end{array}
\end{array}\right\} Pr^1_{LL} \\
\quad M(6): \\
\quad \#i : 1, n, 1 \\
\quad \left[\begin{array}{l}
\underline{path}\ deposit(i), (getmessage; stopsystem; startsystem)\ \underline{end} \\
\underline{path}\ remove(i), (getmessage; stopsystem; startsystem)\ \underline{end}
\end{array}\right] \\
\\
\quad \begin{array}{ll}
P(1): & \underline{path}\ readline, (getmessage; stopsystem; startsystem)\ \underline{end} \\
P(2): & \underline{path}\ printline, (getmessage; stopsystem; startsystem)\ \underline{end}
\end{array} \\
\underline{endprogram}
\end{array}
$$

Figure 3.5. Linereader/lineprinter with n-frame ring buffer and interrupt handler

$$
\begin{array}{l}
MPr_{CS,rep} = \\
\underline{program} \\
\quad \begin{array}{ll}
C(1): & \underline{array}\ request\ granted\ skip\ (k)\ \underline{endarray} \\
C(2): & \underline{array}\ transfer\ (k+1)\ \underline{endarray} \\
M(1): & \#i : 1, k, 1[\underline{path}\ transfer(i); transfer(i+1)\ \underline{end}] \\
P(1): & \underline{path}\ transfer_{k+1}; criticalsection; signal\ \underline{end} \\
M(2): & \#i : 1, k, 1[\underline{path}\ skip(i), (request(i); granted(i))\ \underline{end}] \\
M(3): & \underline{path}\ \#i : 1, k, 1[skip(i), (granted(i); transfer(1))\ ; @]\ \underline{end} \\
M(4): &
\end{array} \\
\quad \#i : 1, k, 1 \\
\quad \left[\begin{array}{l}
\underline{process}\ request(i); granted(i); \\
\#j : 1, k, 1[transfer(j)\ ; @]; criticalsection; signal\ \underline{end}
\end{array}\right] \\
\\
\underline{endprogram}
\end{array}
$$

Figure 3.6. Critical section with k processes

The distributor equivalent to the above concatenator is simply:

$$\underline{path}\ ; [deposit]\ \underline{end}$$

assuming that the collective name *deposit* has been previously defined by the collectivisor:

$$\underline{array}\ deposit\ (n)\ \underline{endarray}$$

Using this notation we can now rewrite MPr_{LL} to obtain $MPr_{LL,dist}$ which is depicted in Fig. 3.7, and we can rewrite $MPr_{CS,rep}$ to obtain $MPr_{CS,dist}$ which is depicted in Fig. 3.8. The new specifications are much less cumbersome than the old. They also permit a very natural reading of the specification. For example, in $MPr_{LL,dist}$ macro paths $M(4)$ and $M(5)$ may be read as:

> "repeatedly, begin by reading a line and then deposit it into any frame"

> "repeatedly, begin by removing a line from any frame and then print it"

for the two paths respectively.
Similarly, $M(2)$ reads:

> "repeatedly, deposit into successive frames".

If we consider $M(4)$ of $MPr_{CS,dist}$ we can read this to say:

> "any process may repeatedly request access followed by being granted access to the pipeline, after which it is transferred through the pipeline, admitted to the critical section, and required to send a signal upon having left the critical section."

Imbricators

There are some important structures which cannot be expressed by concatenators. These structures arise in the specification of *stack* or *last in first out* disciplines, and one also appears in path R of $Pr^{k,m}_{CS,proc}$. To express these structures we need copies to be generated from both sides, and not just appended to the right as is the case for concatenators.

The so called *imbricator* serves to specify the generation of regularities which are nested within each other, and is of the form:

$$\#index : lower, upper, step[p(index)\ @t@\ q(index)]$$

where $p(index), t$ and $q(index)$ denote text patterns. The expansion of the imbricator looks like:
$p(lower)p(lower + step) \ldots p'(upper')tq'(upper') \ldots q(lower + step)q(lower)$

$$
\begin{array}{l}
MPr_{LL,dist} = \\
\underline{program} \\
\quad \left.\begin{array}{l}
\left.\begin{array}{ll}
C(1): & \underline{array}\ deposit\ remove\ (n)\ \underline{endarray} \\
M(1): & \#i:1,n,1[\underline{path}\ deposit(i); remove(i)\ \underline{end}] \\
M(2): & \underline{path}\ ;[deposit]\ \underline{end} \\
M(3): & \underline{path}\ ;[remove]\ \underline{end}
\end{array}\right\} Pr_B \\
\begin{array}{ll}
M(4): & \underline{path}\ readline;,[deposit]\ \underline{end} \\
M(5): & \underline{path}\ ,[remove]; printline\ \underline{end}
\end{array}
\end{array}\right\} Pr^1_{LL} \\
\quad M(6): \\
\quad \#i:1,n,1 \\
\quad \left[\begin{array}{l}
\underline{path}\ deposit(i),(getmessage; stopsystem; startsystem)\ \underline{end} \\
\underline{path}\ remove(i),(getmessage; stopsystem; startsystem)\ \underline{end}
\end{array}\right] \\
\\
\quad P(1): \ \underline{path}\ readline,(getmessage; stopsystem; startsystem)\ \underline{end} \\
\quad P(2): \ \underline{path}\ printline,(getmessage; stopsystem; startsystem)\ \underline{end} \\
\underline{endprogram}
\end{array}
$$

Figure 3.7. Linereader/lineprinter with n-frame ring buffer and interrupt handler

$$
\begin{array}{l}
MPr_{CS,dist} = \\
\underline{program} \\
\quad \begin{array}{ll}
C(1): & \underline{array}\ request\ granted\ skip\ (k)\ \underline{endarray} \\
C(2): & \underline{array}\ transfer\ (k+1)\ \underline{endarray} \\
M(1): & \#i:1,k,1[\underline{path}\ transfer(i); transfer(i+1)\ \underline{end}] \\
P(1): & \underline{path}\ transfer_{k+1}; criticalsection; signal\ \underline{end} \\
M(2): & \#i:1,k,1[\underline{path}\ skip(i),(request(i); granted(i))\ \underline{end}] \\
M(3): & \underline{path}\ \#i:1,k,1[skip(i),(granted(i); transfer(1))\ ;@]\ \underline{end} \\
M(4): &
\end{array} \\
\quad \#i:1,k,1[\underline{process}\ request(i); granted(i);;[transfer]; criticalsection; signal\ \underline{end}] \\
\underline{endprogram}
\end{array}
$$

Figure 3.8. Critical section with k processes

$MPr^{k,m}_{CS,imbr} =$

$\underline{program}$

$C(1): \ \underline{array}\ deposit\ remove\ (k, m)\ \underline{endarray}$

$C(2): \ \underline{array}\ criticalsection\ skip\ (m)\ \underline{endarray}$

$M(1):$

$\#j : 1, m, 1$

$$\left[\begin{array}{l} \#i : 1, k, 1[\underline{path}\ deposit(i, j); remove(i, j)\ \underline{end}] \\ \underline{path}\ ; [deposit(\ , j)]\ \underline{end}\ \underline{path}\ ; [remove(\ , j)]\ \underline{end} \end{array}\right]$$

$M(2): \ \#j : 1, m, 1[\underline{path}\ skip(j), (, [remove(\ , j)]; criticalsection(j))\ \underline{end}]$

$M(3): \ \underline{path}\ \#j : 1, m, 1[(skip(j);\ @; @\ , criticalsection(j))]\ \underline{end}$

$M(4):$

$\#j : 1, m, 1$

$$\left[\ \#i : 1, k, 1[\underline{process}\ , [(deposit(\ , j); remove(\ , j))]; criticalsection(j)\ \underline{end}]\ \right]$$

$\underline{endprogram}$

Figure 3.9. $MPr^{k,m}_{CS,imbr}$ Critical section with m distinct priorities

where $upper'$ denotes the final value of the range of the index (which might be different from $upper$, for instance in the case "$i : 1, 6, 2$" we have $upper' = 5$. p' and q' are the same as p and q respectively, but with any trailing separator of p and any leading separator of q respectively, removed.

In this way path R of $Pr_{CS,proc}$ can be expressed as:

$$\underline{path}\ \#j : 1, m, 1[(skip(j);\ @; @\ , criticalsection(j))]\ \underline{end}$$

Now we can depict the succinct form $MPr^{k,m}_{CS,imbr}$ of $Pr^{k,m}_{CS,proc}$ in Fig. 3.9. The reader should compare the relative sizes of the two specifications. Note also that in $M(4)$ there is the distributor

$$, [(deposit(\ , j); remove(\ , j))]$$

which indicates that one can distribute a separator over more complex expressions than simple collective names. The separator is assumed to distribute over the leftmost distributable dimension which is indicated by the blank field in the list of indices. Each collective name involved in such a more complex expression must have the same number of distributable dimensions, and each corresponding distributable dimension in the collective names must have the same number of elements.

Left and Right Sequence Replicators

Finally, to succinctly express path P_{3n+m+1} from $Pr^{n,m}_{RW2}$ (Figure 2.67)

$$\underline{path}\ write_1, skip_2, skip_3, \ldots, skip_m; write_2, skip_1, skip_3, \ldots, skip_m;$$
$$\ldots; write_i, skip_1, \ldots, skip_{i-1}, skip_{i+1}, \ldots, skip_m;$$
$$\ldots; write_m, skip_1, \ldots, skip_{m-1}\ \underline{end}$$

we will need to use so called *left concatenators*.

Consider the following attempt to formulate part of the path above using concatenators:

$$\underline{path}\ \#i : 1, m, 1[write(i), \#j : 1, i-1, 1[skip(j)\ ,@],$$
$$\#j : i+1, m, 1[skip(j)\ ,@]\ ;@]\ \underline{end}$$

The path program $Pr^{n,m}_{RW2}$ assumes that $m \geq 2$. If $m = 1$, obviously no starvation of writers by writers is possible, so $skip_i$'s are not needed.

For $m = 2$ we initially obtain the empty string for the first concatenator, but not for the second, and we obtain:

$$\underline{path}\ write(1), , skip(2); write(2), skip(1),\ \underline{end}$$

which constitutes a collision of commas and a collision of a comma with a terminal <u>end</u>.

For $m = 3$ we get:

$$\underline{path}\ write(1), , skip(2), skip(3); write(2), skip(1), skip(3);$$
$$write(3), skip(1), skip(2),\ \underline{end}$$

where only the middle part "$write(2), skip(1), skip(3)$" is correct. Generally, it works correctly for $i \neq 1$ and $i \neq m$, i.e., when there is no expansion to empty strings.

To avoid such problems we assume that sequence replicators should not expand to empty strings, and introduce some special kind of replicators which may expand to empty strings, but they will not produce collisions whether expanded to empty or not. To avoid collisions of the type illustrated above, the context of these new replicators should differ from the context of the old replicators in that *a separator should be missing either on their left or on their right*. Their expansion has to provide the extra separator. If the separator is missing from their left context they are called *left replicators*, and if the separator is missing from the right context they are called *right replicators*.

We will only mention left concatenators here. Later we will also be defining right concatenators, as well as left and right imbricators.

A left concatenator has the form:

$$\#index : lower, upper, step[\ sep \mid p(index)\ sep'@]$$

$MPr^{n,m}_{RW2} =$

$\underline{program}$

$C(1): \ \underline{array}\ readrequest\ readbegin\ readend\ (n)\ \underline{endarray}$

$C(2): \ \underline{array}\ writerequest\ write\ skip\ (m)\ \underline{endarray}$

$\underline{priority}\ (,[write],,[skip]) > (,[readbegin]),$

$(,[readend]) > (,[write],,[skip]),$

$(,[write]) > (,[skip])\ \underline{end}$

$M(1):$

$\#j: 1, n, 1$

$$\left[\begin{array}{l} \underline{path}\ readbegin(i), readend(i),, [write],, [skip]\ \underline{end} \\ \underline{path}\ readrequest(i); readbegin(i); readend(i)\ \underline{end} \end{array}\right]$$

$M(2): \ \#i: 1, m, 1[\underline{path}\ writerequest(i); skip(i)*; write(i)\ \underline{end}]$

$M(3):$

$\underline{path}\ \#i: 1, m, 1[write(i)\#j: 1, i-1, 1[\ ,\ |\ skip(j)\ ,@]$

$\#j: i+1, m, 1[\ ,\ |\ skip(j)\ ,@]\ ;@]\ \underline{end}$

$\underline{endprogram}$

Figure 3.10. $MPr^{n,m}_{RW2}$ Second reader-writer solution starvation free for writers

where $index, lower, upper, step, sep, sep'$ and $p(index)$ are as before. If their index range is empty, the strings generated by their expansion will be empty as well. Otherwise, the strings generated by their expansion will be the same as the strings generated by the expansion of the sequence replicators obtained from them by removing "$sep|$", and prefixing the resultant sequence replicators by "sep".

In this notation the part of the path in question can be written as:

$$\underline{path}\ \#i: 1, m, 1[write(i)\#j: 1, i-1, 1[\ ,\ |\ skip(j)\ ,@] \\ \#j: i+1, m, 1[\ ,\ |\ skip(j)\ ,@]\ ;@]\ \underline{end}$$

which for $m = 2$ expands to

$$\underline{path}\ write(1), skip(2); write(2), skip(1)\ \underline{end}.$$

But it is no longer easy to obtain a natural intuitive reading for the resulting macro expression and one needs to know something about the nature of the expansion to understand what the path means.

Now we are in a position to state a concise formulation $MPr^{n,m}_{RW2}$ of $Pr^{n,m}_{RW2}$, which is depicted in Fig. 3.10.

Note the use of distributors in the priority specification where we write:

$$\underline{priority}\ \ (,[write],,[skip]) > (,[readbegin]), (,[readend]) > (,[write],,[skip]), \\ (,[write]) > (,[skip])\ \underline{end}$$

which we take to be equivalent to:

$$\begin{array}{l} \underline{priority} \\ \#i:1,m,1[\\ \quad \#j:1,n,1[\\ \quad write(i)>readbegin(j), skip(i)>readbegin(j), \\ \quad readend(j)>write(i), readend(j)>skip(i)\ ,\ @]\ , \\ write(i)>skip(i)\ ,@] \\ \underline{end} \end{array}$$

Our notation can be read very naturally as stating that:

> "*write*'s and *skip*'s have priority over *readbegin*'s, and *write*'s have priority over *skip*'s and *readend*'s have priority over *write*'s and *skip*'s".

Predicate Distributors

Unlike the previous concepts, predicate distributors are not yet a part of MACRO COSY as defined in this chapter. Rather they are a possible extension and we consider them very useful.

In formulating $M(3)$ of $MPr^{n,m}_{RW2}$ above using left concatenators we lamented the fact that we did not obtain a natural reading for the resulting path. We obtain a very natural reading with so called predicate distributors, which also allow us to reduce the two uses of left concatenators to a single use of such a distributor.

The general form of a *predicate distributor* is:

$$sep\#index : predicate[mseq]$$

where *sep* and *index* are as for concatenators, *mseq* represents a macro sequence, which for the present purposes may be any macro we have explained so far, and *pred* represents any computable predicate over the indices defined at the point where the predicate occurs. For example, if we have:

$$\begin{array}{c} \underline{array}\ integers(n)\ \underline{endarray} \\ \underline{path}\ ,\#i:i\neq 3[integers]\ \underline{end} \end{array}$$

for $n = 5$ the predicate distributor in the path expands to:

$$\underline{path}\ integers(1), integers(2), integers(4), integers(5)\ \underline{end}$$

which shows that the predicate distributor distributes over all elements of the array *integers* except that it filters out those for which the predicate is not true. It is also implicitly assumed that the set of indices is totally ordered.

The next example illustrates the nesting of such distributors and the ability of using predicates involving more than one index, where i and j refer to the first and second dimension of the array respectively:

$$\underline{array}\ integers(n,m)\ \underline{endarray}$$
$$\underline{path}\ ,\#j : j \neq 3[,\#i : i \neq 3[integers]]\ \underline{end}$$

which, for $n = m = 5$, expands to:

$$\begin{aligned}\underline{path}\ & integers(1,1), integers(2,1), integers(4,1), integers(5,1),\\ & integers(1,2), integers(2,2), integers(4,2), integers(5,2),\\ & integers(1,4), integers(2,4), integers(4,4), integers(5,4),\\ & integers(1,5), integers(2,5), integers(4,5), integers(5,5)\ \underline{end}\end{aligned}$$

which indicates that the path renders all of the operations of the array *integers* mutually exclusive, except for all operations of collumn 3 and the 3rd element in each of the other collumns. As an exercise the reader should formulate, e.g., a two dimensional array of operations whose constituent operations are rendered mutually exclusive, except for the operations in the diagonal, which are sequentialized in the increasing order of indices, and other structures that might form efficient synchronizations of highly concurrent strategies for e.g. arithmetical or sorting processes.

Using this notation we can elegantly rewrite the path involving left concatenators in $MPr^{n,m}_{RW2}$ as:

$$\underline{path}\ \#i : 1, m, 1[write(i), , \#j : j \neq i[skip]\ ;@]\ \underline{end}$$

which for $m = 2$ expands to

$$\underline{path}\ write(1), skip(2); write(2), skip(1)\ \underline{end}.$$

Finally, we can now state the whole of $MPr^{n,m}_{RW2}$ very concisely and also quite readable without expansion as $Mpr^{n,m}_{RW2,\,pdist}$ in Fig. 3.11.

System Notation

To complete our development of notation supporting greater economy and generality of system specifications we will next introduce the system notation. This notation will only be used to illustrate our points in this general section on high-level notation, and we will not give a formal definition of it in this book. However, anyone with experience in using a programming language with a macro facility to express modules (C^{++}) should be able to follow the development of our examples.

Every one of the more complex examples from Chap. 2 involved a number of concepts. Thus Pr_{LL} involved the notions of buffer frame, ring buffer, line reader, line printer and interrupt handler. In the basic notation (see Fig. 2.45) we have

$MPr^{n,m}_{RW2,\,pdist} =$
$\underline{program}$
$C(1): \ \underline{array}\ readrequest\ readbegin\ readend\ (n)\ \underline{endarray}$
$C(2): \ \underline{array}\ writerequest\ write\ skip\ (m)\ \underline{endarray}$
$\underline{priority}\ (,[write],,[skip]) > (,[readbegin]),$
$(,[readend]) > (,[write],,[skip]),$
$(,[write]) > (,[skip])\ \underline{end}$
$M(1):$
$\#j : 1, n, 1$
$$\left[\begin{array}{l}\underline{path}\ readbegin(i), readend(i),,[write],,[skip]\ \underline{end}\\ \underline{path}\ readrequest(i); readbegin(i); readend(i)\ \underline{end}\end{array}\right]$$

$M(2): \ \#i : 1, m, 1[\underline{path}\ writerequest(i); skip(i)*; write(i)\ \underline{end}]$
$m(3): \ \underline{path}\ \#i : 1, m, 1[write(i),,\#j : j \neq i[skip]\ ;@]\ \underline{end}$
$\underline{endprogram}$

Figure 3.11. $MPr^{n,m}_{RW2,\,pdist}$ Second reader-writer solution starvation free for writers

indicated some of the groupings of paths corresponding to some of these concepts. Thus Pr_B brackets the paths specifying the ring buffer, and Pr^1_{LL} brackets both the ring buffer and the paths specifying the line reader/line printer spooling system. But there are a number of paths whose association is not clear without an examination of the whole program. In fact, in basic COSY programs information about interconnections between paths is potentially distributed throughout the entire program. There is no way to indicate locality of concepts.

If we consider the corresponding macro program $MPr_{LL,\,dist}$ (Fig. 3.7) we see the same grouping by means of Pr_B and Pr^1_{LL}, but the macro paths are much more concise, and make regularities explicit which had to be deduced from an examination of larger subsections of the original basic COSY program. This improves readability of specifications, but modularization is often dictated by inclusion in common replicators, and conceptual modularization is still only indicated by informal conventions.

But the explicit expression of the *conceptual modularization* of systems is crucial for achieving comprehensible specifications of complex systems. Furthermore, specifications should be *parametrizable* to allow the definition of generic strategies from which specific strategies can be generated by instantiation with specific parameters. That is, we need the ability to *abstract* from a basic or macro COSY program to a *class* of programs (including the original program), where specific instances of these programs are generated by specific instantiations (reifications) of the class definition.

The system notation we will be introducing here serves to overcome the drawbacks of the basic and macro COSY notations with regard to support of *concep-*

tual modularization, and gives a means for expressing *locality* (or accessibility, independence, protection, relevance, information-hiding), and *hierarchy*, as well as supporting *program development by abstraction and concretion*, and *reuse*. Note however that basic and macro COSY programs become potential "bodies" of definitions of abstractions called **systems** in the new notation, and hence they play a crucial role in the development of new classes of systems. In a sense, one can formulate any strategy which is stated in macro or system notation using only the basic notation. But the former two notations allow one to explicitly represent much information about interconnectivity, hierarchy, and other structures, which is largely implicit and scattered throughout the entire corresponding basic COSY program.

Let us reconsider program $MPr_{LL,\,dist}$ from the standpoint of conceptual modularization. We have not said much about "scopes" of collectivisors because in the macro notation there is only one notion of locality, and that is the whole program. Groupings are only by informal convention and have no effect on the meaning of a program. In fact, from the formal theory of basic COSY from Chap. 2 we can easily deduce that:

> the order of occurrence of the paths and processes in a basic COSY program does not affect its meaning.

This result can easily be extended to macro COSY programs.

In the system notation we will have to distinguish scopes of operations (events) and collective names, which may be restricted only to certain modules.

Since we are here concerned with notational support for the design processes, let us reconsider $MPr_{LL,\,dist}$ from its most basic conceptual elements. As we said before, the basic component of the line reader/line printer system is the notion of a buffer frame, repeatedly capable of having some information deposited into it and then removed from it. Using the system notation we can now abstract from the typical path specifying a single frame to the class of such frames by writing the following **system definition**:

$$\begin{array}{l} SPr_{frame} = \\ \underline{system}\ frame\ ; \\ P(1):\ \underline{path}\ deposit; remove\ \underline{end} \\ \underline{operation}\ deposit\ remove\ \underline{endoperation} \\ \underline{endsystem} \end{array}$$

which indicates that one can create and name systems of type $frame$ and give them names by means of an **instantiation** of the form:

$$frame\ f;$$

which upon expansion gives rise to the path:

$$\underline{path}\ f \circ deposit; f \circ remove\ \underline{end}$$

using the conventional "dot" notation for naming instances of records, structures and classes in programming languages like Pascal, PL/I and Simula, respectively. Processes external to the system definition may only use the operations mentioned in the operation definition part, indicated by $\underline{operation}$ and $\underline{endoperation}$ parentheses.

To understand what the legal behaviours of such a frame are one only needs to consider what is specified by $P(1)$ in SPr_{frame}, namely:

$$FS(P(1)) = Pref(Cyc(P(1))^*) = Cyc(P(1))^* Pref(Cyc(P(1)))$$
$$= \{deposit.remove\}^* \{\varepsilon, deposit\}$$

or in other words:

> "A legal behaviour of *frame* is : any finite sequence of strictly alternating executions of *deposit* and *remove*, beginning with an execution of *deposit*."

Of course we also know that any operation mentioned in at least one path may be executed for at most one process at a time, and in case of conflict, one process is chosen arbitrarily to execute the operation. The strict alternation can be thought of as expressing the fact that information is only preserved until the next removal. That legal behaviours must begin with a deposit can be taken to express the fact that frames are initially empty, or that any transaction involving the frame must first overwrite anything that might be left over from a past transaction. We will first define a pool of mutually unsynchronized buffer frames from which we will obtain a ring buffer and various other types of buffer by a technique of **prefixing** similar to that used in Simula class definitions. To begin with, we define a system called SPr_{pool} whose instantiation will produce a collection of distinct frames, where frames may be individually filled and emptied in arbitrary order. Processes will have access to specific *deposit* or *remove* operations of any frame. A solution is:

$SPr_{pool} =$
$frame$
$\underline{system}\ pool(n : \underline{integer})$;
$C(1) :\ \underline{array}\ (frame)\ F\ (n)\ \underline{endarray}$
$\underline{operation}$
$\underline{array}$
$\#i : 1, n, 1[pdeposit(i) = F(i) \circ deposit]$
$\#i : 1, n, 1[premove(i) = F(i) \circ remove]$
$\underline{endarray}$
$\underline{endoperation}$
$\underline{endsystem}$

Prefixing the system name *frame* to the definition of *pool* indicates that all of the definition of *frame* will replace the occurrence of the prefix *frame*. Hence, SPr_{pool} is equivalent to writing:

```
system frame ;
P(1) : path deposit; remove end
operation deposit remove endoperation
endsystem
system pool(n : integer) ;
C(1) : array (frame) F (n) endarray
 operation
 array
 #i : 1, n, 1[pdeposit(i) = F(i) ∘ deposit]
 #i : 1, n, 1[premove(i) = F(i) ∘ remove]
 endarray
 endoperation
endsystem
```

The formal integer parameter in the definition of *pool* must be replaced by an integer constant in the instantiation of *pool* and supplies the upper bound to the collectivisor as in:

$$pool(5)\ P;$$

which expands to:

```
path P ∘ F(1) ∘ deposit; P ∘ F(1) ∘ remove end
path P ∘ F(2) ∘ deposit; P ∘ F(2) ∘ remove end
path P ∘ F(3) ∘ deposit; P ∘ F(3) ∘ remove end
path P ∘ F(4) ∘ deposit; P ∘ F(4) ∘ remove end
path P ∘ F(5) ∘ deposit; P ∘ F(5) ∘ remove end
```

Note that in order to understand the meaning of Spr_{pool} we need less prefixes in the dot notation to obtain the general expansion pattern below:

```
path F(1) ∘ deposit; F(1) ∘ remove end
path F(2) ∘ deposit; F(2) ∘ remove end
path F(3) ∘ deposit; F(3) ∘ remove end
path F(4) ∘ deposit; F(4) ∘ remove end
path F(5) ∘ deposit; F(5) ∘ remove end
```

Hence, to understand or verify complex specifications one does not need to keep all the prefixes and the notation does not become as heavy as might seem at first sight. Furthermore, since one can largely read high level COSY programs without expansion, one is only involved with prefixes to determine that systems are independent of each other, or to determine where in the hierarchic nesting of subsystems one is presently located.

Note also that we have used replicators in the operation definition part to define and make available to processes and internal paths, arrays of operations for potential use individually or in replicators and distributors. We will be using this ability extensively later on.

What can we say about SPr_{pool} based on what we know about SPr_{frame}? First of all, the expansion of:

$$pool(5)\ P$$

in SPr_{pool} depicted above indicates that all paths involved have no shared operations (are disjoint), which follows from the fact that the expansion of a system instantiation is very similar to that for body replicators. It produces a copy of the path (process) patterns, contained in the system definition with the name "system_name", for each array element corresponding to the collectivisor, prefixing the name of the current array element to every operation occurring in the patterns. As far as the operation definitions are concerned, they do not add any synchronization constraints on operations. Hence we know that Spr_{pool} is adequate.

Based on the notion of a pool of mutually unsynchronized buffer frames, which allows processes to determine which frame they will use, we next introduce a so called *free* buffer SPr_{free}, which will not allow processes to determine which frame they will use. This is accomplished by defining free buffer operations $fdeposit$ and $fremove$ which hide the internal structure of the buffer from processes.

$$
\begin{array}{l}
SPr_{free} = \\
pool \\
\quad \underline{system}\ free(n : \underline{integer})\ ; \\
\quad pool(n)\ P; \\
\quad\quad \underline{operation} \\
\quad\quad fdeposit = (, [P \circ pdeposit]) \\
\quad\quad fremove = (, [P \circ premove]) \\
\quad\quad \underline{endoperation} \\
\quad \underline{endsystem}
\end{array}
$$

If we are a little more precise, we should say that paths and processes external to the free buffer system are only able to use information about operations which is on the left hand side of the equality in operation definitions. So in Spr_{pool}, processes had the information that the operations *pdeposit* and *premove* were arrays with specific structure as defined by the replicators, and hence, they could either use singly subscripted operations $pdeposit(i)$ or $remove(i)$, or replictors or distributors using the collective names. In the case of SPr_{free}, however, the situation is different. There processes only have information about operations *fdeposit* and *fremove*, which they do not see as being mapped onto array operations, since this fact is hidden on the right hand side of the equality. In this case, processes may only request the execution of an arbitrary element of the buffer, and they will never know which frame was finally used to accomodate the data written or read.

Let us consider what we can deduce about the adequacy of Spr_{free} based on what we know about SPr_{pool}. Lemma 2.26 about the basic COSY theory says that:

> A path consisting solely of operations separated by commas may be added to (or deleted from) any program without affecting its adequacy.

Hence, no path or process just consisting of an invocation of the operation *fdeposit* or *fremove* only, can affect the adequacy of the free buffer. But a buffer system can only know about these two buffer operations in processes, and hence every occurrence of these buffer operations in a process, possibly surrounded by contexts involving other operations not known to the buffer system, will appear as one single sequence of buffer operations separated by commas. But the process does not have the ability to place constraints on the internal events of the buffer system since it does not know about them, and hence cannot affect the adequacy of the buffer system.

Next we define a ring buffer SPr_{ring} whose instantiation will produce a collection of distinct frames, where frames can only be filled and emptied cyclically and in strict sequence. A solution is:

$$
\begin{array}{l}
SPr_{ring} = \\
pool \\
\quad \underline{system}\ ring(n : \underline{integer})\ ; \\
\quad pool(n)\ P; \\
\quad M(1) : \underline{path}\ ; [P \circ pdeposit]\ \underline{end} \\
\quad M(2) : \underline{path}\ ; [P \circ premove]\ \underline{end} \\
\quad\quad \underline{operation} \\
\quad\quad rdeposit = (, [P \circ pdeposit]) \\
\quad\quad rremove = (, [P \circ premove]) \\
\quad\quad \underline{endoperation} \\
\quad \underline{endsystem}
\end{array}
$$

What can we say about the ring buffer based on what we know about the pool and frame systems used in its definition? Note that prefixing a system definition with a system name whose definition itself involves a prefix etc., will cause prefixes to be expanded successively starting with the prefix which does not itself have a prefix in its definition.

The paths $M(1)$ and $M(2)$ of the ring buffer definition impose the ring discipline on the collection of frames and constitute a synchronizing connection between distinct frames, which previously were unsynchronized. Omitting the P prefix, we know that:

$$Cyc(M(1)) = \{F(1) \circ deposit.F(2) \circ deposit. \cdots .F(n) \circ deposit\}$$

and

$$Cyc(M(2)) = \{F(1) \circ remove.F(2) \circ remove. \cdots .F(n) \circ remove\}$$

and, hence, legal use of the ring buffer must repeatedly deposit into successive frames, and repeatedly remove from succesive frames. Deposits into and removals from distince frames are not globally synchronized, but solely determined by a finite segment of their own past local behaviour.

It is easy to convince oneself that the vector firing sequence:

$$\underline{F(1) \circ pdeposit}.\underline{F(2) \circ pdeposit}. \cdots .\underline{F(n) \circ pdeposit}.$$

$$\underline{F(1) \circ premove}.\underline{F(2) \circ premove}. \cdots .\underline{F(n) \circ premove}$$

satisfies the conditions of Theorem 2.20 and hence the ring buffer paths are adequate. The operations *rdeposit* and *rremove* are again defined as an arbitrary choice of frames for depositing into and removing from. Which one will actually be assigned to the processes invoking the operations will depend on the buffer discipline (in this case first-in-first-out or fifo) and the current state of individual frames reached due to past usage of the system. An analogous argument to that used in the discussion of the operations of the free buffer ensures the adequacy of all of Spr_{ring}.

To bring home the point about independence of processes using *rdeposit* and *rremove* from specific buffer disciplines, consider the following definition of a stack (or last-in-first-out or lifo) discipline:

```
SPr_stack =
pool
 system stack(n : integer) ;
 pool(n) P;
 M(1) : path #i : 1,n,1[(P(i) ∘ pdeposit; @;@ ; P(i) ∘ premove)*] end
  operation
  sdeposit = (, [P ∘ pdeposit])
  sremove = (, [P ∘ premove])
  endoperation
 endsystem
```

A proof of the adequacy of this system is left to the reader as a non-trivial exercise.

Now that we have specified Pr_B as the ring buffer system SPr_{ring} we need to specify the line reader/ line printer Pr^1_{LL} which involves $M(4)$ and $M(5)$ from $MPr_{LL,\,dist}$. We have assumed that a buffer frame can hold one line of information and hence a reader process will repeatedly read one line of information and then deposit into a frame of the buffer, and a printer process will repeatedly remove information from a frame and print a line, which is all expressed by $M(4)$ and $M(5)$. Continuing our development we can now rewrite the definition of Pr^1_{LL} as SPr^1_{LL}:

$$
\begin{array}{l}
SPr^1_{LL} = \\
ring \\
\ \underline{system}\ spooler(n : \underline{integer})\ ; \\
\ ring(n)\ R; \\
\ P(1) :\ \underline{path}\ readline; R \circ rdeposit\ \underline{end} \\
\ P(2) :\ \underline{path}\ R \circ rremove; printline\ \underline{end} \\
\quad \underline{operation} \\
\quad sread = (readline; R \circ rdeposit) \\
\quad sprint = (R \circ rremove; printline) \\
\quad \underline{endoperation} \\
\ \underline{endsystem}
\end{array}
$$

What do we know about SPr^1_{LL} based on our knowledge about SPr_{free}? There is a lemma (Lemma 2.23) in the theory of basic COSY which states that:

> If a belongs to only one path, say P_j, and there are $c_1, \ldots, c_k \in Ev(P_j)$, $\alpha \geq 1$, $\beta \geq 0$, such that for every $x \in FS(P_j)$, $\#_a(x) \geq \alpha(\#_{c_1}(x) + \cdots + \#_{c_k}(x)) - \beta$, then Pr is adequate if and only if the program Pr with all instances of a deleted and all operators made redundant by the deletion removed.

In this way, *readline* and *printline* can be eliminated by using Lemma 2.23 (as shown in Section 2.4.3) from SPr^1_{LL}. Furthermore, Lemma 2.26 can be used to eliminate both the modified paths corresponding to $P(1)$ and $P(2)$. Similar reasoning applies to any paths or processes using the operations *sread* and *sprint*. Hence, SPr^1_{LL} is adequate if and only if SPr_{free} is, which it is as we have shown. Hence, SPr^1_{LL} is adequate.

When we next consider the role of $M(6)$, $P(1)$ and $P(2)$ of $MPr_{LL,\,dist}$ we see that their function is to make all the operations of the buffer and the, spooler "interruptible". We can make frames interruptible by writing:

$SPr_{interruptible\_frame} =$
$frame$
$\underline{system}\ interruptible\_frame(s : \underline{element})$;
$framef;$
$P(1) :\ \underline{path}\ f \circ deposit, s\ \underline{end}$
$P(2) :\ \underline{path}\ f \circ remove, s\ \underline{end}$
$\underline{operation}$
$i\_deposit = f \circ deposit$
$i\_remove = f \circ remove$
$\underline{endoperation}$
$\underline{endsystem}$

Notice that here we have a new type of formal parameter specification *element*, which indicates that at instantiation of a frame an actual parameter of type element,as specified by the basic COSY grammar, may be used. The effect is that the actual parameter is textually substituted for all occurrences of the corresponding formal parameter in the definition when the instantiation is expanded.

Having defined an interruptible frame we obtain an interruptible pool of frames by simply writing:

$SPr_{interruptible\_pool} =$
$interruptible\_frame$
$\underline{system}\ interruptible\_pool(n : \underline{integer} | s : \underline{element})$;
$C(1) :\ \underline{array}\ (interruptible\_frame(s))\ IF\ (n)\ \underline{endarray}$
$\underline{operation}$
$\underline{array}$
$\#i : 1, n, 1[i\_pdeposit(i) = IF(i) \circ i\_deposit]$
$\#i : 1, n, 1[i\_premove(i) = IF(i) \circ i\_remove]$
$\underline{endarray}$
$\underline{endoperation}$
$\underline{endsystem}$

Now all that remains is to define an interruptible ring as:

$SPr_{interruptible\_ring} =$
interruptible_pool
system *interruptible_ring*(n : *integer*|s : *element*) ;
interruptible_pool($n|s$) IP;
$P(1)$: *path* ;[$IP \circ$ *i_pdeposit*] *end*
$P(2)$: *path* ;[$IP \circ$ *i_premove*] *end*
operation
i_rdeposit = (,[$IP \circ$ *i_pdeposit*])
i_rremove = (,[$IP \circ$ *i_premove*])
endoperation
endsystem

But since we also need to be able to interrupt the spooler operations *sread* and *sprint* we need to define an interruptible spooler:

$SPr^1_{LL,interrupt} =$
interruptible_ring
system *interruptible_spooler*(n : *integer*|s : *element*) ;
interruptible_ring($n|s$) IR;
$P(1)$: *path* *readline*; $IR \circ$ *i_rdeposit* *end*
$P(2)$: *path* $IR \circ$ *i_rremove*; *printline* *end*
$P(3)$: *path* *readline*, s *end*
$P(4)$: *path* *printline*, s *end*
operation
i_sread = (*readline*; $IR \circ$ *i_rdeposit*)
i_sprint = ($IR \circ$ *i_rremove*; *printline*)
endoperation
endsystem

Now we can define the whole of $MPr_{LL,\,dist}$ in the system notation:

$SPr_{LL,interrupt} =$
program
interruptible_spooler
system *interrupt_handler*(n : *integer*) ;
interruptible_spooler(n|(*stopsytem*; *startsystem*)) IS;
$P(1)$: *path* *getmessage*; *stopsytem*; *startsystem* *end*
endsystem
endprogram

where $P(1)$ indicates that *getmessage* may happen anytime, concurrently with all other operations in the buffer and spooler, and after its occurrence all these other operations are in conflict with *stopsystem*, because of the argument (*stopsystem*; *startsystem*) sent to all interruptible subsystems.

Now that we have introduced these conflict possibilities, we need to use the priority notation to give the interrupt handler operations *stopsystem* and *startsystem* priority over all other operations as expressed in Pr_{LL}^{new} (2.74). This requires rewriting the interruptible frame and the interruptible spooler definitions, which contain the conflicts necessary for applicability of priority, as follows:

$SPr^{priority}_{interruptible\_frame}$ =
frame
<u>*system*</u> *p_interruptible_frame*(*s* : <u>*element*</u>) ;
framef;
<u>*priority*</u> *f* ∘ *deposit* < *s* , *f* ∘ *remove* < *s* <u>*end*</u>
P(1) : <u>*path*</u> *f* ∘ *deposit*, *s* <u>*end*</u>
P(2) : <u>*path*</u> *f* ∘ *remove*, *s* <u>*end*</u>
<u>*operation*</u>
i_deposit = *f* ∘ *deposit*
i_remove = *f* ∘ *remove*
<u>*endoperation*</u>
<u>*endsystem*</u>

and

$SPr^{1,priority}_{LL,interrupt}$ =
p_interruptible_ring
<u>*system*</u> *p_interruptible_spooler*(*n* : <u>*integer*</u>|*s* : <u>*element*</u>) ;
p_interruptible_ring(*n*|*s*) *IR*;
<u>*priority*</u> *readline* < *s* , *printline* < *s* <u>*end*</u>
P(1) : <u>*path*</u> *readline*; *IR* ∘ *i_rdeposit* <u>*end*</u>
P(2) : <u>*path*</u> *IR* ∘ *i_rremove*; *printline* <u>*end*</u>
P(3) : <u>*path*</u> *readline*, *s* <u>*end*</u>
P(4) : <u>*path*</u> *printline*, *s* <u>*end*</u>
<u>*operation*</u>
i_sread = (*readline*; *IR* ∘ *i_rdeposit*)
i_sprint = (*IR* ∘ *i_rremove*; *printline*)
<u>*endoperation*</u>
<u>*endsystem*</u>

where we assume that the definition of *p_interruptible_ring* is now prefixed by *p_interruptible_pool*, which in turn is prefixed by *p_interruptible_frame*. Having this we can equivalently write SPr_{LL}^{new} for Pr_{LL}^{new}:

SPr_{LL}^{new} =
<u>*program*</u>
p_interruptible_spooler
<u>*system*</u> *interrupt_handler(n :* <u>*integer*</u>*)* ;
p_interruptible_spooler(n|(stopsytem; startsystem)) PIS;
P(1) : <u>*path*</u> *getmessage; stopsytem; startsystem* <u>*end*</u>
<u>*endsystem*</u>
<u>*endprogram*</u>

The same program with prefixes replaced by system definitions is depicted in Fig. 3.12 and Fig. 3.13 for comparison with the original $MPr_{LL,\,dist}$ in macro notation, and Pr_{LL} in the basic notation. Of course the system program is longer than both the other two programs but it also includes explicitly the conceptual structure of the design. Sometimes it is useful to view system programs in prefix form, since programs become much shorter and do not require repetition of already understood definitions. At other times it is more useful to view system programs with expanded prefixes. Any computer supported design system based on COSY should make such views available to the designer at different points in his conceptualization of the system.

3.2 The Process Notation

In this section we provide a formal syntax and semantics for basic COSY programs, i.e., COSY programs with both paths and processes. A **basic COSY program** is a string derived from the production rules given below. The same meta-language conventions have been used in the syntax rules as used in Sect. 2.1.

The syntax of a basic COSY program is given by the following rules:

B1. *basicprogram* = <u>*program*</u> *basicprogrambody* <u>*endprogram*</u>

B2. *basicprogrambody* = $\{path\}^+\{process\}^+|\{path\}^+|\{process\}^+$

B3. *path* = <u>*path*</u> *sequence* <u>*end*</u>

B4. *process* = <u>*process*</u> *sequence* <u>*end*</u>

B5. *sequence* = $orelement\{;\ orelement\}^*$

B6. *orelement* = $starelement\{,\ starelement\}^*$

B7. *starelement* = $element \mid element*$

B8. *element* = $event \mid (sequence)$

B9. *event* = *name*

$SPr_{LL}^{new} =$
$\underline{program}$
$\underline{system}\ frame\ ;$
$P(1):\ \underline{path}\ deposit; remove\ \underline{end}$
$\underline{operation}\ deposit\ remove\ \underline{endoperation}$
$\underline{endsystem}$

$\underline{system}\ p\_interruptible\_frame(s : \underline{element})\ ;$
$frame f;$
$\underline{priority}\ f \circ deposit < s\ ,\ f \circ remove < s\ \underline{end}$
$P(1):\ \underline{path}\ f \circ deposit, s\ \underline{end}$
$P(2):\ \underline{path}\ f \circ remove, s\ \underline{end}$
$\underline{operation}$
$i\_deposit = f \circ deposit$
$i\_remove = f \circ remove$
$\underline{endoperation}$
$\underline{endsystem}$

$\underline{system}\ p\_interruptible\_pool(n : \underline{integer} | s : \underline{element})\ ;$
$C(1):\ \underline{array}\ (p\_interruptible\_frame(s))\ IF\ (n)\ \underline{endarray}$
$\underline{operation}$
$\underline{array}$
$\#i : 1, n, 1[i\_pdeposit(i) = IF(i) \circ i\_deposit]$
$\#i : 1, n, 1[i\_premove(i) = IF(i) \circ i\_remove]$
$\underline{endarray}$
$\underline{endoperation}$
$\underline{endsystem}$

Figure 3.12. Part 1: Linereader/lineprinter with n-frame ring buffer and interrupt handler

```
system p_interruptible_ring(n : integer|s : element) ;
p_interruptible_pool(n|s) IP;
P(1) : path ;[IP ∘ i_pdeposit] end
P(2) : path ;[IP ∘ i_premove] end
 operation
 i_rdeposit = (,[IP ∘ i_pdeposit])
 i_rremove = (,[IP ∘ i_premove])
 endoperation
endsystem

system p_interruptible_spooler(n : integer|s : element) ;
p_interruptible_ring(n|s) IR;
priority readline < s , printline < s end
P(1) : path readline; IR ∘ i_rdeposit end
P(2) : path IR ∘ i_rdeposit; printline end
P(3) : path readline, s end
P(4) : path printline, s end
 operation
 i_sread = (readline; IR ∘ i_rdeposit)
 i_sprint = (IR ∘ i_rremove; printline)
 endoperation
endsystem

system p_interrupt_handler(n : integer) ;
p_interruptible_spooler(n|(stopsytem; startsystem)) IS;
P(1) : path getmessage; stopsytem; startsystem end
endsystem

endprogram
```

Figure 3.13. Part 2: Linereader/lineprinter with n-frame ring buffer and interrupt handler

where "name" is a sequence of letters, digits and eventually some special symbols beginning with a letter, with possible indices if necessary. Assume additionally that we may write some comments immediately before $\underline{program}$, $\underline{path}$ and $\underline{process}$. We shall use this possibility to write names of programs, paths and processes.

The standard formal semantics of path programs was expressed by means of **vectors of strings.**

In general, a basic program Pr is a string of the form:

$$\underline{program}\ P_1 \dots P_n Q_1 \dots Q_m\ \underline{endprogram}$$

or more concisely:

$$Pr = P_1 \dots P_n Q_1 \dots Q_m$$

where P_i for $i = 1, \dots, n$ are paths, and Q_j for $j = 1, \dots, m$ are processes, $m, n \geq 0$ and $m + n \geq 1$. For $j = 1, \dots, m$, let $Ev(Q_j)$ denote the set of all event occurring in Q_j, and let $Ev(Pr) = Ev(P_1) \cup \dots \cup Ev(P_n) \cup Ev(Q_1) \cup \dots \cup Ev(Q_m)$.

The semantics of a basic COSY program Pr including processes is given by means of the vector firing sequences of an equivalent basic program Pr' involving paths only.

The conversion of Pr into Pr', denoted by $Path(Pr)$, is obtained by the following rules:

(1) For every $a \in Ev(Pr)$ construct a set $i^Q(a) = \{i | a \in Ev(Q_i) \wedge i = 1, \dots, m\}$, (note that $i^Q(a)$ differs from $i(a)$ of Chap. 2).

If $i^Q(a) = \{i_1, \dots, i_k\} \neq \emptyset$, where $i_1 < \dots < i_k$ if $i^Q(a) \neq \emptyset$, then replace every occurrence of the event name a in each path it occurs in by the element $(a\&i_1, \dots, a\&i_k)$.

(2) For every process Q_j, $j = 1, \dots, m$, and every event name $a \in Ev(Q_j)$, replace every occurrence of the event name a in process Q_j by $a\&j$.

(3) Replace all occurrences of $\underline{process}$ by $\underline{path}$.

Then the semantics of Pr, both in terms of vector firing sequences and Petri nets are given in terms of $Path(Pr)$. Note that the assumption that $i_1 < \dots < i_k$ is not absolutely necessary, but without it the above construction is not unique, i.e., $Path(Pr)$ then represents a set of path programs, all of them generate the same set of vector firing sequences.

For example, if

Example 3.2.1

$$Pr_{43} = \begin{array}{l} \underline{program} \\ P(1) : \underline{path}\ a\ \underline{end} \\ Q(1) : \underline{process}\ a\ \underline{end} \\ Q(2) : \underline{process}\ a\ \underline{end} \\ \underline{endprogram} \end{array}$$

then

$$Path(Pr_{43}) = \begin{array}{l} \underline{program} \\ P(1) : \underline{path}\ (a\&1, a\&2)\ \underline{end} \\ P(2) : \underline{path}\ a\&1\ \underline{end} \\ P(3) : \underline{path}\ a\&2\ \underline{end} \\ \underline{endprogram} \end{array}$$

The relation between the basic path program $Path(Pr)$ and its vector firing sequences $VFS(Path(Pr))$ has already been defined in Sect. 2.2.3.

Alternative definition of VFS semantics for process programs.

The set of histories $VFS(Path(Pr))$ may be obtained without Pr having to undergo any conversion, that is, it may be obtained directly from Pr. In the method which follows the definition of firing sequences of paths and vector events of programs defined previously is modified, and firing sequences for processes are defined.

We shall use the concept of *substitution*, as meant in formal language theory, to define this alternative semantics. In principle, a *substitution* f is a mapping of an alphabet A onto a subset of B^*, for some alphabet B. The mapping f is extended to strings of A^* in the standard way, and to languages by defining $f(L) = \bigcup_{x \in L} f(x)$. (See Appendix B).
For every process $Q = \underline{process}\ E\ \underline{end}$, define $Cyc(Q) = Cyc(P)$ where $P = \underline{path}\ E\ \underline{end}$. For instance $Cyc(\underline{process}\ a, b; c\ \underline{end}) = \{ac, bc\}$.
Let $Pr = P_1 \ldots P_n Q_1 \ldots Q_m$ be a basic COSY program with paths $P_1, \ldots, P_n$ and processes $Q_1, \ldots, Q_m$. We shall use substitutions to express the fact that for every process Q_j, every $a \in Ev(Q_j)$ denotes in fact an event $a\&j$, and for every path P_i, if $a \in Ev(P_i)$ and $i^Q(a) = \{i_1, \ldots, i_k\} \neq \emptyset$ then a denotes one of the events $a\&i_1, \ldots, a\&i_k$.

Let sub, $sub_1, \ldots, sub_m$ be the following substitutions:

$$(\forall a \in Ev(Pr))\ sub(a) = \begin{cases} \{a\} & if\ i^Q(a) = \emptyset \\ \{a\&i_1, \ldots, a\&i_k\} & if\ i^Q(a) = \{i_1, \ldots, i_k\}, k \geq 1 \end{cases}$$

$$(\forall a \in Ev(Q_j))\ sub_j(a) \ = \ \{a\&j\},\ j = 1, \ldots, m$$

Now, for every P_i, $i = 1, \ldots, n$, and every Q_j, $j = 1, \ldots, m$, let us define:

$$\begin{aligned} Ev^Q(P_i) &= \{a | a \in Ev(P_i) \wedge i^Q(a) = \emptyset\} \cup \{a\&j | a \in Ev(P_i) \wedge j \in i^Q(a)\}, \\ Ev^Q(Q_j) &= \{a\&j | a \in Ev(Q_j)\}, \\ Cyc^Q(P_i) &= sub(Cyc(P_i)), \\ Cyc^Q(Q_j) &= sub_j(Cyc(Q_j)), \\ FS^Q(P_i) &= Pref(Cyc^Q(P_i)), and \\ FS^Q(Q_j) &= Pref(Cyc^Q(Q_j)). \end{aligned}$$

Let Vev^Q be the set of *vector events defined by* the following sets of events: $Ev^Q(P_1), \ldots, Ev^Q(P_n), Ev^Q(Q_1), \ldots, Ev^Q(Q_m)$.
Define:

$$VFS^Q(Pr) = (FS^Q(P_1) \times \ldots \times FS^Q(P_n) \times FS^Q(Q_1) \times \ldots \times FS^Q(Q_m)) \cap (Vev^Q)^*.$$

The set $VFS^Q(Pr)$ is called the set of *vector firing sequences* of Pr and it represents the behaviour of Pr. Note that if $m = 0$ then $VFS(Pr)$ is correctly defined and then $VFS(Pr) = VFS^Q(Pr)$.
It turns out that in the general case the set $VFS^Q(Pr)$ is the same as the set $VFS(Path(Pr))$.

Theorem 3.1
For a basic program $Pr = P_1 \ldots P_n Q_1 \ldots Q_m$, *where* P_i, $i = 1, \ldots, n$ *are paths, and* Q_j, $j = 1, \ldots, m$ *are processes:*

$$VFS^Q(Pr) = VFS(Path(Pr)).$$

Proof. Suppose that $Path(Pr) = P'_1 \ldots P'_n P'_{n+1} \ldots P'_{n+m}$. From the transformation of Pr into $Path(Pr)$ it follows immediately that $Ev(P') = Ev^Q(P_i)$ for $i = 1, \ldots, n$, and $Ev(P'_{n+j}) = Ev^Q(Q_j)$ for $j = 1, \ldots, m$. Hence $Vev^Q = Vev(Path(Pr))$. To finish the proof it suffices to show that $Cyc(P'_i) = Cyc^Q(P_i)$, for $i = 1, \ldots, n$, and that $Cyc(P'_{n+j}) = Cyc^Q(Q_j)$ for $j = 1, \ldots, m$. But due to the definition of Cyc (see Chapter Sect. 2.2.3) and the fact that for every substitution f we have $f(L_1 \cup L_2) = f(L_1) \cup f(L_2)$, $f(L_1 L_2) = f(L_1) f(L_2)$, $f(L^*) = f(L)^*$ (see Lemma B.2). We need only to show that: $Cyc((a\&i_1, \ldots, a\&i_k)) = sub(Cyc(a))$ if $i^Q(a) = \{i_1, \ldots, i_k\} \neq \emptyset$, and $Cyc(a\&j) = sub_j(Cyc(a))$ for $j = 1, \ldots, m$. But this results immediately from the definitions of sub and sub_j, $j = 1, \ldots, m$. □

3.3 Macro Generators for COSY Notation

In this section we shall analyse some criteria for macro notation we feel should be considered when defining a macro notation. Since some of these criteria can be defined in a formal way we shall first introduce some simple notation.
Let BT, BNT, MT, MNT stand for basic notation terminal symbols, basic notation non-terminal symbols, macro notation terminal symbols and macro notation non-terminal symbols respectively. Define $\sigma_B = basicprogram$, i.e., σ_B is the initial non-terminal of basic notation, and let σ_M denote the initial non-terminal of macro notation. Let $x \overset{M}{\mapsto} y$ mean that there exists a production of

macro notation grammar (including context-sensitive constraints) which transforms a string x into a string y, where $x, y \in (MT \cup MNT)^*$. Let $x \overset{B}{\mapsto} y$ denote the similar relation for basic notation.
Define:

$$\begin{aligned} MPROGRAM &= \{s \in MT^+ | \sigma_M \overset{M}{\mapsto}{}^+ s\}, \\ BPROGRAM &= \{s \in BT^+ | \sigma_B \overset{B}{\mapsto}{}^+ s\}, \\ MGEN &= \{s \in MT^+ | (\exists \alpha \in MNT)\alpha \overset{M}{\mapsto}{}^+ s\}, \\ BEX &= \{s \in BT^+ | (\exists \alpha \in BNT)\alpha \overset{B}{\mapsto}{}^+ s\}, \end{aligned}$$

where $\overset{M}{\mapsto}{}^+$ is a transitive closure of $\overset{M}{\mapsto}$ and $\overset{B}{\mapsto}{}^+$ is a transitive closure of $\overset{B}{\mapsto}$, i.e., $x \overset{M}{\mapsto}{}^+ y$ (and $x \overset{B}{\mapsto}{}^+ y$ respectively) means that there exists a production sequence of length one or more starting from x and resulting in y (see Appendix B).

The elements of $MPROGRAM$ are called **macroprograms**, the elements of $BPROGRAM$ are **basicprograms**, the elements of $MGEN$ are called **macro expressions** or **macro generators**, and the elements of BEX are called **(well-formed) basic expressions.**

Consider for instance a macroprogram MPr_{LL} of Fig. 3.5. MPr_{LL} is a correct Macro COSY program, but it clearly cannot be expanded. To make MPr_{LL} expandable we have to replace every occurrence of a **range variable** in range definitions by any integer greater than zero. The expansion of macro generators depends on the range of values of the generator index. Such a range is defined as soon as the lower and upper bounds, and the incremental step are defined for the generator. A **range variable** of a macro generator is any name occurring in any range definition such that it may be replaced by an appropriate integer without violating syntactic rules of macro notation. For every $g \in MGEN$, let $rvar(g)$ denote the set of all range variables of g. Of course $rvar(g)$ may be empty. For every $g \in MGEN$, a mapping r which assigns some values to elements of $rvar(g)$ is called a **g–assignment** if and only if $g^r \in MGEN$, where g^r is a string obtained from g by replacing every $n \in rvar(g)$ by $r(n)$.
Let $assgn(g)$ denote the set of all $g - assignments$. Clearly if $rvar(g) = \emptyset$ then $assgn(g) = \emptyset$. If a macro generator $g \in MGEN$ does not contain range variables, i.e., if $varv(g) = \emptyset$, then it can be expanded. Let $expand(g)$ stands for the result of expanding the macro generator g according to some expansion rule such as we define later and we illustrated informally in Section 3.1.
The first requirement which obviously every macro notation must satisfy is that its grammar must not posses **useless symbols.** A symbol $a \in MT \cup MNT$ is **useful** if there is a derivation $\sigma_M \overset{M}{\mapsto}{}^* xay \overset{M}{\mapsto}{}^* w$, for some x, y, w, where $w \in MT^*$. Otherwise a is useless.
Clearly a macro notation of any practical use must not have useless symbols, so we shall assume that any notation mentioned in this section has this property.

The second criterion which our macro notation must satisfy is **expansion correctness.** We will distinguish collective expansion correctness and individual expansion correctness, and characterize properties generators must have to guarantee these kinds of expansion correctness. **Collective expansion correctness** means that **every well formed macro specification gives rise to a well formed basic specification upon full expansion.**

Individual expansion correctness means that **every macro generator of the notation gives rise to a well formed basic expression when fully expanded on its own.**

Collective expansion correctness is also a property which must be satisfied by a macro notation of any practical use.
More formally, we state **collective expansion correctness** as

$$(\forall g \in MPROGRAM)(\forall r \in assgn(g))\ expand(g^r) \in BPROGRAM,$$

where $g^r = g$ if $assgn(g) = \emptyset$.
Individual expansion correctness is in this style defined as

$$(\forall g \in MGEN)(\forall r \in assgn(g))\ expand(g^r) \in BEX.$$

As one may conclude from Section 3.1, neither left nor right replicators satisfy the individual expansion correctness criterion. Consider the left replicator:

$$\#i : 1,3,1[\ ;\ \mid a(i)\ ,@].$$

It expands to the string

$$;a(1), a(2), a(3)$$

which does not belong to BEX. However, when we put this left replicator in a proper context, we may obtain a well formed basic expression after expansion. For instance:

$$b; \#i : 1,3,1[\ ;\ \mid a(i)\ ,@]$$

expands to the string:

$$b; a(1), a(2), a(3)$$

which is a proper element of BEX. This leads us to the following definition.
A macro generator $g \in MGEN$ is **context dependent** if and only if

(1) $(\exists r \in assgn(g)) expand(g^r) \notin BEX$,

(2) $(\exists xgy \in MGEN)(\forall r \in assgn(xgy)) expand((xgy)^r) \in BEX$.

We shall say that a macro notation is **context independent** if no macro generator $g \in MGEN$ is context dependent.
The following theorem characterizes collective and individual expansion correctness.

Theorem 3.2 *If a macro notation does not contain useless symbols then the following are equivalent:*

(1) a macro notation satisfies individual expansion correctness
(2) a macro notation is context independent and satisfies collective expansion correctness.

Proof.
($\Rightarrow$) Individual expansion correctness implies collective expansion correctness and negates the condition (1) of context dependency.
($\Leftarrow$) If (1) of context dependency is not satisfied then a macro notation satisfies individual expansion correctness from the definition of it. Let $g \in MGEN$ and (2) of context dependency is not satisfied. Suppose that $\alpha \overset{M}{\mapsto}^{+} g$ where $\alpha \in MNT$. Since a macro notation does not have useless symbols, we have $\sigma_M \overset{M}{\mapsto}^{+} x\alpha y \overset{M}{\mapsto}^{+} vgw$, where $vgw \in MPROGRAM$. Since the collective expansion criterion is satisfied, we have $(\forall r \in assgn(vgw))expand((vgw)^r) \in BEX$. But this means that (2) of context dependency is satisfied (as $MPROGRAM \subseteq MGEN$), a contradiction. □

The third criterion is **maximal expressiveness** which means that as many basic specifications should be concisely representable as macro specifications as possible. Obviously, since basic specifications form a subset of macro specifications, we can represent any basic specification by itself, and hence the macro notation satisfies the third criterion trivially. But regularities distributed throughout a basic specification should be more explicitly and concisely represented in the macro notation to achieve our goal of rendering explicit synchronic structure which is implicit in basic notation. Hence, maximal expressiveness must be more narrowly defined as: **every regularity distributed throughout a basic specification should be concisely representable in the macro notation by the use of generators.**

In the sequel we will define macro notations which will try to maximize expressiveness while guaranteeing strong collective expansion correctness. We will say that such notations satisfy **optimal expressiveness with regard to collective expansion correctness.**

Basic specifications can exhibit these kinds of regularity in two ways. First, at the **inter system level**, when there are separate occurrences of more than one subsystem of identical structure, possibly differing in indices of identical event names, or differing in event names, or involving event names where another system involves a compound expression. To express such regularities concisely we need generators which represent all these subsytems up to isomorphism of some kind. Second, at the **intra system level**, when there are separate occurrences of more than one part of a sequential subsystem of identical structure as characterized above.

Generators at the inter system level will involve patterns consisting of sequences of paths and/or processes, which themselves may contain generators at the intra system level, but provided each macro path and macro process gives rise to a well formed basic path or process upon expansion, the expansion of the whole system will obviously give rise to a well formed basic system, as can be seen from the syntax rule for basic COSY specifications. Hence inter system level patterns satisfy collective expansion correctness, individual expansion correctness and maximal expressions if we assume that intra sytem level generators they involve satisfy these ctriteria.

Generators at the intra system level will involve patterns of strings possibly involving additional generators. To satisfy the criteria of expansion correctness full expansion must result in a basic level expression.

The fourth criterion which our macro notation must satisfy is the existence of a **high level semantics consistent with the low level semantics**, which means that it must be possible to interpret (read, understand) any macro specification directly in terms of the macro generators it involves without requiring previous expansion, and in a way consistent with the interpretation of the corresponding expanded specification.

A final criterion which our macro notation should satisfy is to be **close fitting** which means that **formal context free syntax rules should be given requiring the weakest meta restrictions to ensure collective expansion correctness.**

Some of these criteria are in conflict, for example, the criterion of maximal expressiveness and the criterion of direct high level semantics. The criteria defined above are general guidelines for the macro notation which is defined precisely in the next section.

3.3.1 The Macro Program

A macro COSY program will consist of collectivisors, paths, processes and bodyreplicators appearing between the word pair $\underline{program}$ and $\underline{endprogram}$.

Since after expansion of a macro program all its collectivisors disappear, the macro program should include at least one path or process or bodyreplicator, for the body of the basic program obtained by expansion to be non-empty. The syntax for macro programs is given by:

$$M1. \quad mprogram \;=\; \underline{program}\; mprogrambody\; \underline{endprogram}$$
$$M2. \quad mprogrambody \;=\; \{\{collectivisor\}^*\{mpath|mprocess|bodyreplicator\}\}^+$$

In the above rules, and henceforth, non-terminals of macro COSY which correspond to non-terminals of basic COSY have been obtained by prefixing the latter by m standing for *macro*. In this way it is easier to see in what way the macro notation is uniform with the basic notation.

According to the above rules, collectivisors, macro paths, macro processes and bodyreplicators may appear in any order, with the exception that no collectivisor may appear after all the paths, processes and bodyreplicators. Note that here we do not require macropaths to precede macroprocesses.

3.3.2 The Collectivisors

The macro notation permits declarations of collective names corresponding to rectangular arrays and to arrays of other shapes.
When declaring rectangular arrays two conventions have been followed: either the lower bound of indices in their dimensions are implicitly considered as having the value one, or the lower bound is explicitly specified.

To shorten some otherwise lengthy expressions in our macro grammar we shall adopt the following convention:

$$\{\mathit{item}\ \mathit{separator}\ \underline{@}\}^+ = \mathit{item}\ \{\mathit{separator}\ \mathit{item}\}^*$$

This convention is used throughout this Chapter and in Appendix H.

The syntax for collectivisors is given by[1]:

$M2.1.\ \mathit{collectivisor} = \underline{\mathit{array}}\ \{\{\mathit{simpleardecl}|\mathit{replardecl}\}\ \mathit{space}\ \underline{@}\}^+\ \underline{\mathit{endarray}}$
$M2.2.\ \mathit{simpleardecl} = \{\mathit{name}\ \mathit{space}\ \underline{@}\}^+(\{\{\mathit{iexpr} : |\}\mathit{iexpr}\underline{@}, \}^+)$
$M2.3.\ \mathit{replardecl} = \mathit{range\_spec}[\{\mathit{replardecl}|\mathit{subsevents}\}]$
$M2.4.\ \mathit{subsevents} = \{\{\mathit{name}\ \mathit{space}\ \underline{@}\}^+(\{\mathit{iexpr}\ ,\ \underline{@}\}^+)\mathit{space}\ \underline{@}\}^+$
$M2.5.\ \mathit{range\_spec} = \#\mathit{indexvariable} : \mathit{iexpr}, \mathit{iexpr}, \mathit{iexpr}$
$M2.6.\ \mathit{iexpr} = \mathit{integer}|\mathit{rangevariable}|\mathit{indexvariable}|\mathit{iexpr} + \mathit{iexpr}|\mathit{iexpr} * \mathit{iexpr}|$
$\mathit{iexpr} \div \mathit{iexpr}|\mathit{iexpr} ** \mathit{iexpr}|(\mathit{iexpr})$
$M2.7.\ \mathit{space} = \mathit{the\ space\ symbol}$
$M2.8.\ \mathit{indexvariable} = \mathit{name}$
$M2.9.\ \mathit{rangevariable} = \mathit{name}$

In the syntax rules given above, *simpleardecl* stands for a list of collective names which correspond to simple rectangular arrays together with their bounds, and *replardecl* stands for the replicator generating admissible sets of subscripted events. The non-terminal *range_spec* stands for the range specification part of a replicator, *iexpr* for an integer expression, and *subsevents* for a sequence of subscripted events.

When subscripts in the dimensions of arrays are consecutive positive integers starting from 1, the lower bound in these dimensions may be implicitly assumed to have the value one and only the upper bound has to be specified. We permit collective names with a different number of dimensions and/or different bounds

[1] The rule for *iexpr* leads to an ambiguous grammar, but it is very abstract and readable. It may be easily replaced by a more concrete and unambiguous grammar, as that for arithmetic expression from the Pascal programming language, for instance.

in their dimensions to be declared by the same collectivisor.
For example the two *simple array declarations* in $C1$:

$$C1: \ \underline{array}\ a(k)\ \underline{endarray}$$
$$\underline{array}\ b\ c(5)\ d(3,m)\ \underline{endarray}$$

declare the subscripted events:

$$a(1)\dots a(k),$$
$$b(1)\dots b(5),$$
$$c(1)\dots c(5),$$
$$d(1,1)\ d(2,1)\ d(3,1)$$
$$d(1,2)\ d(2,2)\ d(3,2)$$
$$\dots$$
$$d(1,m)\ d(2,m)\ d(3,m)$$

Furthermore the following **total orderings** are defined:

$$a(1) \prec \dots \prec a(k),$$
$$b(1) \prec \dots \prec b(5),$$
$$c(1) \prec \dots \prec c(5),$$
$$d(1,1) \prec d(2,1) \prec d(3,1) \prec d(1,2) \prec d(2,2) \prec d(3,2) \prec \dots$$
$$\prec d(1,m) \prec d(2,m) \prec d(3,m).$$

In general, the order generated by simple array declaration is **antilexicographical**, i.e., the significance of indices is from right to left. Formally, for two indexed events: $a(i_1,\dots,i_k)$ and $a(j_1,\dots,j_k)$, we have

$$a(i_1,\dots,i_k) \prec a(j_1,\dots,j_k) \Leftrightarrow (\exists r \leq k) i_r < j_r \wedge (\forall s > r) i_s = j_s.$$

This will be the same order in which simple distributors will be expanded as discussed in a later section.

If the lower bound in some dimensions of collective names is not 1 but some other fixed integer n we may specify it explicitly. To specify for example that the single dimension of collective name e has lower bound n and upperbound k, and that the first dimension of the two dimensional collective name f has lower bound m and upper bound n and its second dimension lower bound 1 and upper bound k, we write the simple array declarations:

$$C2: \ \underline{array}\ e(n:k)\ f(m:n,k)\ \underline{endarray}$$

We may also combine the declarations in $C1$ and $C2$ into one declaration:

$$C3: \ \underline{array}\ a(k)\ b\ c(5)\ d(m,3)\ e(n:k)\ f(m:n,k)\ \underline{endarray}$$

For collectivisors to be well-formed we shall require all the declarations to satisfy the collectivisor restriction Crest1:

(Crest1) If the upperbound and lowerbound of dimensions of the collective names are integers then the upperbound has to be greater than or equal to their corresponding implicit or explicit lowerbound.

We also permit declaration of subscripted events the indices of which are not consecutive integers, or depend on the index of some other dimension, by means of *replicator array declarations.* For example, the index in the first dimension of the subscripted events:

$$\begin{array}{l} s(1,1) \\ s(3,1)\ s(3,2)\ s(3,3) \\ s(5,1)\ s(5,2)\ s(5,3)\ s(5,4)\ s(5,5) \end{array}$$

takes values 1, 3, 5 and the index in their second dimension takes consecutive values from 1 to the current value of the index of their first dimension. We shall use replicators to generate the set of *admissible subscripted events.* The subscripted events corresponding to s may be declared by the collectivisor C4:

$$C4: \ \underline{array}\ \#i:1,5,2[\#j:1,i,1[s(i,j)]]\ \underline{endarray}$$

using replicators. A general replicator used in a collectivisor has the form:

$$\underline{array}\ldots\#i:i_1,i_2,i_3,[p(i)]\ldots\underline{endarray}.$$

where $p(i)$ is some regularity containing i.

The following collectivisor restriction must also hold:

(Crest2) An array identifier may only occur once in collectivisors.

The index specification part "$\#i : i_1 i_2 i_3$" may be read as "for i from i_1 to t_2 in steps i_3". The set of indexed events declared by the replicator array $\#i : i_1, i_2, i_3[p(i)]$ is defined recursively by:

$$\begin{array}{l} set(a(i_1,\ldots,i_k)) = \{a(i_1,\ldots,i_k)\}, \text{ for every indexed event } a(i_1,\ldots,i_k), \\ set(\#i:i_1,i_2,i_3[p(i))] = \bigcup_{j=1}^{r} set(p(i_1+(j-1)\cdot i_3), \text{ where } r = \frac{i_2-i_1}{i_3}. \end{array}$$

We assume that the elements of the set $set(\#i : i_1, i_2, i_3[p(i)])$ are ordered, and the order is defined as:

$$\begin{array}{l} (\forall a(i_1,\ldots,i_k), a(j_1,\ldots,j_k) \in set(\#i:i_1,i_2,i_3[p(i)]) \\ (a(i_1,\ldots,i_k) \in set(p(l_1)) \setminus set(p(l_2)) \wedge a(j_1,\ldots,j_k) \in set(p(l_2)) \setminus set(p(l_1)) \wedge l_1 < l_2) \\ \Rightarrow\ a(i_1,\ldots,i_k) \prec a(j_1,\ldots,j_k). \end{array}$$

The above order is partial as there is no order between $a(\ldots)$ and $b(\ldots)$ if $a \neq b$. However, if $p(i)$ contains no identical array identifiers (see Crest2), then this order restricted to indexed events having the same name is total. For instance

the collectivisor C4 generates the following total order of $s(i,j)$'s:
$s(1,1) \prec s(3,1) \prec s(3,2) \prec s(3,3) \prec s(5,1) \prec s(5,2) \prec s(5,3) \prec s(5,4) \prec s(5,5)$.
In this notation for replicator array declaration we have separated the range specification part i.e., "$\#i : i_1 i_2 i_3$", which is common to all replicators no matter what they generate, from the regularities they generate. For the replicators to be well-formed they should obey a third collectivisor restriction Crest3:

(Crest3) Each replicator without range variables must specify a non empty range for its index.

Restriction Crest3 says that replicators in collectivisors are supposed to declare at least one subscripted event corresponding to each collective name.

Note that a simple array declaration of an n-dimensional array may now be seen as a shorthand for a replicator array declaration with n nested replicators making the standard expansion of simple array declarations explicit. For example the declaration of d in $C1$ can be written:

$$\#j : 1, m, 1[\#i : 1, 3, 1[d(i,j)]].$$

Subscripted events with the same subscripts in all their dimensions may be declared in the same replicator, and are separated by space. Also subscripted events with the same subscripts in some of their dimensions, may be grouped together in the same replicator, e.g.

$$C5 : \quad \underline{array}\ \#i : 1, 5, 2[t(i)\ \#j : 1, i, 1[s(i,j)\ u(i,j)]]\ \underline{endarray}$$

where the collective name t corresponds to the events

$$t(1)\ t(3)\ t(5)$$

the collective name u to the events

$$\begin{array}{l} u(1,1) \\ u(3,1)\ u(3,2)\ u(3,3) \\ u(5,1)\ u(5,2)\ u(5,3)\ u(5,4)\ u(5,5) \end{array}$$

and the collective name s to the same events as the collectivisor C4. The subscripted events in replicators may be indexed by expressions involving replicator indices.

We also permit grouping together subscripted events, indexed by expressions depending on replicator indices indexing other subscripted events as shown in C6:

$$C6 : \quad \underline{array}\ \#i : 1, 5, 2[v(i+3)\ \#j : 1, i, 1[s(i,j)]]\ \underline{endarray}$$

where v corresponds to the events $v(4)\ v(6)\ v(8)$.

It might happen that some subscripted events are "declared many times" as for instance:

$$\underline{array}\ \#i : 1,5,2[\#j : 1,i,1[t(i)\ \ s(i,j)]]\ \underline{endarray}$$

where $t(3) \in set(\#j : 1,i,1[t(i)\ s(i,j)])$ for $i = 3,4,5$ and $t(5) \in set(\#j : 1,i,1[t(i)\ s(i,j)])$ for $i = 1,2,3,4,5$.

This is acceptable since the part:

$$\text{“}a(i_1,\ldots,i_k) \in set(p(l_1)) \setminus set(p(l_2)) \wedge a(j_1,\ldots,j_k) \in set(p(l_2)) \setminus set(p(l_1))\text{”}$$

in the definition of the order defined by the replicator array, guarantees that a situation like $t(3) \prec t(3)$ does not happen. Of course, we do not encourage anyone to use this kind of side effects as it may clearly worsen the readability of a specification.

We may also combine the $C3$ and $C6$ types of declaration in a single declaration. Also note that one may use *negative* integers both in simple array declarations and replicator array declarations, as illustrated below:

$$\underline{array}\ a(-2:5)\ \#i : 5,-2,-2[b(i)]\ \underline{endarray}$$

which defines:

$$a(-2)\ a(-1)\ a(0)\ a(1)\ a(2)\ a(3)\ a(4)\ a(5)$$
$$b(5)\ b(3)\ b(1)\ b(-1)$$

Identifiers used for replicator indices must satisfy the index restriction Irest1:

(Irest1) Identifiers for replicator indices and range variables should be distinct from any identifiers used for simple events. Identifiers for replicator indices and range variables should also be distinct.

as well as the restriction Irest2:

(Irest2) Replicator index variables are only defined inside the [ ] of the replicators with which they are associated. In the scope of a replicator index variable no other replicator index variable having the same name is permitted.

The restrictions on replicator indices $Irest1$ and $Irest2$ apply to all replicators, including those to be introduced formally later.

3.3.3 The Body-Replicators

So far we have encountered a simple form of replicator in replicator array declarations. Bodyreplicators are another kind of replicator and generate consecutive regularities consisting of paths and/or processes, or rather macro paths and/or processes to be more precise. Nesting of bodyreplicators will also be allowed. Their syntax is formally given by:

$$M2.10. \quad bodyreplicator \;=\; range\_spec[\{mpath|mprocess|bodyreplicator\}^{+}]$$

Note the similarity of this rule to the rule $M2$ for *mprogrambody*.

No meta-restriction is needed in this case. If each of the paths and processes generated is well-formed then the whole expansion is well-formed, which will be formally demonstrated later.

Using the above rule we can, for example, specify m pipelines of size k each, associated with a mechanism controlling access to a critical section similar to that in $MPr^{k,m}_{CS,imbr}$:

$$\begin{array}{l} C(1): \underline{array}\; tr(m,k+1)\; criticalsection(m)\; \underline{endarray} \\ NP1: \\ \;\#i:1,m,1 \\ \;\left[\begin{array}{l} \#j:1,k,1[\underline{path}\; tr(i,j);tr(i,j+1)\; \underline{end}] \\ \underline{path}\; tr(i,k+1);criticalsection(i)\; \underline{end} \end{array}\right] \end{array}$$

We impose the following restriction *BRrest* on bodyreplicators:

(BRrest) The range of the bodyreplicator indices should be non empty,

ensuring that bodyreplicators generate at least one regularity upon expansion. This is important, for a macro program body could consist of just bodyreplicators which upon expansion should generate a non-empty basic program body.

3.3.4 Macro Paths and Macro Processes

The syntax of macro paths and processes is similar to the syntax of paths and processes of basic COSY:

$$\begin{array}{lcl} M3.1. \quad mpath & = & \underline{path}\; msequence\; \underline{end} \\ M3.2. \quad mprocess & = & \underline{process}\; msequence\; \underline{end} \end{array}$$

We have used *msequence* instead of *bsequence* to stand for "macro sequence" since it may contain replicators and distributors as constituents. Similarly, in the syntax rules below, *morelement* will stand for "macro orelement":

$M4.$ $msequence = \{morelement; \underline{@}\}^+$
$M5.$ $morelement = \{gelement, \underline{@}\}^+$
$M5.1.$ $gelement =$
$\{rreplicator\}^*\{mstarelement|sreplicator|distributor\}\{lreplicator\}^*$
$M6.$ $mstarelement = melement|melement*$
$M7.$ $melement = mevent|(msequence)$
$M8.$ $mevent = name\{\ |(\{\{iexpr|\ \}, \underline{@}\}^+)\}$

In the above rules, *gelement* stands for "generalized element" since it can be any of *mstarelement*, *sreplicator* or *distributor*, possibly preceded by right replicators or succeded by left replicators, respectively. We have used *sreplicator* to indicate another kind of replicator which expands to basic COSY sequences, and which we call **sequence replicators.** The difference between the above rules and corresponding ones in basic notation is that here we allow six new types of elements. These are, sequence replicators and distributors, produced by *sreplicator* and *distributor* respectively, which cannot be starred, right and left replicators, produced by *rreplicator* and *lreplicator*, respectively, and indexed events and event slices[2], produced by the second alternative of *mevent.*

3.3.5 Sequence Replicators

Replicators generating basic COSY sequences are of two types:
Concatenators
which generate consecutive regularities are of the form:

$$(Conc) : \#i : in, fi, inc[p(i)\ sep@]$$

and
Imbricators
which generate regularities nested within each other are of the form:

$$(Imbr) : \#i : in, fi, inc[p(i)\ @t@\ q(i)]$$

where p, t, q denote arbitrary strings of macro terminal symbols, and *sep* denotes the separators ";" or ",".

The above definition is quite general and many regularities could be generated by using such replicators, not only those forming well-formed basic strings, so some other rules are necessary.

In the concatenator the @ strips the separator in front of it in the last copy of the regularity $p(i)\ sep$. The expansion of the concatenator therefore is:

$$(concexp) : p(in)\ sep\ p(in + inc)\ sep \ldots sep\ p(fi')$$

[2] Event slices, represented by mevents like $a(3,,)$, are discussed in Chapter 3.3.7 which is devoted to distributors.

where fi' denotes the final value in the range of the index which may be different from fi, more precisely $fi' = max\{j|(\exists k \geq 0)j = in + k * inc \leq fi\}$ if $inc > 0$, or $fi' = min\{j|(\exists k \geq 0)j = in + k * inc \geq fi\}$ if $inc < 0$.

In the imbricator the separators immediately before the first @ and immediately after the second @ in the last copy of the regularity $p(\ )\ q(\ )$ will be replaced by t. The expansion of the imbricator is:

$$\begin{array}{l}(imbrexp):\\ p(in)\ p(in+inc)\ \ldots\ p'(fi')\ t\ q'(fi')\ \ldots\ q(in+inc)\ q(in)\end{array}$$

In the above expression fi', p, q, t respectively are the same as in $(Imbr)$, and p', q' are the same as p, q, respectively, but with any trailing separator of p and any leading separator of q, respectively, removed. The reason we have specified a string t to occcur between the two @s instead of just a separator, is that we would like our grammar to permit paths such as $MP(1)$:

$$\begin{array}{l}MP(1):\\ \underline{path}\ empty, \#i:1,n,1[(up(i);\ @; full*; @\ ; down(i))*]\ \underline{end}\end{array}$$

which specifies a stack of size n with tests for empty and full. When $MP(1)$ is expanded for $n = 3$, for example, the path $BP(2)$ is obtained:

$$BP(2): \quad \begin{array}{l}\underline{path}\ empty\\ \quad ,(up(1)\\ \quad ;(up(2);(up(3);full*;down(3))*;down(2))*\\ \quad ;down(1)\\ \quad)*\ \underline{end}\end{array}$$

in which the starred event $full*$ appears only once, in the innermost regularity. In general we would permit any string to appear at that position as long as it forms a well-formed string in the basic notation with the rest of the expansion.

Having specified what general kind of replicators we will permit, and having decided on the notation of sequence replicators, we proceed to obtain their formal syntax rules. The approach we follow here is not to leave p, t, q as arbitrary strings of macro terminal symbols, but to specify more precisely what these may be syntactically. The syntax of the two types of replicators, concatenators and imbricators, will be considered separately. The non-terminal *sreplicator* producing sequence replicators is defined as follows:

$$M10.\quad sreplicator \;=\; range\_spec[\{concseq|imbrseq\}]$$

where the non-terminals *concseq* and *imbrseq* produce the string inside [ ] of concatenators and imbricators respectively, whose syntax we give next.

The non-terminal *concseq*.

Before we give the syntax rule for the non-terminal *concseq*, let us examine informally what $p(i)$ in *Conc* should be syntactically. Its expansion (*concexp*) was schematically given above. For the string resulting from expansion to be a well-formed basic notation sequence, each of the $p(\ )$ may in general be an *msequence*, as we shall formally prove later. Therefore we may define *concseq* as:

$$M11.\ concseq \ = \ msequence\ sep\ @$$

where $sep = \ ;|$, as the rule M9.1 states (see Appendix H or the beginning of Section 3.3.7).
Examples of replicators according to this rule are $R(1)$ and $R(2)$:

$$\begin{array}{ll} R(1): & \#i:1,n,1[deposit(i)\ ;@] \\ R(2): & \#i:1,n,1[\#j:1,k,1[A(i,j)\ ,@]\ ;@] \end{array}$$

Later we shall formally prove that each of the replicators produced by these rules expands to macro sequences in general, and that their complete expansion forms a basic notation sequence.

The non-terminal *imbrseq*.

Before we give syntax rules for *imbrseq* let us examine informally what imbricators should generate upon complete expansion to always generate well-formed basic notation sequences. First consider just the outermost regularity of their expansion (*imbrexp*):

$$p(in)\ p(in+inc)\ldots\ p'(fi')\ t\ q(fi')\ldots q(in+inc)\ q(in)$$

Since (*imbrexp*) on the whole forms a sequence, $p(in)$ must be a legal head of an *msequence*, and should start with either an *mevent* or "(". Similarly, $q(in)$ must be a legal tail of an *msequence*, and must terminate with an *mevent* or ")" or ")*". Next examine the first and the second regularity of the expansion (*imbrexp*). The strings $p(in)$ and $p(in+inc)$ must be legally connected if they together are to form a legal head of a sequence. Since these two strings differ only in the integer expressions they involve, they start with the same symbols, which implies that for the expansion of the imbricator (*imbrexp*) to be well-formed $p(i)$ must terminate with ";" or "," or "(". Applying a similar argument to $q(in)$ and $q(in+inc)$ we determine that they must start with ";" or "," or ")". Furthermore, the number of unmatched opening parentheses in $p(i)$ should match with closing parentheses in $q(i)$.

The following simple context-free grammar satisfies the requirements specified above:

$$
\begin{array}{ll}
M12 & imbrseq \ = \\
(a) & msequence\ sep\ @\ sep\ \{\ |msequence\}\ @\ sep\ msequence\ | \\
(b) & (@\ \{\ |msequence\ sep\}\ @\ sep\ msequence\ \{)|)*\}\ | \\
(c) & (msequence\ sep\ @\ \{\ |sep\ msequence\}\ @\ \{)|)*\}\ | \\
(d) & (@\ msequence\ @\ \{)|)*\}\ | \\
(e) & (\{\ |msequence\ sep\}\ imbrseq\ \{\ |sep\ msequence\}\{)/)*\}\ | \\
(f) & msequence\ sep\ imbrseq\ |\ imbrseq\ sep\ msequence
\end{array}
$$

We shall formally prove later that the above rules always produce replicators which when expanded yield macro sequences in general.

Let us apply these rules to derive the strings inside [ ] of imbricators $R(3), R(4)$ and $R(5)$:

$$
\begin{array}{ll}
R(3): & \#i:1,n,1[(up(i);reset(i),\ @\ @\ )] \\
R(4): & \#i:1,n,1[(up(i);\ @;@\ ;down(i))*] \\
R(5): & \#i:1,n,1[(skip(i);\ @\ @\ ),v(i)]
\end{array}
$$

The symbol $\mapsto$ in the derivations which follow, indicates that the leftmost non-terminal to the left of $\mapsto$ is replaced using a rule of the grammar to yield a string on the right of $\mapsto$. The symbol $\mapsto^+$ denotes the transitive closure of $\mapsto$ (see Appendix B). For example the derivation

$$gelement \mapsto mstarelement \mapsto melement \mapsto mevent \mapsto up(i)$$

may be abbreviated to

$$gelement \mapsto^+ up(i).$$

Sometimes we shall write $\underset{\text{M0}}{\mapsto}$ to denote explicitly the rule used, as for instance

$$msequence \ \underset{\text{M5}}{\mapsto}\ morelement\ ;\ morelement$$

Leftmost derivations of the strings inside [ ] of replicators R(3), R(4) and R(5) are:

$R(3)$:

$$
\begin{array}{lll}
imbrseq & \underset{\text{M12.c}}{\mapsto} & (msequence\ sep\ @@) \mapsto^+ (up(i);reset(i)\ sep@@) \\
 & \underset{\text{M9.1}}{\mapsto} & (up(i);reset(i),@@)
\end{array}
$$

$R(4)$:

$$
\begin{array}{lll}
imbrseq & \underset{\text{M12.e}}{\mapsto} & (imbrseq)* \underset{\text{M12.a}}{\mapsto} (msequence\ sep@sep@sep\ msequence)* \\
 & \mapsto^+ & (up(i)\ sep@sep@sep\ msequence)* \\
 & \underset{\text{M9.1}}{\mapsto^3} & (up(i);@;@;msequence)* \mapsto^+ (up(i);@;@;down(i))*
\end{array}
$$

$R(5)$:

$$\begin{aligned}
imbrseq &\overset{\mapsto}{\text{M12.f}} \ imbrseq\ sep\ msequence \\
&\overset{\mapsto}{\text{M12.c}} \ (msequence\ sep@@)\ sep\ msequence \\
&\mapsto^{+} \ (skip(i)\ sep@@)\ sep\ msequence \overset{\mapsto^2}{\text{M9.1}} (skip(i);@@), msequence \\
&\mapsto^{+} \ (skip(i);@@), v(i)
\end{aligned}$$

As we have already indicated the expansion rules for sequence replicators will guarantee not to expand to empty strings. (See Section 3.3.8 for details).

3.3.6 Left and Right Sequence Replicators

All replicators in the notation introduced so far *may not expand to empty strings.* A replicator may generate empty strings for two reasons, either because its regularity is the empty string, or because the values of *in*, *fi*, *inc* are such that the range of the index is empty. The former situation cannot occur in replicators produced by the grammar introduced so far since empty regularities are not permitted by the syntax rules. Such replicators are not useful anyway. The latter situation is excluded to avoid collision of terminal symbols as we have shown in Section 3.1.

In our motivation for the concept of left and right sequence replicators in Section 3.1 we concluded that in fact we need some replicators to generate empty strings, but these replicators must differ from the old ones. To avoid collision of separators when the replicators generate empty strings, as in the macro path above, their context should not be the same as that of other generalized elements. A separator should be *missing* either on their left or on their right. *Their expansion has to provide the extra separator.* If the separator on their left is missing they will be called **left replicators**, and will be produced by the non-terminal *lreplicator*, and if the separator on their right is missing, they will be called **right replicators**, and will be produced by the non-terminal *rreplicator*. For their expansion to bind correctly, right replicators should precede, and left replicators should follow starelements, sequence replicators and distributors. This is what the rule M5.1 for *gelement* states.

The replicators produced by the non-terminals *lreplicator* and *rreplicator* will generate sequences with a separator preceding and following respectively. We define their syntax by the rules:

$$\begin{aligned}
M13.\quad lreplicator &= range_spec[\{;|,\}|\{concseq|imbrseq\}] \\
M14.\quad rreplicator &= range_spec[\{concseq|imbrseq\}|\{;|,\}]
\end{aligned}$$

and therefore a replicator produced by the first rule will have the forms:

$$\begin{aligned}
(Lconc)\quad &\#i : in, fi, inc[\ sep \mid p(i)\ sep@] \\
(Limbr)\quad &\#i : in, fi, inc[\ sep \mid p(i)\ @t@\ q(i)]
\end{aligned}$$

and by the second rule the forms:

$$(Rconc): \quad \#i : in, fi, in[p(i)\ sep@|\ sep\ ]$$
$$(Rimbr): \quad \#i : in, fi, inc[p(i)\ @t@\ q(i)\ |\ sep\ ]$$

If their index range is empty the strings generated by their expansion will be empty as well. Otherwise the strings generated by the expansion of *lconc* or *limbr*, and *rconc* or *rimbr* will be the same as the strings generated by the expansion of the sequence replicators obtained from them by removing *sep*| and |*sep* respectively, preceded and respectively followed by *sep*. In the above notation the path corresponding to P_{3n+m+1} of $Pr^{n,m}_{RW2}$ (see Section 3.1) above would be written as:

$$\underline{path}\ \#i : 1, m, 1[write(i)\#j : 1, i-1, 1[\ ,\ |\ skip(j)\ ,@]$$
$$\#j : i+1, m, 1[\ ,\ |\ skip(j)\ ,@]\ ;@]\ \underline{end}$$

which for $m = 2$ expands as illustrated below.

Let us set

$$p(i) = write(i)\#j : 1, i-1, 1[\ ,\ |\ skip(j)\ ,@]\#j : i+1, m, 1[\ ,\ |\ skip(j)\ ,@]$$

so that the above macro path can be written as $\underline{path}\ \#i : 1, m, 1[p(i)\ ;@]\ \underline{end}$. Clearly $\underline{path}\ \#i : 1, 2, 1[p(i)\ ;@]\ \underline{end}$ expands to $\underline{path}\ p(1); p(2)\ \underline{end}$.
Since

$$p(1) = write(1)\underbrace{\#j : 1, 0, 1[\ ,\ |\ skip(j)\ ,@]}_{expands\ to\ empty\ string}\underbrace{\#j : 2, 2, 1[\ ,\ |\ skip(j)\ ,@]}_{expands\ to\ ',skip(2)'}$$

we have $p(1)$ expands to $write(1), skip(2)$.
Similarly,

$$p(2) = write(2)\underbrace{\#j : 1, 1, 1[\ ,\ |\ skip(j)\ ,@]}_{expands\ to\ ',skip(1)'}\underbrace{\#j : 3, 2, 1[\ ,\ |\ skip(j)\ ,@]}_{expands\ to\ empty\ string}$$

so $p(2)$ expands to $write(2), skip(1)$.
Hence $p(1); p(2)$ expands to $write(1), skip(2); write(2), skip(1)$ as it should.

3.3.7 The Distributors

Consider the following macro path involving a concatenator:

$$\underline{path}\ \#i : 1, n, 1[deposit(i)\ ;@]\ \underline{end}$$

As we said earlier, concatenators of the form used above are so regular and occur so frequently that they might be represented by another simpler mechanism, called distributors. The distributor equivalent to the above concatenator is simply:

$$\underline{path}\ ;\ [deposit]\ \underline{end}$$

assuming that the collective name *deposit* has been previously declared, for example by the collectivisor *array deposit remove(n) endarray*. The distributors do not generate indices explicitly, like the concatenators, but generate indices defined by the corresponding collectivisors.

The syntax for the distributor is:

$$
\begin{array}{lll}
M9. & distributor & = sep\ dim\ range\ [\ msequence\ ] \\
M9.1 & sep & = ,\ |\ ; \\
M9.2 & dim & = \ |\ iexpr \\
M9.3 & range & = \ |\ \#\ iexpr, iexpr, iexpr
\end{array}
$$

Recall that every collectivisor defines not only a set of indexed events but also a total order between them. For example the collectivisor:

$$C7: \ \underline{array}\ c(0:3)\ b(2,2,2)\ \underline{endarray}$$

states that the collective name c corresponds to the ordered indexed events set

$$c(0) \prec c(1) \prec c(2) \prec c(3)$$

and the collective name b corresponds to the ordered indexed events set

$$b(1,1,1) \prec b(2,1,1) \prec b(1,2,1) \prec b(2,2,1) \prec b(1,1,2) \prec b(2,1,2) \\ \prec b(1,2,2) \prec b(2,2,2).$$

For every collective name a declared by a collectivisor, let A_a denote the total order of indexed events corresponding to a. More formally $A_a = (S_a, \prec_a)$ where S_a is a set of corresponding indexed events, and $\prec_a \subseteq S_a \times S_a$ is a sharp total order (i.e. $\prec_a \cup =$ is a total order). For instance, for the collective name c declared by the collectivisor C7 we have $S_c = \{c(i) | i = 0, \ldots, 3\}$, and $c(i) \prec_c c(j) \Leftrightarrow i < j$.

Suppose that a collective name a has k dimensions, i.e., every indexed event defined by a is of the form $a(i_1, \ldots, i_k)$, for some $i_1, \ldots, i_k$.

A total order $Asl = (S, \prec)$ is called an **array-slice of** a if and only if either $Asl = A_a$, or $(S, \prec)$ satisfies the following conditions

$$(1)\ (\exists r \geq 1)(\exists j_1, \ldots, j_r \in \{1, \ldots, k\})(\exists l_1, \ldots, l_r \in Integers) \\ S = \{a(i_1, \ldots, i_k) \in S_0 | i_{j_1} = l_1 \wedge \ldots \wedge i_{j_r} = l_r\} \neq \emptyset$$

$$(2)\ \prec = \prec_a \cap S \times S$$

In other words, S is either S_a or a subset of S_a consisting of indexed events, the indices of which are identical for some dimensions. For instance the total order:

$$b(1,2,1) \prec b(2,2,1) \prec b(1,2,2) \prec b(2,2,2)$$

is an array slice of b from the collectivisor C7. In this case we have $r = 1$, $j_r = 2$, $i_2 = l_1 = 2$.

Array slices will be represented by **slice events.** Slice events will be represented like indexed events, but leaving blank the index fields corresponding to the indices which are not the same for all elements of the corresponding array slice.

In particular, the slice event corresponding to the whole set of indexed events defined by a collective name a, will be represented by either a, or $a(\ ,\ldots,\ )$, or $a(\ )$.
In the case of the collective name c defined by the collectivisor C7 we have only one array slice, the whole set of indexed events:

$$c(0) \prec c(1) \prec c(2) \prec c(3) \prec c(4)$$

which may be represented either by slice event c or by slice event $c(\ )$. The collective name b defined by the same collectivisor C7 has 3 dimensions, and we may define 19 array slices and corresponding slice events in this case. For example $b(\ ,\ ,\ )$ or b correspond to: $b(1,1,1) \prec b(2,1,1) \prec b(1,2,1) \prec b(2,2,1) \prec b(1,1,2) \prec b(2,1,2) \prec b(1,2,2) \prec b(2,2,2)$
$b(\ ,2,\ )$ corresponds to $b(1,2,1) \prec b(2,2,1) \prec b(1,2,2) \prec b(2,2,2)$
$b(2,1,\ )$ corresponds to $b(2,1,1) \prec b(2,1,2)$ (in this case $r = 2$, $j_1 = 1$, $j_2 = 2$, $l_1 = 2$, $l_2 = 1$), etc.
The dimensions corresponding to blank fields of a slice event are called the **distributable dimensions** of the corresponding array slice. For instance $b(\ ,2,\ )$ has 2 distributable dimensions: first and third.

The only syntactic difference between macro sequences in distributors and macro sequences anywhere else, is that some of the index fields of the events occurring in distributors may be empty. Hence we have already defined *mevent* in such a way that it would be syntactically valid for them to have some empty index fields. We further restrict the *mevents* involving blank fields to distributors by meta-restriction rules. Recall that the non-terminal *mevent* is defined as:

$$mevent \;=\; name\,\{\ |(\{\{iexpr|\ \},\underline{@}\}^{+})\}$$

Alternatively we could specify rules to distinguish the two macro sequences but this would almost double the number of our syntax rules. We impose the context sensitive restriction:

(Drest1) For every distributor *sep dim range*[*msequence*], *msequence* must contain at least one slice event.

In principle, the slice events are *strings* of the form $a(\alpha_1,\ldots,\alpha_k)$ where some α_i's are blank. In particular a alone may be regarded as a shorthand for $a(,\ldots,)$. If α_i is blank we shall write $\alpha_i = \varepsilon$. Let $a(\alpha_1,\ldots,\alpha_k)$ be a slice event with $\alpha_j = \varepsilon$. Hence j is a distributable dimension of $a(\alpha_1,\ldots,\alpha_k)$.

A slice event or indexed event $a(\alpha_1,\ldots,\alpha_{i-1},\beta,\alpha_{j+1},\ldots,\alpha_k)$ is called a **section of slice event** $a(\alpha_1,\ldots,\alpha_k)$ **generated by the distributable dimension** j if $\beta \neq \varepsilon$.

If a slice event has only one distributable dimension then all its sections are indexed events, otherwise they are slice events with one distributable dimension less. A distributable dimension j of a slice event $a(\alpha_1,\ldots,\alpha_k)$ may, and usually does, generate more than one section, for instance, for the collective name b of the collectivisor C7, the distributable dimension number 1 of $b(\ ,2,\ )$ generates two sections: $b(1,2,\ )$ and $b(2,2,\ )$. The set of all sections generated by a distributable dimension j of $a(\alpha_1,\ldots,\alpha_k)$ is totally ordered in a natural manner by:

$$a(\alpha_1,\ldots,\alpha_{j-1},\beta,\alpha_{j+1},\ldots,\alpha_k) \prec a(\alpha_1,\ldots,\alpha_{j-1},\beta',\alpha_{j+1},\ldots,\alpha_k) \Leftrightarrow \beta < \beta'.$$

For $b(\ ,2,\ )$ and the distributable dimension number 1, we have

$$b(1,2,\ ) \prec b(2,2,\ ).$$

Suppose that the set of all sections generated by the distributable dimension j of $a(\alpha_1,\ldots,\alpha_k)$ consists of n elements, i.e., it is of the form

$$a(\alpha_1,\ldots,\alpha_{j-1},\beta^1,\alpha_{j+1},\ldots,\alpha_k) \prec \ldots \prec a(\alpha_1,\ldots,\alpha_{j-1}\beta^n,\alpha_{j+1},\ldots,\alpha_k)$$
$$\text{with } \beta^1 < \ldots < \beta^k.$$

We may now refer to $a(\alpha_1,\ldots,\alpha_{j-1},\beta^i,\alpha_{j+1},\ldots,\alpha_k)$ as to the **ith section** generated by the dimension j of $a(\alpha_1,\ldots,\alpha_k)$.
A distributor operates on a specific distributable dimension of each slice event in its macro sequence, which after the expansion of the distributor ceases to be distributable. We shall refer to them as the **current distributable dimensions of slice events.** The slice events will be replaced upon the expansion of the distributor by corresponding *sections of slice events.*

The simplest distributor, called the *simple distributor*, has the form:

$$sep\ [\ msequence\ ]$$

In this case it is assumed that for every slice event occurring in *msequence*, the **current distributable dimension is that distributable dimension which is furthest to the right**, For instance in the distributor D1

$$D1:\ \underbrace{;}_{sep}\ [\underbrace{,[b(\ ,2,\ )]}_{msequence}\,]$$

the 3rd dimension of b is the current distributable dimension.

Let us impose the following restriction on simple distributors:

> The current distributable dimensions on which the simple distributor operates all contain the same number of sections

This is an informal statement of the context sensitive restriction Drest2.1 which is precisely discussed in Section 3.3.8. A simple distributor is well-formed and may be expanded only if it satisfies restriction (Drest2.1). A number of context sensitive restrictions for distributors will be clarified in Sect. 3.3.8 where we group all these restrictions together. A distributor in which all the current distributable dimensions on which it operates contain n sections, may be expanded as follows:

> "n copies of the macro sequence in the distributor will be concatenated, separated by the separator associated with the distributor. In the first copy each slice event will be replaced by the first section generated by the current distributable dimension of this slice event. In general, the ith copy $(1 \leq i \leq n)$ of each slice event will be replaced by the ith section generated by the current distributable dimension of this slice event."

According to this scheme the distributor D1 where b is declared by C7 expands to:

$$\underbrace{,[b(\ ,2,1)]}_{simple}\ ;\underbrace{,[b(\ ,2,2)]}_{distributors}$$

and next, according to the same scheme applied twice, to:

$$b(1,2,1), b(1,2,2); b(1,2,2), b(2,2,2).$$

The general syntax for distributors (see M.9) is *sep dim range*[*msequence*]. So far we considered the case where both *dim* and *range* are blanks. Let us now consider the case where *range* is not blank. This case extends the class of regularities which the distributors may generate by permitting them to distribute not only over the whole range of array-slices but over a subrange of them as well. So now we have:

$$D2: \quad sep\ \#i_1, i_2, i_3[msequence]$$

in which i_1, i_2, i_3 denote integer expressions, representing the subrange over which *msequence* is to be distributed. Using the subrange option, we may restrict the expansion of a distributor to some selected "copies" of its regularity. The subrange defines which copies should be selected. The integer expressions i_1, i_2, i_3 specify the first copy to be selected, the upper limit of the copies to be selected and the step by which the upper limit should be reached from i_1, respectively.

Suppose that the distributor: *sep* [*msequence*], where *sep* and *msequence* are the same as in D2 expands to:

$$msequence(1)\ sep\ msequence(2)\ sep\ \ldots\ sep\ msequence(n)$$

where *msequence*(i) results from *msequence* by replacing all slice events by the appropriate ith sections of current distributable dimensions. Then distributor D2 will expand to:

$$msequence(i_1)\ sep\ msequence(i_1 + i_3)\ sep\ msequence(i_1 + 2 \cdot i_3)\ \ldots$$
$$sep\ msequence(i_1 + m \cdot i_3)$$

where $m = (i_2 - i_1) \div i_3 + 1$, i.e., m is the greatest integer such that $i_1 + m \cdot i_3 < i_2$, where n is the number of copies to be selected.
For example, the distributor D3:

$$D3: \quad ;\#1,3,1[c]$$

where c is declared by C7 would expand to:

$$c(0); c(1); c(2)$$

selecting out of all copies of c in the string generated by the expansion of $;[c]$ only the first, second and third. The distributor D4:

$$D4: \quad ;\#1,3,2[c]$$

would expand to:

$$c(0); c(2)$$

selecting the first and the third copies of c in the expansion of $;[c]$.

We require the expansion of D2 to be non-empty, and the indices of the events to be in the range defined by the corresponding collectivisors. In view of the semantics of the subrange, we require now that:

> When a subrange is defined, the slice events are required to contain at least as many sections of current distributable dimensions as specified by the subrange.

This is a slightly informal formulation of the context sensitive restriction Drest5.1 which is precisely discussed later in Sect. 3.3.8. For example, the distributor D5:

$$D5: \quad ;\#1,2,1[c(\ ), b(1,2,\ )]$$

where c and b are defined by C7 satisfies this restriction although $c(\ )$ has four sections and $b(1,2,\ )$ has two, and may be expanded to:

$$c(0), b(1,2,1); c(1), b(1,2,2)$$

It is clear from the syntax of distributors that they may be nested. Each of these distributors must apply to a different distributable dimension of each slice event. We shall say that a distributor is 1-nested if it is the outermost distributor, i.e. it is not within an *msequence* of any other distributor, and a distributor is k-nested, $k > 1$, if it is within an *msequence* of a $k-1$ nested distributor. The following restriction is imposed:

(Drest3) inside a k-nested distributor there must be *at least one* array slice with k distributable dimensions, and no array slices with with more than k distributable dimensions.

Equivalently we may say that after the expansion of the outermost distributor, the rest of the distributors must obey the syntax rules. For example, D7

$$D7: \quad ;[,[c]]$$

where c is defined by C7 is not valid, since after the expansion of the outermost distributor a non-valid distributor is generated:

$$,[c(0)];,[c(1)];,[c(2)];,[c(3)].$$

The reason for this is that the macro sequences inside [ ] of the above expansion do not consist of slice events but of indexed events.

We must specify which of the nested distributors applies to which of the distributable dimensions of array slices. The rule adopted so far is that the outermost distributor will apply to the rightmost distributable dimension of each slice; the second outermost to the rightmost not allocated distributable dimension, etc. A possible relaxation of the above rule would be to consider it as the default rule, and specify explicitly which separator applies to which distributable dimension, which is the case when *dim* of rule M9 is not a blank. The distributor D8, for example:

$$D8: \quad ;[,[b]]$$

where b is defined by C8, where $C8$: $\underline{array}$ $b(4,3)$ $\underline{endarray}$, would expand according to either rules to:

$$\begin{array}{l} b(1,1),b(2,1),b(3,1),b(4,1) \\ ;b(1,2),b(2,2),b(3,2),b(4,2) \\ ;b(1,3),b(2,3),b(3,3),b(4,3) \end{array}$$

with "," applying to the first dimension of b and ";" to the second. But D9:

$$D9: \quad ;1[,[b]]$$

would expand to :

$$\begin{array}{l} b(1,1),b(1,2),b(1,3) \\ ;b(2,1),b(2,2),b(2,3) \\ ;b(3,1),b(3,2),b(3,3) \\ ;b(4,1),b(4,2),b(4,3) \end{array}$$

since it is explicitly specified that ";" applies to the first dimension of b, and implicitly that "," applies to the rightmost unallocated dimension of b, according to the default rule.

The feature for selection of distributable dimensions is very useful when both D8 and D9 are required in the same program. Without it we have to use the equivalent concatenator R8:

$$R8: \quad \#i: 1,4,1[, [b(i, \ )] ; @]$$

instead of D9. If only one of D8 or D9 were required then we could define the collectivisor in such a way as to conform to the default rule. Let us consider now the following array:

$$C9: \quad \underline{array}\ \#i: 1,5,2[\#j: 1,i,1[s(i,j)\ t(j,i)]]\ \underline{endarray}$$

where the indices in the second dimension of s depend on the indices of its first, and the indices of the first dimension of t on the indices in its second dimension. According to the expansion rules the distributor D10:

$$D10: \quad ;[,[t]]$$

expands to :

$$\begin{array}{l} t(1,1) \\ ;t(1,3),t(2,3),t(3,3) \\ ;t(1,5),t(2,5),t(3,5),t(4,5),t(5,5) \end{array}$$

while

$$D11: \quad ;[,[s]]$$

expands to:

$$\begin{array}{l} s(1,1),s(3,1),s(5,1) \\ ;s(3,2),s(5,2) \\ ;s(3,3),s(5,3) \\ ;s(5,4) \\ ;s(5,5) \end{array}$$

Note that it is not immediately obvious which replicator is equivalent to D11, more details on this subject can be found in Sect. 3.3.8. Of course in this case we may just write the expansion of D11 in explicit form, but we cannot do this when for instance 5 is replaced by a range variable m. On the other hand, the expansion of the distributor D12:

$$D12: \quad ;1[,[s]]$$

and the replicator R9:

$$R9: \quad \#i: 1,5,2[, [s(i, \ )] ; @]$$

both result in:

$$\begin{array}{l} s(1,1) \\ ;s(3,1),s(3,2),s(3,3) \\ ;s(5,1),s(5,2),s(5,3),s(5,4),s(5,5) \end{array}$$

So D12 generates the same pattern for s as D10 does for t. To ensure a correct expansion the following restriction needs to be imposed on dimension selectors:

(Drest4) if $dim = j$ then every array slice with **maximal** number of distributable dimensions **must** have the dimension number j as the distributable component.

Furthermore, the case where *dim* is different from a blank requires conditions Drest2.2 and Drest5.2 which correspond to Drest2.1 and Drest5.1 for *dim* equal to a blank, and are precisely defined in Sect. 3.3.8. To demonstrate the use of the two new features of distributors, the subrange and the facility of specifying distributable sections, let us consider two more "realistic" examples. In the first we specify a pipeline, using just replicators, which may be written as:

$$\#i : 1, n, 1[\underline{path}\ transfer(i); transfer(i+1)\ \underline{end}]$$

where array *transfer* is declared by:

$$\underline{array}\ transfer(n+1)\ \underline{endarray}$$

We may replace the sequence in the above path by a distributor obtaining:

$$\#i : 1, n, 1[\underline{path}\ ; \#i, i+1, 1[transfer]\ \underline{end}]$$

In the second example we specify a system containing a square matrix which is initially empty. Processes may read or write to any element of the array asynchronously, but write's and read's on any element should alternate, and no read's should occur before the initial write. These constraints may be specified by:

$$\underline{array}\ write\ read\ (n, n)\ \underline{endarray}$$
$$\#i : 1, n, 1[\#j : 1, n, 1[\underline{path}\ write(i,j); read(i,j)\ \underline{end}]]$$

A writer process which updates the elements of the matrix by columns may be specified by any of the following processes:

$$\underline{process}\ \#j : 1, n, 1[\#i : 1, n, 1[write(i,j)\ ; @]\ ; @]\ \underline{end}$$
$$\underline{process}\ \#j : 1, n, 1[; [write(\ ,j)]\ ; @]\ \underline{end}$$
$$\underline{process}\ ; [\#i : 1, n, 1[write(i,\ )\ ; @]]\ \underline{end}$$
$$\underline{process}\ ; [; [write]]\ \underline{end}$$

We now give a number of processes, each specifying reading from selected elements of the matrix. A process reading all the elements of the matrix by rows may be specified most concisely by:

$$\underline{process}\ ; 1[; [read]]\ \underline{end}$$

A process reading the elements of the first r $(1 \leq r \leq n)$ rows by columns may be specified by:

$$\underline{process}\ ;[;\#1,r,1[read]]\ \underline{end}$$

A process reading the lower left triangular matrix may be specified by:

$$\underline{process}\ \#i:1,n,1[\#j:1,i,1[read(i,j)\ ;@]\ ;@]\ \underline{end}$$

or by:

$$\underline{process}\ \#i:1,n,1[;\#1,i,1[read(i,\ )]\ ;@]\ \underline{end}$$

Finally a process reading the elements of the matrix forming the upper right triangular matrix by rows may be specified by:

$$\underline{process}\ \#j:1,n,1[\#k:j,n,1[read(j,k)\ ;@]\ ;@]\ \underline{end}$$

or by:

$$\underline{process}\ \#j:1,n,1[;\#j,n,1[read(j,\ )]\ ;@]\ \underline{end}$$

3.3.8 Expansion of Macro COSY Programs

In the previous sections the expansion of replicators and distributors was only indicated schematically. In this section we define the formal rules for the expansion of complete macro programs, replicators and distributors.
We shall start with introducing some auxiliary notions. Consider the following three mappings:

- *as* - a mapping which assigns some integers to range variables,
- *dec* - the mapping which gives a decimal representation of integer values,
- *val* - the mapping which computes the value of an integer expression without variables

The domains of *as* and *val* are some special kind of strings, while their co-domains are integer values, the domain of *dec* is the set of integers and the co-domain is the set of decimal representations, i.e., some kind of strings.
We now extend *as* and *val* to: $as^* : MGEN \longrightarrow MGEN$ and $val^* : MGEN \longrightarrow MGEN$. Let $IEXPR_0$ denote the set of all integer expressions of our macro notation without range variables, i.e.,

$$IEXPR_0 = \{S \in MT^+ | iexpr \longrightarrow^+ s \wedge s \text{ contains no range variables}\}$$

and let $RVAR$ denote the set of all range variables, i.e.,

$$RVAR = \{s \in MT^+ | rangevariable \longrightarrow^+ s\}.$$

Every $g \in MGEN$ can be represented as

$$g = u_1 x_1 u_2 \dots u_n x_n u_{n+1},$$

where $n \geq 0$, $x_i \in IEXPR_0$, $u_i \in (MT \cup MNT)^*$ and no substring of u_i belongs to $IEXPR_0$. We define:

$$val^*(g) = u_1 dec(val(x_1))u_2 \dots u_n dec(val(x_n))u_{n+1}$$

For example,
$val^*(a(3*5-3);b(5+2)) = a(dec(val(3*5-3)));b(dec(val(5+2))) = a(12);b(7)$.
Similarly, every $g \in MGEN$ can be represented as

$$g = u_1 v_1 u_2 \dots u_n v_n u_n,$$

where $n \geq 0$, $v_i \in RVAR$, $u_i \in (MT \cup MNT)^*$ and no substring of u_i belongs to $RVAR$. The context-sensitive restriction Irest1 guarantees uniqueness of the above representation. We define:

$$as^*(x) = val^*(u_1 dec(as(v_1))u_2 \dots u_n dec(as(v_n)u_{n+1}).$$

In the sequel we shall use the symbol as to denote both as and as^*, and val to denote both val and val^*.
We may now precise the definition of the set $assgn(y)$ which has already been used in the definition of collective expansion correctness. We say that for every macro generator $g \in MGEN$, an assignment as is a g-assignment, i.e., $as \in assgn(g)$, if and only if $as(g) \in MGEN$. Clearly, for every $as \in asgn(g)$, $as(g)$ is range variable free, so it may be expanded. In general, as we have already mentioned, **only range variable free macro programs can be expanded.** We shall show that in such a case they expand to well-formed basic programs, i.e., our notation satisfies the collective expansion correctness criterion.
Throughout the following sections we will be representing macro programs schematically to facilitate the statement of definitions, theorems and their proofs. Capitalized names of basic or macro non-terminals will stand for the set of strings which may be generated from that non-terminal. Thus:

$$MPROGRAMBODY = \{s \in MT^+ | mprogrambody \mapsto^+ s\}, \text{ etc.}$$

The subscript "0" will indicate that all strings of the set contain no range variable. Thus:

$MPROGRAMBODY_0 =$
$\{s \in MPROGRAMBODY | s$ contains no range variable etc.$\}$
To obtain the schematic representations of macro programs we now define:

$$
\begin{array}{lll}
MPBODY & = & CPQBR^+ \\
CPQBR & = & CL\ PQBR \\
CL & = & COLLECTIVISOR_0^* \\
PQBR & = & \{MPATH_0 \cup MPROCESS_0 \cup BODYREPLICATOR_0\} \\
MSEQ & = & MOR;\ldots;MOR \\
MOR & = & GEL,\ldots,GEL \\
GEL & = & RR^*\ MSD\ LR^* \\
RR & = & RIGHTREPLICATOR_0 \\
LR & = & LEFTREPLICATOR_0 \\
MSD & = & MSTEL \cup SR \cup DST \\
SR & = & SREPLICATOR_0 \\
DST & = & DISTRIBUTOR_0 \\
MSTEL & = & MEL \cup \{MEL*\} \\
MEL & = & MEV \cup \{(MSEQ)\} \\
MEV & = & MEVENT_0
\end{array}
$$

The complete expansion of a macro program *mprog* is given by *expand*(*mprog*), where the lower case names corresponding to the upper case names defined above represent arbitrary elements of the sets denoted by the corresponding upper case name. The function *expand* is defined as follows:

1. $expand(mprog) = \underline{program}\ pathsfirst(expand(mpbody))\ \underline{endprogram}$
2. $expand(cpqbr_1 \ldots cpqbr_n) = expand(cpqbr_1) \ldots expand(cpqbr_n)$
3. $expand(cl\ \ pqbr_1 \ldots pqbr_w)$
   $= expand(pqbr_1 \ldots pqbr_w) = expand(pqbr_1) \ldots expand(pqbr_w)$
4. $expand(\#i : in, fi, inc[pqbr_1 \ldots pqbr_k])$
   $= expand(exprepl(\#i : in, fi, inc[pqbr_1 \ldots pqbr_k]))$
5. $expand(\underline{path}\ mseq\ \underline{end}) = \underline{path}\ expand(mseq)\ \underline{end}$
6. $expand(\underline{process}\ mseq\ \underline{end}) = \underline{process}\ expand(mseq)\ \underline{end}$
7. $expand(mor_1;\ldots;mor_r) = expand(mor_1);\ldots;expand(mor_r)$
8. $expand(gel_1,\ldots,gel_s) = expand(gel_1),\ldots,expand(gel_s)$
9. $expand(rr_1 \ldots rr_u\ msd\ lr_1 \ldots lr_v)$
   $= expand(exprepl(rr_1) \ldots exprepl(rr_u)\ msd\ exprepl(lr_1) \ldots exprepl(lr_v))$
10. $expand(srepl) = expand(exprepl(srepl))$
11. $expand(distr) = expand(expdistr(distr))$
12. $expand(mel*) = expand(mel)*$
13. $expand(mev) = val(mev)$
14. $expand((mseq)) = (expand(mseq))$

An auxiliary mapping *pathsfirst* works in the following way. If a string x is a sequence of paths and processes, say $x = P_1 \ldots P_{n+m}$, where $n, m \geq 0$, $n+m \geq 1$, $P_{i_1} \ldots P_{i_n}$ are paths, $P_{i_{n+1}} \ldots P_{i_{n+m}}$ are processes, then

$$paths first(x) = P_{i_1} \ldots P_{i_n} P_{i_{n+1}} \ldots P_{i_{n+m}}$$

This means that it moves all paths before processes as required by the basic COSY grammar from Section 3.2. If the string x is not a sequence of paths and processes then $pathsfirst(x)$ is not defined. To simplify our reasoning a little bit, we shall assume that from now throughout the rest of this section, all **names** (of arrays, slice-events, range variables, events, indices, etc.) **are terminals**, i.e., $NAME \subseteq MT$.

For every string x, let $x\{y/t\}$, where y is a string and $t \in MT$, denote the string obtained from x by replacing each occurrence of t by y. If x contains no t then $x\{y/t\}$ equals to x.

To define *exprepl* and *expdist* let us first define the operator $COPY_{j=k}^{m}(x)$ which transforms strings into strings, where j, k, m are parameters, j is the name (of an integer variable), k, m are integer values, x is an argument and x is a string from $(MT \cup MNT)^*$. The result of this operator is the following:

$$COPY_{j=k}^{m}(x) = \begin{cases} val(x\{dec(k)/j\})COPY_{j=k+1}^{m}(x) & if\ m \geq k \\ \varepsilon & if\ m < k \end{cases}$$

For example:

$$\begin{aligned} &COPY_{i=1}^{3}(a(2*i-1);) = val(a(2*1-1);)COPY_{i=2}^{3}(a(2*i-1);) = \\ &= a(1); val(a(2*2-1);)COPY_{i=3}^{3}(a(2*i-1);) = \\ &= a(1); a(3); val(a(2*3-1);)COPY_{i=4}^{3}(a(2*i-1);) = a(1); a(2); a(3); \end{aligned}$$

We need also two operators $LSTRIP$ and $FSTRIP$ which strip off the last and first separators of a given string respectively. We assume that $LSTRIP(\varepsilon) = FSTRIP(\varepsilon) = \varepsilon$, and for every $x \in (MT \cup MNT)^*$ and $t \in MT \cup MNT$, let:

$$LSTRIP(xt) = \begin{cases} x & if\ t =;\ \ or\ t =, \\ xt & otherwise \end{cases}$$

$$FSTRIP(tx) = \begin{cases} x & if\ t =;\ \ or\ t =, \\ tx & otherwise \end{cases}$$

For instance:

$$\begin{aligned} &LSTRIP(COPY_{i=1}^{3}(a(2*1-1);) = a(1); a(2); a(3) \\ &FSTRIP(COPY_{i=1}^{3}(a(2*1-1);) = a(1); a(2); a(3); \end{aligned}$$

The Expansion of Replicators

We shall start with defining *exprepl* for bodyreplicators. Let us consider a bodyreplicator $\#i : in, fi, inc[spqbr]$, where $spqbr = pqbr_1 \ldots pqbr_k$, $pqbr_i \in PQBR$, $i \geq 1$. We point out that, for all replicators considered, *exprepl*(*replicator*) means that only *replicator* is expanded and not any other replicator which may be generated by the expansion of *replicator*. We define:

$$exprepl(\#i : in, fi, inc[spqbr]) = COPY_{j=1}^{m}(spqbr\{(in + (j-1) * inc)/i\}),$$

where $m = (fi - in) \div inc + 1$, and the name j does not occur in *spqbr*.
Let us apply this formalism to the following bodyreplicator:

$$\#i : 1,4,1[\underline{path}\ deposit(i); remove(i)\ \underline{end}]$$

In the above symbolism: $spqbr = pqbr_1 = \underline{path}\ deposit(i); remove(i)\ \underline{end}$, $\underline{path}\ deposit(i); remove(i)\ \underline{end}\{(1 + (j-1) * 1/i\} = \underline{path}\ deposit((1 + (j - 1) * 1)); remove((1+(j-1)*1))\ \underline{end}$, $m = (4-1) \div 1 + 1 = 4$, so the expansion is given by:

$$COPY_{j=1}^{4}(\underline{path}\ deposit((1 + (j-1) * 1)); remove((1 + (j-1) * 1))\ \underline{end})$$

which yields:

$$\begin{array}{l} \underline{path}\ deposit(1); remove(1)\ \underline{end} \\ \underline{path}\ deposit(2); remove(2)\ \underline{end} \\ \underline{path}\ deposit(3); remove(3)\ \underline{end} \\ \underline{path}\ deposit(4); remove(4)\ \underline{end} \end{array}$$

The expansion of a concatenator $\#i : in, fi, inc[mseq\ sep@]$ where $mseq \in MSEQ$, *sep* is either ";" or ",", is defined as:

$$\begin{array}{l} exprepl(\#i : in, fi, inc[mseq\, sep@]) = \\ LSTRIP(COPY_{j=1}^{m}(mseq\{(in + (j-1) * inc)/i\}sep)) \end{array}$$

where $m = (fi - in) \div inc + 1$, and the name j does not occur in *mseq*.
For example consider $\#i : 4, 1, -2[deposit(2 * i - 1)\ ;@]$. Its expansion is given by:

$$\begin{array}{l} LSTRIP(COPY_{j=1}^{2}(deposit(2 * (4 + (j-1) * (-2)) + 1);) = \\ LSTRIP(deposit(9); deposit(5);) = deposit(9); deposit(5) \end{array}$$

The second kind of sequence replicators, an imbricator $\#i : in, fi, inc[p\ @t@\ q]$ is expanded as follows:

$$exprepl(\#i : in, fi, inc[p\ @t@\ q]) = \begin{cases} LtR & if\ m \geq 1 \\ LSTRIP(FSTRIP(t)) & if\ m < 1 \end{cases}$$

where L, R are the following strings:

$$L = LSTRIP(COPY_{j=1}^{m}(p\{(in + (j-1) * inc)/i\})),$$
$$R = FSTRIP(COPY_{j=1}^{m}(q\{(dec(m) - (j-1) * inc/i\}))$$

where $m = (fi - in) \div inc + 1$, and the name j occurs in neither p nor q nor t. In the case of the inbricator:

$$\#i : 1,3,1[(up(i);\ @; full^*; @\ ; down(i))^*]$$

we have:

$$p\ =\ (up(i);$$
$$t\ = ; full^*;$$
$$q\ = ; down(i))^*$$

and the expansion is given by

$$\begin{aligned} &LSTRIP(COPY_{j=1}^{3}((up((1 + (j-1) * 1));)); full^*; \\ &\quad FSTRIP(COPY_{j=1}^{3}(; down((3 - (j-1) * 1)))^*)) = \\ &= LSTRIP((up(1); (up(2); (up(3);); full^*; \\ &\quad FSTRIP(; down(3))^*; down(2))^*; down(1))^*) = \\ &= (up(1); (up(2); (up(3); full^*; down(3))^*; down(2))^*; down(1))^* \end{aligned}$$

The expansion of both left- and right-replicators can be defined in a similar way. Let $\#i : in, fi, inc[sep|pattern]$ and $\#i : in, fi, inc[pattern|sep]$, where either $pattern = mseq\ sep'\ @$, or $pattern = p@t@q$, be a left- and right-replicator respectively.

Let $m = (fi - in) \div inc + 1$. We define:

$$exprepl(\#i : in, fi, inc[sep|pattern]) = \begin{cases} sep\ exprepl(\#i : in, fi, inc[pattern]) & if\ m \geq 1 \\ \varepsilon & if\ m < 1 \end{cases}$$

$$exprepl(\#i : in, fi, inc[pattern|sep]) = \begin{cases} exprepl(\#i : in, fi, inc[pattern])sep & if\ m \geq 1 \\ \varepsilon & if\ m < 1 \end{cases}$$

In the previous section we restricted concatenator, imbricator and bodyreplicator to always expand to non-empty string. This restriction may formally be expressed as:

(Rrest1.1) For every replicator if inc is an integer then $inc \neq 0$.

(Rrest1.2) For every concatenator and bodyreplicator if in, fi, inc are integers then $(fi - in) \div inc + 1 > 0$.

(Rrest1.3) For every imbricator, if in, fi, inc are integers then $(fin - in) \div inc + 1 \leq 0 \Rightarrow FSTRIP(LSTRIP(t)) \neq \varepsilon$.

A final restriction which is imposed on replicators not in collectivisors is:

(Rrest2) The replicators should generate indexed events permitted by their corresponding collectivisors.

It is obvious that different replicators may expand to identical strings. Note that in general every replicator can be represented by

$$\#i : in, fi, inc[pattern]$$

and, depending on the form of *pattern*, it is either a bodyreplicator, or a concatenator, or an impricator, or left- or right-replicator.
A replicator in which $in = inc = 1$ and fi is an integer is said to be in **normal form.** The following result shows that if in, inc, fi are all integers in some replicator, then there is a normal form for the replicator which expands to the same basic expression.

Theorem 3.3
Let in, fi, inc be integers, $m = (fi - in) \div inc + 1$.
For every replicator $\#i : in, fi, inc[pattern]$,
$exprepl(\#i : in, fi, inc[pattern]) =$
$\quad exprepl(\#j : 1, m, 1[pattern\{(in + (j-1) * inc)/i\}])$

Proof. This follows immediately from the fact that in *exprepl* for every replicator, the argument of *COPY* is always subject to substitution by $\{in + (j-1) * inc/i\}$ or $\{(dec(m) - (j-1) * inc/i)\}$. □

The Expansion of Distributors

Let $a(\alpha_1, \ldots, \alpha_k)$ be a slice event representing an array slice $Asl(S, \prec)$. Define:

$$Indices(a(\alpha_1, \ldots, \alpha_k)) = \{(i_1, \ldots, i_k) | a(i_1, \ldots, i_k) \in S\}.$$

Consider the following collectivisor

$$C11 : \underline{array}\ \#i : 1, 10, 1[\#j : 1, 10, i[a(i, j)]]\ \underline{endarray}$$

The sets $Indices(a)$, $Indices(a(,4))$, $Indices(a(,5))$, $Indices(a(,7))$ and $Indices(a(,9))$ are depicted in Figure 3.14.

For every dimension j of $a(\alpha_1, \ldots, \alpha_k)$, let $Indices_j(a(\alpha_1, \ldots, \alpha_k))$ denote the projection of $Indices(a(\alpha_1, \ldots, \alpha_k))$ onto the jth coordinate, i.e.,

$$Indices_j(a(\alpha_1, \ldots, \alpha_k)) =$$
$$\{l | (\exists i_1, \ldots, i_{j-1}, i_{j+1}, \ldots, i_k)(i_1, \ldots, i_{j-1}, l, i_{j+1}, \ldots, i_k) \in Indices(a(\alpha_1, \ldots, \alpha_k))\}$$

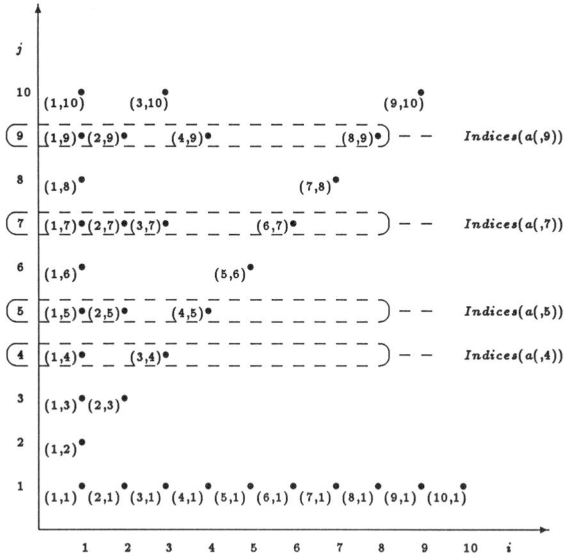

Figure 3.14. *Indices*(a) defined by C11

Suppose that j is a distributable dimension of $a(\alpha_1, \ldots, \alpha_k)$, $Indices_j(a(\alpha_1, \ldots, \alpha_k)) = \{l_1, \ldots, l_n\}$ and $l_1 < \ldots < l_n$. This means that the jth dimension of $a(\alpha_1, \ldots, \alpha_k)$ generates n sections which are ordered as follows:

$$a(\alpha_1, \ldots, \alpha_{j-1}, l_1, \alpha_{j+1}, \ldots, \alpha_k) \prec \ldots \prec a(\alpha_1, \ldots, \alpha_{j-1}, l_n, \alpha_{j+1}, \ldots, \alpha_k).$$

Let $index_{a(\alpha_1,\ldots,\alpha_k)}(i,j)$ denote the following integer expression with the variable i:

$$index_{a(\alpha_1,\ldots,\alpha_k)}(i,j) = dec(l_1) * (i \div 1) * (1 \div i) + \ldots + dec(l_n) * (i \div n) * (n \div i).$$

Lemma 3.1 *For every* $i = 1, \ldots, n$*:*

$$val(index_{a(\alpha_1,\ldots,\alpha_k)}(i,j)) = l_i.$$

Proof. Clearly

$$val(index_{a(\alpha_1,\ldots,\alpha_k)}(i,j)) = \sum_{r=1}^{n} l_r (i \div r)(r \div i) = l_i$$

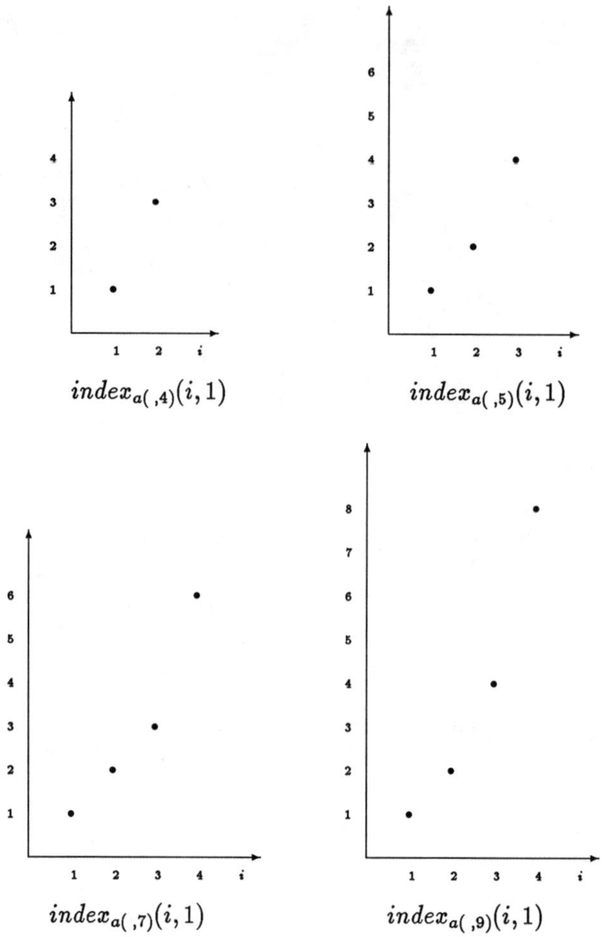

Figure 3.15. $index_{a(\ ,4)}(i,1)$, $index_{a(\ ,5)}(i,1)$, $index_{a(\ ,7)}(i,1)$, $index_{a(\ ,9)}(i,1)$, where a is defined by the collectivisor C11

since

$$(i \div r)(r \div i) = \begin{cases} 0 & if\ r \neq i \\ 1 & if\ r = i \end{cases}$$
$$for\ i = 1, \ldots, n. \qquad \square$$

The above lemma gives the following interpretation of $index_{a(\alpha_1,\ldots,\alpha_k)}(i,j)$: $index_{a(\alpha_1,\ldots,\alpha_k)}(i,j)$ computes the value of the *index* of the *i*th section of $a(\alpha_1,\ldots,\alpha_k)$ along the dimension j. Fig. 3.15 illustrates this concept.
Suppose now that i_1, i_2, i_3 are some integers. Define: $m = (i_2 - i_1) \div i_3 + 1$, and let $\xi(l)$ be the following integer function: $\xi(l) = i_1 + (l-1) * i_3$. Suppose also that $i_2 \leq n$. Define $Indices_j^{i_1,i_2,i_3}(a(\alpha_1,\ldots,\alpha_k))$ to be the following subset of $Indices_j(a(\alpha_1,\ldots,\alpha_k)) = \{l_1,\ldots,l_n\}$:

$$Indices_j^{i_1,i_2,i_3}(a(\alpha_1,\ldots,\alpha_k)) = \{l_{\xi(l)} | l = 1,\ldots,m\}.$$

The above definition is correct because $\xi(m) \leq i_2 \leq n$.
For example if a is declared by the collectivisor C11 (see Fig. 3.14) then $Indices_1(a(,9)) = \{1,2,4,8\}$, and for instance $Indices_1^{2,4,2}(a(,9)) = \{2,8\}$.
Suppose that $Indices_j^{i_1,i_2,i_3}(a(\alpha_1,\ldots,\alpha_k)) = \{l'_1,\ldots,l'_m\}$.
Let $index^{i_1,i_2,i_3}_{a(\alpha_1,\ldots,\alpha_k)}(i,j)$ denote the following integer expression with the variable i:

$$index^{i_1,i_2,i_3}_{a(\alpha_1,\ldots,\alpha_k)}(i,j) = dec(l'_1) * (i \div 1) * (1 \div i) + \ldots + dec(l'_m) * (i \div n) * (n \div i).$$

Lemma 3.2 *For every* $i = 1,\ldots,m$*:* $val(index^{i_1,i_2,i_3}_{a(\alpha_1,\ldots,\alpha_k)}(i,j)) = l'_i$.

Proof. As for Lemma 3.1. $\square$

We may now say that $index^{i_1,i_2,i_3}_{a(\alpha_1,\ldots,\alpha_k)}(i,j)$ computes the value of the index of the *i*th section of $a(\alpha_1,\ldots,\alpha_k)$ along a subrange defined by i_1, i_2, i_3 over the dimension j.
Let $SECTION^{i,d}_{i_1,i_2,i_3}(x)$ be an operator where i, d, i_1, i_2, i_3 are parameters, i is the name (of an integer variable), $d \in \{1,2,\ldots\}$, i_1, i_2, i_3 are integer values, x is an argument which is a slice event. It is assumed that the name i does not occur in x, and if it is not a distributable dimension of x the operator simply does nothing. The operator $SECTION^{i,d}_{i_1,i_2,i_3}(x)$ produces the following slice event or indexed event:

$$SECTION^{i,d}_{i_1,i_2,i_3}(a(\alpha_1,\ldots,\alpha_k)) =$$
$$\begin{cases} a(\alpha_1,\ldots,\alpha_{d-1}, index^{i_1,i_2,i_3}_{a(\alpha_1,\ldots,\alpha_k)}(i,d), \alpha_{d+1},\ldots,\alpha_k) & if\ \alpha_d = \varepsilon \\ a(\alpha_1,\ldots,\alpha_k) & if\ \alpha_d \neq \varepsilon \end{cases}$$

Similarily, let $SECTION^i_{i_1,i_2,i_3}(x)$, $SECTION^{i,d}(x)$, $SECTION^i(x)$ be operators, where i, d, i_1, i_2, i_3, x are as above, defined as follows:

$SECTION^{i}_{i_1,i_2,i_3}(a(\alpha_1,\ldots,\alpha_k)) = SECTION^{i,r}_{i_1,i_2,i_3}(a(\alpha_1,\ldots,\alpha_k))$, where r is an integer satisfying: $\alpha_r = \varepsilon$ and for each j, $r < j \leq k$, $\alpha_j \neq \varepsilon$.
In other words, r is the number of the rightmost distributable dimension.

$$SECTION^{i,d}(a(\alpha_1,\ldots,\alpha_k)) = \begin{cases} a(\alpha_1,\ldots,\alpha_{d-1},index_{a(\alpha_1,\ldots,\alpha_k)}(i,d),\alpha_{d+1},\ldots,\alpha_k) & if\ \alpha_d = \varepsilon \\ a(\alpha_1,\ldots,\alpha_k) & if\ \alpha_d \neq \varepsilon \end{cases}$$

and
$SECTION^{i}(a(\alpha_1,\ldots,\alpha_k)) = SECTION^{i,r}(a(\alpha_1,\ldots,\alpha_k))$, where r is an integer satisfying: $\alpha_r = \varepsilon$ and for each j, $r < j \leq k$, $\alpha_j \neq \varepsilon$.
Recall that the general pattern for a distributor is:

$$sep\ dim\ range[mseq]$$

where *sep* is either ";" or ",", *dim* is either blank or a positive integer, *range* is either blank or has the form $\#i_1, i_2, i_3$
Let $GSECTION^{i}_{dim\ range}(x)$ be an operator where i, *dim*, *range* are parameters, i is the name (of an integer variable), *dim* and *range* are strings as in the syntax for distributor, x is an argument and x is a slice event and the name i does not occur in x. The operator $GSECTION^{i}_{dim\ range}(x)$ produces the following slice event or indexed event:

$$GSECTION^{i}_{dim\ range}(x) = \begin{cases} SECTION^{i}(x) & \text{if both } dim \text{ and } range \text{ are blanks} \\ SECTION^{i,d}(x) & \text{if } range \text{ is blank but } dim = d \\ SECTION^{i}_{i_1,i_2,i_3}(x) & \text{if } dim \text{ is blank but } range = \#i_1,i_2,i_3 \\ SECTION^{i,d}_{i_1,i_2,i_3}(x) & \text{if } dim = d \text{ and } range = \#i_1,i_2,i_3 \end{cases}$$

Hence $GSECTION^{i}_{dimrange}$ may be regarded as a generalization of $SECTION$'s.
Every macrosequence $mseq \in MSEQ$ may be represented as:

$$mseq = u_1 s_1 u_2 \ldots u_m s_m u_{m+1}$$

where s_j are slice events, $u_j \in (MT \cup MNT)^*$, no substring of u_j is a slice event.
If the name i does not occur in $mseq$, we may apply $GSECTION$ to $mseq$ by:

$$GSECTION^{i}_{dim\ range}(mseq) = u_1 GSECTION^{i}_{dim\ range}(s_1) u_2 \ldots u_m GSECTION^{i}_{dim\ range}(s_m) u_{m+1}.$$

We shall now define all context sensitive restrictions mentioned and partially defined in Section 3.3.7. A well-defined distributor of the form *sep dim range*[*p*] must satisfy the following:

(Drest1) p contains at least one slice event.

(Drest2.1) If both *range* and *dim* are blank then the number of sections generated by the rightmost distributable (current) dimension must be the same for all slice events occurring in p.

(Drest2.2) If range is blank and $dim = j$ and j is an integer, then the number of sections generated by the jth dimension must be the same for all slice events occurring in p and having the jth dimension distributable.

(Drest3) Inside a k-nested distributor there must be at least one slice event with k distributable dimensions, and no array slices with more than k distributable dimensions.

(Drest4) If $dim = j$ then every slice event with maximal number of distributable dimensions must have the jth dimension as a distributable dimension.

(Drest5.1) If $range = \#i_1, i_2, i_3$, dim is blank and i_1, i_2, i_3 are integers, then $i_2 \geq i_1 \geq 1$, $i_3 \geq 1$ and i_2 must be smaller than or equal to the number of sections generated by the rightmost distributable (current) dimension of any array slice event occurring in p.

(Drest5.2) If $range = \#i_1, i_2, i_3$, $dim = j$ and i_1, i_2, i_3 are integers, then $i_2 \geq i_1 \geq 1$, $i_3 \geq 1$ and i_2 must be smaller than or equal to the number of sections generated by the jth dimension of any slice event occurring in p which has the jth dimension distributable.

We can now define *expdistr* as follows:

$$expdistr(sep\ dim\ range[mseq]) =$$
$$LSTRIP(COPY_{i=1}^{m}(GSECTION_{dim\ range}^{i}(mseq)sep))$$

where the name i does not occur in *mseq* and *range*, $m = (i_2 - i_1) \div i_3 + 1$ if $range = i_1, i_2, i_3$, or m is the number of sections generated by jth dimension of some slice event $a(\alpha_1, \ldots, \alpha_k)$ occurring in *mseq* with $\alpha_j = \varepsilon$ if *range* is blank but $dim = j$, or m is the number of sections generated by the rightmost distributable dimension of some slice event occurring in *mseq* if both *range* and *dim* are blank. The conditons Drest2.1, 2.2, 5.1 and 5.2 make the definition of m correct.
We show formally that the distributors can be replaced by concatenators.

Theorem 3.4 *For every well-defined distributor sep dim range[mseq]:*
$expand(sep\ dim\ range[mseq]) =$
$expand(\#i : 1, m, 1[GSECTION_{dim\ range}^{i}\ (mseq)sep@]),$

where m is as in the definition of expdistr.

Proof. It is sufficient to show that

$$expdistr(sep\ dim\ range[mseq]) =$$
$$exprepl(\#i : 1, m, 1[GSECTION_{dim\ range}^{i}\ (mseq)sep@]).$$

From the definition of *exprepl* it follows that:

$$exprepl(\#i:1,m,1[GSECTION^{i}_{dim\ range}(mseq)\ sep@] =$$
$$= LSTRIP(COPY^{m'}_{j=1}(GSECTION^{i}_{dim\ range}(mseq)[1+(j-1)*1/i]sep))$$
where $m' = (m-1) \div 1 + 1 = m$.

Since $1+(j-1)*1 = j$ then the substitution $[1+(j-1)*1/i]$ is equivalent to replacing a variable i by another variable j. Thus

$$LSTRIP(COPY^{m'}_{j=1}(GSECTION^{i}_{dim\ range}(mseq)[1+(j-1)*1/i]sep))$$
$$= LSTRIP(COPY^{m}_{j=1}(GSECTION^{i}_{dim\ range}(mseq)sep))$$
$$= expdistr(sep\ dim\ range[mseq]). \qquad \square$$

To illustrate this theorem let us consider the following example. Let a be as described by $C11$.

$;[a(\ ,9)]$ is equivalent to
$\#i:1,4,1[a(1*(i\div 1)*(1\div i)+2*(i\div 2)*(2\div i)$
$\quad +4*(i\div 3)*(3\div i)+8*(i\div 4)*(4\div i),9)\ ;\ @]$

since both expand to: $a(1,9);a(2,9);a(4,9);a(8,9)$.
Note that $GSECTION^1(a(\ ,9)) =$
$a(1*(i\div 1)*(1\div i)+2*(i\div 2)*(2\div i)+4*(i\div 3)*(3\div i)+8*(i\div 4)*(4\div i),9)$, but for instance $\#j:1,4,1[a(j+(j\div 3)*(3\div j)+4*(j\div 4),9)\ ;\ @]$ also expands to $a(1,9);a(2,9);a(4,9);a(8,9)$, and so is equivalent to $;[a(\ ,9)]$.

Expansion Theorem

Recall that a capitalized name of a basic or macro-nonterminal stands for the proper set of strings which may be generated from that non-terminal, i.e. for instance:

$MPROGRAM = \{s \in MT^{+} \mid mprogram \mapsto^{+} s\}$,
$BPROGRAM = \{s \in BT^{+} \mid bprogram \mapsto^{+} s\}$
and so on.

Furthermore, the subscript "0" indicates range variable freeness. Assume additionally that $BT \subset MT$, i.e. all basic-terminals are macro-terminals. Of course by the definition $BPROGRAM = BPROGRAM_0$, etc.

The following theorem states **strong collective expansion correctness for the general macro notation.**

Theorem 3.5 (Expansion Theorem)
$(\forall mprog \in MPROGRAM_0) \quad expand(mprog) \in BPROGRAM. \qquad \square$

As one might expect, the proof of Theorem 3.5 is by induction on the definition of the expansion function. The first obstacle we have to overcome is the fact that it is *not true* that $(\forall mev \in MEV) expand(mev) \in BEVENT$, for instance $expand(A(i)) = A(i) \notin BEVENT$ because index **variables** are not allowed in $BEVENT$.

Lemma 3.3
Let $mprog \in MPROGRAM_0$, $s = expand(mprog)$.
Then s contains no index variables.

Proof. Suppose that $s' = exprepl(\#i : in, fi, inc[pattern])$ From the definitions of *exprepl* and $COPY$ it follows immediately that the variable (symbol) "i" does not occur in s', i.e. $\#_i(s) = 0$. This ends the proof since index variables result from collectivisors and replicators, and collectivisors expand to the empty string (see (3) of the definition of *expand*). □

Since the function *expand* uses the functions *exprepl* and *expdistr*, its correctness requires their correctness as well. In fact, due to Theorem 3.4, we need to show the correctness of *exprepl* only.

Lemma 3.4 *Let* $SEP = \{;\} \cup \{,\}$.
(1) $(\forall repl \in SR)\ \ exprepl(repl) \in MSEQ$.
(2) $(\forall rr \in RR)\ \ exprepl(rr) \in MSEQ\ SEP$.
(3) $(\forall lr \in LR)\ \ exprepl(lr) \in SEP\ MSEQ$.

Proof. (1) We have to consider two cases: *repl* is a concatenator or an imbricator. Suppose that *repl* is a concatenator, i.e. $repl = \#i : in, fi, inc[p(i)\ sep@]$ and $concseq \mapsto^+ p(i)\ sep@$. The grammar rule is: $concseq = msequence\ sep@$ which means that $p(i) \in MSEQ$. From the definition of *exprepl* we have that:
$exprepl(repl) = p(in)\ sep\ p(in + inc) \ldots sep\ p(fi')$, where
$fi' = max\{j | (\exists k) j = in + k * inc \leq fi\}$ if $inc > 0$ and
$fi' = min\{j | (\exists k) j = in + k * inc > fi\}$ if $inc < 0$. Since $p(i) \in MSEQ$ then also $p(j) \in MSEQ$ for $j = in, in + inc, \ldots, fi'$, but this means that
$p(in)\ sep\ p(in + inc) \ldots sep\ p(fi') \in MSEQ$.
(We left to the reader the trivial proof that $s_1, s_2 \in MSEQ \Rightarrow s_1\ sep\ s_2 \in MSEQ$).
Suppose now that *repl* is an imbricator, i.e.

$$repl = \#i : in, fi, inc[p(i)\alpha\ @t@\ \beta q(i)]$$

where $\alpha \in SEP \cup \{(\}$, $\beta \in SEP \cup \{),)*\}$, and $imbrseq \mapsto^+ p(i)\alpha@t@\beta q(i)$.

The grammar rule is:

$$\begin{array}{lll}(a) & imbrseq = & msequence\ sep\ @\ sep\ \{\ |msequence\}\ @\ sep\ msequence| \\ (b) & & (@\ \{\ |msequence\ sep\}\ @\ sep\ msequence\ \{)|)*\}| \\ (c) & & (msequence\ sep\ @\ \{\ |sep\ msequence\}\ @\ \{)|)*\}| \\ (d) & & (@\ msequence\ @\ \{)|)*\}| \\ (e) & & (\{\ |msequence\ sep\}\ imbrseq\ \{\ |sep\ msequence\}\{)|)*\}| \\ (f) & & msequence\ sep\ imbrseq\ |\ imbrseq\ sep\ msequence\end{array}$$

In the derivation $imbrseq \mapsto^{+} p(i)\alpha @ t @ \beta q(i)$ we use both rules for *macrosequence* and *imbrseq*, and as far as the rules for *imbrseq* are concerned the derivation always satisfies the following order (since *imbrseq* occurs on the right hand side only in (e) and (f)):

1. the rules (e) and (f) can be used any finite number of times, or not at all, but all their uses must precede any use of $(a-d)$,
2. after k (eventually $k=0$) applications of (e) and/or (f), one and only one of the rules $(a-d)$ is applied.

Define:

$$HMSEQ(k) = \{s|s\underbrace{))\ldots)}_{k\ times} \in MSEQ\},$$

$$TMSEQ(k) = \{s|\underbrace{((\ldots(}_{k\ times}s \in MSEQ\}$$

Let $exprepl(repl) = s$. From the definition of *exprepl* it follows that in this case:

$$s = p(in)\alpha p(in+inc)\alpha\ldots\alpha p(fi')\alpha' t\beta' q(fi')\beta\ldots\beta q(in+inc)\beta q(in)$$

where $fi' = max\{j|(\exists k)j = in+k*inc \le fi\}$ if $inc > 0$ and $fi' = min\{j|(\exists k)j = in+k*inc > fi\}$ if $inc < 0$,

$$\alpha' = \begin{cases}\alpha & \alpha = (\\ \varepsilon & \alpha \in SEP,\end{cases}$$

$$\beta' = \begin{cases}\beta & \beta \in \{),)*\} \\ \varepsilon & \beta \in SEP.\end{cases}$$

Suppose we used the rule (e) k times, the rule (f) k' times, and $k > 1$, $k' \ge 0$. Now we have to consider 4 cases (recall that $s_1, s_2 \in MSEQ \Rightarrow s_1; s_2$ and $s_1, s_2 \in MSEQ$).

Case 1. We applied (a) next. In this case $p(j) \in HMSEQ(k)$, $q(j) \in TMSEQ(k)$, all j, $\alpha, \beta \in SEP$, $\alpha' t \beta' \in SEP \cup SEP\ MSEQ\ SEP$, so clearly $s \in MSEQ$.

Case 2. We used (b) next. In this case $p(j) \in HMSEQ(k) \cup \{(\}^k$,

$\alpha = (, \beta \in SEP, q(i) \in TMSEQ(1)(TMSEQ(k) \cup \{),)*\}^k) \subseteq TMSEQ(k+1)$, $\alpha' t \beta' = (t$, where $t \in SEP \cup MSEQ\ SEP$. Let $n = (fi - in) \div inc + 1$. We have:

$$s' = \underbrace{p(in)\alpha p(in+inc)\alpha \dots \alpha p(fi'-1)\alpha}_{n-1 \ times \ p(j)\alpha} p(fi') \in HMSEQ((n-1)*(k+1)+k)$$
$$= HMSEQ((k+1)*n-1)$$
$$s'' = q(fi')\beta q(fi'-1)\beta \dots \beta q(in+inc)\beta q(in) \in TMSEQ((k+1)*n).$$

Thus $s = s'\alpha' t \beta' s'' = s'(ts''$, but this means that $s \in MSEQ$.
Case 3. We used (c) next. This case is symmetric to (b), so we proceed similarly.
Case 4. We used (d) next. In this case $p(i) \in HMSEQ(k) \cup \{(\}^k$, $q(i) \in TMSEQ(k) \cup \{),)*\}^k$, $\alpha = (, \beta \in \{),)*\}$, $\alpha' = \alpha$, $\beta' = \beta$, $t \in MSEQ$, $p(i)\alpha \in HMSEQ(k+1) \cup \{(\}^{k+1}$, $q(i) \in TMSEQ(k+1) \cup \{),)*\}^{k+1}$, so clearly $s \in MSEQ$. Define $HMSEQ(0) = TMSEQ(0) = MSEQ$. For the case when we do not use the rule (e) at all, we proceed identically as above but with $k = 0$. Thus, $exprepl(repl) \in MSEQ$ for all imbricators.
(2) and (3) of the lemma are immediate consequences of (1) and the definition of *exprepl*. □

The proof of Theorem 3.5

First, in the proof we shall follow the definition of *expand*.
$mprog = \underline{program}\ mpbody\ \underline{endprogram}$, so clearly:

$$expand(mprog) \in BPROGRAM_0 \Leftrightarrow$$
$$pathsfirst(expand(mpbody)) \in BPROGRAMBODY.$$

Define the nonterminal: $bprogrambody' = \{path|process\}^+$. Clearly, $pathsfirst(expand(mpbody)) \in BPROGRAMBODY \Leftrightarrow expand(mpbody) \in BPROGRAMBODY'$.
But $mpbody = cpqbr_1 \dots cpqbr_n$, so:

$$expand(mpbody) \in BPROGRAMBODY' \Leftrightarrow$$
$$expand(cpqbr_i) \in BPROGRAMBODY',$$

for $i = 1, \dots, n$.
Every $cpqbr = cl\ pqbr_1 \dots pqbr_w$, so:

$$expand(cpqbr) \in BPROGRAMBODY' \Leftrightarrow$$
$$expand(pqbr_i) \in BPROGRAMBODY',$$

for $i = 1, \dots, w$. For every $pqbr$ we have either $pqbr \in MPATH_0$, or $pqbr \in MPROCESS_0$, or $pqbr =$
$\#i : in, fi, inc[pqbr'_1(i) \dots pqbr'_k(i)] \in BODYREPLICATOR_0$. Consider first the case $pqbr = \#i : in, fi, inc[pqbr'_1(i) \dots pqbr'_k(i)]$. From the definition of *expand* we have that in this case $expand(pqbr) = expand(exprepl(pqbr))$. From

the definition of *exprepl* and bodyreplicators we have that: $exprepl(pqbr) = pqbr''_1 \dots pqbr''_r$, where $pqbr''_i \in PQBR$ for $i = 1, \dots, n$.

If $pqbr''_i \in BODYREPLICATOR_0$, we have to apply *exprepl* to it and repeat the above reasoning, but in any case after a finite number of steps we obtain the case where there are no bodyreplicators left. Thus we can say that there always is a sequence $pqbr'''_1 \dots pqbr'''_l$ such that $pqbr'''_i \in MPATH_0 \cup MPROCESS_0$, $i = 1, \dots, l$, and:

$$expand(pqbr) = expand(pqbr'''_1 \dots pqbr'''_l) = expand(pqbr'''_1) \dots expand(pqbr'''_l).$$

From the condition $Irest2$ it follows that $pqbr'''_i$ does not contain body-replicator index variables, which together with the above means that:

$$expand(pqbr) \in BPROGRAMBODY' \Leftrightarrow$$
$$expand(pqbr'''_i) \in \begin{cases} PATH & pqbr'''_i \in MPATH_0 \\ PROCESS & pqbr'''_i \in MPROCESS_0 \end{cases}$$

for $i = 1, \dots, l$.
If $pqbr = \underline{path}\ mseq\ \underline{end}$ then:

$$expand(pqbr) \in PATH \Leftrightarrow expand(mseq) \in SEQUENCE.$$

If $pqbr = \underline{process}\ mseq\ \underline{end}$ then:

$$expand(pqbr) \in PROCESS \Leftrightarrow expand(mseq) \in SEQUENCE.$$

We show now that for every $mseq \in MSEQ$ there is an $mseq' \in MSEQ$ such that $expand(mseq) = expand(mseq')$ and $mseq'$ contains no replicator and no distributor. Since:

$$mseq = mor_1; \dots; mor_r, \text{ and}$$
$$mor = gel_1, \dots, gel_s,$$

it is sufficient to show that for every $gel \in GEL$ there is an $mseq' \in MSEQ$ such that $expand(gel) = expand(mseq')$ and $mseq'$ contains no replicators and no distributors. $gel = rr_1 \dots rr_u\ msd\ lr_1 \dots lr_v$, where $u = 0 \Rightarrow rr_1 \dots rr_u = \varepsilon$, $v = 0 \Rightarrow lr_1 \dots lr_v = \varepsilon$, and

$$expand(gel) =$$
$$expand(\underbrace{exprepl(rr_1) \dots exprepl(rr_u)\ msd\ exprepl(lr_1) \dots exprepl(lr_v)}_{s}).$$

Lemma 3.4 (2) and (3) guarantee that $s \in MSEQ$, i.e. s is a macrosequence itself, and s contains one layer of left and right replicators less than gel. If $gel = msd$, then either $gel \in MSTEL$, or $gel \in SR$ or $gel \in DST$. If $gel \in MSTEL$, gel does not contain outside replicators or distributors, if $gel \in$

SR then $expand(gel) = expand(exprepl(gel))$, and Lemma 3.4(1) guarantees that $s = exprepl(gel) \in MSEQ$, while if $gel \in DST$ then $expand(gel) = expand(expdist(gel))$, and by Theorem 3.4 and Lemma 3.4(1) we have that $s = expdist(gel) \in MSEQ$. In both cases s contains one layer of macro generators less than gel. Repeating this reasoning as many times as necessary we always can obtain $mseq'' \in MSEQ$, such that $expand(gel) = expand(mseq'')$ and $mseq''$ contains neither replicators nor distributors. Thus we proved that for every $mseq \in MSEQ$ there is an $mseq' \in MSEQ$ such that $mseq'$ contains no replicator and no distributor, and additionally $expand(mseq) = expand(mseq')$.

Let us now consider our $mseq'$. First observe that in this case every general element occurring in $mseq'$ is simply a starelement, but this means that $expand(mseq') = mseq'$ (since $expand(mev) = mev$, and then by induction). Second, for this case the rules of the grammar for macro notation and the rules of the grammar for basic notation are almost identical. The only difference is that indexed events in the macro notation may contain variables. But since $expand(mseq') = mseq'$, then $expand(mprog) = s_1\ mseq'\ s_2$, and Lemma 3.3 guarantees that indexed events in $mseq'$ contain no variables. Thus $mseq' \in SEQUENCE$, consequently $expand(mseq) \in SEQUENCE$, and consequently $mprog \in BPROGRAM$. □

3.4 The Semantics of Macro COSY Programs

Theorem 3.5 allows us to define the semantics of macro-programs in terms of the vector firing sequences of the basic programs obtained by their expansion. The semantics of a range variable free macro program MPr, which does not include any macro process will therefore be given in terms of $VFS(expand(MPr))$, and that of a program including processes will be given in terms of $VFS(Path(expand(MPr)))$.

We may also define the semantics of MPr in terms of Petri nets associated to $expand(MPr)$, or $Path(expand(MPr))$ if processes are involved.

The essential idea of the above approach was to allow us to specify a system consisting of event like points in any n-dimensional discrete Cartesian space, which is the task of collectivisors combined with the power of replicators. In particular, arithmetic expressions of the usual kind used in mathematics can be used to characterize and generate the index sets involved in a concise manner.

Once such a system of independent indexed events has been generated, sequential constraints can be imposed on it by combining subsets of events into sequential subsystems by means of paths and/or processes. Such subsystems will be synchronized by means of the events they share with other subsystems.

3.5 The COSY Environment

The practical use of the formal apparatus, presented in this book, during the design, analysis, and development of example systems has to be done manually on paper. Often analysis based on the COSY formalism involves extensive mechanical transformations of programs and their corresponding semantical objects (e.g. operation sets, cycle sets, vector firing sequences, etc.), which by its tediousness tends to dampen the designer's enthusiasm for exploring thoroughly more than one or two alternative designs. This of course is unfortunate and can be avoided by judicious use of the computer for performing these mechanical tasks.

Since we have always been concentrating on the task of discovering analytic criteria which are efficiently mechanisable, and since we have been fortunate enough to obtain a respectable number of such criteria, we have mechanised much of the COSY formalism. This makes it possible to experiment extensively with larger and more varied example systems.

We have devoted considerable effort to the writing of computer programs intended to reduce the mechanical labour of manipulating COSY specifications of example systems. At present we have two loosely coupled packages which perform functions as follows:

The first package is the Basic COSY System or BCS for short, which emphasizes the analysis and simulated execution of concurrent and distributed systems specified in the Basic COSY notation. BCS will accept a program written int his notation, check its syntax for correctness, and provides an internal representation of the program as well as object code. More importantly, BCS facilitates a simulated execution of the program thus allowing the user to study the dynamic properties of the specification, and to explore various alternative designs in a managed environment. BCS has a flexible command-driven interface with extensive multi-level on-line help, to allow the user to concentrate on the problem at hand which, in many cases, is quite complex.

BCS is currently implemented in C^{++} in Unix, as well as in SIMULA in both Unix and Michigan Terminal System (MTS). The version in C^{++} makes it possible to choose either the net semantics or the vector firing sequence semantics during simulation. The two semantics differ for programs involving ambiguous paths, they are the same for unambiguous programs.

The second package is the COSY System called CS for short, which is an interactive program which is complementary to BCS in the sense that it allows the development of COSY programs involving an earlier version of the high-level COSY notation discussed in this chapter. CS will accept a program written in this notation, check its syntax for correctness, and provides and internal representation of the high-level program which can be expanded to yield an equivalent basic COSY program, which in turn can be input to BCS for analysis and simulation. The part of CS dealing with the macro notation is also implemented in SIMULA in Unix and MTS, and in C^{++} in Unix.

Now that we have briefly characterized the packages we will sketch the organization and functionality of BCS in more detail.

The basic COSY machine itself on its own would be rather useless. In order to work, it needs to communicate with several satellite systems. In the next section, we provide a general overview of these satellites.

3.5.1 The Architecture of BCS

Fig. 3.16 depicts the overall architecture of BCS. The conceptual framework underlying the environment concerns three components: the interface, which oversees the interaction between BCS and the user, the semantic transformer, which generates an internal representation for the source specification, and the analyser, which exercises the specification to generate its behaviours. These components together facilitate the input of a source specification written in the basic COSY notation, its compilation , the generation of histories or traces of the concurrent behaviour permitted by the specification, and the display of analysis results.

The BCS Analysis Mechanism

The main element of the analysis component is the simulator whose structure is derived from the COSY formalism. The simulation mechanism carries out the execution of a COSY specification by implementing what is effectively a COSY machine based on the principles developed as parts of the COSY theory. In essence, the simulator is an algorithmic description of how to carry out instructions implicit in a COSY specification in order to generate its behaviour. The implementation of the simulator has the characteristic of a *virtual multiprocessor* (referred to as machines) in that each of the sequential subsystems could in principle be assigned to a different physical processor.

Simulation is initiated by initializing the internal representation of the specification, establishing all the structural conditions necessary to start a simulation run, and determining the events that may next be executed. The simulator runs interactively in two modes: free and manual. In the manual mode, the simulator pauses each time a set of events is fired, displaying the next set of enabled events. In the free mode, the simulation advances without interruption and only vectors (or scalars) of fired events are displayed if desired. The simulation can be terminated anywhere during the run. This leaves the model in a valid state from which the simulation can validly be resumed.

During the simulation run, the events fired are recorded in a history which may be (i) *undone* i.e. the last recorded event may be deleted and the state of

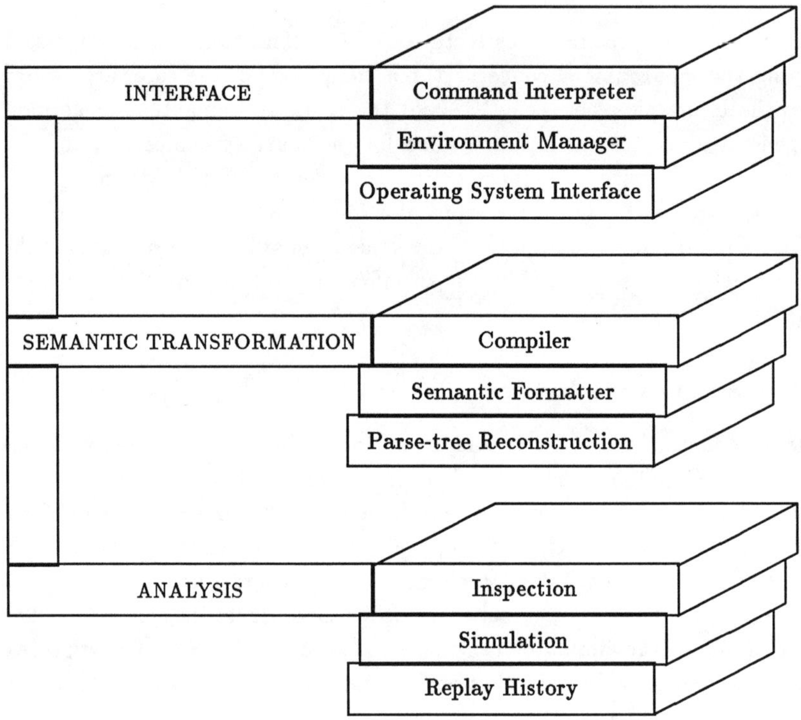

Figure 3.16. BCS Architecture

the simulation restored accordingly (a useful feature for backtracking during a manual exploration of the allowable behaviour) (ii) stripped of completed cyclic behaviours, leaving only prefixes of cycles, (iii) replayed in one of four styles, namely independent replay, sequential replay, concurrent replay and verbatim replay .

BCS is more than just a means of constructing a simulation model. It provides a representation of a COSY specification which can be manipulated in a variety of ways. Such manipulation includes behavioral analysis whereby properties of the behaviour of a system are deduced by examining its structure. The inspector provides the mechanism for such examination. It operates on the individual paths or processes of a compiled COSY specification without regard for the synchronisation constraints imposed by the coincidence of operations in different paths. The inspection mechanism is mainly useful for checking the semantics of more complicated paths which are otherwise difficult to read.

The BCS Semantic Transformation Mechanism

At the heart of the transformation level is the BCS compiler which works much like a traditional compiler: parsing the source program into an *intermediate code*, *optimizing* that intermediate code, and then translating the intermediate code into *object code*. The source program, in our case, is a specification written in basic COSY notation, while the intermediate code is equivalent to the internal representation of the program. The representation, in the form of a parse-tree together with an operation table, is optimized by pruning redudant nodes from the parse tree and compacting the operation table. The compiler accepts a source specification written in the basic COSY notation, describing one or more paths and from that source, constructs a parse tree representation of the specification. During the construction of the parse tree it checks the syntax and static semantics of the specification for correctness, generating appropriate error messages embedded in a listing of the source specification. If there are no errors, the compiler optimizes the internal representation and finally generates the object code which is written out to a file. The internal representation may be reconstructed from this object code.

The semantic formatter displays the specification formatted in such a way that lines which are too long for the page width currently in force are split only at semantically significant points. Each of the subsystems are given unique names $P_1, \ldots, P_n$, where n is the number of subsytems. Furthermore, levels of nesting are indicated by appropriate indentations.

The BCS Interface

At the interface level, BCS's functionality revolves around the command interpreter. It invokes the appropriate parts of the other levels and presents analysis results. The command interpreter itself has a multi-level structure reflecting the operations that can be performed at each level, together with context-sensitive on-line documentation . Each command may be entered in an abbreviated form and may require parameters.

Each BCS problem under study requires its own named *environment* which comprises a collection of files (source, object, listing, etc) associated with a COSY specification. The environment manager facilitates construction of families of specifications describing variants of the same problem. Specifications are given a generic name structure with a system supplied suffix (version number) so that they can be identified as representing alternative specifications of the same problem. An automatic versioning system keeps track of the versions as they are developed, enabling one to develop a specification in stages until a syntactically and semantically correct solution is obtained.

The interface is also responsible for most of the interactions between the environment and the operating system.

3.5.2 The Macro COSY Environment

The COSY system (CS for short) extends BCS by allowing for the interactive translation of high level COSY specifications involving macro generators to equivalent basic COSY specifications. The resulting specifications can then be studied by using the facilities of BCS.

The implementation of the macro COSY notation involved the incorporation of a compiler, a semantic formatter and an expander for the macro COSY notation in BCS, without repeating what had been achieved in BCS, to obtain the COSY System (CS for short). Facilities are also provided for incremental expansion to allow:

the specification of the values of integer variables as the expansion proceeds,

the resetting of values of integer variables after each expansion permitting the expansion of the same macro program repeatedly using different sets of integer values without requiring recompilation,

error detection and recoveries to enforce the context sensitive restrictions,

linking the output from the macro expander to the compiler of BCS allowing the simulation of expanded macro programs.

The syntax checker of the CS compiler serves the usual function of that part of any compiler. The context conditions of the macro grammar are partly checked in the CS compiler, and these checks are completed by the CS expander. Apart from the role of the expander to complete static semantic checking of the CS compiler the CS expander plays another important role. In applications of COSY to specific synchronization problems COSY programs are usually written in macro COSY, i.e. in terms of path patterns and generators over these patterns. In the process of learning the macro COSY notation it is often necessary to manually expand the macro program for different values of the integer parameters to ensure that one has written a program with the expected semantics as expressed by the corresponding expanded basic COSY program. The expander automates this process and hence makes it possible to study the structure of many possible expansions without much labour.

The following is from a sample session in CS using the expander.

```
mcosy > unix
Enter Unix Command: cat m_sysource

program
array a(4) b(0:1,2) c(0:2,4:5,-1:1) endarray
array #i:1,5,2[#j:2,4,2[d(i,j) #k:1,2,1[e(i,j,k) f(i,j,k)]]] endarray
array #i:1,5,2[#j:1,i,1[s(i,j) t(j,i)]] endarray
array #i:1,5,1[#j:5,i,-2[ss(j,i) tt(i,j)]] endarray
endprogram

mcosy > compile
No errors.
mcosy > expand

total_order_of_indexed_events

{
a (1) (2) (3) (4)
b (0,1) (1,1) (0,2) (1,2)
c (0,4,-1) (1,4,-1) (2,4,-1) (0,5,-1) (1,5,-1) (2,5,-1) (0,4,0) (1,4,0)
 (2,4,0) (0,5,0) (1,5,0) (2,5,0) (0,4,1) (1,4,1) (2,4,1) (0,5,1) (1,5,1)
 (2,5,1)
}
{
 d(1,2) d(1,4) e(1,2,1) e(1,2,2) f(1,2,1) f(1,2,2)
 e(1,4,1) e(1,4,2) f(1,4,1) f(1,4,2)
```

```
 d(3,2) d(3,4) e(3,2,1) e(3,2,2) f(3,2,1) f(3,2,2)
 e(3,4,1) e(3,4,2) f(3,4,1) f(3,4,2)

 d(5,2) d(5,4) e(5,2,1) e(5,2,2) f(5,2,1) f(5,2,2)
 e(5,4,1) e(5,4,2) f(5,4,1) f(5,4,2)
}
{
 s(1,1) t(1,1)
 s(3,1) s(3,2) s(3,3) t(1,3) t(2,3) t(3,3)
 s(5,1) s(5,2) s(5,3) s(5,4) s(5,5) t(1,5) t(2,5) t(3,5) t(4,5) t(5,5)
}
{
 ss(5,1) ss(3,1) ss(1,1) tt(1,5) tt(1,3) tt(1,1)
 ss(5,2) ss(3,2) tt(2,5) tt(2,3)
 ss(5,3) ss(3,3) tt(3,5) tt(3,3)
 ss(5,4) tt(4,5)
 ss(5,5) tt(5,5)
}
```

This example shows how one can determine that expressions used in macro generators give the expected indexed events and their order.

Two versions of CS, were coded in SIMULA and one was implemented in the Michigan Terminal System on an IBM/370 and the other in Unix on a Vax/780, and both support an earlier version of the macro COSY notation. Recently we rewrote the SIMULA version for MTS in C^{++} for Unix. Presently we are engaged in the development of an implementation of the generalized COSY notation introduced in this book.

3.6 The COSY System Dossier

Analysis based on the COSY formalism often involves extensive mechanical transformations of programs and their corresponding semantical objects, a laborious process which inhibits thorough exploration of more than one or two alternative designs. Judicious use of the BCS environment to perform these mechanical tasks greatly extends the practical utility of the method.

Some of the uses one can make of the COSY formalism are:

1. **Specification:** To specify the synchronisation aspects of systems of congreable (cooperating) concurrent behaviours.
2. **Abstraction:** To abstract to the synchronisation aspects of programs (implementations). That is, rather than expressing "how" a set of con-

current behaviours is enforced by means of the synchronizational elements of a program, one expresses "what" the set of synchronised concurrent behaviours is.

3. **Implementation:** To translate a specification of synchronized concurrent behaviours into a program involving synchronisation primitives in such a way as to enforce the behaviour specified when the program is executed. That is, we have the inverse of abstraction, rather than expressing "what" the set of synchronised concurrent behaviours is, we are to express "how" a set of synchronized concurrent behaviours is enforced by the synchronizational elements of a program of some implemented programming language.

4. **Verification:**

   (a) Implementation (program) satisfies specification.

   (b) Specification is implemented by program.

5. **Transformation:** To translate one specification into another which is more or less concurrent and/or distributed but defines the same interleaved or pseudo–concurrent behaviours.

6. **Comparison:** To compare two different programs implementing some synchronization strategy by comparing their corresponding abstract specifications in COSY. Comparisons may be made with regard to their degree of concurrency and distribution. However, the abstractions must preserve concurrency and distribution.

7. **Evaluation:** To evaluate programming languages with respect to their ability to support concurrency and distribution. Hence, a specification may be transformed to a more concurrent and/or distributed specification, but it may not be possible to implement it by means of a program which has the same degree of concurrency and distribution.

8. **Realisation:** To obtain a means of "executing" specifications by either:

   (a) direct implementation of specification in VLSI; or

   (b) implementation in the sense of point 3 above; or

   (c) combining the specification grammar with the grammar for a more conventional algorithmic programming language.

Full system COSY notation including the distinction between specification and implementation was developed to support the processes of *abstraction* and *implementation* (programming, coding, translating). **Verification** requires that there exist formal methods for determining the "meaning" of both the specification part and the implementation part. If there is a formal semantics associated with the implementation part then its meaning can be determined and must be shown to determine the same behaviours in some sense as the specification.

Given the BCS environment it is possible to produce a lot of information about a system formulated in COSY and it becomes necessary to devise a means for organising this information in a way which facilitates the analysis of the system formulated and allows ready comparison between different system designs. For this purpose, we introduced the notion of a **system dossier** as a framework for organising one's information about a system as well as organizing the processes generating this information [Lau83]. This notion is particularly helpful if one has several complementary but differing viewpoints supported by formalisms and automated procedures with the help of which one is gathering and analysing this information.

In the process of developing the dossier notion it became apparent that it would be presumptuous if we fixed the possible viewpoints constituting a dossier a priori. Hence, the dossier notation was introduced as an extendable notation. In this notation it is possible to introduce new **viewpoints** provided each viewpoint is adequately defined particularly with respect to its interface with other viewpoints which have already been defined.

Furthermore, since the dossier is a rather general notion it should not only be applicable to systems developed in COSY but should also serve as a means for describing the COSY formalism itself, the BCS environment and even the dossier concept itself.

Finally, we envisage the dossier notation as eventually being implemented, and if it were then "compiled" this would mean, for example:

1. certain communication paths are established between, for example, the user and maintainer or designer of the system;
2. interconnections between suite components are made which allow, for instance:
   (a) output of one component to become input to another component;
   (b) systems descriptions written in a combination of notations from different components, e.g. C^{++} programs and COSY specifications, to be interpreted using the semantics of the components involved. It might even be possible to run a program written, e.g. in a combination of C^{++} and Concurrent Pascal, by linking the respective compilers in appropriate ways.

Hence, the dossier concept is intended to combine aspects of a programming language construct, an operating system construct, a communication system construct and even a text processing construct.

3.7 Bibliographical Notes

The path/process notation originates from [LC75] and it also forms the main notation of Roy Campbell's Ph.D thesis [Cam76], although the latter also includes various other forms of path notation.

The simple forms of collectivisors and distributors were developed by Campbell and Lauer in [CL75].

Collaboration between Lauer, H. Genrich, and P.R. Torrigiani led to the development of a very wide grammar for collectivisors, distributors, replicators and system notation which were published in [TL77], [LT78] and [Tor78], the latter of which also contains translation to high-level Petri nets as introduced in [LC75], [Lau75b] and later in [LJJ83] and [LJ84]. This system notation was wide in the sense that it allowed all entities in the system to be parameters of system and operation definitions. R. Devillers extended the power of replicators to allow formulation of various general solutions of the hyperfast Banker (see Chapter 4.2) and the resulting notation was published in [LTD80], [Dev79b] and [Dev79a].

The initial replicator, distributor and system notation incorporated in COSY was published for the first time in [LTS79]. In [CL81] and particularly in J.Y. Cotronis' Ph.D thesis [Cot82], which contains a detailed analysis of the properties of earlier notations and motivates the design of the new notation, this macro notation was reexamined and revised to eliminate drawbacks of earlier notations and grammars; some aspects of programming methodology were investigated, such as when replicators and distributors may be replaced by other replicators and distributors without affecting the results of expansion; and finally, direct vector firing sequence semantics for macro programs was given, rather than indirectly in terms of the corresponding basic programs resulting from expansion. This permits the comprehension of macro programs directly without expansion. This same notation is presented in [LJ87] which also traces a more detailed history of the development of the macro notation in analogy to similar generators in Petri net theory. This indicates that not only the basic notation has a corresponding standard semantics in terms of Petri nets, but that the generators of the macro notation have corresponding generators in net theory, making COSY a very convenient complementary language theoretic formalism consistent with Petri net theory.

Finally, the macro notation presented in this book is a simplification and generalization of the notation in Cotronis' thesis. However, Cotronis' grammar is LR(k), which is not the case for our grammar, and hence allows more effective implementation as it stands. However, we consider simplification of reader comprension to be more important in the context of this book, and our grammar could easily be modified to be LR(k).

Theorems 3.1 and 3.3, Corollaries 3.3 and 3.4, Theorems 3.4 and 3.5 were all proven in [Cot82], but for the macro notation used there. Our generalizations and simplification of the grammar however greatly simplifies our proofs.

The first environment for basic COSY (BCS) was developed by Brian Hamshere and Lauer in [LH81] and its SIMULA implementation is described in [Ham83]. The implementation of the macro notation from [Cot82] and [LJ87] was initiated by J.Y. Cotronis and completed by Phesany Wong [Won85] on an IBM/780 under Michigan Terminal System, and ported to Unix by Mike Milanetti [Mil86]. The current C++ implementation of BCS, including both net semantics and vector firing sequence semantics, is due to S.J. Okech [Oke91].

The dossier concept was introduced in [Lau83] where it is explained in detail and its use illustrated by defining a dossier on COSY itself, including a sub-dossier on BCS, as well as by defining a dossier specifying the Smokers' problem, its solution and its verification. The dossier concept was incorporated in a general framework of requirements for an evolutionary computer based information processing environment called RECIPE in [CL84].

Chapter 4

COSY Applications

This chapter is devoted to various applications of the COSY theory and techniques. In general we can find two classes of applications. The first kind is the application of COSY, its theory and techniques to specify and/or prove properties of some realistic non-sequential systems or their parts. In fact, we already used COSY in Chap. 2 and 3 to analyse specifications of a linereader-lineprinter system, a specification mechanism for controlling access to a critical section, a mechanism regulating the use of a critical section by processes with differing priorities, and a reader-writer system.

In this chapter we additionally specify a two-way channel (part of a communication system), and a Banker's strategy (part of an operating system), and we will emphasize the ways of translating real world rules and constraints into COSY specifications. The specification of the Banker's problem also shows how some limitations of COSY can easily be removed in a very natural way. We shall also apply the COSY approach to the Cigarette Smoker's Problem, and show how the COSY simulator together with maximally concurrent semantics can be used to prove that a specification is adequate. The last problem of this first class of applications considered in this chapter is the Merlin-Randell Problem of Train Journeys, one of the most interesting synchronization problems. In this case COSY is used to implement a graph theoretical solution, and the correctness of this implementation is proven.

The second kind of application of COSY involves the use of the theory as a tool which aids in the development of another theory. We shall present two such applications. One is presentation of some formal bases for an alternative methodology for representing non-sequential systems, which is more suited for specifying systems in which concurrency is not the primary object of specification, but where concurrency is mainly used to speed up the performance of algorithms which may also be specified sequentially. In this approach the programmer starts with a sequential solution together with a characterization of independent events, from which a concurrent implementation can be derived. The second application in this latter class shows how COSY can be used in the theory of N-Modular

Redundancy in the area of reliability of systems.

4.1 Two-Way Channel with Disconnect

The problem we shall specify in this section can be described as follows:

> "The channel between endpoints a and b can pass messages in both directions simultaneously until it receives a 'disconnect' message from one end, after which it neither delivers nor accepts messages at that end. It continues to deliver and accept messages at the other end until the 'disconnect' message arrives, after which it can do nothing. The order of messages sent in a given direction is preserved."

This is one of the problems in communication which is very often met in practice. Two parties can communicate with each other until one of them disconnects.

In order to specify the two-way channel with disconnect we need two queues through which messages from node a to node b and, respectively, from node b to node a are sent. We assume that sizes of the queues are bounded by arbitrary numbers. We shall use a ring buffer discipline to meet the channel behaviour described by:

> "The channel between endpoints a and b can pass messages in both directions simultaneously. The order of messages sent in a given direction is preserved."

Let $(a \longrightarrow b)$-queue be the queue by which node a sends messages to node b. Assume that the ring buffer at $(a \longrightarrow b)$-queue consists of n frames. The queue has to distinguish between depositing ordinary messages and disconnect messages. By distinguishing we mean that if an ordinary message is deposited in a frame, then the ordinary message will be removed from that frame and if a disconnect message is deposited in a frame, then the disconnect message will be removed from that frame. Additionally, when a node accepts a disconnect message it should empty its incoming queue. Thus, each frame should be capable of being reset to empty as an alternative to removing a message, and the queue also has to allow testing for emptiness to verify that it has been reset for all frames.

Let us now define the following events for the $(a \longrightarrow b)$-queue:

$abmd_i$ –message is deposited in frame i of $(a \longrightarrow b)$-queue,
$abmr_i$ –message is removed from frame i of $(a \longrightarrow b)$-queue,
abe_i –test whether frame i of $(a \longrightarrow b)$-queue is empty
abF –test whether all frames of $(a \longrightarrow b)$-queue are full

$abdd_i$ —disconnect message is deposited in frame i of $(a \longrightarrow b)$-queue,
$abdr_i$ —disconnect message is removed from frame i of $(a \longrightarrow b)$-queue,
$abres_i$ —reset frame i of $(a \longrightarrow b)$-queue to empty,
abE —test whether all frames of $(a \longrightarrow b)$-queue are empty,

and similarly, by replacing 'ab' by 'ba', we obtain the events for $(b \longrightarrow a)$-queue. For $n = 3$, the $(a \longrightarrow b)$-queue can be specified by the following paths:

$\underline{path}\ abe_1, abE, (abmd_1; abF*; abmr_1, abres_1), (abdd_1; abF*; abdr_1, abres_1)\ \underline{end}$
$\underline{path}\ abe_2, abE, (abmd_2; abF*; abmr_2, abres_2), (abdd_2; abF*; abdr_2, abres_2)\ \underline{end}$
$\underline{path}\ abe_3, abE, (abmd_3; abF*; abmr_3, abres_3), (abdd_3; abF*; abdr_3, abres_3)\ \underline{end}$
$\underline{path}\ abmd_1, abdd_1; abmd_2, abdd_2; abmd_3, abdd_3\ \underline{end}$
$\underline{path}\ abmr_1, abdr_1, abres_{1;}abmr_2, abdr_2; abres_2; abmr_3, abdr_3, abres_3\ \underline{end}$
$\underline{path}\ abe_1; abe_2; abe_3\ \underline{end}$

and symmetrically for the $(b \longrightarrow a)$-queue.

Since some of the actions of each node depend on the state of the other node, which may be connected or disconnected, we need to specify the conditions for change, and a way of determining the state of each node. The two nodes must:

(1) be connected together, and

(2) after accepting or delivering a disconnect message become disconnected.

To specify this we need events:

CB —both node a and node b connected,
aC —node a becomes connected (symmetrically bC),
aD —node a becomes disconnected (symmetrically bD),
adA —node a accepts disconnect message (symmetrically bdA),
adD —node a delivers disconnect message (symmetrically bdD).

Now the paths specifying the conditions for connection and disconnection, and also the state of the nodes a and b are:

$$\underline{path}\ CB; aC*; adA, adD; aD*\ \underline{end},$$

for the state of node a, and

$$\underline{path}\ CB; bC*; bdA, bdD; bD*\ \underline{end},$$

for the state of node b, respectively. Since event CB occurs in both the above paths, both paths must permit it to occur, after which both nodes are in their connected state. After accepting a disconnect message or delivering a disconnect message, a node becomes disconnected and remains disconnected. Only after the other node gets disconnected, may both nodes be connected again.

Let us now describe in detail the actions of node a (the actions of node b are symmetric). This can be expressed as:

"connect together with b, then

repeat zero or more times:

either acknowledge request for accepting an ordinary message, then

if b is connected then

if $(a \longrightarrow b)$-queue is not full, accept and deposit into $(a \longrightarrow b)$-queue

if $(a \longrightarrow b)$-queue is full, do not accept

if b is disconnected then ignore message

or remove ordinary message from $(b \longrightarrow a)$-queue and deliver it

until

either accept a disconnect message and disconnect, followed by

if b is connected then deposit disconnect message into $(a \longrightarrow b)$-queue

if b is disconnected do nothing, then empty $(b \longrightarrow a)$-queue taking care of any pending message

or removing a disconnect message, deliver it and disconnect."

We believe that the above description of the actions of node a are precise, apart from the phrase "taking care of any pending message" which needs some explanation. Pending messages are those which are to be deposited in the $(a \longrightarrow b)$-queue after node b has been found connected, but before they are actually deposited node b accepts a disconnect message and disconnects itself. Since node b must reset the $(a \longrightarrow b)$-queue and the $(a \longrightarrow b)$-queue must be left empty, care must be taken to reset any pending messages as well.

Let us introduce the last events we need:

$amRA$	–node a is requested to accept a message (symmetrically bmRA),
amA	–node a accepts a message (symmetrically bmA),
amD	–node a delivers a message (symmetrically bmD),
aI	–node a ignores a message (symmmetrically bI),
$abNP$	–test that there is no pending message in $(a \longrightarrow b)$-queue (symmetrically $baNP$).

Now the activities of node a for $n = 3$ may be specified by the path below which follows directly from the above English description:

$$
\begin{array}{l}
\underline{path}\ CB;\\
((amRA; (bC; (abe_1, abe_2, abe_3; amA; abmd_1, abmd_2, abmd_3), abF), (bD; aI)),\\
\quad (bamr_1, bamr_2, bamr_3; amD))*;\\
(adA; (bC; abe_1, abe_2, abe_3; abdd_1, abdd_2, abdd_3), bD;\\
\quad (bares_1, bares_2, bares_3)*; baNP; (bares_1, bares_2, bares_3)*; baE),\\
(badr_1, badr_2, badr_3; adD)\ \underline{end}
\end{array}
$$

The above path resets pending messages. The resetting of the $(b \longrightarrow a)$-queue is done in two stages. During the first stage we guarantee that there is space in the $(a \longrightarrow b)$-queue for any pending messages to be deposited. After testing that there are no pending messages, the resetting is completed. We may specify the actions of node b symmetrically, by replacing all prefixes 'a', 'b', 'ab' and 'b' of events in the above path by 'b', 'a', 'ba' and 'ab', respectively.

Finally, we must specify the conditions under which a message is pending to be deposited into a queue. The conditions and the state of pending messages to the $(a \longrightarrow b)$-queue are:

(1) initially no message is pending;

(2) after b is found connected, indicating the intention of node a to deposit a message into the $(a \longrightarrow b)$-queue, a deposit is pending;

(3) after a deposit or when the $(a \longrightarrow b)$-queue is found to be full (in the case where node b is connected and there is no room for depositing), no deposit is pending.

The path specifying the above behaviour, i.e. for emptying the $(a \longrightarrow b)$-queue completely is:

$$\underline{path}\ abNP*; bC; abdd_1, abdd_2, abdd_3, abmd_1, abmd_2, abmd_3, abF\ \underline{end}$$

Symmetrically, specify the conditions and the state of pending messages for the $(b \longrightarrow a)$-queue (i.e. for emptying the $(b \longrightarrow a)$-queue completely). The whole program for 3-frame queues is presented in Fig. 4.1. The general case, where $(a \longrightarrow b)$-queue has length n, and $(b \longrightarrow a)$-queue has length m, is specified by means of macro COSY in Fig. 4.2.

It was proved in [Siu90] that the above solution is adequate. The proof is lengthy (19 pages) and uses the "custom tailored" techniques of adequacy-preserving transformations which were described in Section 2.4.3 of this book. The solution is transformed ten times and Theorem 2.20 is used to prove adequacy of the result of last transformation.

```
program
  path abe1, abE, (abmd1; abF*; abmr1, abres1),              } (a → b) – queue
    (abdd1; abF*; abdr1, abres1) end
  path abe2, abE, (abmd2; abF*; abmr2, abres2),
    (abdd2; abF*; abdr2, abres2) end
  path abe3, abE, (abmd3; abF*; abmr3, abres3),
    (abdd3; abF*; abdr3, abres3) end
  path abmd1, abdd1; abmd2, abdd2; abmd3, abdd3 end
  path abmr1, abdr1, abres1; abmr2, abdr2, abres2;
    abmr3, abdr3, abres3 end
  path abe1; abe2; abe3 end
  path bae1, baE, (bamd1; baF*; bamr1, bares1),              } (b → a) – queue
    (badd1; baF*; badr1, bares1) end
  path bae2, baE, (bamd2; baF*; bamr2, bares2),
    (badd2; baF*; badr2, bares2) end
  path bae3, baE, (bamd3; baF*; bamr3, bares3),
    (badd3; baF*; badr3, bares3) end
  path bamd1, badd1; bamd2, badd2; bamd3, badd3 end
  path bamr1, badr1, bares1; bamr2, badr2, bares2;
    bamr3, badr3, bares3 end
  path bae1; bae2; bae3 end
path CB; aC*; adA, adD; aD * end}  state of node a
path CB; bC*; bdA, bdD; bD * end}  state of node b
 path CB;                                                     } node
 ((amRA; (bC; (abe1, abe2, abe3; amA; abmd1, abmd2, abmd3), abF), (bD; aI)),
   (bamr1, bamr2, bamr3; amD))*;
 (adA; (bC; abe1, abe2, abe3; abdd1, abdd2, abdd3), bD;
   (bares1, bares2, bares3)*; baNP; (bares1, bares2, bares3)*; baE),
 (badr1, badr2, badr3; adD) end
 path CB;                                                     } node
 ((bmRA; (aC; (bae1, bae2, bae3; bmA; bamd1, bamd2, bamd3), baF), (aD; bI)),
   (abmr1, abmr2, abmr3; bmD))*;
 (bdA; (aC; bae1, bae2, bae3; badd1, badd2, badd3), aD;
   (abres1, abres2, abres3)*; abNP; (abres1, abres2, abres3)*; abE),
 (abdr1, abdr2, abdr3; bdD) end
 path abNP*; bC;                                              } empty (a → b) – queue
   abdd1, abdd2, abdd3, abmd1, abmd2, abmd3, abF end
 path baNP*; aC;                                              } empty (b → a) – queue
   badd1, badd2, badd3, bamd1, bamd2, bamd3, baF end
endprogram
```

Figure 4.1. Two-way Channel with Disconnect (3-frame queues)

```
program
  array abe abmd abmr abres abdd abdr (1 : n)
        bae bamd bamr bares badd badr (1 : m) endarray
  #i : 1, n, 1[ path abe(i), abE, (abmd(i); abF*; abmr_i, abres_i),
                     (abdd_i; abF*; abdr_i, abres_i) end ]              }
  path ; [abmd, abdd] end                                               } (a → b) − queue
  path ; [abmr, abdr, abres] end                                        }
  path ; [abe] end                                                      }
  #i : 1, m, 1[ path bae(i), baE, (bamd(i); baF*; bamr_i, bares_i),
                     (badd_i; baF*; badr_i, bares_i) end ]              }
  path ; [bamd, badd] end                                               } (b → a) − queue
  path ; [bamr, badr, bares] end                                        }
  path ; [bae] end                                                      }
 path CB; aC*; adA, adD; aD * end} state of node a
 path CB; bC*; bdA, bdD; bD * end} state of node b
  path CB;
  ((amRA; (bC; (, [abe]; amA; , [abmd], abF), (bD; aI)), (, [bamr]; amD))*;
  (adA; (bC; , [abe]; , [abdd]), bD; (, [bares])*; baNP; (, [bares])*; baE),   } node a
     (, [badr]; adD) end
  path CB;
  ((bmRA; (aC; (, [bae]; bmA; , [bamd]), baF), (aD; bI)), (, [abmr]; bmD))*;
  (bdA; (aC; , [bae]; , [badd]), aD; (, [abres])*; abNP; (, [abres])*; abE),   } node b
     (, [abdr]; bdD) end
 path abNP*; bC; , [abdd, abmd], abF end} empty (a → b) − queue
 path baNP*; aC; , [badd, bamd], baF end} empty (b → a) − queue
endprogram
```

Figure 4.2. Macro Notation Two-way Channel with Disconnect

4.1.1 Bibliographic Notes

The basic ideas of the above solution follow from [CL83]. The solution from J.Y. Cotronis and P.Lauer [CL83] was refined and an error was corrected by S. Siu in [Siu90]. The solution presented above is in principle that of [Siu90]. The reader interested in different solutions to this particular problem, is referred to [DJW85], where various solutions to this problem and different specification tools including Temporal Logic are presented.

4.2 The Hyperfast Banker

The Banker's problem and its connection with the allocation of non shareable, non premptive and reusable resources in operating systems have been considered in computer science since the late sixties.

The version we shall consider here is a banker with a single currency, say dollars. He is supposed to give loans to customers of his bank, who may apply for them as many times as they please, requesting in each application no more than the whole capital of the banker, and committing themselves to give back the loan and not to apply for new loans before having returned the previous one. The banker wants to avoid any deadlock or starvation (see Chap. 2.6.4) situations and to spread his money as much as possible.

To deal with this problem we shall use the following strategy. The banker builds in his bank as many windows as he has money and he lets the customers queue at a window depending on the amount of money they ask for. Cyclically, he looks at how much money he may loan and opens the windows corresponding to loans not greater than this amount. At each open window, he serves a single customer, if any is waiting, and then sending him to the next window, i.e. the window corresponding to a loan one less, and then he closes the window. Moreover, in order to avoid starvation, he counts the customers applying for a new loan and blocks them when a certain counter amount is reached, until he gets all his money back, at which time he resets the counter.

We shall specify the above strategy using a slightly extended macro COSY notation and priorities. To make the reading easier we shall also present a specification of a simple case in terms of basic COSY and priorities. The specification will consist of nine parts, namely: bank director, vault, doors, fair doormen, counters, clerks, fair reimbursers, customers, and some declarations and priority constraints.

4.2.1 The Bank Director

As stated above, the banker cyclically examines the state of his vault. If he has some money he will open a set of windows, and if he has all his money he will first reset the counting locks on all doors to the bank to allow new customers to apply for a loan.

The vault containing the money will be modelled by a buffer, each dollar corresponding to a single frame. Initially, each frame is full, meaning that all the money is in the vault. A dollar may be borrowed by a $get(\ldots)$ event: the corresponding frame is then empty. It is paid back by a $put(\ldots)$ event: the frame is then full again. The frames will be numbered from 1 to n, from right to left, where n is the capacity of the vault. We shall present the detailed specification of this free buffer when we discuss the vault.

In order to avoid possible conflicts between customers it will be useful to know not only the current amount in the vault but also the state, borrowed or not, of each dollar, i.e. the state, empty or full, of each frame. Such a **configuration** will be represented by an integer c, ranging from 0 to 2^n-1, whose binary representation reflects the state of the various frames. More precisely, if we number the bits of a binary representation from 1 to n, from right to left, and if $bit(b,c)$ is read "bit b of c", $bit(f,c)$ will be 1 if frame f is full in the considered configuration, 0 otherwise. The banker will check the current configuration with a set of events $conf(c)$, where c ranges from 1 to 2^n-1, since the empty configuration does not interest him.

The banker will not open the appropriate windows himself sequentially, he will use the services of the clerks, working concurrently. Their service will be activated by a $pgetb(c)$ event and the completion of the job will be controlled by a $pgete(c)$ event, where $pgetb$ and $pgete$ stand for "parallel get begin" and "paralell get end", respectively.

Similarly, when the configuration is full, i.e., when it is represented by the integer 2^n-1, the counters on the doors will reset themselves in parallel, as a result of an activation event $resb$, until the occurrence of a completion event, $rese$, where $resb$ and $rese$ stand for "reset-begin" and "reset-end", respectively.

The banker will not be concerned with the collection of borrowed money. This will be done independently and in parallel. It may then happen that the content of the vault is increased while the banker is controlling a money distribution. This will not affect him until his next cycle.

For $n=2$ the banker path is the following (here $2^n-1=3$):

$$\underline{path}\ (conf(1); pgetb(1); pgete(1)), (conf(2); pgetb(2); pgete(2)),$$
$$(conf(3); resb; rese; pgetb(3); pgete(3))\ \underline{end}$$

In general the banker has the form:

$$\underline{path}\ \#c : 1, 2^n-2, 1[conf(c); pgetb(c); pgete(c))\ , @]$$
$$, (conf(2^n-1); resb; rese; pgetb(2^n-1); pgete(2^n-1))\ \underline{end}$$

where $conf$ and $pgetb, pgete$ are declared by:

$$\underline{array}\ conf\ pgetb\ pgete\ (1 : 2^n-1)\ \underline{endarray}$$

4.2.2 The Vault

The vault of the bank will be represented by the n frames of a free buffer. Since the distribution of money and its return may be done in parallel for different customer processes p, at the various windows w, for the various frames f, we use three-dimensional $get(p,w,f)$ and $put(p,w,f)$ events. However, a frame f will never be associated with a window w with a higher number; only the cases where $w \leq f$ have to be considered. In the general solution this is taken into account in the collectivisor declared arrays *put* and *get*. Since the *put* and *get* events regulate the configuration changes, they are associated with the configuration events $conf$.

The vault paths where $n = 2$ and the number of customers is $m = 2$ are shown below. We also note that $2^n - 1 = 3$ and define the mapping *bit* by the following table:

c	$bit(2,c)$	$bit(1,c)$
1	0	1
2	1	0
3	1	1

The vault consists of two paths, namely:

$$\underline{path}\ (conf(1), conf(3))*; get(1,1,1), get(2,1,1);$$
$$conf(2)*; put(1,1,1), put(2,1,1)\ \underline{end}$$

which represents frame number 1, and

$$\underline{path}\ (conf(2), con(3))*; get(1,1,2), get(1,2,2), get(2,1,2), get(2,2,2);$$
$$conf(1)*; put(1,1,2), put(1,2,2), put(2,1,2), put(2,2,2)\ \underline{end}$$

representing frame number 2.

The general case requires a slight extension of the macro COSY notation. We need to introduce the concept of **predicate concatenator**. The standard concatenator has the form:

$$\#i : in, fi, inc[p(i)\ sep@]$$

and expands to:

$$p(in)\ sep\ p(in + inc)\ sep\ \ldots\ sep\ p(fi')$$

where fi' denotes the final value in the range of the index i and it might differ from fi (see Chap. 3.3.5).

The predicate concatenator has the form:

$$\#i : in, fi, predicate(i)[p(i)\ sep@].$$

Suppose that

$$\{i | in \leq i \leq fi \wedge predicate(i)\} = \{i_1, \ldots, i_k\} \text{ and } i_1 \leq i_2 \leq \ldots \leq i_k.$$

Then the predicate concatenator expands to:

$$p(i_1)\ sep\ p(i_2)\ sep\ \ldots\ sep\ p(i_k)\ .$$

For instance if $n = 2$ then

$$\#c : 1, 2^n - 1, bit(1, c) = 1[conf(c), @]$$

expands to:

$$conf(1), conf(3)$$

since

$$\{c | 1 \leq c \leq 2^2 - 1 \wedge bit(1, c) = 1\} \ = \ \{1, 3\},$$

and

$$\#c : 1, 2^n - 1, bit(2, c) = 1[conf(c), @]$$

expands respectively to:

$$conf(2), conf(3)$$

Clearly in the same style we may introduce predicate imbricators and predicate body replicators (predicate distributors were already mentioned in Chap. 3.1), but for the banker specification we only need predicate concatenators or predicate distributors.

Using predicate concatenators we can now give the specification of the general case for the vault as:

$$\begin{array}{l} \#f : 1, n, 1[\\ \underline{path}\ (\#c : 1, 2^n - 1, bit(f, c) = 1[conf(c), @])*;, [, [get(, , f)]]; \\ \qquad (\#c : 1, 2^n - 2, bit(f, c) = 0[conf(c), @])*;, [, [put(, , f)]]\ \underline{end}] \end{array}$$

where *conf*, *put*, *get* are declared by the following collectivisors:

$$\begin{array}{l} \underline{array}\ conf(1 : 2^n - 1)\ \underline{endarray} \\ \underline{array}\ \#p : 1, m, 1[\#w : 1, n, 1[\#f : w, n, 1[put(p, w, f)\ \ get(p, w, f)]]]\ \underline{endarray} \end{array}$$

where m is the number of customers. Note that the last collectivisor *guarantees* that $w \leq f$ for any valid $get(p, w, f)$ and $put(p, w, f)$.

The general vault can also be specified using **predicate distributors** instead of predicate concatenators. The standard distributor is defined as (see Chap. 3.3, and 3.3.7, 3.3.8 in particular for details):

$$sep\ dim\ range\ [\ msequence\ ]$$

where *range* is either empty or of the form to $\#in, fi, inc$. In the latter case *msequence* is distributed over the subrange $\{in, in+inc, \ldots, fi\}$ which must be a subset of an appropriate range of array-slices.

The predicate distributor, as we mentioned at the beginning of Chap. 3.3, may be defined by allowing the following form of range specification:

$$\#index : predicate(index)$$

and assuming that in such a case *msequence* is distributed over the subrange being the set of all indices from an appropriate range of array-slices which additionally satisfy a given predicate.

For example, if $n = 2$, and *conf* is defined as above, then

$$, \#c : bit(1, c) = 1[conf]$$

expands to '$conf(1), conf(3)$', and

$$, \#c : bit(2, c) = 1[conf]$$

expands to '$conf(2), conf(3)$', since in both cases the range of array-slices *conf* is $\{1, 2, 3\}$. Thus, using predicate distributors, we may specify the general vault as:

$$\begin{array}{l} \#f : 1, n, 1[\\ \underline{path}\ (, \#c : bit(f, c) = 1[conf])*;, [, [get(\ , \ , f)]]; \\ \quad (, \#c : bit(f, c) = 0 \wedge c \neq 2^n - 1[conf])*;, [, [put(\ , \ , f)]]\ \underline{end}] \end{array}$$

with *put*, *get* and *conf* being defined as before.
We remark that we did not provide formal expansion rules for both predicate concatenators and predicate distributors. This may be done in a more or less straightforward way following the approach of Sect. 3.3.8, but it requires extending the mapping $COPY$ from $COPY_{i=k}^{m}$ to $COPY_{predicate(j)}$.

4.2.3 The Doors

Access to the windows will be made through various doors. In front of each window, there will be a set of m doors, one for each customer. When a customer wants to go to a window, he will first knock at his corresponding door using a 2-dimensional $rq(p, w)$ event, where rq stands for "request" (a dollar at window w for customer process p). Each door will cyclically look if there is a request pending for it and then wait until the corresponding process has got a frame. If there is no pending request, this will be reflected by a 2-dimensional $rqsk(p, w)$ event, where $rqsk$ stands for "request-skip".

For $n = 2$, and $m = 2$ the door paths are the following:

$$\underline{path}\ rqsk(1,1),(rq(1,1);get(1,1,1),get(1,1,2))\ \underline{end}$$
$$\underline{path}\ rqsk(2,1),(rq(2,1);get(2,1,1),get(2,1,2))\ \underline{end}$$
$$\underline{path}\ rqsk(1,2),(rq(1,2);get(1,2,2))\ \underline{end}$$
$$\underline{path}\ rqsk(2,2),(rq(2,2);get(2,2,2))\ \underline{end}$$

The general doors are specified by:

$$\#p:1,m,1[\#w:1,n,1[\underline{path}\ rqsk(p,w),(rq(p,w);,[get(p,w,\ )])\ \underline{end}]]$$

with:

$$\underline{array}\ rqsk\ rq\ (1:m,1:n)\ \underline{endarray}$$

and *get* declared as for the vault.

4.2.4 The Fair Doormen

At each window, a single customer may be served at a time. A doorman will then be used to select a customer, if any, among the ones who knocked at their doors, i.e. who performed an *rq* event, thus invalidating the corresponding request-skip event. In order to avoid starvation of one door by other ones, the doorman will inspect them cyclically: he will thus select the first knocking customer he finds after the one previously served, and then he waits until this customer is served, i.e. gets the requested frame. During this period, the other knocking customers, if any, will be delayed. For $n = 2$ and $m = 2$ the doormen paths are the following:

$$\underline{path}\ rqsk(1,1),get(1,1,1),get(1,1,2),rqsk(2,1),get(2,1,1),get(2,1,2)\ \underline{end}$$
$$\underline{path}\ rqsk(1,2),get(1,2,2),rqsk(2,2),get(2,2,2)\ \underline{end}$$

and the general doormen specification is:

$$\#w:1,n,1[\underline{path}\ ,[rqsk(\ ,w),,[get(\ ,w,\ )],@]\ \underline{end}]$$

with *rqsk* and *get* declared as above.

4.2.5 The Counters

The admittance counters, which will sometimes temporarily block some customers from applying for a new loan, might be attached to the whole system, to each customer, to each window or to each customer-window pair. In order to avoid unnecessary conflicts between customers and to get a simpler specification, we shall here associate a counter with each customer.

Each counter has a **capacity** k and will be modelled by a free buffer with k frames which are initially empty. The frames are filled by 2-dimensional $cnt(p,i)$

events, which will be performed by the customers before each loan transaction. They are emptied by 2-dimensional $rset(p,i)$ events, activated by the resetting mechanism, which is controlled by the $resb-rese$ pair. The 2-dimensional events $leave(p,i)$ are introduced to allow the resetting of frames that are already empty.

To avoid unnecessary and possibly time consuming choices for the customers when they count themselves, we also use sequentializing paths, which will impose a cyclical order on the way each customer accesses his counting frames.

For $n = 2$, $m = 2$, $k = 2$ the paths for the counters, for the resetting mechanism, and for the sequentializers are the following.
The counters:

$$\underline{path}\ leave(1,1),(cnt(1,1);rset(1,1))\ \underline{end}$$
$$\underline{path}\ leave(1,2),(cnt(1,2);rset(1,2))\ \underline{end}$$
$$\underline{path}\ leave(2,1),(cnt(2,1);rset(2,1))\ \underline{end}$$
$$\underline{path}\ leave(2,2),(cnt(2,2);rset(2,2))\ \underline{end}$$

the resetting mechanism:

$$\underline{path}\ resb;leave(1,1),rset(1,1);rese\ \underline{end}$$
$$\underline{path}\ resb;leave(1,2),rset(1,2);rese\ \underline{end}$$
$$\underline{path}\ resb;leave(2,1),rset(2,1);rese\ \underline{end}$$
$$\underline{path}\ resb;leave(2,2),rset(2,2);rese\ \underline{end}$$

and the sequentializers:

$$\underline{path}\ cnt(1,1);cnt(1,2)\ \underline{end}$$
$$\underline{path}\ cnt(2,1);cnt(2,2)\ \underline{end}.$$

In the general case, we have:
the counters:

$$\#p:1,m,1[\#i:1,k,1[\underline{path}\ leave(p,i),(cnt(p,i);rset(p,i))\ \underline{end}]]$$

the resetting mechanism:

$$\#p:1,m,1[\#i:1,k,1[\underline{path}\ resb;leave(p,i),rset(p,i);rese\ \underline{end}]]$$

and the sequentializers:

$$\#p:1,m,1[\underline{path}\ ;[cnt(p,\ )]\ \underline{end}]$$

with a collectivisor of the form:

$$\underline{array}\ cnt\ leave\ rset\ (1:m,1:k)\ \underline{endarray}.$$

4.2.6 The Clerks

The clerks are in charge of actually giving the money to customers at the various windows. Their task is controlled by the *pgetb* – *pgete* events (see the bank manager). To achieve the highest degree of parallelism, one has to avoid competitions between paths and/or processes whenever possible. To ensure this was obviously not possible for the doormen, but as far as the clerks are concerned, we are simply faced with the problem which has sometimes been called the "Argentinian party problem". Briefly, this is a problem where a set of consumers apply for certain resources, and one knows that there are always enough resources to satisfy everyone, and one wants the distribution to be made without conflict. To solve the problem one may specify a strategy beforehand which says which resource goes to which consumer, if any. In our case we know that the number of windows opened from the right is equal to the number of full frames observed by the bank manager. Hence, one may simply associate the window w with the w-th full frame from the right. We shall then associate a clerk with each window w and with each frame $f \geq w$. This clerk will only be activated if frame f is the w-th full frame from the right. If a customer has been selected by the doormen for this window, the clerk will grant his request and give him frame f. Otherwise the clerk performs a 2-dimensional $grsk(w, f)$ event, where $grsk$ stands for "grant-skip".

The paths for clerks for $n = 2$ and $m = 2$ are the following:

$\underline{path}\ pgetb(1), pgetb(3); grsk(1,1), get(1,1,1), get(2,1,1); pgete(1), pgete(3)\ \underline{end}$
$\underline{path}\ pgetb(2); grsk(1,2), get(1,1,2), get(2,1,2); pgete(2)\ \underline{end}$
$\underline{path}\ pgetb(3); grsk(2,2), get(1,2,2), get(2,2,2); pgete(3)\ \underline{end}$

In the general case the clerks may be specified as:

$$\begin{array}{l}\#w: 1, n, 1[\#f: w, n, 1[\\ \underline{path}\ \#c: 1, 2^n-1, bit(f,c) = 1 \wedge \sum_{i<f} bit(i,c) = w-1[pgetb(c), @];\\ \qquad grsk(w,f),,[get(\ ,w,f)];\\ \qquad \#c: 1, 2^n-1, bit(f,c) = 1 \wedge \sum_{i<f} bit(i,c) = w-1[pgete(c), @]\ \underline{end}]]\end{array}$$

or, using predicate distributors instead of predicate concatenators as:

$$\begin{array}{l}\#w: 1, n, 1[\#f: w, n, 1[\\ \underline{path}\ , \#c: 1, 2^n-1, bit(f,c) = 1 \wedge \sum_{i<f} bit(i,c) = w-1[pgetb];\\ \qquad grsk(w,f),,[get(\ ,w,f)];\\ \qquad \#c: 1, 2^n-1, bit(f,c) = 1 \wedge \sum_{i<f} bit(i,c) = w-1[pgete]\ \underline{end}]]\end{array}$$

with $grsk$ declared as:

$$\underline{array}\ \#w: 1, n, 1[\#f: w, n, 1[grsk(w,f)]]\ \underline{endarray}$$

and the remaining indexed events as before.

4.2.7 The Fair Reimbursers

When a customer p reaches the point where he wants to return the amount j of his loan, instead of doing it himself sequentially, he will simply activate, by a 2-dimensional event $pputb(p, j)$ a set of j reimbursers who will concurrently return one dollar, i.e. fill one frame each. Each reimburser may be seen as associated with a customer and with a window where the customer received a frame. He may then be activated by a $pputb(p, j)$ event, where $j \geq$ window number.

We will also impose the restriction that the reimbursers are 'fair', i.e. that reimburser (p, w) gives back the same dollar that customer p obtained at window w during the preceding loan. We will thus associate the put events with their corresponding get event, and prevent the customer from obtaining a new dollar at a window, during a new loan, if he previously obtained one at that window without returning it. This fair behaviour also avoids any unnecessary competition between the reimbursers to refill the empty frames. For $n = 2$ and $m = 2$ the paths for the reimbursers and for fair reimbursements are the following:
The reimbursers:

$$\underline{path}\ pputb(1,1), pputb(1,2); put(1,1,1), put(1,1,2)\ \underline{end}$$
$$\underline{path}\ pputb(2,1), pputb(2,2); put(2,1,1), put(2,1,2)\ \underline{end}$$
$$\underline{path}\ pputb(1,2); put(1,2,2)\ \underline{end}$$
$$\underline{path}\ pputb(2,2); put(2,2,2)\ \underline{end}$$

and the fair reimbursements:

$$\underline{path}\ get(1,1,1), get(1,1,2); put(1,1,1), put(1,1,2)\ \underline{end}$$
$$\underline{path}\ get(2,1,1), get(2,1,2); put(2,1,1), put(2,1,2)\ \underline{end}$$
$$\underline{path}\ get(1,2,2); put(1,2,2)\ \underline{end}$$
$$\underline{path}\ get(2,2,2); put(2,2,2)\ \underline{end}.$$

In the general case we have, the reimbursers:

$$\#w : 1, n, 1[\#p : 1, m, 1[\underline{path}\ \#j : w, n, 1[pputb(p, j), @];, [put(p, w, \ )]\ \underline{end}]]$$

and, the fair reimbursements:

$$\#w : 1, n, 1[\#p : 1, m, 1[\underline{path}\ , [get(p, w, \ )];, [put(p, w, \ )]\ \underline{end}]]$$

where $pputb$ is declared by the following collectivisor:

$$\underline{array}\ pputb(1 : n, 1 : m)\ \underline{endarray}.$$

4.2.8 The Customers

Each customer is modelled by a process which will cyclically count himself ($cnt(\ldots)$ events) for a new loan, choose a loan amount j, go through a sequence of windows as required by the strategy (knocking at a door, rq, and waiting until he gets a frame due to the occurrence of a get), and finally will return his entire loan ($pputb$).

The customers will thus ask for their dollars one at a time and will not return any until they have received their entire loan.

In the case of $n = 2$, $m = 2$ and $k = 2$, the processes representing the customers are:

$$\begin{array}{l}\underline{process}\\ cnt(1,1), cnt(1,2);\\ (rq(1,1); get(1,1,1), get(1,1,2); pputb(1,1)),\\ (rq(1,2); get(1,2,2); rq(1,1); get(1,1,1), get(1,1,2); pputb(1,2))\\ \underline{end}\\ \underline{process}\\ cnt(2,1), cnt(2,2);\\ (rq(2,1); get(2,1,1), get(2,1,2); pputb(2,1)),\\ (rq(2,2); get(2,2,2); rq(2,1); get(2,1,1), get(2,1,2); pputb(2,2))\\ \underline{end}\end{array}$$

For the general case, the processes for the customers are:

$$\begin{array}{l}\#p : 1, m, 1[\\ \underline{process}\\ ,[cnt(p,\ )];\\ \#j : 1, n, 1[(\#w : j, 1, -1[rq(p,w);, [get(p,w,\ )]; @]; pputb(p,j)), @]\ \underline{end}]\end{array}$$

4.2.9 Priority Constraints

The various events of the specification have been introduced above. All the collective events can be declared by the following collectivisors:

$$\begin{array}{l}\underline{array}\ conf\ pgetb\ pgete\ (1 : 2^n - 1)\\ \qquad rq\ rqsk\ pputb\ (1 : m, 1 : n)\\ \qquad cnt\ leave\ rset\ (1 : m, 1 : k)\ \underline{endarray}\\ \underline{array}\ \#p : 1, m, 1[\#w : 1, n, 1[\#f : w, n, 1[get(p,w,f)\ put(p,w,f)]]]\ \underline{endarray}\\ \underline{array}\ \#w : 1, n, 1[\#f : w, n, 1[grsk(w,f)]]\ \underline{endarray}\end{array}$$

where n is the capacity of the vault, m is the number of customers, and k is the capacity of counters. We also utilized two non-collective events, $resb$ and $rese$.

To obtain fair behaviour and to avoid starvation situations (see Chap. 2.6.3), we will have to add some priority constraints. In general, the skip-like events will have a lower priority than events they are in conflict with, since they are only introduced to reflect the situations where these conflicting events are disabled. Thus:

$rqsk(\ldots) < rq(\ldots)$, to avoid that knocking at a door is indefinitely delayed by the watch round of the corresponding doorman;
$leave(\ldots) < cnt(\ldots)$, to avoid that counting a customer is indefinitely delayed by the resetting mechanism, i.e. by the supervisory actions of the director of reimbursing; and
$grsk(\ldots) < get(\ldots)$, to avoid that a clerk always skips the money distribution.

Note that the rule does not concern the pairs $rqsk - get$ and $leave - rset$, since the examination of the specifications for the doors and the counters shows that they are never in conflict.

It is also necessary to require that $conf(\ldots) < put(\ldots)$, so as to avoid that the examination of the configuration by the bank director indefinitely delays the reimbursements. It is not necessary to impose the same condition on $conf$ and get, since get is controlled by the $pgetb - pgete$ pair, i.e. by the bank director himself, so that they are never in conflict.

For $n = 2$, $m = 2$ and $k = 2$ the above priority constraints can be specified as:

$\underline{priority}$
$rqsk(1,1) < rq(1,1), rqsk(1,2) < rq(1,2), rqsk(2,1) < rq(2,1), rqsk(2,1) < rq(2,2),$
$leave(1,1) < cnt(1,1), leave(1,2) < cnt(1,2), leave(2,1) < cnt(2,1),$
$leave(2,1) < cnt(2,2),$
$grsk(1,1) < (get(1,1,1), get(2,1,1)), grsk(1,2) < (get(1,1,2), get(2,1,2)),$
$grsk(2,2) < (get(1,2,2), get(2,2,2))$
$conf(2) < (put(1,1,1), put(2,1,2)),$
$conf(1) < (put(1,1,2), put(1,2,2), put(2,1,2), put(2,2,2))$ $\underline{end}$

As we already mentioned in Chap. 3.1, we can use distributors and replicators to specify the priority constraints in the general case as follows:

$\underline{priority}$
$\#p : 1, m, 1[\#w : 1, n, 1[rqsk(p,w) < rq(p,w), @]],$
$\#p : 1, m, 1[\#w : 1, n, 1[leave(p,w) < cnt(p,w), @]],$
$\#w : 1, n, 1[\#f : w, n, 1[grsk(w,f) < (, [get(, w, f)]), @]],$
$\#f : 1, n, 1[(\#c : 1, 2^n - 1, bit(f,c) = 0[conf(c), @]) < (, [, [put(, , f)]]), @]$ $\underline{end}$

Note that the last line of the above priority specification **cannot** be replaced by $(, [conf] < (, [, [, [put]]])$ since the definition of the priority relation $<$ requires that this relation can only be defined for events being in local conflict (see Chap. 2.6.2).

The specification of the whole banker for $n = 2$, $m = 2$, and $k = 2$ is given in Fig. 4.3 and 4.4 by means of basic COSY with priorities. The general banker is specified in Fig. 4.5 using our extended macro COSY notation with priorities.

4.2.10 Final Comment and Bibliographical Notes

The Banker's problem originates from E. Dijkstra [Dij68], and some classical computational solutions are given in [Hab69] and [Hov68]. The solution presented here is clearly non-computational and based on [LTD80] by P. Lauer, R. Torrigiani and R. Devillers. An extension of this approach to the case of a Banker with several currencies is given in [Dev79b]. The kernel of the Bank, consisting of the banker, the vault and the clerks was subjected to a careful formal analysis in [SL80] and [LS81] and proved to be adequate and correct.

4.3 Cigarette Smokers

This section requires from the reader some knowledge about Dijkstra's P and V operations on semaphores. For those not familiar with semaphores we recommend [BH73].

The Cigarette Smokers' problem is one of the oldest synchronization problems. It was introduced by Patil in [Pat71] and can be formulated as follows.

Three smokers are sitting at a table. Each smoker has one and only one of the resources he requires for rolling and smoking a cigarette. One has tobacco, another has cigarette papers, and the third has matches. Smokers may not exchange ingredients with each other. On the table in front of them, two out of the three resources will be placed. No new resources will be placed on the table until some smoker has successfully smoked. Smokers may only pick up resources that are on the table sequentially one after the other. Smokers cannot return resources to the table when they discover that the third resource they need to smoke has not been placed on the table. Hence, once the wrong smoker picks up a resource which another smoker could successfully use, no smoker can successfully smoke and the smokers are deadlocked.

Patil gave a program representing this problem which uses binary semaphore operations P and V, and which may deadlock. This program is presented in Fig. 4.6. In this figure and the subsequent figures $**$ stands for some unspecified operation performed by the process. To inform the smokers about the resources which have been placed on the table, three binary semaphores, initialized to 0, representing tobacco, paper and matches are used. On placing the resources on the table the corresponding semaphore is incremented by performing a V operation. On the smokers' side binary semaphores, initialized to 0, *smokert*, *smokerp*, and *smokerm* are used to signal that a cigarette has been smoked. The smoker who finishes smoking a cigarette performs the operation V on the corresponding semaphore.

The **Smokers' Problem**, as formulated by Patil, is to define additional semaphores and processes, and to introduce appropriate P and V statements so as to make the smokers' system deadlock free. No alteration may be made

```
program
 priority
 rqsk(1,1) < rq(1,1), rqsk(1,2) < rq(1,2),
 rqsk(2,1) < rq(2,1), rqsk(2,1) < rq(2,2),
 leave(1,1) < cnt(1,1), leave(1,2) < cnt(1,2),
 leave(2,1) < cnt(2,1), leave(2,1) < cnt(2,2),
 grsk(1,1) < (get(1,1,1), get(2,1,1)), grsk(1,2) < (get(1,1,2), get(2,1,2)),
 grsk(2,2) < (get(1,2,2), get(2,2,2))
 conf(2) < (put(1,1,1), put(2,1,2)),
 conf(1) < (put(1,1,2), put(1,2,2), put(2,1,2), put(2,2,2)) end
 path (conf(1); pgetb(1); pgete(1)), (conf(2); pgetb(2); pgete(2)),     } bank manager
      (conf(3); pgetb(3); pgete(3)) end
  path (conf(1), conf(3))*; get(1,1,1), get(2,1,1);                    }
       conf(2)*; put(1,1,1), put(2,1,1) end                           }
  path (conf(2), con(3))*; get(1,1,2), get(1,2,2), get(2,1,2), get(2,2,2);  } vault
        conf(1)*; put(1,1,2), put(1,2,2), put(2,1,2), put(2,2,2) end  }
 path rqsk(1,1), (rq(1,1); get(1,1,1), get(1,1,2)) end   }
 path rqsk(2,1), (rq(2,1); get(2,1,1), get(2,1,2)) end   } doors
 path rqsk(1,2), (rq(1,2); get(1,2,2)) end               }
 path rqsk(2,2), (rq(2,2); get(2,2,2)) end               }
 path rqsk(1,1), get(1,1,1), get(1,1,2),                 }
      rqsk(2,1), get(2,1,1), get(2,1,2) end              } fair doormen
 path rqsk(1,2), get(1,2,2), rqsk(2,2), get(2,2,2) end   }
  path leave(1,1), (cnt(1,1); rset(1,1)) end   }               }
  path leave(1,2), (cnt(1,2); rset(1,2)) end   } counters      }
  path leave(2,1), (cnt(2,1); rset(2,1)) end   }               }
  path leave(2,2), (cnt(2,2); rset(2,2)) end   }               }
  path resb; leave(1,1), rset(1,1); rese end   }               }
  path resb; leave(1,2), rset(1,2); rese end   } resetting     } counters
  path resb; leave(2,1), rset(2,1); rese end   }               }
  path resb; leave(2,2), rset(2,2); rese end   }               }
  path cnt(1,1); cnt(1,2) end   } sequentializers              }
  path cnt(2,1); cnt(2,2) end.  }                              }
```

Figure 4.3. Banker1 for $n = 2, m = 2, k = 2$ (part 1)

```
path pgetb(1), pgetb(3);                                                    ⎫
      grsk(1,1), get(1,1,1), get(2,1,1);                                    ⎪
      pgete(1), pgete(3) end                                                ⎬ clerks
path pgetb(2); grsk(1,2), get(1,1,2), get(2,1,2); pgete(2) end              ⎪
path pgetb(3); grsk(2,2), get(1,2,2), get(2,2,2); pgete(3) end              ⎭
  path pputb(1,1), pputb(1,2);             ⎫                     ⎫
        put(1,1,1), put(1,1,2) end         ⎪                     ⎪
  path pputb(2,1), pputb(2,2);             ⎬ reimbursers         ⎪
        put(2,1,1), put(2,1,2) end         ⎪                     ⎪
  path pputb(1,2); put(1,2,2) end          ⎪                     ⎪
  path pputb(2,2); put(2,2,2) end          ⎭                     ⎬ fair reimbursers
  path get(1,1,1), get(1,1,2);             ⎫                     ⎪
        put(1,1,1), put(1,1,2) end         ⎪                     ⎪
  path get(2,1,1), get(2,1,2);             ⎬ reimbursements      ⎪
        put(2,1,1), put(2,1,2) end         ⎪                     ⎪
  path get(1,2,2); put(1,2,2) end          ⎪                     ⎪
  path get(2,2,2); put(2,2,2) end.         ⎭                     ⎭
process                                                                      ⎫
cnt(1,1), cnt(1,2);                                                          ⎪
(rq(1,1); get(1,1,1), get(1,1,2); pputb(1,1)),                               ⎪
(rq(1,2); get(1,2,2); rq(1,1); get(1,1,1), get(1,1,2); pputb(1,2))           ⎪
end                                                                          ⎪
process                                                                      ⎬ customers
cnt(2,1), cnt(2,2);                                                          ⎪
(rq(2,1); get(2,1,1), get(2,1,2); pputb(2,1)),                               ⎪
(rq(2,2); get(2,2,2); rq(2,1); get(2,1,1), get(2,1,2); pputb(2,2))           ⎪
end                                                                          ⎭
endprogram
```

Figure 4.4. Banker1 for $n = 2, m = 2, k = 2$ (part 2)

```
program
  array conf pgetb pgete (1 : 2^n − 1)
        rq rqsk pputb (1 : m, 1 : n)
        cnt leave rset (1 : m, 1 : k) endarray
 array #p : 1, m, 1[#w : 1, n, 1[#f : w, n, 1[get(p, w, f) put(p, w, f)]]] endarray
 array #w : 1, n, 1[#f : w, n, 1[grsk(w, f)]] endarray
 priority
 #p : 1, m, 1[#w : 1, n, 1[rqsk(p, w) < rq(p, w), @]],
 #p : 1, m, 1[#w : 1, n, 1[leave(p, w) < cnt(p, w), @]],
 #w : 1, n, 1[#f : w, n, 1[grsk(w, f) < (, [get( , w, f)]), @]],
 #f : 1, n, 1[(#c : 1, 2^n − 1, bit(f, c) = 0[conf(c), @]) < (, [, [put( , , f)]]), @] end
 path #c : 1, 2^n − 2, 1[conf(c); pgetb(c); pgete(c)) , @]                         } bank manager
   , (conf(2^n − 1); resb; rese; pgetb(2_n − 1); pgete(2_n − 1)) end                 }
 #f : 1, n, 1[                                                                       }
 path ( , #c : bit(f, c) = 1[conf])*; , [, [get( , , f)]];                           } vault
     ( , #c : bit(f, c) = 0 ∧ c ≠ 2^n − 1[conf])*; , [, [put( , , f)]] end]          }
#p : 1, m, 1[#w : 1, n, 1[path rqsk(p, w); , [get(p, w, )]) end]]} doors
#w : 1, n, 1[path , [rqsk( , w), , [get( , w, )], @] end]} fair doormen
 #p : 1, m, 1[#i : 1, k, 1[path leave(p, i), (cnt(p, i); rset(p, i)) end} counters      }
        path resb; leave(p, i), rset(p, i); rese end]} resetting                        } counter
        path ; [cnt(p, )] end]} sequentializers                                         }
 #w : 1, n, 1[#f : w, n, 1[                                                          }
 path , #c : 1, 2^n − 1, bit(f, c) = 1 ∧ Σ_{i<f} bit(i, c) = w − 1[pgetb];           } clerks
        grsk(w, f), , [get( , w, f)];                                                }
        #c : 1, 2^n − 1, bit(f, c) = 1 ∧ Σ_{i<f} bit(i, c) = w − 1[pgete]end]]       }
  #w : 1, n, 1[#p : 1, m, 1[                             } reimbursers   }
  path , [get(p, w, )]; , [put(p, w, )] end]]            }               } reimbursers
 path , [get(p, w, )]; , [put(p, w, )] end]]} reimbursements             }
 #p : 1, m, 1[                                                                       }
 process                                                                             }
 , [cnt(p, )];                                                                       } customers
 #j : 1, n, 1[(#w : j, 1, −1[rq(p, w); , [get(p, w, )]; @]; pputb(p, j)), @]         }
 end]                                                                                }
endprogram
```

Figure 4.5. General Banker Solution

```
Rtobacco              Rpaper                Rmatch              )
rt : P(s);            rp : P(s);            rm : P(s);          |
   V(paper);             V(match);             V(tobacco);      > agent
   V(match);             V(tobacco);           V(paper);        |
   goto rt               goto rp               goto rm          )
bt : P(smokert);      bp : P(smokerp);      bm : P(smokerm);
   V(s);                 V(s);                 V(s);
   goto bt               goto bp               goto bm
Smokerwithtobacco     Smokerwithpaper       Smokerwithmatches   )
at : P(paper);        ap : P(match);        am : P(tobacco);    |
   P(match);             P(tobacco);           P(paper);        |
   **                    **                    **               > smokers
   V(smokert);           V(smokerp);           V(smokerm);      |
   goto at               goto ap               goto am          )
```

Figure 4.6. Patil's statement of Smokers' Problem

to the process defining the agent and no conditional statements may be used in the solution. The problem was solved by Parnas in [Par75] and the solution is presented in Fig. 4.8 It is relatively easy to show that the program from Fig. 4.6 may deadlock. For instance the sequence of event occurrences depicted in Fig. 4.7 leads to a deadlock. How can we prove that the solution of Fig. 4.8 is correct, and can we use the COSY approach to do this?

Let us first consider the more fundamental question whether we can model binary semaphores in COSY, and if yes, how? Recall that a binary semaphore is a variable whose values can only be 0 or 1. Operation $P(s)$ decrements the value of the semaphore s by 1, and if the value of the semaphore is 0, the execution of the operation is delayed until its value becomes 1. This provides a way to allow a process to wait until other processes catch up with it. Furthermore, if the value of the semaphore s is 1 and several processes try to execute $P(s)$ then only one of the processes is permitted to complete the execution of $P(s)$ and the other processes are delayed until the value of s becomes 1. Processes are independent (but eventually cooperating) programs and they are all cyclic.

The COSY notation does not contain any data types, so we have to *abstract from the notion of semaphores as a data type*, and just represent the synchronization properties of semaphores instead.

The transformation rules are the following:

(1) a binary semaphore s is represented by <u>path</u> $sP; sV$ <u>end</u>, if $s = 1$ initially,

(2) a binary semaphore s is represented by <u>path</u> $sV; sP$ <u>end</u>, if $s = 0$

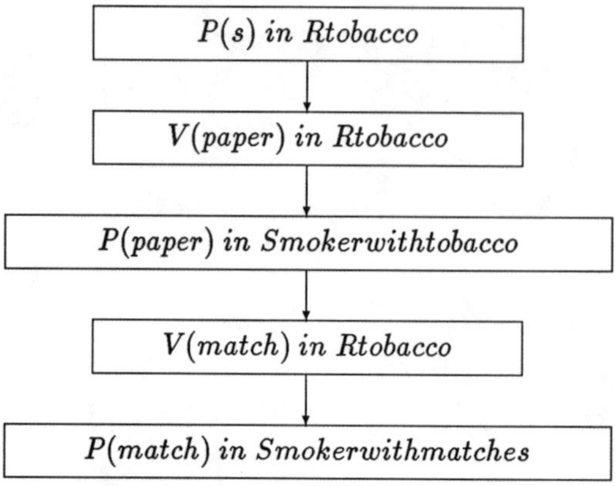

Figure 4.7. Deadlocking sequence of event occurrences for Fig. 4.6

initially,

(3) cyclic processes consisting of sequences of semaphore operations can be directly modelled by COSY processes.

To illustrate how these rules work, we shall consider the classical example of two processes with a critical section. The standard semaphore solution is presented in Fig. 4.9.

Its COSY (path and process) representation is the following:

$$\begin{aligned}
&Pr_{criticalsection} = \\
&semaphore : \underline{path}\ sP; sV\ \underline{end} \\
&process1 : \underline{process}\ sP; criticalsection; sV\ \underline{end} \\
&process2 : \underline{process}\ sP; criticalsection; sV\ \underline{end}
\end{aligned}$$

After replacing all processes by paths (see the transformation rules in Chap. 3.2), writing a_i instead of $a\&i$, and erasing all redundant parentheses, we obtain the following *path* program which describes the meaning of $Pr_{criticalsection}$:

$$\begin{aligned}
&Pr'_{criticalsection} = \\
&semaphore : \underline{path}\ sP_1, sP_2; sV_1, sV_2\ \underline{end} \\
&process1 : \underline{path}\ sP_1; criticalsection_1; sV_1\ \underline{end} \\
&process2 : \underline{path}\ sP_2; criticalsection_2; sV_2\ \underline{end}
\end{aligned}$$

It can easily be seen that:

$$VFS(Pr'_{criticalsection}) = Pref(\{\underline{sP_1}\ \underline{criticalsection_1}\ \underline{sV_1}, \underline{sP_2}\ \underline{criticalsection_2}\ \underline{sV_2}\}^*)$$

```
initially : s = mutex = 1,
        t = tobacco = paper = match = smokert = smokerp = smokerm = 0,
        S[1] = S[2] = S[3] = S[4] = S[5] = S[6].
  Rtobacco              Rpaper               Rmatch          ⎫
  rt : P(s);            rp : P(s);           rm : P(s);      ⎪
    V(paper);             V(match);            V(tobacco);   ⎬ agent
    V(match);             V(tobacco);          V(paper);     ⎪
    goto rt               goto rp              goto rm       ⎭
  bt : P(smokert);      bp : P(smokerp);     bm : P(smokerm);
    V(s);                 V(s);                V(s);
    goto bt               goto bp              goto bm
  Smokerwithtobacco     Smokerwithpaper      Smokerwithmatches  ⎫
  at : P(S[6]);         ap : P(S[5]);        am : P(S[3]);      ⎪
    t := 0;               t := 0;              t := 0;          ⎬ smokers
    * *                   * *                  * *              ⎪
    V(smokert);           V(smokerp);          V(smokerm);      ⎪
    goto at               goto ap              goto am          ⎭
  dt : P(tobacco);      dp : P(paper);       dm : P(match);  ⎫
    P(mutex);             P(mutex);            P(mutex);     ⎪
    t := t + 1;           t := t + 2;          t := t + 4;   ⎬ pushers
    V(S[t]);              V(S[t]);             V(S[t]);      ⎪
    V(mutex);             V(mutex);            V(mutex);     ⎪
    goto dt               goto dp              goto dm       ⎭
  d1 : P(S[1]);         d2 : P(S[2]);        d3 : P(S[4]);
    goto d1               goto d2              goto d3
```

Figure 4.8. Parnas' solution to Smokers' Problem

```
initially : s = 1
  process1              process2
  pr1 : P(s);           pr2 : P(s);
    critical section;     critical section;
    V(s);                 V(s);
    goto pr1              goto pr2
```

Figure 4.9. Semaphore solution for critical section

and this is exactly the behaviour we wanted.

The COSY (path and process) representation of Patil's statement from Fig. 4.6 and its COSY path program equivalent are shown together in Fig. 4.10, and it can be observed that the vector $\underline{sP_1}\ \underline{paperV_1}\ \underline{paperP_7}\ \underline{matchV_1}\ \underline{matchP_9}$ corresponds to the sequence from Fig. 4.7, and deadlocks the whole system.

Parnas' solution from Fig. 4.8 is slightly more difficult to translate into the COSY (path and process) formalism. This solution uses an array of semaphores called S and an index t for indexing the elements of the array. It is not possible to directly translate this solution into COSY paths and processes because the concept of data type is not included in this formalism. However, by considering the role of t in this solution we may represent the solution from Fig. 4.8 in COSY. Using the rules for semaphores and processes we already applied in the critical section problem and for Patil's statement, we can obtain a semi-COSY intermediate specification which is represented in Fig. 4.11 This is not a COSY program since $t := 0$, $t := t+1$, $t := t+2$, $t := t+4$ are disallowed in COSY, however this is a correct COSY specification in all aspects not involving t, so this intermediate specification facilitates our understanding of the role of t in the original solution. The integer t may take values from 0 to 6, and may be incremented twice successively by 1, 2, or 4; it may not, however, be incremented by the same amount nor simultaneously. Note that t is always set to 0 before new resources are supplied. So the effects of the three pusher processes can only be:

$t = 1$ if dt is first, after which $t = 3$ if dp is next, or $t = 5$ if dm is next;

$t = 2$ if dp is first, after which $t = 3$ if dt is next, or $t = 6$ if dm is next;

$t = 4$ if dm is first, after which $t = 5$ if dt is next, or $t = 6$ if dp is next.

This situation can be adequately modelled by the path:

$$\begin{array}{l} \mathit{change\ of\ t}\ : \\ \underline{path}\ (S[1]V; S[3]V, S[5]V), (S[2]V; S[3]V, S[6]V), (S[4]V; S[5]V, S[6]V)\ \underline{end} \end{array}$$

which has to be added to our COSY specification. The above path describes global changes of t, we also have to find what values the variable t may have in a particular process in which it occurs. In process dt, $t = 1$ if dt is first, or $t = 3$ if dt is second and dp is first, or $t = 5$ if dt is second but dm is first. In process dm, $t = 4$ if dm is first, or $t = 5$ if dm is second and dt is first, or $t = 6$ if dm is second but dp is first. Finally, in process dp, $t = 2$ if dp is first, or $t = 3$ if dp is second and dt is first, or $t = 6$ if dp is second but dm is first. To model this situation in terms of COSY it is sufficient to replace:

$$
\begin{array}{ll}
Pr_{smokers1} = & \\
semaphore\ tobacco: & \underline{path}\ tobaccoV; tobaccoP\ \underline{end} \\
semaphore\ paper: & \underline{path}\ paperV; paperP\ \underline{end} \\
semaphore\ match: & \underline{path}\ matchV; matchP\ \underline{end} \\
semaphore\ smokert: & \underline{path}\ smokertV; smokertP\ \underline{end} \\
semaphore\ smokerp: & \underline{path}\ smokerpV; smokerpP\ \underline{end} \\
semaphore\ smokerm: & \underline{path}\ smokermV; smokermP\ \underline{end} \\
semaphore\ s: & \underline{path}\ sP; sV\ \underline{end} \\
Rtobacco: & \underline{process}\ sP; paperV; matchV\ \underline{end} \\
Rmatch: & \underline{process}\ sP; tobaccoV; paperV\ \underline{end} \\
Rpaper: & \underline{process}\ sP; matchV; tobaccoV\ \underline{end} \\
bt: & \underline{process}\ smokertP; sV\ \underline{end} \\
bp: & \underline{process}\ smokerpP; sV\ \underline{end} \\
bm: & \underline{process}\ smokermP; sV\ \underline{end} \\
smokerwithtobacco: & \underline{process}\ paperP; matchP; smokertV\ \underline{end} \\
smokerwithmatches: & \underline{process}\ tobaccoP; paperP; smokermV\ \underline{end} \\
smokerwithpaper: & \underline{process}\ matchP; tobaccoP; smokerpV\ \underline{end}
\end{array}
$$

$$
\begin{array}{l}
Pr'_{smokers1} = \\
\left.\begin{array}{l}
P1: \underline{path}\ tobaccoV_1, tobaccoV_3; tobaccoP_8, tobaccoP_9\ \underline{end} \\
P2: \underline{path}\ paperV_1, paperV_2; paperP_7, paperP_8\ \underline{end} \\
P3: \underline{path}\ matchV_1, matchV_3; matchP_7, matchP_9\ \underline{end} \\
P4: \underline{path}\ smokertV; smokertP\ \underline{end} \\
P5: \underline{path}\ smokerpV; smokerpP\ \underline{end} \\
P6: \underline{path}\ smokermV; smokermP\ \underline{end} \\
P7: \underline{path}\ sP_1, sP_2, sP_3; sV_4, sV_5, sV_6\ \underline{end}
\end{array}\right\}\ \text{semaphores} \\
Q1: \underline{path}\ sP_1; paperV_1; matchV_1\ \underline{end}\}\ \text{Rtobacco} \\
Q2: \underline{path}\ sP_2; tobaccoV_2; paperV_2\ \underline{end}\}\ \text{Rmatch} \\
Q3: \underline{path}\ sP_3; matchV_3; tobaccoV_3\ \underline{end}\}\ \text{Rpaper} \\
Q4: \underline{path}\ smokertP; sV_4\ \underline{end}\}\ \text{bt} \\
Q5: \underline{path}\ smokerpP; sV_5\ \underline{end}\}\ \text{bp} \\
Q6: \underline{path}\ smokermP; sV_6\ \underline{end}\}\ \text{bm} \\
Q7: \underline{path}\ paperP_7; matchP_7; smokertV\ \underline{end}\}\ \text{smoker with tobacco} \\
Q8: \underline{path}\ tobaccoP_8; paperP_8; smokermV\ \underline{end}\}\ \text{smoker with matches} \\
Q9: \underline{path}\ matchP_9; tobaccoP_9; smokerpV\ \underline{end}\}\ \text{smoker with paper}
\end{array}
$$

Figure 4.10. COSY representation of Patil's statement

$$t := t + 1; S[t]V \text{ by } S[1]V, S[3]V, S[5]V \text{ in process } dt,$$

$$t := t + 4; S[t]V \text{ by } S[4]V, S[5]V, S[6]V \text{ in process } dm,$$

$$t := t + 2; S[t]V \text{ by } S[2]V, S[3]V, S[6]V \text{ in process } dp,$$

Finally, since we represented both global and local changes of t, we may remove $t := 0$; from all smokers. All these changes transform our intermediate semi-COSY specification into a true COSY program. This program, called $Pr_{smokers2}$, is represented in Fig. 4.12, and will be regarded as an equivalent COSY representation of Parnas' solution.

The path program, $Pr'_{smokers2} = Path(Pr_{smokers2})$ which is equivalent to $Pr_{smokers2}$ (see Section 3.2) is depicted in Fig. 4.13. To make the program $Pr'_{smokers2}$ more readable and shorter we write $name_i$ instead of $name\&i$, e.g. $mutexV_{10}$ instead of $mutexV\&10$. First we shall reduce the program $Pr'_{smokers2}$ using some results of Section 2.4.3. Lemma 2.26 allows us to delete the paths P_{28}, P_{29} and P_{30}. The new programs satisfies the condition of Lemma 2.23 with $a = S[1]P_{13}, a = S[2]P_{14}$ and $a = S[4]P_{15}$, respectively, and with $\alpha = 1, \beta = 0$ in all three cases. Hence we can delete $S[1]P_{13}$, $S[2]P_{14}$ and $S[4]P_{15}$ from the paths P_9, P_{10} and P_{12}. Suppose that after applying Lemma 2.23 the paths P_9, P_{10}, P_{12} are replaced by P'_9, P'_{10}, P'_{12}. We may now apply Lemma 2.26 again and delete the paths P'_9, P'_{10}, P'_{12} altogether. Let us denote the program obtained by the above transformation by $Pr^{reduced}_{smokers2}$, i.e., $Pr^{reduced}_{smokers2} = P_1P_2 \ldots P_8P_{11}P_{13} \ldots P_{27}$ where P_i are those from Fig. 4.13. Clearly $Pr'_{smokers2}$ is adequate if and only if $Pr^{reduced}_{smokers2}$ is adequate. To prove that $Pr^{reduced}_{smokers2}$ is adequate we shall first show that $CON \approx MAX$ for $Pr^{reduced}_{smokers2}$. From Corollary 2.18 and Theorem 2.33 it follows that in such a case $Pr^{reduced}_{smokers2}$ is adequate if and only if it is MAX-adequate. In analysing maximally concurrent behaviour of $Pr^{reduced}_{smokers2}$ we can use a computer aided design and analysis tool called BCS (Basic COSY Simulator, see Chap. 3.5). From the abstract viewpoint, BCS is an implementation of the relations CON, SEQ and MAX. It is generally not easy for human beings to check if $\underline{x} \in VFS$ or to find $enabled_{CON}(\underline{x})$ or $enabled_{MAX}(\underline{x})$, but it is usually an easy task for BCS. Seeking an analogy, it is generally not easy for human beings to multiply matrices, whilst it is rather an easy task for special programs. We shall show that the maximally concurrent behaviour of $Pr^{reduced}_{smokers2}$ is very regular.

```
semaphore tobacco :     path tobaccoV; tobaccoP end
semaphore paper :       path paperV; paperP end
semaphore match :       path matchV; matchP end
semaphore smokert :     path smokertV; smokertP end
semaphore smokerp :     path smokerpV; smokerpP end
semaphore smokerm :     path smokermV; smokermP end
semaphore s :           path sP; sV end
semaphore mutex :       path mutexP; mutexV end
semaphore S[1] :        path S[1]V; S[1]P end
semaphore S[2] :        path S[2]V; S[2]P end
semaphore S[3] :        path S[3]V; S[3]P end
semaphore S[4] :        path S[4]V; S[4]P end
semaphore S[5] :        path S[5]V; S[5]P end
semaphore S[6] :        path S[6]V; S[6]P end
rt :  process sP; paperV; matchV end
rm :  process sP; tobaccoV; paperV end        } agent
rp :  process sP; matchV; tobaccoV end
bt :  process smokertP; sV end
bp :  process smokerpP; sV end
bm :  process smokermP; sV end
at :  process S[6]P; t := 0; smokertV end
am :  process S[3]P; t := 0; smokermV end     } smokers
ap :  process S[5]P; t := 0; smokerpV end
dt :  process tobaccoP; mutexP; t := t + 1; S[t]V; mutexV end
dm :  process matchP; mutexP; t := t + 4; S[t]V; mutexV end   } pushers
dp :  process paperP; mutexP; t := t + 2; S[t]V; mutexV end
d1 :  process S[1]P end
d2 :  process S[2]P end
d3 :  process S[4]P end
```

Figure 4.11. Semi-COSY representation of Parnas' solution

$$
\begin{array}{ll}
\mathit{semaphore\ tobacco}: & \underline{path}\ tobaccoV; tobaccoP\ \underline{end} \\
\mathit{semaphore\ paper}: & \underline{path}\ paperV; paperP\ \underline{end} \\
\mathit{semaphore\ match}: & \underline{path}\ matchV; matchP\ \underline{end} \\
\mathit{semaphore\ smokert}: & \underline{path}\ smokertV; smokertP\ \underline{end} \\
\mathit{semaphore\ smokerp}: & \underline{path}\ smokerpV; smokerpP\ \underline{end} \\
\mathit{semaphore\ smokerm}: & \underline{path}\ smokermV; smokermP\ \underline{end} \\
\mathit{semaphore\ s}: & \underline{path}\ sP; sV\ \underline{end} \\
\mathit{semaphore\ mutex}: & \underline{path}\ mutexP; mutexV\ \underline{end} \\
\mathit{semaphore}\ S[1]: & \underline{path}\ S[1]V; S[1]P\ \underline{end} \\
\mathit{semaphore}\ S[2]: & \underline{path}\ S[2]V; S[2]P\ \underline{end} \\
\mathit{semaphore}\ S[3]: & \underline{path}\ S[3]V; S[3]P\ \underline{end} \\
\mathit{semaphore}\ S[4]: & \underline{path}\ S[4]V; S[4]P\ \underline{end} \\
\mathit{semaphore}\ S[5]: & \underline{path}\ S[5]V; S[5]P\ \underline{end} \\
\mathit{semaphore}\ S[6]: & \underline{path}\ S[6]V; S[6]P\ \underline{end} \\
\mathit{change\ of\ t}: & \underline{path}\ (S[1]V; S[3]V, S[5]V), (S[2]V; S[3]V, S[6]V), \\
 & \quad (S[4]V; S[5]V, S[6]V)\ \underline{end}
\end{array}
$$

$$
\left.\begin{array}{ll}
rt: & \underline{process}\ sP; paperV; matchV\ \underline{end} \\
rm: & \underline{process}\ sP; tobaccoV; paperV\ \underline{end} \\
rp: & \underline{process}\ sP; matchV; tobaccoV\ \underline{end}
\end{array}\right\}\ \text{agent}
$$

$$
\begin{array}{ll}
bt: & \underline{process}\ smokertP; sV\ \underline{end} \\
bp: & \underline{process}\ smokerpP; sV\ \underline{end} \\
bm: & \underline{process}\ smokermP; sV\ \underline{end}
\end{array}
$$

$$
\left.\begin{array}{ll}
at: & \underline{process}\ S[6]P; smokertV\ \underline{end} \\
am: & \underline{process}\ S[3]P; smokermV\ \underline{end} \\
ap: & \underline{process}\ S[5]P; smokerpV\ \underline{end}
\end{array}\right\}\ \text{smokers}
$$

$$
\left.\begin{array}{ll}
dt: & \underline{process}\ tobaccoP; mutexP; S[1]V, S[3]V, S[5]V; mutexV\ \underline{end} \\
dm: & \underline{process}\ matchP; mutexP; S[4]V, S[5]V, S[6]V; mutexV\ \underline{end} \\
dp: & \underline{process}\ paperP; mutexP; S[2]V, S[3]V, S[6]V; mutexV\ \underline{end}
\end{array}\right\}\ \text{pushers}
$$

$$
\begin{array}{ll}
d1: & \underline{process}\ S[1]P\ \underline{end} \\
d2: & \underline{process}\ S[2]P\ \underline{end} \\
d3: & \underline{process}\ S[4]P\ \underline{end}
\end{array}
$$

Figure 4.12. COSY (path and process) equivalent of Parnas' solution

Let $\underline{x}_1, \underline{x}_2, \underline{x}_3, \underline{y}_1, \underline{y}_2, \underline{y}_3\ \underline{y}_4, \underline{y}_5, \underline{y}_6$ be the following elements of $Vev(Pr^{reduced}_{smokers2})^*$:

$$\underline{x}_1 = \underline{sP_1}\ \underline{paperV_1}\ \underline{paperP_{12}}\ \underline{matchV_1}\ \underline{matchP_{11}},$$
$$\underline{x}_2 = \underline{sP_2}\ \underline{tobaccoV_2}\ \underline{tobaccoP_{10}}\ \underline{paperV_2}\ \underline{paperP_{12}},$$
$$\underline{x}_3 = \underline{sP_3}\ \underline{matchV_3}\ \underline{matchP_{11}}\ \underline{tobaccoV_3}\ \underline{tobaccoP_{10}},$$
$$\underline{y}_1 = \underline{mutexP_{11}}\ \underline{S[4]V_{11}}\ \underline{mutexV_{11}}\ \underline{mutexP_{12}}\ \underline{S[6]V_{12}}\ \underline{mutexV_{12}}\ \underline{S[6]P_7}\ \underline{smokertV_7}\ \underline{smokertP_4}\ \underline{sV_4},$$
$$\underline{y}_2 = \underline{mutexP_{12}}\ \underline{S[2]V_{12}}\ \underline{mutexV_{12}}\ \underline{mutexP_{11}}\ \underline{S[6]V_{11}}\ \underline{mutexV_{11}}\ \underline{S[6]P_7}\ \underline{smokertV_7}\ \underline{smokertP_4}\ \underline{sV_4},$$
$$\underline{y}_3 = \underline{mutexP_{10}}\ \underline{S[1]V_{10}}\ \underline{mutexV_{10}}\ \underline{mutexP_{12}}\ \underline{S[3]V_{12}}\ \underline{mutexV_{12}}\ \underline{S[3]P_8}\ \underline{smokermV_8}\ \underline{smokermP_6}\ \underline{sV_6},$$
$$\underline{y}_4 = \underline{mutexP_{12}}\ \underline{S[2]V_{12}}\ \underline{mutexV_{12}}\ \underline{mutexP_{10}}\ \underline{S[3]V_{10}}\ \underline{mutexV_{10}}\ \underline{S[3]P_8}\ \underline{smokermV_8}\ \underline{smokermP_6}\ \underline{sV_6},$$
$$\underline{y}_5 = \underline{mutexP_{10}}\ \underline{S[1]V_{10}}\ \underline{mutexV_{10}}\ \underline{mutexP_{11}}\ \underline{S[5]V_{11}}\ \underline{mutexV_{11}}\ \underline{S[5]P_9}\ \underline{smokerpV_9}\ \underline{smokerpP_5}\ \underline{sV_5},$$
$$\underline{y}_6 = \underline{mutexP_{11}}\ \underline{S[4]V_{11}}\ \underline{mutexV_{11}}\ \underline{mutexP_{10}}\ \underline{S[5]V_{10}}\ \underline{mutexV_{10}}\ \underline{S[5]P_9}\ \underline{smokerpV_9}\ \underline{smokerpP_5}\ \underline{sV_5}.$$

Define: $\underline{z}_1 = \underline{x}_1\underline{y}_1$, $\underline{z}_2 = \underline{x}_1\underline{y}_2$, $\underline{z}_3 = \underline{x}_2\underline{y}_3$, $\underline{z}_4 = \underline{x}_2\underline{y}_4$, $\underline{z}_5 = \underline{x}_3\underline{y}_5$, $\underline{z}_6 = \underline{x}_3\underline{y}_6$.

Lemma 4.1 *For $Pr^{reduced}_{smokers2}$ we have: $Scope_{MAX} = Pref(MAX^*(\varepsilon)) = Pref(\{\underline{z}_1, \underline{z}_2, \underline{z}_3, \underline{z}_4, \underline{z}_5, \underline{z}_6\})$.*

Proof. By inspection, immediate result if BCS is used. After 13 maximal steps the whole system always returns to the initial state and there are only 6 possibilities of performing these steps, namely left normal forms of $\underline{z}_1, \underline{z}_2, \underline{z}_3, \underline{z}_4, \underline{z}_5, \underline{z}_6$. □

As an immediate consequence of Lemma 4.1 we have:

Corollary 4.1 *$Pr^{reduced}_{smokers2}$ is MAX-adequate.* □

We shall now recall Theorem 2.37 from Section 2.5.2 which characterizes the major result characterizing $CON \approx MAX$ by an element of $MAX^*(\varepsilon)$. Theorem 2.37 states:
"The following properties are equivalent:
(1) $CON \not\approx MAX$
(2) there are $\underline{x} \in MAX^*(\underline{\varepsilon})$, $\underline{x}\,\underline{A_1} \ldots \underline{A_k} \in Scope_{MAX}$ and $A_1, \ldots, A_k \in Ind$, where $k \geq 2$, such that

(a) $\underline{y} = \underline{x}\underline{A_1} \ldots \underline{A_k} \in VFS \wedge \underline{x}\underline{A_1} \ldots \underline{A_{k-1}} \in Scope_{MAX}$

(b) $(\forall a \in Ev)\underline{ya} \in VFS \Rightarrow (\exists b \in A_{k-1} \cup A_k)\ (a, b) \notin ind$

(c) $< A_1, \ldots, A_k >$ is the left-normal form of $\underline{z} = \underline{A_1} \ldots \underline{A_k}$, i.e., $z \doteq \underline{A_1} \ldots \underline{A_k}$.

(d) $(\exists b \in Ev)\underline{xb} \in VFS \wedge (\forall c \in A_1 \cup \ldots \cup A_{k-1})(b,c) \in ind.$ "

The following lemma can be derived from Theorem 2.37.

Lemma 4.2 *Let $\underline{x} \in MAX^*(\underline{\varepsilon})$ and $A_1, \ldots, A_k \in Ind$, where $k \geq 2$ be those from Theorem 2.37 (2). Then:*
$(\exists b \in A_k)(\exists a \in Ev \setminus A_{k-1})A_{k-1} \cup \{a\} \in enabled_{CON}(\underline{x}\underline{A_1} \ldots \underline{A_{k-2}}) \wedge (a,b) \notin ind \wedge a \neq b.$

Proof.
From (2d) of Theorem 2.37 it follows that $(\exists a \in Ev \setminus A_{k-1})A_{k-1} \cup \{a\} \in enablen_{CON}(\underline{x}\underline{A_1} \ldots \underline{A_{k-2}})$ (just take b from (2)). Hence $\underline{x}\underline{A_1} \ldots \underline{A_{k-1}}\underline{a} \in VFS, A_{k-1} \cup \{a\} \in Ind$. Due to (2c) of Theorem 2.37 we have $a \in A_k$, i.e., $a \neq b$. Suppose that $(\forall b \in A_k)(a,b) \in ind$. This means that $\underline{x}\underline{A_1} \ldots \underline{A_{k-1}}\underline{A_k \cup \{a\}} = \underline{x}\underline{A_1} \ldots \underline{A_k}\underline{a} \in VFS$.
From (2b) of Theorem 2.37 it follows that $(\exists b \in A_{k-1} \cup A_k)(a,b) \notin ind$. Since $A_{k-1} \cup \{a\} \in Ind$, we have that $(\exists b \in A_k)(a,b) \notin ind$, a contradiction. Hence $(\exists b \in A_k)(a,b) \notin ind$. □
Let $Conflicts \subseteq VFS$ be the following set of vector firing sequences:

$$\underline{x} \in Conflicts :\Leftrightarrow (\exists a,b \in Ev)\underline{xa} \in VFS \wedge \underline{xb} \in VFS \wedge (a,b) \notin ind \wedge a \neq b.$$

Lemma 4.3 *Let $\underline{x} \in MAX^*(\underline{\varepsilon})$ and $A_1, \ldots, A_k \in Ind$, $k \geq 2$ be those from Theorem 2.37(2). Then: $\underline{x}\underline{A_1} \ldots \underline{A_{k-1}} \in Conflicts \cap Scope_{MAX}$.*

Proof. Let a, b be those from Lemma 4.2. Note that $\underline{x}\underline{A_1} \ldots \underline{A_{k-1}}\underline{a} \in VFS \wedge \underline{x}\underline{A_1} \ldots \underline{A_{k-1}}\underline{b} \in VFS \wedge (a,b) \notin ind \wedge a \neq b$, so $\underline{x}\underline{A_1} \ldots \underline{A_{k-1}} \in Conflicts$. From (2a) of Theorem 2.37 it follows that $\underline{x}\underline{A_1} \ldots \underline{A_{k-1}} \in Scope_{MAX}$. □
For the program $Pr^{reduced}_{smokers2}$ we may check that $Conflicts \cap Scope_{MAX} = Conflicts_1 \cup Conflicts_2$, with $Conflicts_1 \cap Conflicts_2 = \emptyset$ and:

$$\underline{z} \in Conflicts_1 \Rightarrow \underline{z} = \underline{yc} \wedge c \in \{matchP_{11}, paperP_{12}, tobaccoP_{10}\},$$
$$\underline{z} \in Conflicts_2 \Rightarrow \underline{z} = \underline{\varepsilon} \vee (\underline{z} = \underline{yc} \wedge c \in \{sV_4, sV_5, sV_6\}.$$

Consider the case $\underline{z} = \underline{yc} \in Conflicts_1$ and $c = matchP_{11}$. This means that $\underline{z} = \underline{u}\,\underline{sP_i}\,\underline{w}\,\underline{matchP_{11}}$, where $i = 1,3$ and $matchP_{11}$ does not occur in w. Suppose that $\underline{z} = \underline{u}\,\underline{sP_1}\,\underline{w}\,\underline{matchP_{11}}$. We shall show that $\underline{z}$ cannot be decomposed into $\underline{z} = \underline{x}\underline{A_1} \ldots \underline{A_{k-1}}$ such that $\underline{x}, A_1, \ldots, A_{k-1}$ satisfy Lemma 4.2 and Lemma 4.3.
Suppose that it can. Lemma 4.2 implies that $A_{k-1} \notin enabled_{MAX}(\underline{x}\underline{A_1} \ldots \underline{A_{k-2}})$, which means that $\underline{z} \notin MAX^*(\underline{\varepsilon})$, i.e., $\underline{sP_1}\,\underline{w}\,\underline{matchP_{11}} \neq \underline{x}_1$. Hence, because $\underline{z} = \underline{u}\,\underline{sP_1}\,\underline{w}\underline{matchP_{11}} \in Scope_{MAX} = Pref(MAX^*(\underline{\varepsilon}))$, $\underline{sP_1}\,\underline{w}\,\underline{matchP_{11}}$

must be a proper prefix of $\underline{x}_1$ which implies that $A_{k-1} = \{\underline{matchP_{11}}\}$ and $\underline{w} = \underline{paperV_1}\,\underline{matchV_1}$.
Let us now analyse the prefix $\underline{u}$ of $\underline{z} = \underline{u}\,\underline{sP_1}\,\underline{w}\,\underline{matchP_{11}}$. If $\underline{u}$ contains at least one sP_i, $i = 1,2,3$, then it may be represented as $\underline{u} = \underline{u'sP_iu''}$, where $\underline{u''}$ contains no sP_i at all. In such a case either $\underline{u'sP_iu''} \in \{\underline{z}_1, \ldots, \underline{z}_6\}^*$, or $\underline{u'sP_i\,u''}\,\underline{mutexV_j} \in \{\underline{z}_1, \ldots, \underline{z}_6\}^*$, where $j = 10, 12$. If $\underline{u}$ contains no sP_i at all, then $\underline{u} = \underline{\varepsilon} \in \{\underline{z}_1, \ldots, \underline{z}_6\}^*$.
If $\underline{u} \in \{\underline{z}_1, \ldots, \underline{z}_6\}^*$ then $enabled_{CON}(\underline{u}\,\underline{sP_1}\underline{paperV_1}\,\underline{matchV_1}\underline{matchP_{11}}) = \{\{paperP_{12}\}, \{mutexP_{11}\}, \{paper_{12}, mutexP_{11}\}\}$, which means that $\underline{z} = \underline{u}\,\underline{sP_1}\underline{paperV_1}\,\underline{matchV_1}\underline{matchP_{11}} \notin Conflicts$, a contradiction.
If $\underline{u}\,\underline{mutexV_j} \in \{\underline{z}_1, \ldots, \underline{z}_6\}^*$, where $j = 10, 12$,
then $enabled_{CON}(\underline{u}\,\underline{sP_1}\underline{paperV_1}\,\underline{matchV_1}) = \{\{mutexV_j\}\}$, i.e., $\underline{z} = \underline{u}\,\underline{sP_1}\underline{paperV_1}\,\underline{matchV_1}\underline{matchP_{11}} \notin Conflicts$, a contradiction.
For $\underline{z} = \underline{u}\,\underline{sP_3}\,\underline{w}\,\underline{matchP_{11}}$, $\underline{z} = \underline{y}\,\underline{paperP_{12}}$ and $z = \underline{y'}\underline{tobaccoP_{10}}$ we proceed almost identically. Thus no element of $Conflicts_1$, can satisfy Lemma 4.3.
Consider now $\underline{z} = \underline{y}\underline{a} \in Conflicts_2$ and $a = sV_4$. Suppose that $\underline{z}$ can be represented as $\underline{z} = \underline{x}\underline{A}_1 \ldots \underline{A}_{k-1}$, such that $\underline{x}_1 A_1, \ldots, A_{k-1}$ satisfy Lemma 4.2 and Lemma 4.3.
Hence $A_{k-1} \notin enabled_{MAX}(\underline{x}\underline{A}_1 \ldots \underline{A}_{k-2})$ and $\underline{z} \notin MAX^*(\underline{\varepsilon})$, which implies that $A_{k-1} = \{sV_4\}$ and either $\underline{z} = \underline{v}\,\underline{v'}\,\underline{sV_4}$ where $\underline{v'}\,\underline{sV_4}\,\underline{mutexV_{12}} = \underline{y_1}$, or $\underline{z} = \underline{v}\ \underline{v''}\ \underline{sV_4}$ and $\underline{v''}\,\underline{sV_4}\,\underline{mutexV_{11}} = \underline{y_2}$. Suppose that $\underline{z} = \underline{v}\,\underline{v'}\,\underline{sV_4}$ and $\underline{v'}\ \underline{sP_4}\underline{mutexV_{12}} = \underline{y_1}$. We now have: $\underline{x}\underline{A}_1 \ldots \underline{A}_{k-2} = \underline{vv'}$, and $enabled_{CON}(\underline{vv'}) = \{\{mutexV_{12}\}, \{sV_4\}, \{mutexV_{12}, sV_4\}\}$. This means that only $mutexV_{12}$ can play the role of the event a from Lemma 4.2.
Of course $A_k \in enabled_{CON}(\underline{x}\underline{A}_1 \ldots \underline{A}_{k-1})$. We have here $\underline{x}\underline{A}_1 \ldots \underline{A}_{k-1} = \underline{v}\,\underline{v'}\,\underline{sV_4}$, and

$$\begin{aligned}
&enabled_{con}(\underline{v}\,\underline{v'}\,\underline{sV_4}) = \\
&\{\{mutexV_{12}\}, \{sP_1\}, \{sP_2\}, \{sP_3\}, \\
&\{sP_1, mutexV_{12}\}, \{sP_2, mutexV_{12}\}\{sP_3, mutexV_{12}\}\}.
\end{aligned}$$

But since $(sP_i, mutexV_{12}) \in ind$, the conditions of Lemma 4.2 cannot be satisfied. For the cases $\underline{z} = \underline{v}\,\underline{v''}\,\underline{sV_4}$ and $\underline{v''}\,\underline{sV_4}\,\underline{mutexV_{11}} = \underline{y_2}$, $\underline{z} = \underline{y}\underline{sV_5}$, $\underline{z} = \underline{y'}\underline{sV_6}$ we proceed almost identically.

Hence no element of $Conflicts_2$ can satisfy Lemma 4.2. In this way we proved that $CON \approx MAX$ for $Pr^{reduced}_{smokers2}$. Due to Corollary 4.1 (and Corollary 2.18, Theorem 2.33) we may now conclude that $Pr^{reduced}_{smokers2}$ is adequate. Thus $Pr'_{smokers2}$ is also adequate and, of course, deadlock-free.

In this way we proved, using COSY, that Parnas' solution is correct.

So far we have represented semaphores and operations on them abstractly by means of path expressions. However, we may abstract away from the use of semaphores altogether and find a solution to the problem in terms of COSY alone, without any intermediary. To do this we restate the problem in the following

$Pr'_{smokers2} =$

$P_1 :$ $\underline{path}$ $tobaccoV_2, tobaccoV_3; tobaccoP_{10}$ $\underline{end}$

$P_2 :$ $\underline{path}$ $paperV_1, paperV_2; paperP_{12}$ $\underline{end}$

$P_3 :$ $\underline{path}$ $matchV_1, matchV_3; matchP_{11}$ $\underline{end}$

$P_4 :$ $\underline{path}$ $smokertV_7; smokertP_4$ $\underline{end}$

$P_5 :$ $\underline{path}$ $smokerpV_9; smokerpP_5$ $\underline{end}$

$P_6 :$ $\underline{path}$ $smokermV_8; smokermP_6$ $\underline{end}$

$P_7 :$ $\underline{path}$ $sP_1, sP_2, sP_3; sV_4, sV_5, sV_6$ $\underline{end}$

$P_8 :$ $\underline{path}$ $mutexP_{10}, mutexP_{11}, mutexP_{12}; mutexV_{10}, mutexV_{11}, mutexV_{12}$ $\underline{end}$

$P_9 :$ $\underline{path}$ $S[1]V_{10}; S[1]P_{13}$ $\underline{end}$

$P_{10} :$ $\underline{path}$ $S[2]V_{12}; S[2]P_{14}$ $\underline{end}$

$P_{11} :$ $\underline{path}$ $S[3]V_{10}, S[3]V_{12}; S[3]P_8$ $\underline{end}$

$P_{12} :$ $\underline{path}$ $S[4]V_{11}; S[4]P_{15}$ $\underline{end}$

$P_{13} :$ $\underline{path}$ $S[5]V_{10}, S[5]V_{11}; S[5]P_9$ $\underline{end}$

$P_{14} :$ $\underline{path}$ $S[6]V_{11}, S[6]V_{12}; S[6]P_7$ $\underline{end}$

$P_{15} :$ $\underline{path}$ $(S[1]V_{10}; S[3]V_{10}, S[3]V_{12}, S[5]V_{10}, S[5]V_{11}),$
$(S[2]V_{12}; S[3]V_{10}, S[3]V_{12}, S[6]V_{11}, S[6]V_{12}),$
$(S[4]V_{11}; S[5]V_{10}, S[5]V_{11}, S[6]V_{11}, S[6]V_{12})$ $\underline{end}$

$P_{16}(Q_1) :$ $\underline{path}$ $sP_1; paperV_1; matchV_1$ $\underline{end}$

$P_{17}(Q_2) :$ $\underline{path}$ $sP_2; tobaccoV_2; paperV_2$ $\underline{end}$

$P_{18}(Q_3) :$ $\underline{path}$ $sP_3; matchV_3; tobaccoV_3$ $\underline{end}$

$P_{19}(Q_4) :$ $\underline{path}$ $smokertP_4; sV_4$ $\underline{end}$

$P_{20}(Q_5) :$ $\underline{path}$ $smokerpP_5; sV_5$ $\underline{end}$

$P_{21}(Q_6) :$ $\underline{path}$ $smokermP_6; sV_6$ $\underline{end}$

$P_{22}(Q_7) :$ $\underline{path}$ $S[6]P_7; smokertV_7$ $\underline{end}$

$P_{23}(Q_8) :$ $\underline{path}$ $S[3]P_8; smokermV_8$ $\underline{end}$

$P_{24}(Q_9) :$ $\underline{path}$ $S[5]P_9; smokerpV_9$ $\underline{end}$

$P_{25}(Q_{10}) :$ $\underline{path}$ $tobaccoP_{10}; mutexP_{10}; S[1]V_{10}, S[3]V_{10}, S[5]V_{10}; mutexV_{10}$ $\underline{end}$

$P_{26}(Q_{11}) :$ $\underline{path}$ $matchP_{11}; mutexP_{11}; S[4]V_{11}, S[5]V_{11}, S[6]V_{11}; mutexV_{11}$ $\underline{end}$

$P_{27}(Q_{12}) :$ $\underline{path}$ $paperP_{12}; mutexP_{12}; S[2]V_{12}, S[3]V_{12}, S[6]V_{12}; mutexV_{12}$ $\underline{end}$

$P_{28}(Q_{13}) :$ $\underline{path}$ $S[1]P_{13}$ $\underline{end}$

$P_{29}(Q_{14}) :$ $\underline{path}$ $S[2]P_{14}$ $\underline{end}$

$P_{30}(Q_{15}) :$ $\underline{path}$ $S[4]P_{15}$ $\underline{end}$

Figure 4.13. Basic COSY path program equivalent of $Pr_{smokers2}$ from Fig. 4.12

way:

(1) Decide which of the ingredients should be put on the table.

(2) Produce each ingredient and place it on the table.

(3) Choose the correct consumer to take the available ingredients.

(4) Go back to (1).

As a matter of fact, the decision as to which of the ingredients should be put on the table immediately indicates the correct consumer.

A COSY representation of the restated problem may be the following:

$$\begin{array}{l} Pr_{smokers3} = \\ \underline{path}\ supplytm, supplypt; tobacco; m-smoker, p-smoker\ \underline{end} \\ \underline{path}\ supplytm, supplymp; match; t-smoker, p-smoker\ \underline{end} \\ \underline{path}\ supplypt, supplymp; paper; t-smoker, m-smoker\ \underline{end} \end{array}$$

where the interpretation of events is the following:

supplytm – supply tobacco and matches,

supplymp – supply matches and paper,

supplypt – supply paper and tobacco,

tobacco – tobacco on the table,

match – matches on the table,

paper – paper on the table,

$m-smoker$ – smoker with matches smokes,

$p-smoker$ – smoker with paper smokes,

$t-smoker$ – smoker with tobacco smokes,

As we can see the solution is much simpler than the previous ones, and it is almost trivial to check its adequacy by inspection.

Bibliographic Notes

The Cigarette Smokers' Problem originates from S.S. Patil [Pat71] and was analyzed in detail in [Hab73]. The solution in terms of arrays of semaphores is due to Parnas [Par75], and the COSY techniques we used in this section comes from [LC75], [Lau83] and [JLKD86].

4.4 Merlin-Randell Problem of Train Journeys

The Merlin-Randell problem of train journeys can be informally described in the following way:

> "There is a finite set of trains, a finite set of stations and one engine house. Each station can hold at most one train. Each train has a route to follow, starting and ending at the engine house. The route may be repeated one or more times. A train may leave a station (or the engine house) only if its immediate destination is an empty station or the engine house. At the beginning all trains are placed in the engine house. The problem is to find a synchronization among train movements which allows parallel movements of trains where possible, and enables the completion of each train journey."

In fact this is a classic example of resource sharing and mutual exclusion problems encountered in computing systems, although it uses the notions of trains and stations instead of packages, nodes, etc.

4.4.1 Movement Graphs and Minimal Critical Patterns

In the model we shall present, the train system is modelled as a *movement graph*, with vertices representing stations, and arcs representing possible movements of trains. The states of the system are represented by *markings*, which are mappings assigning stations to trains, and the possible changes of system states are given by the *reachability* relation on markings.

In what follows, every finite, labelled, directed graph will simply be called a *graph*. A graph will be denoted by $G = (V, LA, \Pi)$, where V is the *vertex-set*, $LA \subseteq V \times \Pi \times V$ is the *labelled arc set*, and Π is the *label set*. We assume that $V \neq \emptyset$, $V \cap \Pi = \emptyset$, $card(V \cup \Pi) < \infty$, and that for every $a \in \Pi$ there is at least one arc labelled by a. We shall sometimes write $G = (V(G), LA(G), \Pi(G))$ rather than $G = (V, LA, \Pi)$. A graph G' is a *subgraph* of G, denoted $G' \leq G$, if $V(G') \subseteq V(G)$, $LA(G') \subseteq LA(G)$ and $\Pi(G') \subseteq \Pi(G)$. The subgraph of G which is arc-induced by all arcs labelled by $a \in \Pi$ will be denoted by G_a.

A *movement graph* is a graph G such that: $\Pi \neq \emptyset$, no vertex of V is isolated, and for every $a \in \Pi$, $LA(G_a) = \{(s_i, a_i, s_{i+1}) | i = 1, \ldots, k-1\}$ where $k \geq 2$ and $s_i \neq s_j$ for $i \neq j$. We shall later write $next_G(a, s_i) = s_{i+1}$ $(i = 1, \ldots, k-1)$, $next_G(a, EH) = s_1$ and $next_G(a, s_k) = EH$, where $EH \notin V$ is a distinguished symbol representing the engine house.

We remark that in this model the labels of G represent trains, and that $next_G(a, s)$ denotes the station following the station s on the route of train a. The last definition may seem incomplete since the engine house EH is not part of

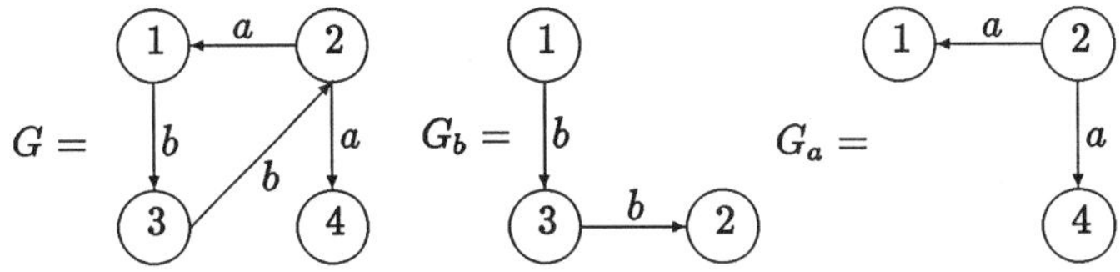

Formally:

$V(G) = \{1, 2, 3, 4\}$

$\Pi(G) = \{a, b\}$

$LA(G) = \{(2, a, 1), (2, a, 4), (1, b, 3), (3, b, 2)\}$

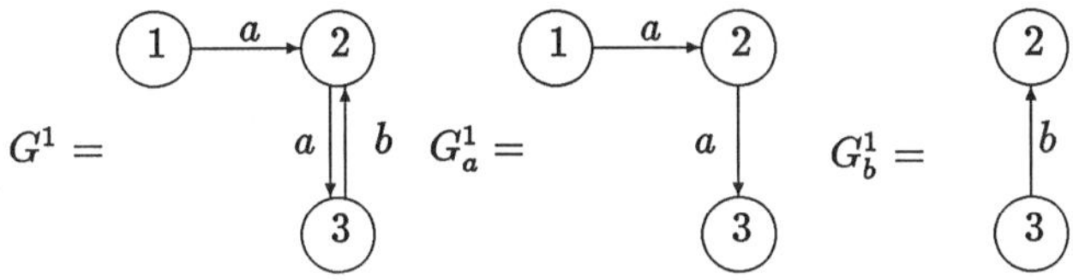

Figure 4.14. Explanation of concept of movement graph

the movement graph. However, such a definition is convenient from the technical point of view as some important subgraphs of a movement graph G are also movement graphs. Furthermore, the $next_G(a, EH) = s_1$ and $next_G(a, s_k) = EH$ provide sufficient information about the role of EH. Fig. 4.14 illustrates the concepts introduced above. In the graphical representation in the figure, each vertex is represented by a small circle and each arc by a labelled arrow joining appropriate circles. The graph G *is not* a movement graph since $LA(G_a) = \{(2, a, 1), (2, a, 4)\}$ is not of the form $\{(s_i, a_i, s_{i+1}) | i = 1, \ldots, k-1\}$, $k = 2$. The graph G^1 is a movement graph.

A **marking**[1] of G, $M \in Mar_G$, is a mapping $M : \Pi \longrightarrow V \cup \{EH\}$ such that for every $a \in \Pi$, $M(a) \in V(G_a) \cup \{EH\}$, and for every $s \in V$, $card(M^{-1}(s)) \leq 1$. Intuitively, $M(a) = s$ means that the train a is at station s, thus the condition $M(a) \in V(G_a) \cup \{EH\}$ represents the constraint that each train can only follow its own route, and $card(M^{-1}(s)) \leq 1$ says simply that each station can hold at most one train. A marking M' is directly reachable from M if there is a set $\Pi_0 \subseteq \Pi$, such that $\Pi_0 \neq \emptyset$, for every $a \in \Pi_0 : next_G(a, M(a)) \notin M(\Pi) - \{EH\}$, and

$$M'(a) = \begin{cases} next_G(a, M(a)) & a \in \Pi_0 \\ M(a) & a \in \Pi - \Pi_0 \end{cases}$$

[1] Although we borrow the concept and notation of marking from Petri net theory, markings used in this subsection differ formally from those in Petri nets.

If M' is directly reachable from M we shall write $M \longrightarrow M'$ or $M \longrightarrow_{\Pi_0} M'$. Also if $M_i \longrightarrow_{\Pi_i} M_{i+1}$, for $i = 1, \ldots, k$, we shall write $M_1 \longrightarrow_{\Pi_1 \ldots \Pi_k} M_{k+1}$. Clearly $\longrightarrow$ can be treated as a relation, so we can extend it standardly to $\longrightarrow^*$.

Let M_G^{EH} denote the marking representing the state in which all the trains are placed in the engine house (the initial state), i.e. $M_G^{EH}(a) = EH$ for every $a \in \Pi$.

A marking M is said to be *admissible* if $M \longrightarrow^* M_G^{EH}$. Otherwise M is *inadmissible*. The set of all inadmissable markings of a movement graph G is denoted by $IMar_G$.

The inadmissible markings represent those states of a train system for which the completion of each train journey is impossible. Thus any solution to the train problem must avoid them at any cost. To avoid inadmissible markings without having to resort to checking the whole reachability graph, we need some alternative, more convenient, characterization of these markings. We shall now give such a characterization by means of so called *minimal critical patterns*, which is a class of movement graphs.

For a movement graph G, let M_G^{start} be a marking defined by:

$$M_G^{start}(a) = next_G(a, EH)$$

for every $a \in \Pi(G)$. Clearly, M_G^{start} is only defined if $next_G(a, EH) \neq next_G(b, EH)$ for $a \neq b$.

Let M be a marking of a movement graph G, and let $G' \leq G$. A movement graph G' is said to be $M - active$ if $M_{G'}^{start}$ exists and for every $a \in \Pi(G')$: $M(a) = M_{G'}^{start}(a)$. Fig. 4.15 illustrates the concepts introduced above.

We first introduce the auxiliary notion of a critical pattern. A *critical pattern* is a movement graph G such that M_G^{start} exists and M_G^{start} is inadmissible. The graphs in Fig. 4.16 are critical patterns.

Lemma 4.4 *A marking M of a movement graph G is inadmissible if and only if there is an $M - active$ critical pattern in G.*

Proof. For every movement graph G and $a \in \Pi(G)$, let $last_G(a)$ denote the vertex $s \in V(G)$ such that $next_G(a, s) = EH$.
($\Rightarrow$) From $M \in IMar_G$ it follows that $Z = \{G' | G' \; is \; M - active \; in \; G\} \neq \emptyset$. Let H be a maximal (w.r.t $\leq$) element of Z. We observe that $last_G(a) = last_H(a)$, for all $a \in \Pi(H)$, and that $M(a) \in \{EH, last_G(a)\}$ for all $a \in \Pi(G) - \Pi(H)$. Suppose that M_G^{start} is admissible. Then there is a sequence $a_1, \ldots, a_n$ ($a_i \in \Pi(H)$) such that $n = card(LA(H)) + card(\Pi(H))$ and $M_H^{start} \longrightarrow_{\{a_1\} \ldots \{a_n\}} M_H^{EH}$. One can see that $M \longrightarrow_{\Pi_0 \{a_1\} \ldots \{a_n\}} M_G^{EH}$, where $\Pi_0 = \{a \in \Pi(G) | M(a) = last_G(a)\}$, contradicting $M \in IMar_G$. Hence H is an $M - active$ critical pattern.
($\Leftarrow$) Let H be an $M - active$ critical pattern in G. Suppose that M is admissible. Then there is a sequence $a_1, \ldots, a_n$ ($a_i \in \Pi(G)$) such that $M \longrightarrow_{\{a_1\} \ldots \{a_n\}} M_G^{EH}$.

$G^1 =$ 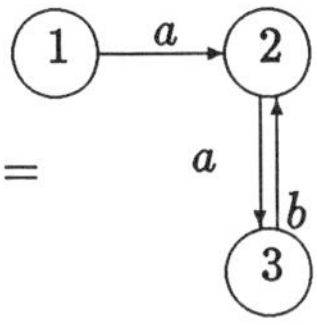

$G^2 =$ 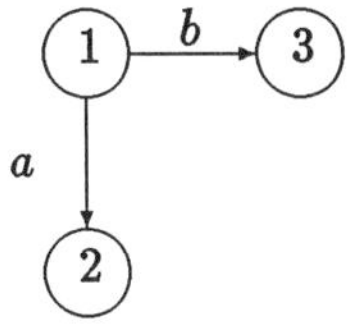

$M_{G^1}^{start}(a) = 1$

$M_{G^1}^{start}(b) = 3$

$M_{G^2}^{start}$ is undefined as

$next_{G^2}(a, EH) = next_{G^2}(b, EH) = 1$

$G^3 =$ 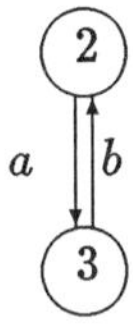

$M_{G^3}^{start}(a) = 2$

$M_{G^3}^{start}(b) = 3$

Let M be the following marking of $G : M(a) = 2,\ M(b) = 3$

G^3 is $M-active$ in G^1 because:

$G^3 \leq G^1$, $M_{G^3}^{start}$ exists and

$M_{G^3}^{start}(a) = M(a),\ M_{G^3}^{start}(b) = M(b)$

Figure 4.15. Example M_G^{start} and $M - active$ movement graph

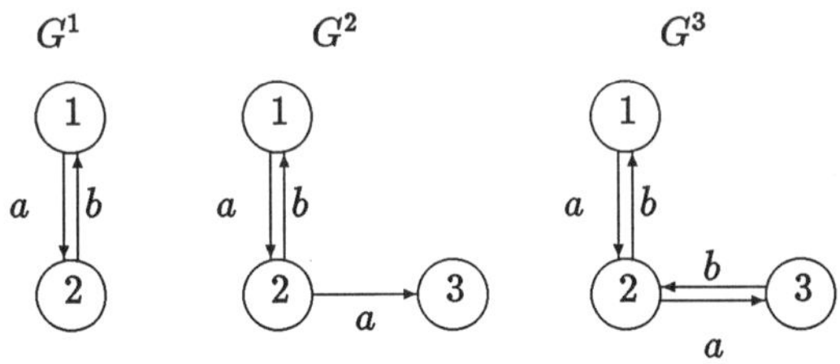

Figure 4.16. Critical patterns

Define $B_1, \ldots, B_n$ as follows. Let k_i denote the number of occurrences of a_i in the sequence $a_1, \ldots, a_i$. If $a_i \in \Pi(H)$ and $k_i \leq card(V(H_{a_i}))$, then $B_i = \{a_i\}$; otherwise $B_i = \emptyset$ $(i = 1, \ldots, n)$. Let $C_1, \ldots, C_m$ be a sequence obtained from $B_1, \ldots, B_n$ by dropping all B_i's which are empty sets. (Note that $m \geq 1$ since H is $M - active$ in G.) One can see that $M_H^{start} \longrightarrow_{C_1 \ldots C_m} M_H^{EH}$, contradicting $M_H^{start} \in IMar_H$. Hence M is inadmissible. □

A *minimal critical pattern* is a critical pattern G which is the only $M_G^{start} - active$ critical pattern in G, i.e.

$$[G' \text{ is a } M_G^{start} - active \text{ critical pattern in } G] \Rightarrow G' = G.$$

The critical patterns G^1 and G^3 in Fig. 4.16 are minimal critical patterns, but G^2 is not, since G^1 is an $M_{G^2}^{start} - active$ critical pattern in G^2.

We can now provide a fundamental characterization of inadmissible markings.

Theorem 4.1 (Admissibility Criterion for Movement Graphs)
A marking M of a movement graph G is inadmissible if and only if there is an $M - active$ minimal pattern.

Proof. ($\Leftarrow$) It follows immediately from Lemma 4.4.
($\Rightarrow$) By Lemma 4.4, $Z = \{G' | G' \textit{ is an } M - \textit{active critical pattern in } G\} \neq \emptyset$. Let H be a minimal (w.r.t. $\leq$) element of Z. Suppose that H' is an $M_H^{start} - active$ critical pattern. Clearly we have $H' \in Z$. Thus, from the minimality of H and $H' \leq H$ it follows that $H' = H$. Hence H is a minimal critical pattern. □

The last theorem provides a basis for a synchronization which intuitively consists in preventing minimal critical patterns from activation. It is defined as a relation of reachability on markings, as follows:

Let $\rightsquigarrow$ be a binary relation defined on markings of a movement graph G such that $M \rightsquigarrow M'$ if $M \longrightarrow M'$ and no minimal critical pattern of G is $M' - active$.

We may observe that the relation $\rightsquigarrow$ satisfies the following two properties:

$$(1)\ (\forall M \in Mar_G)\ M_G^{EH} \rightsquigarrow^* M \Rightarrow M \rightsquigarrow^* M_G^{EH},$$

$$(2)\ (\forall M \in Mar_G)(\forall M' \in Mar_G \setminus IMar_G)\ M \longrightarrow M' \Rightarrow M \rightsquigarrow M'.$$

The first condition says that the synchronization enables the completion of each train journey. The second condition says that the strategy based on $\rightsquigarrow$ allows parallel movements of trains whenever possible. Thus $\rightsquigarrow$ is indeed a solution to the Merlin-Randell problem of train journeys.

In the next section we show how to implement $\rightsquigarrow$ in terms of the COSY model.

4.4.2 COSY Specification of the Synchronization $\rightsquigarrow$

In our COSY specification of the relation $\rightsquigarrow$ we use three kinds of events.

The movements of trains are represented by the set of events:

$$Movements = \{[s_1as_2] | a \in \Pi, s_1 \in V(G_a) \cup \{EH\}, s_2 = next_G(a, s_1)\}.$$

The event $[s_1as_2]$ may be interpreted as follows: "the train a moves from station s_1 to station s_2".

To synchronize train movements we use two sets of auxiliary events:

$$Synch1 = \{[as] | a \in \Pi, s \in V(G_a)\}$$
$$Synch2 = \{[sa] | a \in \Pi, s \in V(G_a)\}.$$

Each $[as]$ represents an event in which the train a is given a station s for exclusive use. Each $[sa]$ represents an event in which the train a releases the station s.

We define three types of simple paths to model the behaviour of trains, stations, and devices preventing the activation of minimal critical patterns. Let $a \in \Pi$ and let $LA(G_a) = \{(s_i, a, s_{i+1}) | i = 1, \ldots, k-1\}$. We define a path $Train(a)$ as follows:

$$Train(a) = \underline{path}\ ST_1; ST_2; \ldots; ST_k; ST_{k+1}\ \underline{end}$$

where:

(1) $ST_1 = [as_1]; ([s_1a]; [as_1])^*; [EHas_1]$
(2) $ST_i = [as_i]; ([s_ia]; [as_i])^*; [s_{i-1}as_i]; [s_{i-1}a]$ for $i = 2, \ldots, k$
(3) $ST_{k+1} = [s_kaEH]; [s_ka]$.

The path $Train(a)$ represents the route of the train a, ST_1 represents a move of train a from the engine house to the station s_1, ST_i for $i = 2, \ldots, k$ represent a move of train a from the station s_{i-1} to the station s_i, and ST_{k+1} represents a move of train a from station s_k to the engine house.

Let $s \in V$ and let $\{a | s \in V(G_a)\} = \{a_1, \ldots, a_n\}$. We define a path $Station(s)$ as follows:

$$Station(s) = \underline{path}\ ([a_1s]; [sa_1]), \ldots, ([a_ns]; [sa_n])\ \underline{end}$$

The path $Station(s)$ represents the behaviour of the station s.

Let H be a minimal critical pattern of G, $\Pi(H) = \{a_1, \ldots, a_n\}$ and $s_i = next_H(a_i, EH)$. We define a path $mcp(H)$ as follows:

$$mcp(H) = \underline{path}\ W_1; (W_2; \ldots; (W_{n-1}; Z_{n-1})*; \ldots; Z_2)*; Z_1\ \underline{end}$$

where:

$$W_1 = \ldots = W_{n-1} = [a_1s_1], [a_2s_2], \ldots, [a_ns_n]$$
$$Z_1 = \ldots = Z_{n-1} = [s_1a_1], [s_2a_2], \ldots, [s_na_n].$$

The path $mcp(H)$ represents a mechanism preventing activation of the minimal critical pattern H.

Suppose that for our movement graph G we have:

$$\Pi(G) = \{a_1, \ldots, a_{k_0}\}, V(G) = \{s_1, \ldots, s_{m_0}\}, \text{ and } MCP(G) = \{H_1, \ldots, H_{n_0}\}$$

where $MCP(G)$ is the set of all minimal critical patterns of G. Then the following path program TJ is a COSY specification of the synchronization relation $\rightsquigarrow$:

$TJ =$
$\underline{program}$
$Train(a_1) \ldots Train(a_{k_0}) Station(s_1) \ldots Station(s_{m_0}) mcp(H_1) \ldots mcp(H_{n_0})$
$\underline{endprogram}$

We shall now analyze the path program TJ in more detail. In what follows we will use:

$$\underline{x}|_{a_k} = pr_k(\underline{x}),\ \underline{x}|_{s_k} = pr_{k_0+m}(\underline{x}),\ \underline{x}|_{H_k} = pr_{k_0+m_0+n}(\underline{x})$$

for every $\underline{x} \in Vev(TJ)^*$ and all $1 \leq k \leq k_0$, $1 \leq m \leq m_0$, $1 \leq n \leq n_0$. From the definition of TJ it follows immediately that if $\underline{x} \in VFS(TJ)$, $t = [sa] \in Synch2$ and $H \in MCP(G)$, then:

(1) $\underline{x}|_a t \in FS(Train(a)) \Rightarrow \underline{xt} \in VFS(TJ)$
(2) $\sum_{w \in Synch1} \#_w(\underline{x}|_H) - \sum_{w \in Synch2} \#_w(\underline{x}|_H) \leq card(\Pi(H)) - 1$

and if (2) is a proper inequality the we also have:

$$(3)\ (\forall t' \in Synch1 \cap Ev(mcp(H)))\ \underline{x}|_H t' \in FS(mcp(H)).$$

With every history $\underline{x}$ of TJ we associate a marking $mar(\underline{x})$ of G giving the position of each train after that history. The mapping

$$mar : VFS(TJ) \longrightarrow Mar_G$$

is defined as follows.

Let $\underline{x} \in VFS(TJ)$ and $a \in \Pi$. If $\underline{x}|_a \in (Synch1 \cup Synch2)^*$ then $mar(\underline{x})(a) = EH$. Otherwise, $\underline{x}|_a$ can be decomposed as $\underline{x}|_a = w[sas']y$, where $y \in (Synch1 \cup Synch2)^*$, and we define $mar(\underline{x})(a) = s'$. In particular, $mar(\underline{\varepsilon}) = M_G^{EH}$.

Note that directly from the definition of TJ it follows that if $mar(\underline{x})(a) = s \in V$, then:

(4) $\#_{[as]}(\underline{x}) = \#_{[sa]}(\underline{x}) + 1$
(5) $(\exists z \in (Synch1 \cup Synch2)^*)\ \underline{x}|_s = z[as]$.

By (5) above, $card((mar(\underline{x}))^{-1}(s))) \leq 1$ for every $s \in V$. Hence $mar(\underline{x})$ is always a correctly defined marking.

To compare the behaviour of TJ and the synchronization $\rightsquigarrow$, we introduce a relation of reachability on histories of TJ, denoted by $\approx>$. This relation will essentially be a kind of step sequence semantics of TJ which does not take into account the events from the sets $Synch1$ and $Synch2$.

Let $\underline{x}, \underline{y} \in VFS(TJ)$. Then:

$$\underline{x} \approx> \underline{y}$$

if $\underline{y} = \underline{x}\underline{z}\underline{A}$, where $z \in (Synch1 \cup Synch2)^*$ and $A \in Ind(TJ) \cap Movements$.

To show that the path program TJ is a correct specification of $\rightsquigarrow$ for a given movement graph G, we will demonstrate that for all $\underline{x}, \underline{y} \in VFS(TJ)$ and every $M \in Mar_G$ the following conditions hold:

$$\underline{x} \approx> \underline{y} \Rightarrow mar(\underline{x}) \rightsquigarrow mar(\underline{y})$$
$$mar(\underline{x}) \rightsquigarrow M \Rightarrow [(\exists \underline{z})\underline{x} \approx> \underline{z} \wedge mar(\underline{z}) = M].$$

In other words, we want to show that all the movements of trains specified by TJ are also specified by $\rightsquigarrow$, and that every train movement specified by $\rightsquigarrow$ is also realizable by the path program TJ. In particular this and $mar(\underline{\varepsilon}) = M_G^{EH}$ would imply that in particular TJ is adequate. Note also that the above conditions are equivalent to saying that:

$$\{mar(\underline{y}) | \underline{x} \approx> \underline{y}\} = \{M | mar(\underline{x}) \rightsquigarrow M\}.$$

Theorem 4.2
For all $\underline{x}, \underline{y} \in VFS(TJ)$ and every $M \in Mar_G$ the following conditions are satisfied:
(1) $\underline{x} \approx> \underline{y} \Rightarrow mar(\underline{x}) \rightsquigarrow mar(\underline{y})$,
(2) $mar(\underline{x}) \rightsquigarrow M \Rightarrow [(\exists \underline{z})\underline{x} \approx> \underline{z} \wedge mar(\underline{z}) = M]$.

Proof. Before proving these properties we shall show some properties of TJ we require for this proof.

$$Property\ 1: \ mar(VFS(TJ)) \cap IMar_G = \emptyset.$$

Suppose that $mar(\underline{x}) \in IMar_G$. Then there is a $mar(\underline{x})-active\ H \in MCP(G)$. Assume that $\Pi(H) = \{b_1, \ldots, b_k\}$ and let $p_i = next_H(b_i, EH)$, $i = 1, \ldots, k$. By the property (4) of TJ we have $\#_{[b_i p_i]}(\underline{x}) = \#_{[p_i b_i]}(\underline{x}) + 1$, $i = 1, \ldots, k$. Thus,

$$\sum_{t \in Synch1} \#_t(\underline{x}|_H) = \sum_{t \in Synch2} \#_t(\underline{x}|_H) + card(\Pi(H)),$$

contradicting the property (2) of TJ.

$$Property\ 2: \ (H \in MCP(G) \wedge M \rightsquigarrow M') \Rightarrow ((\exists a \in \Pi(H)) M_H^{start}(a) \notin \{M(a), M'(a)\}).$$

Suppose that $M_H^{start}(a) \in \{M(a), M'(a)\}$ for every $a \in \Pi(H)$. We are going to show that $M' \in IMar_G$, a contradiction with the definition of $\rightsquigarrow$. Suppose that $M \longrightarrow_{\Pi_1} M'$, and let $\Pi_2 = \{a \in \Pi_1 \cap \Pi(H) | M(a) \neq M_H^{start}(a)\}$. If $\Pi_2 = \emptyset$ the H is $M - active$ and, consequently, $M \in IMar_G$. Hence $M' \in IMar_G$. If $\Pi_2 \neq \emptyset$ then there is M'' such that $M \longrightarrow_{\Pi_2} M''$ and $M'' = M'$ or $M'' \rightarrow M'$. Clearly, H is $M'' - active$, and hence $M' \in Imar_G$. Thus Property 2 holds.

Property 3 : if $\underline{x} \in VFS(TJ)$ then there is $w \in Synch2^*$ such that $\underline{xw} \in VFS(TJ)$, and for every $s \in V - mar(\underline{x})(\Pi(G))$:

$$\sum_{t \in Synch1} \#_t(\underline{xw}|_s) = \sum_{t \in Synch2} \#_t(\underline{xw}|_s).$$

Property 3 follows directly from the definition of the paths representing the behaviour of trains, the definition of $mar(\underline{x})$ and the property (1) of TJ.

Now we shall prove the conditions of the theorem.

Proof of (1) *of the theorem.*

Let $\underline{y} = \underline{xwA}$, where $w \in (Synch1 \cup Synch2)^*$ and $A = \{[p_1 b_1 q_1], \ldots, [p_k b_k q_k]\}$. We will show that $mar(\underline{xw}) \rightsquigarrow mar(\underline{y})$, which suffices to prove the condition (1) of the theorem since $mar(\underline{xw}) = mar(\underline{x})$. Since A contains only independent events, $b_i \neq b_j$ for $i \neq j$. Hence we have:

$$(\forall i = 1, \ldots, k) mar(\underline{xw})(b_i) = p_i \wedge mar(\underline{y})(b_i) = q_i = next_G(b_i, p_i).$$

Moreover, $mar(\underline{xw})(a) = mar(\underline{y})(a)$ for every $a \in \Pi(G) \setminus \{b_1, \ldots, b_k\}$. Suppose that $q = q_i \in mar(\underline{xw})(\Pi(G)) \setminus \{EH\}$. Then, by property (5) of TJ, $\underline{xw}|_q = z[bq]$, where $b \neq b_i$. On the other hand, $mar(\underline{y})(b_i) = q$, so $\underline{xw}|_q = \underline{y}|_q = y'[b_i q]$, a contradiction. Thus $q_i \notin mar(\underline{xw})(\Pi(G)) \setminus \{EH\}$, for $i = 1, \ldots, k$. Therefore $mar(\underline{xw}) \rightarrow_{\{b_1, \ldots, b_k\}} mar(\underline{y})$. This and *Property* 1 from the beginning of this proof yield $mar(\underline{xw}) \rightsquigarrow mar(\underline{y})$.

Proof of (2) *of the theorem.*

Suppose that $mar(\underline{x}) \rightarrow_{\{b_1, \ldots, b_k\}} M$. Let $mar(\underline{x})(b_i) = p_i$, $M(b_i) = q_i$ and $t_i = [p_i b_i q_i]$, for $i = 1, \ldots, k$. Let us take $w \in Synch2^*$ from *Property* 3, and let $v = r_1 \ldots r_k$, where $r_i = [b_i q_i]$ if $q_i \neq EH$ and $r_i = \varepsilon$ if $q_i = EH$. Define $\underline{y} = \underline{xwv}$. We prove that $\underline{y} \in VFS(TJ)$. One can see that from $q_i \neq q_j$ (for $i \neq j$), $q_i \notin mar(\underline{x})(\Pi(G)) \setminus \{EH\}$ and *Property* 3 it follows that $\underline{y}|_a \in FS(Train(a))$ and $\underline{y}|_s \in FS(Station(s))$, for all $a \in \Pi(G)$ and $s \in V$.

Suppose now that $H \in MCP(G)$ and $B = \{r_1, \ldots, r_k\} \cap Ev(mcp(H)) \neq \emptyset$. Without loss of generality we can assume that $B = \{r_1, \ldots, r_m\}$, $1 \leq m \leq k$. By *Property* 2, there is $a \in \Pi(H) \setminus \{b_1, \ldots, b_m\}$ such that $mar(\underline{x})(a) \neq M_H^{start}(a)$. Thus, by *Property* 3:

$$\sum_{t \in Synch1} \#_t(\underline{xw}|_H) - \sum_{t \in Synch2} \#_t(\underline{xw}|_H) \leq card(\Pi(H)) - (m+1).$$

Therefore, by the property (3) of TJ, $\underline{xw}|_H r_1 \ldots r_m \in FS(mcp(H))$, so $\underline{y}|_H \in FS(mcp(H))$. Consequently, $\underline{y} \in VFS(TJ)$. One can easily see that $\underline{yt_i} \in VFS(TJ)$, for $i = 1, \ldots, k$. Moreover, $A = \{t_1, \ldots, t_k\} \in Ind(TJ)$. Hence $\underline{yA} \in VFS(TJ)$. So we have $\underline{x} \approx> \underline{z}$, where $\underline{z} = \underline{yA}$ and $mar(\underline{z}) = M$ which completes the proof of (2) of the theorem. □

To illustrate the technique and results formulated above let us consider the movement graph G from Fig. 4.17. This movement graph has only one minimal critical pattern, namely the graph H from Fig. 4.17, so in this case $MCP(G) = \{H\}$. The path program TJ, also in Fig. 4.17, represents a COSY specification of the solution to this particular set of train journeys.

One might wonder if the regular expressions of the form $([sa]; [as])*$ in the paths representing train routes are necessary? The answer is yes. The reason is that they enable releasing of stations which do not hold trains (see *Property* 3 in the proof of Theorem 4.2), and thus the condition (2) of Theorem 4.2 can be proved. Consider, for instance, the path program TJ from Fig. 4.17. Let us replace the path $Train(a)$ by the following path:

$$Train'(a) = \underline{path}\ [a1]; [EHa1]; [a2]; [1a2]; [2aEH]; [2a]\ \underline{end}.$$

Let $x = [a1]$. Clearly, $\underline{x} \in VFS(TJ)$ and $mar(\underline{x}) = M_G^{EH}$. Thus $mar(\underline{x}) \rightsquigarrow M$, where $M(a) = EH$, $M(b) = 2$ and $M(c) = 3$, however, for every $\underline{z} \in Vev(TJ)^*$, $[a1] \approx> \underline{z} \Rightarrow mar(\underline{z}) \neq M$.

4.4.3 Final Comment and Bibliographic Notes

The solution to the Merlin-Randell problem of train journeys presented here essentially consists in avoiding the activation of minimal critical patterns. The problem of finding all minimal critical patterns of a given movement graph is a crucial point of such an approach. This problem is a non-trivial graph theoretical problem. More results concerning a graph theoretic description of minimal critical patterns can be found in [Kou86]. The whole problem of train journeys as formulated in this section follows from [MR78]. Several attempts have been made to find a solution to and a formal specification of this problem. The COSY model, as a specification tool, has been used in [Shi79b], [Lau85],[Kou84] and [Kou86]. CCS was used in [Shi85], CSP in [Ros85] and Temporal Logic in [Mos85]. The most complete approach is due to M. Koutny [Kou86] and this section is based on that paper.

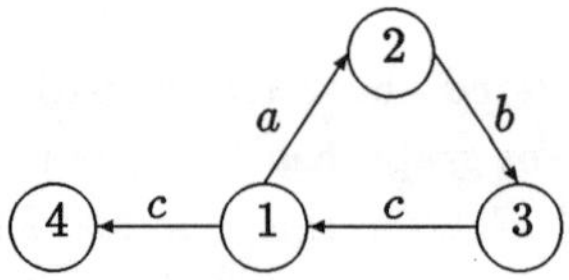

movement graph G

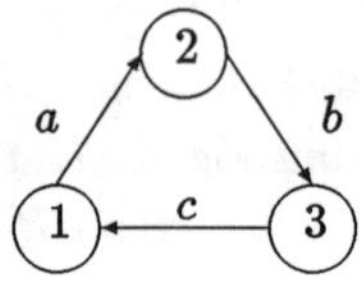

H the only minimal critical pattern of G

$TJ =$

program

Train(*a*) =

path [*a*1]; ([1*a*]; [*a*1])*; [*EHa*1]; [*a*2]; ([2*a*]; [*a*2])*;
[1*a*2]; [1*a*]; [2*aEH*]; [2*a*] *end*

Train(*b*) =

path [*b*2]; ([2*b*]; [*b*2])*; [*EHb*2]; [*b*3]; ([3*b*]; [*b*3])*;
[2*b*3]; [2*b*]; [3*bEH*]; [3*b*] *end*

Train(*c*) =

path [*c*3]; ([3*c*]; [*c*3])*; [*EHc*3]; [*c*1]; ([1*c*]; [*c*1])*;
[3*c*1]; [3*c*]; [*c*4]; 9[4*c*]; [*c*4])*; [1*c*4]; [1*c*]; [4*cEH*]; [4*c*] *end*

Station(1) = *path* ([*a*1]; [1*a*]), ([*c*1]; [1*c*]) *end*

Station(2) = *path* ([*a*2]; [2*a*]), ([*b*2]; [2*b*]) *end*

Station(3) = *path* ([*b*3]; [3*b*]), ([*c*3]; [3*c*]) *end*

Station(4) = *path* [*c*4]; [4*c*] *end*

mcp(*H*) =

path [*a*1], [*b*2], [*c*3]; ([*a*1], [*b*2], [*c*3]; [1*a*], [2*b*], [3*c*])*;
[1*a*], [2*b*], [3*c*] *end*

endprogram

Figure 4.17. COSY solution of problem of train journeys

4.5 Transforming Sequential Systems into Concurrent Systems

The most popular method for specifying non-sequential systems consists in decomposing the problem into parts being sequential by nature (such a decomposition is often ambiguous), specifying each sequential part in a relatively independent way, and then composing all sequential specifiactions to obtain the whole system. The COSY notation indeed supports this method, each single path describes sequential constraints, each process describes how a sequential process can utilize the constrained events, and the composition of paths and processes (or paths only) describes the whole system. In fact, *all widely known* programming languages and specification tools such as CSP, CCS, ADA, *occam*, Path Pascal are based on this method as a principle. This method is convenient for specifying systems with inherent concurrent activities, such as operating systems and/or real time distributed systems, etc. Unfortunately, this method is much less useful for specifying systems (or algorithms) in which concurrent activities are not the primary concern of specification. For example, a system or algorithm might be specified as a sequential one, and concurrency is merely used to speed up performance and to use hardware more effectively. As practical examples we can take sorting, linear algebra procedures, searching routines, algorithms such as the Sieve of Eratostenes, etc. In all these examples concurrency is not a primary concern, however its introduction might substantially improve efficiency. In such cases, the method described above is not very convenient since the existing sequential solutions cannot be explicitly used to obtain a concurrent solution. A method which seems to suit such cases better can be described as follows.

We start with a *sequential specification* (solution). Next, we determine which events (actions) can occur (be performed) *independently*. We might also modify the original sequential specification a bit to make it easier to determine the (potentially) independent events. *The sequential specification (solution) together with the specification of independent events constitutes a fully non-sequential specification (solution).* This specification may then be *implemented* by transforming it into a set of interacting sequential subsystems (processes). The rules for such a transformation should allow for an *automation* and a programmer would need have no knowledge about them (just as he does not have to know the implementation of, for example, "while-do" loops).

The difference between the two methods is that in the first one the programmer specifies synchronization in an explicit way. He has to specify a set of sequential subsystems and a way of synchronizing them. In the second approach, he specifies only which events (actions) are independent, and does not bother about synchronization details. The rest would presumably be done by a compiler-like program.

The first approach is convenient for systems (algorithms) with inherently concurrent activities, the second one is more convenient when concurrency is not in fact the primary issue, but it could substantially improve efficiency.

In this section we shall show how COSY can be used to create some theoretical foundations for the second approach. To illustrate the basic ideas let us consider the sequential, iterative factorial scheme to compute $x!$:

$$z := 1;\ \underline{while}\ (x > 0)\ \underline{do}\ \underline{begin}\ z := z * x; x := x - 1\ \underline{end}$$

Since the body of the above loop consists of two dependent actions (the order of $z := z * x$ and $x := x + 1$ cannot be changed), it is modified to a form in which independent actions can be identified:

$$z := 1;\ \underline{while}\ (x > 0)\ \underline{do}\ \underline{begin}\ y := x; z := z * y; x := x - 1\ \underline{end}$$

We can model the above program by the *extended* path[2]:

$$EP_1 = \underline{epath}\ onetoz; (xgt0; xtoy; ztimesytoz; xminus1tox)^*; xnotgt0\ \underline{end}$$

where the events occuring in the path correspond to program actions and conditions in the following way:

$$\begin{aligned} onetoz &= z := 1 \\ xgt0 &= (x > 0) \\ xtoy &= y := x \\ ztimesytoz &= z := z * y \\ xminus1tox &= x := x - 1 \end{aligned}$$

Additionally the event $xnotgt0$ corresponds to $\neg(x > 0)$, i.e., to $(x \leq 0)$. The "extension" indicated by using $\underline{epath}$ instead of $\underline{path}$ means that the implicit assumption that every sequential subsystem is cyclic is no longer valid. The extended paths will be formally defined in the very next subsection, for now we just say that for example, $\underline{path}\ a, b; c\ \underline{end}$ corresponds to the regular expression $(a, b; c)^*$, while $\underline{epath}\ a, b; c\ \underline{end}$ corresponds to $(a, b; c)$.

For the events of the extended path EP_1 we can define an independency relation by defining two events to be independent if the corresponding program actions have no common variable. Note that this is the definition for EP_1. It may be too weak or too strong or even inadequate for other examples. It is up to the program designer to specify the independency relation. In this approach he or she is relieved from synchronization details but must must specify independent events instead. [JM90] shows how it can be done within a simple prototype

[2] The idea of using regular expressions (formal grammars in general) to model the control structure of various kinds of programs date back to the early seventies (see for instance [GL73] [Maz72], for Algol-like programs, and [Jan76] for programs with co-routines).

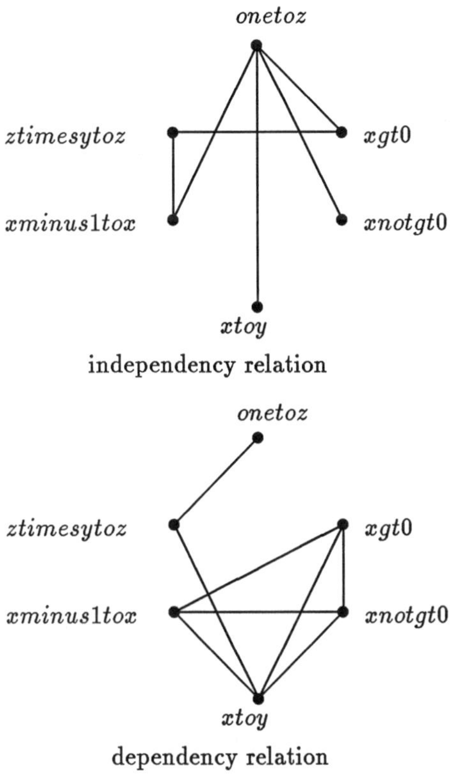

Figure 4.18. Independency relation for the extended path EP_1

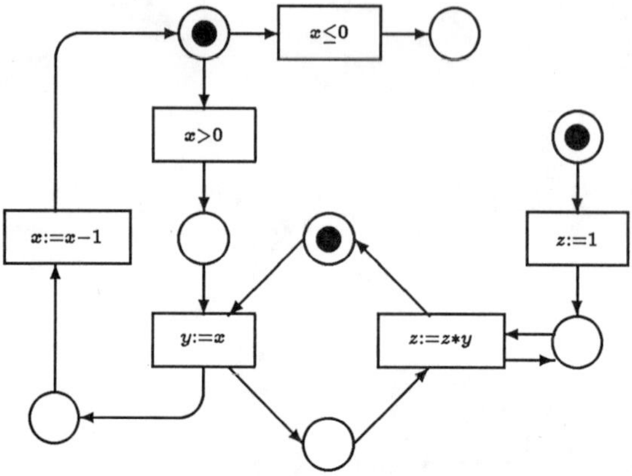

Figure 4.19. Petri net version of EPr_1

programming language. Two events are dependent if they are not independent. Fig. 4.18 shows both the independency and dependency relation for the events of EP_1. From this figure one can obtain the three maximal sets of mutually dependent events:

$$\{xgt0, xnotgt0, xtoy, xminus1tox\},\ \{xtoy, ztimesytoz\},\ \{ztimesytoz, onetoz\}.$$

Now, the three sequential components of the concurrent program being constructed can be obtained by erasing events that do not appear in the above sets together with all superfluous separators. Thus we get an extended path program:

$$\begin{aligned} EPr_1 = \ & \underline{epath}\ (xgt0; xtoy; xminus1tox)^*; xnotgt0\ \underline{end} \\ & \underline{epath}\ (xtoy; ztimesytoz)^*\ \underline{end} \\ & \underline{epath}\ onetoz; (ztimesytoz)^*\ \underline{end} \end{aligned}$$

A Petri net version of the program EPr_1 is shown on Fig. 4.19, and it is easy to prove that this program indeed represents a parallel factorial.

In some cases the above transformations results in a concurrent specification which by no means can be regarded as equivalent to the initial sequential one. Consider the following program which computes the sum of two integers using the successor and predecessor operations (for simplicity denoted by $x + 1$ and $x - 1$, respectively):

$$\underline{while}\ (x > 0)\ \underline{do}\ x := x - 1; y := y + 1\ \underline{end}$$

We can model this program by the extended path:

$$EP_2 : \underline{epath}\ (xgt0; minus1tox; yplus1toy)^*; xnotgt0\ \underline{end}.$$

Repeating the same procedure as for EP_1, we obtain

$$EPr_2 = \begin{array}{l} \underline{epath}\ (xgt0; xminus1tox)^*; xnotgt0\ \underline{end} \\ \underline{epath}\ (yplus1toy)^*\ \underline{end} \end{array}$$

which certainly does not compute the sum $x+y$ as there is absolutely no synchronization between the extended path containing *xminus1tox* and that containing *yplus1toy*.

We shall extend the ideas described above and show that by introducing special events, called *synchronization guards*, we can guarantee that the transformation described above preserves the desired properties of the initial specification. In the sequel we will simply use Pr and P to denote extended path programs and extended paths instead of EPr and EP, respectively.

4.5.1 e-Paths and e-Path Programs

To build a theoretical basis for transforming sequential systems into concurrent systems we have to extend the concept of path program. The implicit assumption that every sequential subsystem is cyclic is no longer desirable. Furthermore we need tools both for defining all histories (like VFS's for path programs), and for defining some histories that might be interpreted as *resulting* histories (like VL's for products of regular expressions). Using the concepts already discussed in this book, we can easily define such a model by extending path programs or by extending products of regular expressions. Since path programs are the kernel of this book, we choose the first alternative.

An **e-path program** (for extended path program) is a collection of **e-paths** in *eprogram* and *endprogram* parentheses. As in the case of path programs we shall omit these parentheses for brevity.

An **e-path** (from extended path) is an expression:

$$P = \underline{epath}\ sequence\ \underline{end}$$

where *sequence* is a string derived from rules $B4 - B8$ of the grammar from Sect. 2.1, with an additional assumption that the special event name "*nil*" might appear in *sequence* (one or more times).

Thus, from the syntactic viewpoint, there is little difference between path programs and e-path programs, however there are more semantical differences. First, single e-paths are no longer implicitly cyclic; second, the behaviour of single e-paths is characterized by *two* sets of event sequences instead of one.

The sets of event sequences generated by an e-path will be defined by the function RFS (from: Resulting Firing Sequences). The function RFS will replace the function Cyc we used in the case of ordinary single paths.

To a string E, being either an $e-path$, $sequence$, $starelement$, $element$ or $event$ (see the rules $B4-B8$ in Sect. 2.1), we associate a set of event sequences $RFS(E)$ defined recursively as:

1. $RFS(\underline{epath}\ E\ \underline{end}) = RFS(E)$
2. $RFS(E_1;\ldots;E_k) = RFS(E_1)\ \ldots\ RFS(E_k)$
3. $RFS(E_1,\ldots,E_k) = RFS(E_1) \cup \ldots \cup RFS(E_k)$
4. $RFS(E*) = RFS(E)^*$
5. $RFS((E)) = RFS(E)$
6. $RFS(a) = \begin{cases} \{a\} & a \text{ a is an event and } a \neq nil \\ \{\varepsilon\} & a = nil. \end{cases}$

Let $P = \underline{epath}\ sequence\ \underline{end}$. Clearly, $RFS(P)$ is a regular language defined by $sequence$ treated as a regular expression, with nil interpreted as ε. We then define, $FS(P)$ as $FS(P) = Pref(RFS(P))$. We shall often write FS and RFS instead of $FS(P)$ and $RFS(P)$ if the name P is understood.

The set of event sequences $RFS(P)$ is called the set of (finite) **resulting firing sequences of** P. From the set $RFS(P)$ we can derive the set of (finite) **firing sequences of** P:

$$FS(P) = Pref(RFS(P)).$$

For instance for $P = \underline{epath}\ a;b,c\ \underline{end}$, $RFS(P) = \{ab, ac\}$, $FS(P) = \{\varepsilon, a, ab, ac\}$ (compare $FS(P_1)$ from Pr_1 of Sect. 2.1).

The set FS represents all behaviours of P, while RFS represents the *resulting behaviours*, i.e. histories which lead either to an end of system activity, or in the case of cyclic systems, lead back to the initial state of the system.

The behaviour of an e-path program $Pr = P_1 \ldots P_n$ is described by the set of all vector firing sequences that might be produced by Pr. We shall denote this set by VFS (or $VFS(Pr)$, and define it by:

$$VFS = FS_1 \times \ldots \times FS_n \cap Vev^*.$$

The set VFS is called the set of all (finite) **vector firing sequences of** Pr.

The resulting behaviour of an e-path program $Pr = P_1 \ldots P_n$ is described by the set of all resulting vector sequences that might be produced by Pr. This set will be denoted by $RVFS$ (or $RVFS(Pr)$), called the set of all (finite) **resulting vector firing sequences**, and defined by:

$$RVFS = RFS_1 \times \ldots \times RFS_n \cap Vev^*.$$

Lemma 4.5
$Pref(RVFS) \subseteq VFS$

Proof. $RVFS \subseteq VFS \Rightarrow Pref(RVFS) \subseteq Pref(VFS) = VFS$ (see Lemma 2.6 (1); its proof also holds when Cyc is replaced by RFS). □

The example below shows that in general the equality $VFS = Pref(RVFS)$ does not hold:

$$Pr_{43} = \frac{\underline{epath}\ a, (b; c; d)\ \underline{end}}{\underline{epath}\ a, (b; d; c)\ \underline{end}}$$

Here we have $VFS(Pr_{43}) = \{\underline{a}, \underline{b}, \underline{\varepsilon}\}$, while $RVFS(Pr_{43}) = \{\underline{a}\}$, i.e. $Pref(RVFS(Pr_{43}) = \{\underline{\varepsilon}, \underline{a}\}$. We could also define the concepts of *deadlock-freeness* and *adequacy* for e-path programs in the same way as for path programs, however, as we drop the assumption about cyclicity of sequential components, these properties are no longer of major interest.

The next difference between paths and e-paths is that for an e-path P, the set $Ev(P)$ *denotes the set of all events occurring in* P *except nil*. The event nil does not belong to $Ev(P)$ as it is transformed into ε by RFS.

Let $Pr = P_1 \dots P_n$ be an e-path program, where the P_i's are e-paths. Let Ev (or $Ev(Pr)$) denote the set of events occurring in Pr except nil, i.e. $Ev = Ev_1 \cup \dots \cup Ev_n$, where Ev_i (or $Ev(P_i)$) is the set of events of P_i for $i = 1, \dots, n$, and let Vev (or $Vev(Pr)$) be the set of vector events defined by $Ev_1, \dots, Ev_n$. Furthermore, let $FS_1, \dots, FS_n$ and $RFS_1, \dots, RFS_n$ be sets of firing sequences and resulting firing sequences generated by $P_1, \dots, P_n$ respectively.

4.5.2 Concurrent e-Paths and Proper Concurrent e-Paths

Our goal is to describe formally transformations of sequential specifications into concurrent specifications. As shown in the introductory section, the required sequential specification consists of two components: a description of sequential behaviour and a declaration of independent events. The first component is in the form of a pure sequential specification, the second component identifies sets of dependent events, which implicitly specifies the independency relation. This small inconsistency follows from the fact that on the level of programs an explicit definition of independency seems to be more convenient and popular, while in the COSY approach, the independency relation *ind* results from the distribution of events $Ev_1, \dots, Ev_n$, and each Ev_i is a set of dependent events. To model these concepts formally we shall introduce the concept of concurrent e-path.
By a **concurrent e-path** we mean a tuple

$$CP = (P, Ev_1, \dots, Ev_n)$$

where $P = \underline{epath}\ sequence\ \underline{end}$ is an e-path, $n \geq 1$ and $Ev_1, \ldots, Ev_n$ are sets of events satisfying:

$$Ev(P) = Ev_1 \cup \ldots \cup Ev_n.$$

Let Vev (or $Vev(CP)$ if needed) be the set of vector events defined by $Ev_1, \ldots, Ev_n$. The set $Vev(CP)$ defines a concurrent environment generated by CP, it plays the same role in the theory of concurrent e-paths, as $Vev(Pr)$ plays in the theory of e-path programs and path programs. In our model, an e-path P represents a pure sequential specification, and $CP = (P, Ev_1, \ldots, Ev_n)$ represents a sequential specification to be implemented concurrently.

To explain the intuitive meaning of the next definitions we consider the following example. Let $P = \underline{epath}\ a; b; c\ \underline{end}$, $Ev_1 = \{a, b\}$, $Ev_2 = \{a, c\}$. In this case we have: the behaviour of P,

$$FS(P) = \{\varepsilon, a, ab, abc\},$$

and the resulting behaviour of P,

$$RFS(P) = \{abc\}.$$

Let us reflect what kind of sequence vectors may be interpreted as a *behaviour* or a *resulting behaviour* of $CP_1 = (P, Ev_1, Ev_2)$. The resulting behaviour is easier to define. Note that

$$\{\underline{x} | x \in RFS(P)\} = \{\underline{abc}\}$$

may be regarded as a *resulting behaviour* defined by CP_1. The problem with the *behaviour* is somewhat more complex. The set $\{\underline{x} | x \in FS(P)\}$ does not seem to be a proper candidate . For instance, $\underline{abc} \in \{\underline{x} | x \in FS(P)\}$, $\underline{ab} \in \{\underline{x} | x \in FS(P)\}$, but $\underline{ac} \notin \{\underline{x} | x \in FS(P)\}$ although $\underline{abc} = \underline{acb}$! From the notion of the behaviour we demand that an initial part of each history is also a history, which in terms of this model means that the behaviour should be closed under the operation $Pref$. On the other hand, the behaviour defined by CP_1 should not be much bigger than the set $\{\underline{x} | x \in FS(P)\}$, because $FS(P)$ defines the behaviour of P and the operation "_" is an operation which forgets about superfluous sequentializations. This means that the least set containing $\{\underline{x} | x \in FS(P)\}$ and closed under $Pref$ is a good candidate to represent the behaviour of CP_1. One can easily prove that this set is equal to $Pref(\{\underline{x} | x \in RFS(P)\}$.
The above discussion leads to the following definitions:
The **nonsequential behaviour** of CP is characterized by the set

$$VFS(CP) = Pref(\{\underline{x} \in Vev^* | x \in RFS(P)\}$$

i.e. the set of all (finite) *vector firing sequences of* CP.
The **resulting non-sequential behaviour** of CP is characterized by the set

$$RVFS(CP) = \{\underline{x} \in Vev^* | x \in RFS(P)\}$$

i.e. the set of all (finite) *resulting vector firing sequences* of CP.

For example for $CP_1 = (\underline{epath}\ a; b; c\ \underline{end}, \{a,b\}, \{a,c\})$ we have

$$RVFS(CP_1) = \{\underline{abc}\}, \quad VFS(CP) = \{\underline{\varepsilon}, \underline{a}, \underline{b}, \underline{ab}, \underline{abc}\}.$$

Now we describe a transformation $\mathcal{C}$ of concurrent e-path into e-path programs. We have used this transformation informally in the introductory section of this chapter.

Consider the concurrent e-path $CP = (P, Ev_1, \ldots, Ev_n)$. Let $\mathcal{C}(CP)$ be the set of e-path programs such that:

$$Pr = P_1 \ldots P_n \in \mathcal{C}(CP) :\Leftrightarrow (\forall i = 1, \ldots, n) RFS(P_i) = h_i(RFS(P)).$$

There may be more than one element of $\mathcal{C}(CP)$. To derive e-paths P_i satisfying the above condition, one possible algorithm is to replace in P those events that do not appear in Ev_i by *nil* and use equivalence rules for regular expressions (interpreting *nil* as the symbol ε) to erase as many *nil*'s as possible. In fact, we used this algorithm in the introductory part of this chapter.

For example for $CP_2 = (\underline{epath}\ a; b, c\ \underline{end}, \{a,b\}, \{a,c\})$, both

$$Pr_{44} = \begin{array}{l} \underline{epath}\ a; b, nil\ \underline{end} \\ \underline{epath}\ a; nil, c\ \underline{end} \end{array}$$

and

$$Pr_{45} = \begin{array}{l} \underline{epath}\ (a; b), a\ \underline{end} \\ \underline{epath}\ a, (a; c)\ \underline{end} \end{array}$$

belong to $\mathcal{C}(CP_2)$.

For every concurrent e-path CP and $Pr \in \mathcal{C}(CP)$, let

$$\begin{array}{l} VFS(\mathcal{C}(CP)) = VFS(Pr), and \\ RVFS(\mathcal{C}(CP)) = RVFS(Pr). \end{array}$$

From the definition of $\mathcal{C}$ it follows that for any $Pr', Pr'' \in \mathcal{C}(CP)$, we have $VFS(Pr') = VFS(Pr'')$ and $RVFS(Pr') = RVFS(Pr'')$, so both $VFS(\mathcal{C}(CP))$ and $RVFS(\mathcal{C}(CP))$ are defined in an unambiguous way, that is they are independent of a particular choice of $Pr \in \mathcal{C}(CP)$.

We say a concurrent e-path $CP = (P_1, Ev_1, \ldots, Ev_n)$ is **proper** if and only if:

(1) $VFS(CP) = VFS(\mathcal{C}(CP))$,

(2) $RVFS(CP) = RVFS(\mathcal{C}(CP))$.

From the preceding discussion it follows that the above two equalities are well-defined. The following examples show that we cannot leave out any of the conditions (1) and (2) of the above definition.

Let $CP_3 = (\underline{epath}\ (a;b)^*\ \underline{end}, \{a\}, \{b\})$, and let

$$Pr_{46} = \frac{\underline{epath}\ a^*\ \underline{end}}{\underline{epath}\ b^*\ \underline{end}}$$

Clearly $Pr_{46} \in \mathcal{C}(CP_3)$ and $VFS(CP_3) = VFS(Pr_{46}) = \{\underline{a}, \underline{b}\}^*$, while $RVFS(CP_3) = \{\underline{x} \in \{a,b\}^* \mid \#_a(\underline{x}) = \#_b(\underline{x})\} \neq \{\underline{a}, \underline{b}\}^* = RVFS(Pr_{46})$, so (1) does not always imply (2). Next, let

$$CP_4 = (\underline{epath}\ (a;c;e),(b;d;f)\ \underline{end}, Ev_1, Ev_2)$$

where $Ev_1 = \{a,e,b,f\}$, $Ev_2 = \{c,e,d,f\}$, and

$$Pr_{47} = \frac{\underline{epath}\ (a;e),(b;f)\ \underline{end}}{\underline{epath}\ (c;e),(d;f)\ \underline{end}.}$$

Note that $Pr_{47} \in \mathcal{C}(CP_4)$, $RVFS(CP_4) = RVFS(Pr_{47}) = \{\underline{ace}, \underline{bdf}\}$, but $\underline{ad} \in VFS(Pr_{47}) \setminus VFS(CP_4)$, so $VFS(Pr_{47}) \neq VFS(CP_4)$.
The above examples are also examples of concurrent e-paths which are not proper. The concurrent e-path CP_3 is not proper because of the way in which $\mathcal{C}$ interprets the independency. In this case it maps a and b into two different e-paths, while the semantics of the loop $(a;b)^*$ in CP_3 assumes that the number of occurrences of a and b is the same in each resulting vector firing sequence. So $\mathcal{C}$ interprets the independency in a much less sophisticated way than the semantics of loops in concurrent e-paths implies.
The concurrent e-path CP_4 is not proper for similar reasons. Now $\mathcal{C}$ interprets the independency differently from the way the choice operator in the concurrent e-path does. In CP_4 the two independent events occurring in branches of the alternative, namely a and d, are mapped by $\mathcal{C}$ to different e-paths allowing both these events to occur.

Let us explain why *proper* concurrent e-paths are so important in our approach. Recall that in our model, an e-path P represents a pure sequential specification, and $CP = (P, Ev_1, \ldots, Ev_n)$ represents a sequential specification to be implemented concurrently. Since $\mathcal{C}$ preserves behavioural properties of *proper* concurrent e-paths only, when CP is proper, it can be implemented by any $Pr \in \mathcal{C}(Pr)$. Necessary and sufficient conditions stating what properties CP must posses to be proper have not been found yet for the general case. But in fact, what we are really looking for is a transformation Π such that for every concurrent e-path CP, $\Pi(CP)$ can be regarded as behaviourally equivalent to CP, and $\Pi(CP)$ is proper.

4.5.3 Synchronized e-Path Programs

The transformation Π mentioned at the end of the previous section will be defined by introducing *synchronization guards* which are to be inserted into the original concurrent e-paths. These guard events will help to control the concurrent execution. Recall that in the above examples, concurent e-paths were *not* proper because of the confusion between the choice and loop operations in the way in which $\mathcal{C}$ interprets the independency relation. For example, independent events occurring in branches of the choice were mapped by $\mathcal{C}$ to different paths of the resulting e-path program. To recover this, synchronization guards will be inserted into alternatives and loops and, by definition, the synchronization guards must not be independent of any event from the bodies of the constructs they are inserted in.

The synchronization guards (denoted by $\bigtriangledown$'s), are suppose to occur in two contexts:

(1) in alternatives, where they are inserted as the first event of one of the branches of the alternatives: $branch_1, (\bigtriangledown; branch_2)$

(2) in loops, where they are inserted at the beginning of the body: $(\bigtriangledown; body)^*$.

Additionally we shall require the synchronization guards to be *unique*, and to restrict concurrency of other events as little as possible. We shall now develop the necessary formalism to present the ideas described above. We first define the class $SEPATH$ of synchronized e-paths.

Let $Ev_1, \ldots, Ev_n$ be sets of events, $Ev = Ev_1 \cup \ldots \cup Ev_n, Ev_{\bigtriangledown} \subseteq Ev$. Elements of $Ev_{\bigtriangledown}$ are called **synchronization guards** and will be denoted by the symbol $\bigtriangledown$ with indices if necessary. Recall that for every $a \in Ev$, $i(a) = \{i \mid a \in A_i\}$. Furthermore, for every regular expression R built from the elements of Ev, i.e., $R \in REX(Ev)$ (see Appendix B, B.4), let $Ev(R)$ denote the set of all elements of Ev occurring in R. For instance, if $Ev = \{a, b, c\}$, $Ev(a; b, a) = \{a, b\}$.

The class $SEPATH(Ev_1, \ldots, Ev_n, Ev_{\bigtriangledown})$ of **synchronized e-paths** is defined by the following grammar:

$$sepath = \underline{epath}\ sequence_{\bigtriangledown}\ \underline{end}$$
$$sequence_{\bigtriangledown} = element_{\bigtriangledown} \mid (sequence_{\bigtriangledown}; element_{\bigtriangledown})$$
$$element_{\bigtriangledown} = event \mid (\bigtriangledown; sequence_{\bigtriangledown}) * \mid (sequence_{\bigtriangledown}, (\bigtriangledown; sequence_{\bigtriangledown}))$$

where *event* stands for any element of $Ev \setminus Ev_{\bigtriangledown}$ (i.e., $event \neq nil$), and $\bigtriangledown$ stands for an element of $Ev_{\bigtriangledown}$, and the following conditions are satisfied:

(1) for each loop $(\bigtriangledown; sequence_{\bigtriangledown})* : (\forall b \in Ev(sequence_{\bigtriangledown}))i(b) \subseteq i(\bigtriangledown)$

(2) for each alternative $(sequence^1_{\bigtriangledown}, (\bigtriangledown; sequence^2_{\bigtriangledown}))$:

$$(\forall b \in Ev(sequence^1_{\bigtriangledown}) \cup Ev(sequence^2_{\bigtriangledown}))i(b) \subseteq i(\bigtriangledown)$$

(3) synchronization guards are unique, i.e., for each

$P \in SEPATH(Ev_1, \ldots, Ev_n)$, each $\nabla \in Ev_\nabla$ may occur at most once.

We shall write just $SEPATH$ rather than $SEPATH(Ev_1, \ldots, Ev_n, Ev_\nabla)$ if it does not lead to confusion.
A concurrent e-path $CP = (P, Ev_1, \ldots, Ev_n)$ is called a **synchronized concurrent e-path**, $CP \in SCEPATH$, if there is $Ev_\nabla \subseteq Ev_1 \cup \ldots \cup Ev_n$ such that $P \in SEPATH(Ev_1, \ldots, Ev_n, Ev_\nabla)$.
For example $CP_3' = (\underline{epath}\ (\nabla; a; b)^*\ \underline{end}, \{\nabla, a\}, \{\nabla, b\})$,
$CP_4' = (\underline{epath}\ (a; c; e), (\nabla; b; d; f)\ \underline{end}, \{\nabla, a, e, b, f\}, \{\nabla, c, e, d, f\})$ both belong to the class $SCEPATH$.
The following theorem characterizes the most important property of synchronized concurrent e-paths.

Theorem 4.3 (Properness of Synchronized Concurrent e-Paths)
Every synchronized concurrent e-path from the class $SCEPATH$ is proper. □

The proof by induction on the structure of e-paths from the class $SEPATH$ is given in Appendix G.
Now we may define the transformation Π. The mapping Π is from the set of all *nil-free* concurrent e-path into $SCEPATH$ and it inserts synchronization guards as described above. Formally, this mapping is defined as follows.
Let $CP = (P, Ev_1, \ldots, Ev_n)$ be a concurrent e-path such that P can be derived from the following grammar:

$$\begin{aligned}
epath &= \underline{epath}\ sequence\ \underline{end}\\
sequence &= orelement | (orelement; orelement)\\
orelement &= starelement | (starelement, starelement)\\
starelement &= element | element*\\
element &= event | (sequence)
\end{aligned}$$

where *event* differs from *nil*.
The difference is that now neither "$a; b; c$" are valid sequences, they must be replaced by "$(a; b); c$" or "$a; (b; c)$", and by "$(a, b), c$" or "$a, (b, c)$" respectively. The other restriction is that the event *nil* is not allowed in P. Clearly every *nil*-free e-path can be represented in this manner, so the restriction is rather symbolic. Let G_∇ be an infinite set of events such that $G_\nabla \cap Ev = \emptyset$, where $Ev = Ev_1 \cup \ldots \cup Ev_n$. The elements of G_∇ are synchronization guards to be inserted and they are denoted by ∇. Let $SEQUENCE$ denote the set of all strings which can be derived from the non-terminal sequence of the above grammar. Elements of $SEQUENCE$ can be interpreted as some kind of regular expressions.
First we define the mapping Φ from the set $SEQUENCE$ into $SEPATH$ and the mappings $G_1, \ldots, G_n$ from the set $SEQUENCE$ into subsets of G_∇. Let

E, E_i, $i = 1,2$, denote strings which can be derived from the non-terminals *sequence*, *orelement*, *starelement* and *element* of the above grammar.
The mappings Φ and $G_1, \ldots, G_n$ are defined inductively:

(1) for every event a, $\Phi(a) = a$, $G_i(a) = \emptyset$, $i = 1, \ldots, n$.

(2) for every $(E_1; E_2) \in SEQUENCE$, $\Phi((E_1; E_2)) = (\Phi(E_1); \Phi(E_2))$ and $G_i((E_1; E_2)) = G_i(E_1) \cup G_i(E_2)$ for $i = 1, \ldots, n$.

(3) for every $(E_1, E_2) \in SEQUENCE$,
$\Phi((E_1, E_2)) = (\Phi(E_1), (\bigtriangledown; \Phi(E_2)),$
and

$$G_i((E_1, E_2)) = \begin{cases} G_i(E_1) \cup G_i(E_2) \cup \{\bigtriangledown\} & \text{if } (Ev(E_1) \cap Ev(E_2)) \cap Ev_i \neq \emptyset \\ G_i(E_1) \cup G_i(E_2) & \text{otherwise} \end{cases}$$

for $i = 1, \ldots, n$, where $\bigtriangledown$ is any element of $G_{\bigtriangledown}$ such that
$\bigtriangledown \notin \bigcup_{i=1}^{n} (G_i(E_1) \cup G_i(E_2))$.

(4) for every $E* \in SEQUENCE$, $\Phi(E*) = (\bigtriangledown; \Phi(E))*$
and

$$G_i(E*) \begin{cases} G_i(E) \cup \{\bigtriangledown\} & \text{if } Ev(E) \cap Ev_i \neq \emptyset \\ G_i(E) & \text{otherwise} \end{cases}$$

for $i = 1, \ldots, n$, where $\bigtriangledown$ is any element of $G_{\bigtriangledown}$ such that $\bigtriangledown \notin \bigcup_{i=1}^{n} G_i(E)$.

The set $G_i(E)$ contains synchronization guards which will be added to Ev_i for $i = 1, \ldots, n$.
Finally we define $\Phi(\underline{epath}\ E\ \underline{end}) = \underline{epath}\ \Phi(E)\ \underline{end}$, and $G_i(\underline{epath}\ E\ \underline{end}) = G_i(E)$ for $i = 1, \ldots, n$. The concurrent e-path $\Pi(CP)$ is defined as follows:

$$\Pi(CP) = (\Phi(P), Ev_1 \cup G_1(P), \ldots, Ev_n \cup G_n(P))$$

For example for $CP_5 = (\underline{epath}\ (a, b), c\ \underline{end}, \{a\}, \{b\}, \{c\})$, we have

$$\Pi(CP_5) = (\underline{epath}\ (a, (\bigtriangledown_1; b)), (\bigtriangledown_2; c)\ \underline{end}, \{a, \bigtriangledown_1, \bigtriangledown_2\}, \{b, \bigtriangledown_1, \bigtriangledown_2\}, \{c, \bigtriangledown_2\})$$

The transformation Π is partial as it requires the e-path P in $CP = (P, Ev_1, \ldots, Ev_n)$ to have some special form. However, it may easily be extended to any *nil*-free concurrent e-path. If P is not in the required special form, it may be transformed into such a form by adding some additional parentheses. To guarantee nonambiguity of this auxiliary transformation, we may assume that the parentheses are added starting from the left. For instance $\underline{epath}\ a, b, c\ \underline{end}$ will be transformed into $\underline{epath}\ (a, b), c\ \underline{end}$. Thus we may say that the transformation Π is defined for all *nil*-free concurrent e-paths.
From the definition of Π it immediately follows that $\Pi(CP)$ is a synchronized concurrent e-path.

Corollary 4.2 *For every concurrent nil-free e-path CP we have:*
(1) $\Pi(CP) \in SCEPATH$.
(2) $\mathcal{C}(\Pi(CP))$ *is proper.* □

Recall that in Section 2.2.1 we have defined the erasing homomorphism h_B which erases all occurrences of events which do not belong to the set B. Now we extend h_B to vector sequences by defining $h_B(\underline{x}) = \underline{h_B(x)}$. The following result follows immediately from Theorem 4.1.

Corollary 4.3 *Let* $CP = (P, Ev_1, \ldots, Ev_n)$ *be a concurrent e-path with* $Ev = Ev_1 \cup \ldots \cup Ev_n$. *Then:* $VFS(CP) = h_{Ev}(VFS(\mathcal{C}(\Pi(CP))))$ *and* $RVFS(CP) = h_{Ev}(RVFS(\mathcal{C}(\Pi(CP))))$. □

Therefore, for a sequential specification of a system in the form of a concurrent e-path CP, we can first apply the transformation Π to get a proper concurrent e-path $\Pi(CP)$, and then apply the transformation $\mathcal{C}$ to get an equivalent concurrent specification $Pr \in \mathcal{C}(\Pi(CP))$.
Applying this procedure to the e-path EP_2 from the introductory part of this chapter we obtain:

$$\begin{array}{ll} \Pi(EP_2) = & \underline{epath}\ (\triangledown; xgt0; xminus1tox; yplus1toy)^*; xnotgt0\ \underline{end}, \\ \text{and } EPr'_2 = & \underline{epath}\ (\triangledown; xgt0; xminus1tox)^*; xnotgt0\ \underline{end} \\ & \underline{epath}\ (\triangledown; yplus1toy)^*\ \underline{end} \\ \text{where} & EPr'_2 \in \mathcal{C}(\Pi(CP_6)), \\ & CP_6 = (EP_2, \{xgto0, xminus1tox, xnotgto0\}, \{yplus1toy\}). \end{array}$$

One may easily prove that EPr'_2 indeed represents the addition $x + y$. For the concurrent e-path $CP_5 = (\underline{epath}\ (a, b), c\ \underline{end}, \{a\}, \{b\}, \{c\})$, as we have already shown, $\Pi(CP_5) = (\underline{epath}\ (a, (\triangledown_1; b)), (\triangledown_2; c)\ \underline{end}, \{a, \triangledown_1, \triangledown_2\}, \{b, \triangledown_1 \triangledown_2\}, \{c, \triangledown_2\})$, and for example

$$Pr_{48} = \begin{array}{l} \underline{epath}\ (a, \triangledown_1), \triangledown_2\ \underline{end} \\ \underline{epath}\ nil, (\triangledown_1; b), \triangledown_2\ \underline{end} \\ \underline{epath}\ nil, (\triangledown_2; c)\ \underline{end} \end{array}$$

belongs to $\mathcal{C}(\Pi(CP_5))$.

4.5.4 Final Comment and Bibliographic Notes

Until recently, the problem of deriving an equivalent concurrent system from a sequential specification has not been very popular in the computing community, and this still seems to be the case among theoreticians. However, there have been many recent research projects on this subject, see [Bac89, Bos88, Die88, Hey89, SA88] and others. To explain the reason for the interest in this kind of research we quote [Hey89]:

> "...there is an economic fact of life that cannot be ignored, namely that large companies with many hundreds of man-years invested in sequential software will not invest substantially in parallel hardware until there is a convincing 'migration route' to such machines...".

Unfortunately, most of those results are concerned with very specific hardware or software, so their general use is rather limited.

Some ideas concerning the approach presented here can be traced to [LS79], however the first explicit formulation of the problem and some early solutions originate from R. Janicki and C. Lengauer ([Jan81b, Jan81a, LH82] and [Len82]). Petri nets were used as a basic formal tool in [Jan81b] and [Jan81a], and the weakest preconditions ([Dij76]) were used in [LH82] and [Len82]. The COSY model, as a formal tool, was applied to find a transformation of sequential specifications into concurrent specifications in [Jan84a] and [Jan85b]. The notion of synchronized concurrent e-paths, Theorem 4.3 and the transformation Π follow from R. Janicki and T. Müldner [JM90]. Concepts similar to e-path and e-path program were considered in [Bes87]. The prototype programming language Banach, which is mainly due to T. Müldner, has been defined and analyzed in [Jan88, JM89, JM90]. Banach supports the form of specification of non-sequential systems we advocate in this chapter. It has been designed in such a way that the programmer does not have to be concerned about the synchronizatioon details. In Banach, the basic unit of specification is called an event. A Banach program consists of two parts; a description of the behaviour of events and a declaration of independent events. The semantics of Banach programs is based on the theory presented in this chapter.

4.6 Modelling N-Modular Redundancy

Modular redundancy is a form of replication of processing modules with majority voting which is one of the best known techniques for tolerating failures in concentrated processing. The need to ensure correct input-output behaviour and a high level of fault-masking in the case of real time systems, has led designers to consider the application of N-Modular Redundancy (NMR for short) in the construction of software.

In this subsection we focus on the application of the NMR technique in the area of distributed systems composed of a number of nodes, each node consisting of several modules performing the same computational task, connected by means of communication channels. One kind of major problem encountered in the design of such systems is the so-called sequencing failure, namely a failure of the modules within a node to process input requests in the same order. In this section we shall show how the COSY model can be used to specify and prove correct a fully distributed protocol allowing the establishment of synchronization points among replicated nondeterministic computations in such a way that sequencing failures do not occur. We shall first present in some detail the structure of an NMR system which we then discuss, and next we shall introduce a formal specification of the proposed synchronization mechanism for the NMR system.

4.6.1 Replicated Computation in Networks of Computing Nodes

Let us consider a network of computing nodes connected by communicating channels and communicating (amongst each other and the external environment) only by synchronous message passing over those channels. The communication channels are established by input and output ports, i.e. each node owns some input ports on which it can receive messages sent on corresponding output ports owned by other nodes; and it owns some output ports on which it can send messages to be received on corresponding input ports owned by other nodes. The behaviour of each node is cyclic and each particular single cycle can be described in the following way:

> "After receiving a message (computation request) on one of its input ports the node performs the required computational task and then sends result-messages on some of its output ports (one message per output port)."

The overall structure of the network (and its environment) can be represented by a pair $(G, æ)$ called a *communication graph*, where

$$G = (V \cup \{env\}, A)$$

is a *finite labelled directed* graph with the set of vertices $V \cup \{env\}$, the set of labelled arcs A, and æ is a *request-response* mapping. Elements of V represent the nodes of a network, env represents the external environment, and $env \notin V$. Let Π denote the set of all channel names. Each labelled arc from A is a triple (v, p, v') where $v, v' \in V \cup \{env\}$, and $p \in \Pi$, i.e. we can say that $A \subseteq (V \cup \{env\}) \times \Pi \times (V \cup \{env\})$. Let $\Pi' \subseteq \Pi$ be defined as $\Pi' = \{p | (\exists v \in V \cup \{env\})(\exists v' \in V)(v, p, v') \in A\}$, i.e. elements of Π' are labels of arcs which lead to elements of V.

The *request-response mapping* æ is defined as: $æ : \Pi' \longrightarrow 2^{\Pi} \setminus \{\emptyset\}$. It is additionally required that the pair $(G, æ)$ satisfies the following conditions:

(1) for every $p \in \Pi$ there is exactly one arc labelled with p, and if $(v, p, v') \in A$ then $v \neq v'$.

(2) for each $v \in V \cup \{env\}$ the number of outgoing arcs is greater than zero, and the number of incoming arcs is also greater than zero.

$$æ(a) = \{b, c\}$$
$$æ(b) = \{d\}$$
$$æ(c) = \{e, f\}$$
$$æ(d) = \{e, f\}$$

$$\Pi = \{a, b, c, d, e, f\}$$
$$\Pi' = \{a, b, c, d\}$$

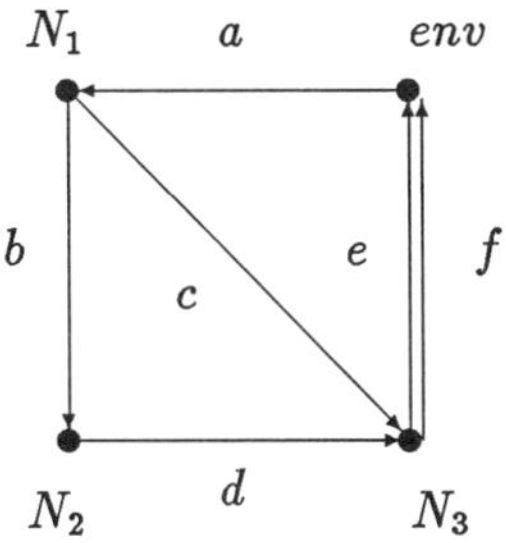

Figure 4.20. Example of pair $(G, æ)$

(3) for every $N \in V$ and $p \in In_N$, $æ(p) \subseteq Out_N$, where $In_N = \{q|(v, q, N) \in A\}$, $Out_N = \{q|(N, q, v) \in A\}$.

(4) for every $N \in V$ and $q \in Out_N$ there is $p \in In_N$ such that $q \in æ(p)$.

The above conditions are interpreted as follows:

(1) says that in the system each communication channel joins two distinct nodes,

(2) guarantees that each node has at least one input and at least one output port,

(3) means that a node sends results on the output ports it posesses, and

(4) says that there are no idle output ports.

Finally, a $p \in \Pi$ is an *external* input (output) port/channel if $(env, p, N) \in A$ (or $(N, p, env) \in A$) for some $N \in V$. A simple example of a communication graph is presented in Fig. 4.20.

The communication graph represents the structure of the non-replicated system of computing nodes. Its replicated version is obtained by replacing each node $N \in V$ (see Fig. 4.21 (a)) by n modules $M1, \ldots, Mn$ where $n \geq 2$ is a fixed integer (see Fig. 4.21 (b)), each module having the same set of ports as N and performing the same computation as N. The result is an NMR-node composed of n autonomously operating modules. The idea is to replicate each single computation carried out by N by n independent computations, the results of which can then be compared by a decision algorithm (residing in the nodes receiving the results from N) giving a single, more reliable result. In most applications the decision algorithm is a kind of majority voting on the n results. Due to the replication we no longer have a single request-message being received by a module, but rather n result messages sent by the modules of a replicated node. The generic module Mi (Fig. 4.21 (c)) is thus further composed of a pair

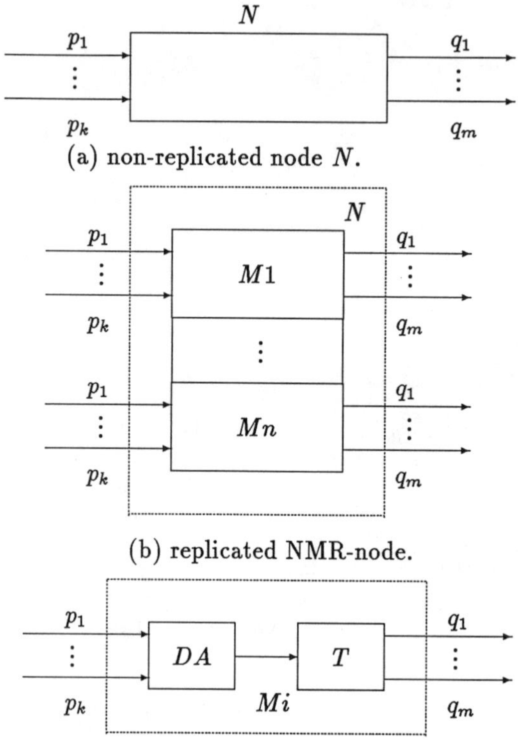

(a) non-replicated node N.

(b) replicated NMR-node.

(c) internal structure of module in replicated node.

Figure 4.21. Detailed structure of a node N

of subcomponents DA and T, corresponding to the decision algorithm and the computation performed by the module. We also assume that the message passing communication between the modules is synchronous and one-to-many, i.e. a message sent on an output port labelled by p is simultaneously and instantaneously received on all input ports labelled with the same p.

The behaviour of a module with the internal structure shown in Fig. 4.21 (c) can be expressed in the following way:

> "After receiving n messages at an input port p_i, the decision algorithm DA produces a single result message which is processed by the task T."

After the completion of the computational task, M sends result messages to the output ports æ(p_i). A correct replicated system must behave (in terms of the output messages sent to the external environment in response to the request messages received from it) like the original non-replicated system working without any faults. It turns out that a sufficient condition for this (in addition to appropriate assumptions about the reliability of modules) is that every NMR node in the network satisfies the following requirement:

> **REQ:** All modules of a replicated process input requests in an identical order.

In a distributed system, meeting requirement REQ is a surprisingly difficult task, and this difficulty is basically the result of non-determinism present in distributed message passing systems. In the sequel we will present a suitable distributed synchronization mechanism which will guarantee the satisfaction of the above requirement.

4.6.2 Formal Specification

Let us consider an NMR system represented by a communication graph $(G, æ)$. The basic kind of event we will use is:

m_{ip}- a message sent at output port p by the i-th module of the NMR node N such that $p \in Out_N$.

If p is an external input port, then m_{ip} denotes the i-th request sent by the external environment. We remark that in order to maintain the consistency of the whole system, the environment sends n replicated request messages and receives n result messages.

Suppose that M is a module from Fig. 4.21 (c). To specify its behaviour we need the following other types of events, where $p \in \{p_1, \ldots, p_k\}$:

$accept_{M:p}$- acceptance test performed by the decision algorithm DA in M on n messages received at input port p,

$result_{M:p}$- single decision result message (derived from n messages received at p) sent by DA to the task T in M,

$set_{M:p}$- the setting of a flag which allows M to receive the n-th message at p (this is a synchronization event),

$reset_{M:p}$- the resetting of the flag set by $set_{M:p}$ (another synchronization event),

$task_{M:p}$- computation in T in response to messages received at p.

The behaviour of M is then specified by the following sequence of single paths:

$$Module(M) =$$
$$IN_{M:p_1} \ldots IN_{M:p_k} FLAG_{M:p_1 \ldots p_k} TASK_{M:p_1 \ldots p_k} OUT_{M:in(q_1)q_1} \ldots OUT_{M:in(q_m)q_m}$$

where

$IN_{M:p} =$
$\underline{path}\ \Lambda_{1p}; \ldots; \Lambda_{n-1p}; set_{M:p}; (reset_{M:p}; set_{M:p})*;$
$\Lambda_{np}; accept_{M:p}; result_{M:p}; reset_{M:p}\ \underline{end}$
$FLAG_{M:p_1 \ldots p_k} =$
$\underline{path}\ (set_{M:p_1}, \ldots, set_{M:p_k}); (reset_{M:p_1}, \ldots, reset_{M:p_k})\ \underline{end}$
$TASK_{M:p_1 \ldots p_k} =$
$\underline{path}\ (result_{M:p_1}; task_{M:p_1}), \ldots, (result_{M:p_k}; task_{M:p_k})\ \underline{end}$
$OUT_{M:in(q)q} =$
$\underline{path}\ (task_{M:r_1}, \ldots, task_{M:r_l}); m_{iq}\ \underline{end}$
$in(q) = æ^{-1}(q) = \{r_1, \ldots, r_l\},$
$\Lambda_{1p} = \ldots = \Lambda_{n-1p} = \Lambda_{np} = (m_{1p}, \ldots, m_{np}).$

To make the notation shorter we shall also write $TASK_M$ and $FLAG_M$ instead of $TASK_{M:p_1 \ldots p_k}$ and $FLAG_{M:p_1 \ldots p_k}$, respectively.

The intuitive meaning of the single paths specifying module M might be expressed as follows:

> $IN_{M:p}$ – (Input): specifies the behaviour of the input port p and additionally that after the receipt of n messages DA performs the acceptance test and sends the single result to T,
>
> $FLAG_{M:p_1 \ldots p_k}$ – (Flags Manager): allows at most one of the k flags (each one assigned to a different input port p_i) to be set at any time; this prevents the module M from receiving two sets of n messages at different input ports at any time,
>
> $TASK_{M:p_1 \ldots p_k}$ – (Computation Task): specifies the behaviour of T,
>
> $OUT_{M:in(q)q}$ – (Output): specifies the behaviour of the output port $q \in \{q_1, \ldots, q_m\}$.

Having provided the specification of a single module, we define the specification of an NMR node N, which is composed out of the modules $M1, \ldots, Mn$, as the following sequence of single paths:

$$Node(N) = Module(M1) \ldots Module(Mn).$$

Finally assuming that:

$$V = \{N_1, \ldots, N_s\}, \text{ and } \bigcup_{N \in V} In_N = \{p_1, \ldots, p_t\},$$

our COSY specification of the NMR-system $(G, æ)$ is the following path program:

$NMRSYST =$
$\underline{program}\ Node(N_1) \ldots Node(N_s) Syn(p_1) \ldots Syn(p_t)\ \underline{endprogram}$

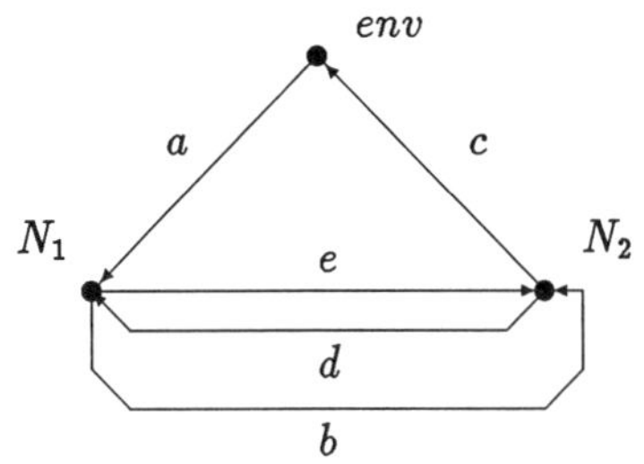

$$\text{æ}(a) = \{b\}$$
$$\text{æ}(b) = \{c, d\}$$
$$\Pi = \{a, b, c, d, e\} \qquad \text{æ}(d) = \{e\}$$
$$\Pi' = \{a, b, d, e\} \qquad \text{æ}(e) = \{c\}$$

Figure 4.22. Simple communication graph

where each path $Syn(p)$ specifies that on each input port p messages are received in sequences of n messages, each being sent by a different module owning the output port p. For every p,

$$Syn(p) = \underline{path}\ Expr(m_{1p}, \ldots, m_{np})\ \underline{end}$$

where $Expr(m_{1p}, \ldots, m_{np})$ is any regular expression written in COSY syntax which generates the language of all strings $a_1 a_2 \ldots a_n$ such that $a_i \in \{m_{1p}, \ldots, m_{np}\}$ and $a_i \neq a_j$ for $i \neq j$. For example $Expr(m_{1p}, m_{2p})$ might be defined as $(m_{1r}; m_{2r}), (m_{2r}; m_{1r})$.

To illustrate the concepts introduced above, let us analyze the following example. Let $(G, \text{æ})$ be as in Fig. 4.22 and let $n = 2$. Moreover, let A and B denote respectively the first and second module of N_1, and let C and D denote respectively the first and second module of N_2. The formal specification of $(G, \text{æ})$ is given by the following path program:

$$
\begin{array}{l}
NMRSYST = \\
\underline{program} \\
Syn(a)Syn(b)Syn(c)Syn(d)Syn(e) \\
IN_{A:a}IN_{A:d}FLAG_{A:ad}TASK_{A:ad}OUT_{A:\{a\}b}OUT_{A:\{d\}e} \\
IN_{B:a}IN_{B:d}FLAG_{B:ad}TASK_{B:ad}OUT_{B:\{a\}b}OUT_{B:\{d\}e} \\
IN_{C:b}IN_{C:e}FLAG_{C:be}TASK_{C:be}OUT_{C:\{b,e\}c}OUT_{C:\{b\}d} \\
IN_{D:b}IN_{D:e}FLAG_{D:be}TASK_{D:be}OUT_{D:\{b,e\}c}OUT_{D:\{b\}d} \\
\underline{endprogram}
\end{array}
$$

We shall now show that the specification presented above satisfies the fundamental requirement REQ introduced previously. For this we need some auxiliary notions.

Let $Pr = NMRSYST = P_1 \ldots P_m$, and let *Result* denote the set of all events of $Ev(Pr)$ which represent messages sent from the decision algorithms to the computational tasks, i.e. those which have the form: $result_{M:p}$.

For all $\underline{x} \in VFS(Pr)$ and $1 \leq i \leq m$, let $\underline{x}|P_i$ denote $pr_i(\underline{x})$.

Let $\vartheta : Ev(Pr)^* \longrightarrow \Pi^*$ denote the following homomorphism:

$$(\forall a \in Ev(Pr))\vartheta(a) = \begin{cases} \varepsilon & a \notin Result \\ p & a = result_{M:p} \in Result. \end{cases}$$

We remark that $\vartheta(\underline{x}|TASK_M)$ captures the sequence of input requests processed by module M throughout the history $\underline{x}$. This leads to the following formulation of the requirement REQ.

Formal definition of REQ: A history $\underline{x} \in VFS(Pr)$ satisfies REQ if for every NMR node and for every pair M, M' of modules of that node, **one** of the following conditions is satisfied:

(1) $\vartheta(\underline{x}|TASK_M) = \vartheta(\underline{x}|TASK_{M'})$,

(2) $(\exists p \in \Pi)\vartheta(\underline{x}|TASK_M) = \vartheta(\underline{x}|TASK_{M'})p$ or

$$\vartheta(\underline{x}|TASK_{M'}) = \vartheta(\underline{x}|TASK_M)p.$$

In other words, if a history satisfies the above definition then either all the n modules of each particular NMR node performed the same sequences of requests, or those sequences differ in that there is one request (represented by p) which has not been processed by a subset of modules of that node. In such a case p is the next request to be performed by the modules in that subset.

Theorem 4.4

Each history $\underline{x} \in VFS(Pr)$ satisfies REQ.

Proof. Let $\underline{x} \in VFS(Pr)$, and let M and M' ($M \neq M'$) be modules of the same NMR node. For every $K \in \{M, M'\}$, and for every $p \in In_K$ we define $\Xi_{K:p}(\underline{x})$ as follows. Let $y = \underline{x}|IN_{K:p}$. One may unambiguously decompose y as $y = wz$, where w is the maximal prefix of y which belongs to $Cyc(IN_{K:p})^*$. Then we can define $\Xi_{K:p}(\underline{x})$ to be the number of occurences of events representing messages (i.e. $m_{1p}, \ldots, m_{np}$) within z.

Clearly, $0 \leq \Xi_{K:p}(\underline{x}) \leq n$. Intuitively, $\Xi_{K:p}(\underline{x})$ is the number of messages waiting for the acceptance test after being received at p. One may observe that the single path $FLAG_K$ guarantees that:

$$(1)\ (\forall q \neq p)\Xi_{K:p}(\underline{x}) = n \Rightarrow \Xi_{K:q}(\underline{x}) < n$$

i.e., K cannot receive n messages at two different input ports at any time. Furthermore, for every $p \in In_M = In_{M'}$ we have:

$$\begin{array}{ll} (2) & -1 \leq \#_p(\vartheta(\underline{x}|TASK_M)) - \#_p(\vartheta(\underline{x}|TASK_{M'})) \leq 1 \\ (3a) & \#_p(\vartheta(\underline{x}|TASK_M)) - \#_p(\vartheta(\underline{x}|TASK_{M'})) = 1 \Rightarrow \Xi_{M':p}(\underline{x}) = n \\ (3b) & \#_p(\vartheta(\underline{x}|TASK_{M'})) - \#_p(\vartheta(\underline{x}|TASK_M)) = 1 \Rightarrow \Xi_{M:p}(\underline{x}) = n \end{array}$$

Note that the above conditions $(2), (3a)$ and $(3b)$ follow essentially from the fact that single paths $IN_{M:p}$ and $IN_{M':p}$ share events $m_{1p}, \ldots, m_{np}$. We can now prove the theorem by induction on $length(\underline{x})$. Clearly, $\vartheta(\underline{\varepsilon}|TASK_M) = \varepsilon = \vartheta(\underline{\varepsilon}|TASK_{M'})$. Assume that $\underline{x} \in VFS(Pr)$ satisfies REQ. Let $\underline{xa} \in VFS(Pr)$, $a \in Ev(Pr)$. We have to prove that $\underline{xa}$ satisfies REQ. Clearly, if $a \notin B = (Result \cap (Ev(TASK_M) \cup Ev(TASK_{M'})))$, then $\underline{xa}$ satifies REQ, so let us assume that $a \in B$. If $\vartheta(\underline{x}|TASK_M) = \vartheta(\underline{x}|TASK_{M'})$, then clearly $\underline{xa}$ satisfies REQ. Otherwise, by the induction hypothesis, we may assume without loss of generality that $\vartheta(\underline{x}|TASK_M) = \vartheta(\underline{x}|TASK_{M'})p$, where each $p \in In_M$. Hence, by the condition $(3a)$ above, $\Xi_{M':p}(\underline{x}) = n$.

We now consider two cases.

Case 1: $a = result_{M:q}$. By condition (2), $p \neq q$. Thus, by $(3a)$, $\Xi_{M':q}(\underline{xa}) = n$. On the other hand, $\Xi_{M':p}(\underline{xa}) = \Xi_{M':p}(\underline{x}) = n$, which contradicts condition (1).
Case 2: $a = result_{M':q}$. By condition (1), $\Xi_{M':r}(\underline{x}) < n$ for every $r \neq p$. Thus, $q = p$.
Hence, we have shown that $a = result_{M':p}$. Consequently, $\vartheta(\underline{xa}|TASK_{M'}) = \vartheta(\underline{x}|TASK_{M'})p = \vartheta(\underline{xa}|TASK_M)$. Therefore, $\underline{xa}$ satisfies REQ. □

Another property of the specification presented is that if the flow of messages in the replicated part of the communication graph can be represented by an acyclic graph, then $NMRSYST$ is an adequate path program.

We say that a communication graph $(G, æ)$ is **acyclic** if the directed graph $T_{(G,æ)} = (\Pi, A)$, where $A = \{(p,q) | p \in \Pi' \wedge q \in æ(p)\}$, does not contain a sequence of vertices $p_1, \ldots, p_s$ such that $(p_1,p_2), (p_2,p_3), \ldots, (p_{s-1},p_s), (p_s,p_1)$ are all arcs in $T_{(G,æ)}$.

For example, the NMR-system $(G, æ)$ from Fig. 4.22 is acyclic, and the corresponding graph $T_{(G,æ)}$ is shown in Fig. 4.23.

Theorem 4.5

If $(G, æ)$ is acyclic, then $NMRSYST$ is an adequate path program.

Proof. (Sketch) Let $NMRSYST = Pr = P_1 \ldots P_m$. One can prove this theorem in two steps. First we show that:

$$(1)\ (\forall \underline{x} \in VFS(Pr))(\exists \underline{xz} \in VFS(Pr))(\forall i) pr_i(\underline{xz}) \in Cyc(P_i)^*,$$

i.e. $\underline{xz}$ is a history leading to the initial state of the system. The crucial points in proving (1) are:

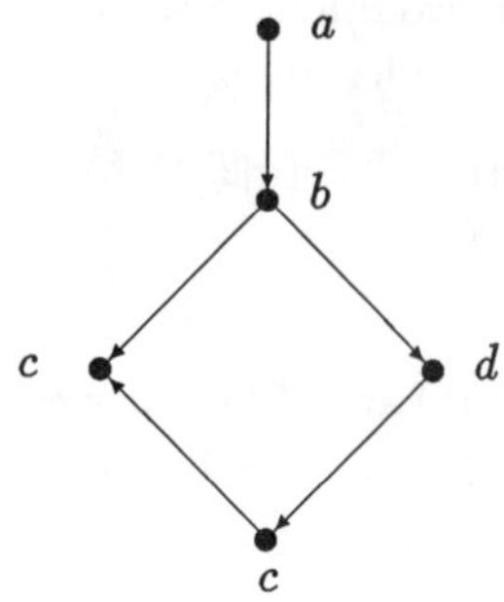

Figure 4.23. Message flow for system of Fig. 4.22

(i) $(G, æ)$ is acyclic,

(ii) the presence of $(reset_{M:p}; set_{M:p})*$ between $set_{M:p}$ and Λ_{np} in each single path $IN_{M:p}$ allows the flag $set_{M:p}$ to be released (if set) after each history $\underline{w}$ satisfying $\Xi_{M:p}(\underline{w}) = n-1$, and avoids any artificial deadlock situation.

In the second step we prove that starting from the "initial state" it is possible to perform each event, i.e.

$$(2)\ (\forall a \in Ev(Pr))(\exists \underline{v} \in Vev(Pr)^{*})\underline{va} \in VFS(Pr).$$

Notice that the above condition (2) follows essentially from $(G, æ)$ being acyclic, which implies that for every $p \in \Pi$ there is an external input port q such that there is a directed path from q to p in $T_{(G,æ)}$. Finally, from (1) and (2) we obtain adequacy of Pr. □

4.6.3 Bibliographical Notes

The use of COSY to model NMR systems originates from M. Koutny and L. Mancini [MK86, KM89] and the present subsection is based on their approach. An application of NMR techniques to NASA's Space Shuttle is described in [Skl76], and another important application can be found in [Gol80]. The COSY specification of NMR systems is not founded on, and does not suggest, any specific implementation. [MK86] presents two implementation techniques, one based on atomic actions, and the second using decentralized agreement. Some general information concerning the NMR technique can be found in [Avi85].

Chapter 5

Comparison of COSY with Other Models

In this chapter we compare COSY with other models of non-sequential systems. Besides Petri Nets, the Calculus of Communicating Systems (CCS) and Communicating Sequential Processes (CSP) have been developed to a great extent and sophistication. We devote the first two sections of this chapter to the analysis of the relationship between COSY and, respectively, CCS and CSP. In the third section we show how the process notation and the concept of priority can be modelled by appropriate classes of Petri nets. The fourth section shows the equivalence between vector sequences and Mazurkiewicz traces. The last section is an analysis of the relationship between COSY and Synchronized Behaviours.

5.1 COSY and CCS

The earliest and still one of the most complete models of sequential computations is the λ-calculus (which existed long before any notion of implementing a programming language). The Calculus of Communicating Systems (CCS for short) is an attempt to provide an analogous model for concurrent computations, carried out by communication among independent agents. The syntax of CCS contains operators expressing behavioural notions of sequentialization, nondeterminism and communicating concurrent behaviours. The theory of CCS is primarily concerned with an algebraic study of various derived notions of behavioural equivalence including a set of proof techniques.

The syntax of CCS may be defined in the following way. Let Δ be a set of **names**, $\overline{\Delta} = \{\overline{a} | a \in \Delta\}$ be a set of **co-names**, $\Delta \cap \overline{\Delta} = \emptyset$, and let $\Lambda = \Delta \cup \overline{\Delta}$. Elements of Λ are called **labels**. It is also assumed that $\overline{\overline{a}} = a$ for all $a \in \Lambda$. Basically labels are used to name actions (events) which may be performed. Let τ denote a special, invisible action (event), and assume that $\tau \notin \Lambda$. The set $\Lambda \cup \{\tau\}$ is called the set of **actions** (events). Finally, let rl denote any mapping $rl : \Lambda \cup \{\tau\} \longrightarrow \Lambda \cup \{\tau\}$ satisfying $rl(\overline{a}) = \overline{rl(a)}$ and $rl(a) = \tau \Leftrightarrow a = \tau$ for any

$a \in \Lambda \cup \{\tau\}$, and let $\mathcal{V}$ denote the set of all variables.

Now the abstract syntax for CCS expressions can be defined by the following little grammar:

$$P ::= nil \mid a.P \mid P + P \mid fix_i < X >< P > \mid P \,|\, P \mid P[c/b] \mid P \setminus A \mid X$$

where $a \in \Lambda \cup \{\tau\}$, $c, b \in \Lambda$, $A \subseteq \Delta$, $X \in \mathcal{V}$. The expression $fix_i < X >< P >$ is said to bind X, and a **CCS agent** is defined as any CCS expression without free variables.

A CCS agent is intuitively a device which may perform certain *sequences* of named *actions* (events). A CCS expression is a presentation of one such agent, in the same sense as a COSY program may be viewed as a presentation of a device which may perform certain sequences of vector events. The CCS expression *nil* is the CCS agent which simply does nothing. The CCS operator "." can be viewed as an operation expressing the performance of actions "in sequence". A CCS expression $a.P$, where P is an agent, defines a new agent which performs the action a first and the acts as P. For instance the CCS expression:

$$a.b.nil$$

represents the agent capable of "performing a first, b next and then nothing more". In terms of COSY this may be represented either by a sequence:

$$a; b$$

if this is part of a sequential component, or by an e-path (see Sect. 4.5.2):

$$\underline{epath}\ a; b\ \underline{end}$$

if this is the complete description of a sequential component.

Nondeterminism is expressed in CCS by the operator "+" (*summation*), so the new agent $P + Q$ can act as the agent P or the agent Q. Whether it acts as P or Q is determined non-deterministically. The CCS expression:

$$a.nil\ +\ b.nil$$

describes an agent capable of "*either* prforming a or b", and it corresponds to a COSY orelement (if this is part of a sequential component):

$$a, b$$

or to the e-path (if it is the whole sequential component):

$$\underline{epath}\ a, b\ \underline{end}.$$

The fix operator represents a recursion in CCS and is used to describe agents with infinite and unbounded behaviours. $fix_i < X >< P >$ is a CCS expression where $P = \{P_i | i \in \mathcal{I}\}$ is an indexed family of expressions, $\mathcal{I}$ is a set of indices, $X = \{X_i | i \in \mathcal{I}\}$ is an indexed family of distinct variables. Intuitively $fix_i < X >< P >$ stands for the i-th component of the solution (i.e. fixed point) of the equations: $X_i = P_i$, $i \in \mathcal{I}$.

In COSY mutual recursion does not appear, iteration is used to represent infinite behaviour instead. On the other hand, iteration does not appear explicitly in CCS. But since iteration can be regarded as a kind of simple form of recursion, it can easily be represented in CCS, and this also means that some recursion is implicitly representable in COSY. When $\mathcal{I} = \{1\}$, we shall omit the index and simply write $fix < X >< P >$. For instance the path:

$$\underline{path}\ a; b\ \underline{end}$$

which corresponds to the labelled Petri net:

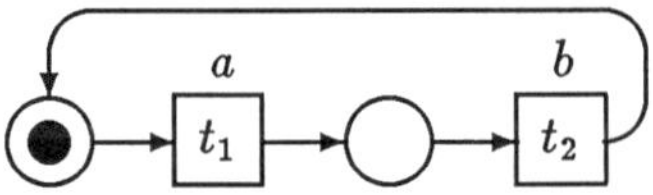

corresponds to the CCS agent:

$$fix < x >< a.b.x > .$$

In this case $fix < x >< a.b.x >$ may be thought of as the definition and call of a recursive agent named by the variable x.

So far we have only described a language for expressing non-deterministic behaviours. Parallelism is represented by an operator "$\mid$", called *composition*. Intuitively, $P \mid Q$ represents the behaviour of agents P and Q computing independently in parallel, and with *communication through complementary actions*. Without complementary actions P and Q are regarded as completely independent, hence the CCS agent:

$$fix < x >< a.b.x > \mid fix < x >< a.b.x >$$

corresponds to the following labelled Petri net:

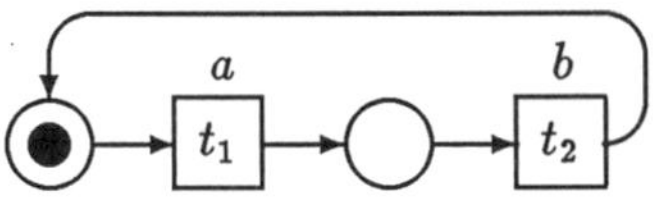

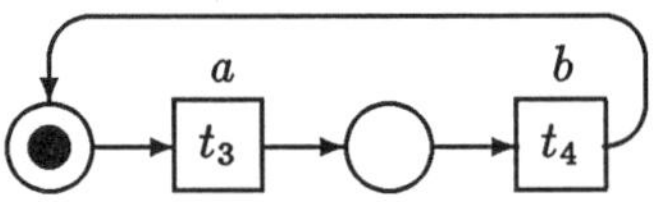

and, depending on the context, it may be represented by two paths:

$$\underline{path}\ a_1; b_1\ \underline{end} \quad \underline{path}\ a_2; b_2\ \underline{end}$$

or by two processes:

$$\underline{process}\ a; b\ \underline{end}\quad \underline{process}\ a; b\ \underline{end}.$$

The communication through complementary actions means that whenever P and Q are capable of performing actions "a" and "$\overline{a}$" respectively, then in $P \mid Q$ they *may* perform these actions together in one synchronized action, and as far as the behaviour of $P \mid Q$ is concerned, this joint action becomes an invisible internal action, represented by $\tau \notin \Lambda$.

Thus, the CCS agent:

$$fix < x >< a.c.nil > \mid fix < x >< b.\overline{c}.x >$$

corresponds to the following COSY paths:

$$\underline{path}\ a; c, \tau\ \underline{end}\quad \underline{path}\ b; \overline{c}, \tau\ \underline{end}$$

with the additional assumption that τ is a special synchronizing event.

Now one may ask how to represent in CCS the behaviours defined by the paths:

$$\underline{path}\ a; c\ \underline{end}\quad \underline{path}\ b; c\ \underline{end}$$

or, if you prefer synchronization to be invisible, by something like:

$$\underline{path}\ a; \tau\ \underline{end}\quad \underline{path}\ a; \tau\ \underline{end}.$$

To do this we can use the CCS operator called *restriction*. It has the form: $P \backslash A$, where $A \subseteq \Delta$, and the behaviour of this is exactly that of P, but forbidding any action from $A \cup \overline{A}$. Thus:

$$(fix < x >< a.c.x > \mid fix < x >< b.\overline{c}.x >) \setminus \{c\}$$

corresponds to:

$$\underline{path}\ a; c\ \underline{end}\quad \underline{path}\ b; c\ \underline{end}.$$

The last of the CCS operators is called *relabelling*. The intention of relabeling is nothing but changing the labels of actions of agents. The agent $P[c/b]$ represents the behaviour of P with all actions b renamed to c. For instance:

$$(fix < x >< a.b.x >)[c/a]$$

corresponds to the path:

$$\underline{path}\ c; b\ \underline{end}.$$

The standard semantics of CCS is defined in an operational manner. Let *Agents* denote the set of all CCS agents. Let us introduce a family of **action relations**, indexed by $\Lambda \cup \{\tau\}$, over *Agents*.

Intuitively the term

$$P \xrightarrow{a} P'$$

means that P may do a, and become P' in doing so. The action relations $\xrightarrow{a}$ are defined to be the smallest relations satisfying the rules in Fig. 5.1.

From the operational rules for CCS, the entire behaviour of any agent can be inferred and presented as a tree. For example consider:

$$P = fix < x >< a.c.x > \,|\, fix < x >< b.\bar{c}.x > .$$

Its derivation tree and action tree are represented in Fig. 5.2.

The CCS approach emphasizes the proper definition of an equivalence relation over agents. Intuitively, two agents are equivalent when their action trees are sufficiently "similar", and in the spirit of the CCS approach, the invisible actions, represented by arcs labelled τ, should be ignored to a certain extent in defining this similarity. These τ labelled arcs cannot be ignored entirely, since an invisible τ-action represents the ability to preempt alternatives. The two agents $a.nil + b.nil$ and $a.nil + \tau.b.nil$ define trees:

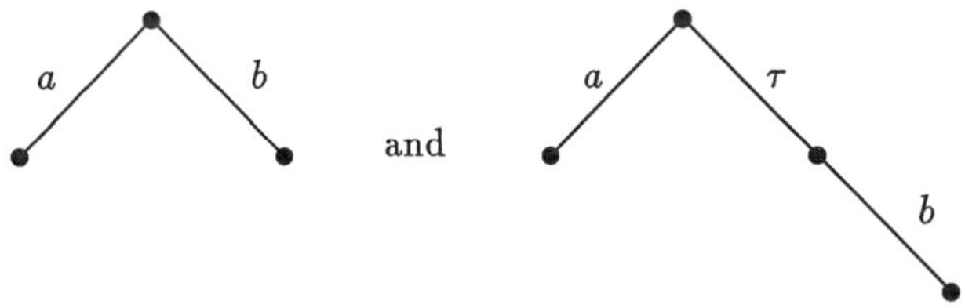

but these trees are not considered similar in CCS. The reason is that $a.nil + b.nil$ offers a and b as alternative actions, while $a.nil + \tau.b.nil$ can discard the b alternative by performing the invisible τ-action.

Roughly speaking the equivalence in CCS is expressed in terms of visible actions ($\neq \tau$) using the concept of **bisimulation**. For every $s = a_1 a_2 \ldots a_n \in \Lambda^*$, let $\stackrel{s}{\Rightarrow} \subseteq Agents \times Agents$ be the relation defined as follows:

$$\begin{array}{l}(\forall P, Q \in Agents): \\ P \stackrel{s}{\Rightarrow} Q \;:\Leftrightarrow P \xrightarrow{\tau}{}^* \xrightarrow{a_1} \xrightarrow{\tau}{}^* \xrightarrow{a_2} \ldots \xrightarrow{\tau}{}^* \xrightarrow{a_n} \xrightarrow{\tau}{}^* Q.\end{array}$$

In particular $P \stackrel{\epsilon}{\Rightarrow} Q$ if and only if $P \stackrel{\tau}{\Rightarrow}{}^* Q$.

A relation $BIS \subseteq Agents \times Agents$ is said to be a **bisimulation** if and only if, for all $(P, Q) \in BIS$ and $s \in \Lambda^*$:

(a) whenever $P \stackrel{s}{\Rightarrow} P'$ then, for some Q', $Q \stackrel{s}{\Rightarrow} Q'$ and $(P', Q') \in BIS$,

(b) whenever $Q \stackrel{s}{\Rightarrow} Q'$ then, for some P', $P \stackrel{s}{\Rightarrow} P'$ and $(P', Q') \in BIS$.

P and Q are said to be **observationally equivalent**, and we write $P \sim_{BIS} Q$ if and only if there exists a bisimulation BIS such that $(P, Q) \in BIS$.

$$\textit{action}: \qquad a.P \xrightarrow{a} P$$

$$\textit{summation}: \qquad \frac{P \xrightarrow{a} P'}{P+Q \xrightarrow{a} P'} \qquad \frac{Q \xrightarrow{a} Q'}{P+Q \xrightarrow{a} Q'}$$

$$\textit{fix}: \qquad \frac{P\{fix_i <X><P>/X\} \xrightarrow{a} Q}{fix_i <X><P> \xrightarrow{a} Q} \quad i \in \mathcal{I}$$

$$\textit{composition}: \qquad \frac{P \xrightarrow{a} P'}{P \,|\, Q \xrightarrow{a} P' \,|\, Q} \qquad \frac{Q \xrightarrow{a} Q'}{P \,|\, Q \xrightarrow{a} P \,|\, Q'}$$

$$\frac{P \xrightarrow{a} P' \,,\; Q \xrightarrow{\bar{a}} Q' \,,\; a \neq \tau}{P \,|\, Q \xrightarrow{\tau} P' \,|\, Q'}$$

$$\textit{relabelling}: \qquad \frac{P \xrightarrow{a} Q}{P[rl] \xrightarrow{rl(a)} Q[rl]}$$

$$\textit{restriction}: \qquad \frac{P \xrightarrow{a} Q \,,\; a \notin A \cup \overline{A}}{P \setminus A \xrightarrow{a} Q \setminus A}$$

Note: In the rule for recursion $P\{E/X\}$ denotes the result of substituting expressions E for variables X simultaneously in P, with changes of bound variables as necessary to avoid clashes.

Figure 5.1. Operational rules for CCS

$$P_1 = fix < x >< a.c.x >$$
$$P_2 = fix < x >< b.\overline{c}.x >$$
$$P = P_1 \,|\, P_2$$

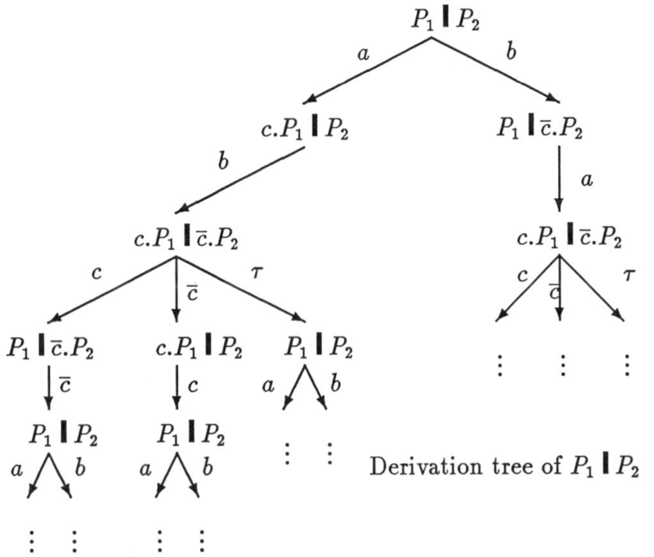

Derivation tree of $P_1 | P_2$

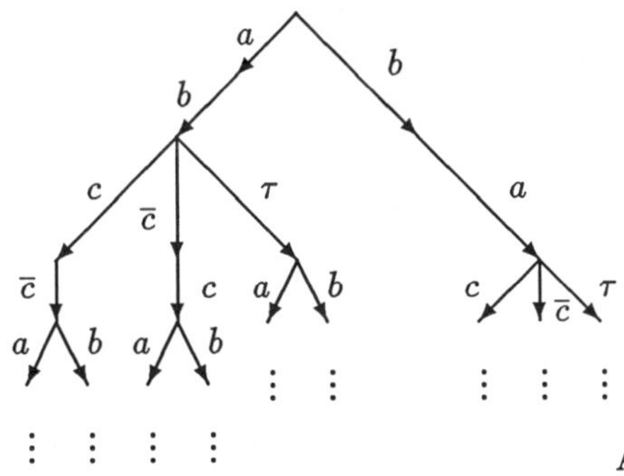

Action tree of $P = P_1 | P_2$

Figure 5.2. Derivation- and action-tree of $P = fix < x >< a.c.x > | fix < x >< b.\overline{c}.x >$

The bisimulations and observational equivalence turn out to have a number of pleasant properties, some of which are: bisimulations are closed under composition and inversion; observational equivalence is an equivalence relation and it is the largest bisimulation, i.e. $\sim_{BIS} = \bigcup\{R | R \text{ is a bisimulation}\}$.

Although both COSY and CCS start conceptually from the notion of *uninterpreted atomic event* (called action in CCS), which is capable of occurring whenever the synchronization constraints of its immediate context allows, their main concerns are different. First of all, standard CCS with observational equivalence is an interleaving theory, i.e. it represents parallelism in terms of nondeterminism. For instance, one can easily verify that:

$$a.nil \,|\, b.nil \;\sim_{BIS}\; a.b.nil + b.a.nil$$

while their COSY counterparts:

$$\underline{path}\ a\ \underline{end}\quad \underline{path}\ b\ \underline{end}$$

and:

$$\underline{path}\ (a;b),(b;a)\ \underline{end}$$

generate different behaviours, one "truly concurrent" the other "truly sequential".

The CCS agents can only be synchronized via complementary actions which when synchronized become an invisible action τ, and the concept of observational equivalence in CCS ignores (to a certain extent) the invisible actions. This means that this concept of equivalence is deliberately abstracting away from internal synchronization structure, which is *one of the main concerns* of the COSY theory.

COSY focusses on expressing and consequently proving properties of non-sequential systems such as absence of deadlock, adequacy (liveness), absence of starvation, etc.; on expressing, in an explicit way, the execution semantics, and it provides a treatment of "true concurrency" as Petri nets do also. CCS emphasizes the structuring of systems in terms of subsystems, the proper definition of an equivalence relation over agents, finding sound and complete axiomatizations, and applying modal logic to characterize this approach. Basically, our feeling is that the interleaving semantics of CCS renders the concerns listed above easier of accomplishment, however some of the "spirit of concurrency" is lost.

Finally, as we have already pointed out, in CCS, parallel composition " | " is connected with hiding. Whenever two communication actions synchronize, they yield an invisible internal action τ which cannot synchronize any further, i.e. every communication action a has a unique matching communication action $\overline{a}$, with $\overline{\overline{a}} = a$. As the philosophy of CCS is that only matching communication actions a and $\overline{a}$ can synchronize to yield τ, it allows only for binary synchronization.

CCS was originally presented in [Mil80] as a kind of programming language based on a notion of asynchronously computing agents with synchronous communication including value passing. Since then the theory of CCS has been developed considerably, mainly around more basic calculi, adopting the concept of an uninterpreted atomic event (action) as the most primitive component, and the concept of bisimulation as the main tool to develop equivalence relations. The version of CCS used here is based on [Mil84]. The concept of a bisimulation relation originates from [Par81]. A more theoretical basis for this kind of calculi can be found in [Mil83], where a synchronous calculus is presented first, and next it is shown how an asynchronous calculus, similar to, but more general than standard CCS, can be derived from the synchronous one. In [Nie86] the relationship between CCS and Petri nets is analyzed and used to define a partial order semantics of a subset of CCS. The complete description of CCS together with many examples can be found in the monograph [Mil89].

5.2 COSY and CSP

The original, version of CSP ([Hoa78a]) was a programming notation for distributed computing in which guarded commands ([Dij75]) were extended by parallel composition and handshake communication. The parallel composition operator had the form:

$$[P_1 :: C_1 || \ldots || P_n :: C_n]$$

where the P_i are unique identifiers for the C_i. The C_i are basically sequential programs representing processes with *disjoint* sets of local variables. The only allowed way of communication between the processes was via two specially designed communication commands: $P_i?x$ and $P_j!expr$, where x is a variable, $expr$ is an expression which may occur inside the C_j's. The command $P_i?x$ expresses a request for input from process P_i. The command $P_j!expr$ expresses a request to output the value of expression $expr$ to process P_j. These requests are granted only if simultaneously:

> $P_j!expr$ occurs in process P_i and P_i is ready to perform $P_j!expr$, and
>
> $P_i?x$ occurs in process P_j and P_j is ready to perform $P_i?x$.

The combined effect of the simultaneous execution of $P_j!expr$ in P_i and $P_i?x$ in P_j corresponds to the atomic assignment:

$$x := expr$$

of P_i's value of $expr$ to the local variable x in P_j. This is exactly a synchronized (handshake) communication between the processes P_i and P_j.

This original version of CSP generated a lot of theoretical problems, the most important seem to be difficulties in finding good compositional semantics and verification methods. These types of problems as well as an influence of other models such as COSY and CCS led to a more abstract version of CSP, called Theoretical CSP or TCSP. Instead of a concrete programming syntax with assignments, guarded commands and communication commands $P_i?x$ and $P_j!expr$, the abstract concept of uninterpreted atomic events (actions) became the central concept (as in COSY, CCS and Petri Nets). In the sequel we shall compare the basic concepts of TCSP with those of COSY.

Let Ev be a set of atomic *events*, $\mathcal{V}$ a set of *variables*. An abstract syntax for TCSP expressions can be given by the following grammar:

$$\begin{aligned} P \ ::= \ & stop \mid skip \mid a \ \rightarrow \ P \mid P;P \mid P \ \square \ P \mid \\ & P \ \sqcap \ P \mid P \ \|_A \ P \mid P \setminus A \mid X \mid P[b/a] \mid \mu X.P \end{aligned}$$

where $a, b \in Ev$, $A \subseteq Ev$, $X \in \mathcal{V}$.

The expression $\mu X.P$ is said to bind X, and a (TCSP) *process* is a TCSP expression without free variables.

Intuitively, *stop* represents a deadlocked process, and *skip* represents a process which terminates successfully. The process $a \rightarrow P$ behaves like: "first a then P". For instance the TCSP process

$$a \ \rightarrow \ b \ \rightarrow \ skip$$

represents the process which "performs a first, b next and then terminates", and corresponds to the COSY sequence:

$$a; b$$

if this is a part of a sequential component.

The process *stop* has no explicit equivalent expression in COSY, however it can be easily simulated. For instance the TCSP process:

$$a \ \rightarrow \ stop$$

may be expressed in COSY by two paths:

$$\underline{path}\ a; stop; pots\ \underline{end}\ \underline{path}\ pots; stop\ \underline{end}.$$

In both cases, after an occurrence of a the whole system deadlocks.

The TCSP operator "$\longrightarrow$" can thus be viewed as a way of expressing the performance of events "in sequence". Semicolon is the second TCSP operator which can be used to specify sequential composition. The TCSP process

$$P; Q$$

means "first P and if P terminates successfully then Q". The TCSP process:

$$(a \ \rightarrow \ skip); (b \ \rightarrow \ skip)$$

corresponds to the COSY e-path:

$$\underline{epath}\ a; b\ \underline{end}.$$

The COSY modelling of, for instance:

$$(a \rightarrow stop); (b \rightarrow skip)$$

is more tricky, since an equivalent of *stop* does not occur explicitly in COSY, but the two paths below may be viewed as a COSY equivalent of this process:

$$\underline{path}\ a; stop; pots; b\ \underline{end}\ \underline{path}\ pots; stop\ \underline{end}.$$

In TCSP there are two forms of choice. The first one, represented by:

$$P \square Q$$

and sometimes called "external nondeterminism", may be modelled by an orelement in COSY. For instance, the TCSP process:

$$(a \rightarrow skip)\square(b \rightarrow skip)$$

may either perform a and terminate successfully, or perform b and then terminate successfully, and it corresponds to:

$$a, b$$

if this is part of a sequential component.

The second form of choice in TCSP, represented by:

$$P \sqcap Q$$

sometimes called "internal nondeterminism", defines "uncontrollable" or "purely" nondeterministic choice. The difference between $\square$ and $\sqcap$ is rather subtle, and we shall discuss it later on.

The operator μ defines a recursion in TCSP, $\mu X.P$ represents the solution of the equation $X = P$. COSY does not allow for full recursion, only iteration (as some early versions of CSP), which can be regarded as a special case of recursion. The COSY path:

$$\underline{path}\ a; b\ \underline{end}$$

corresponds to the TCSP process:

$$\mu X.a \longrightarrow b \longrightarrow X.$$

Parallelism is represented in TCSP by the parallel comosition operator "$||_A$" with $A \subseteq Ev$. Intuitively, $P\ ||_A\ Q$ behaves like P and Q running independently of each other, except for all the events from the set A. The events from A are used to synchronize P and Q, and the rule (*handshake* synchronization) is similar to that of COSY. Every $a \in A$ may be performed only if it is allowed by both P and Q. For instance:

$$a \longrightarrow c \longrightarrow skip \, ||_{\{c\}} \, b \longrightarrow c \longrightarrow skip$$

corresponds to:

$$\underline{path}\ a; c\ \underline{end}\ \underline{path}\ b; c\ \underline{end}.$$

By varying its synchronization set A from $A = \emptyset$ to $A = Ev$, parallel composition $||_A$ ranges from complete independency $||_\emptyset$ to full synchrony $||_{Ev}$.

For instance:

$$\mu X.a \longrightarrow b \longrightarrow X \, ||_\emptyset \, \mu X.a \longrightarrow b \longrightarrow X$$

corresponds to the following Petri net:

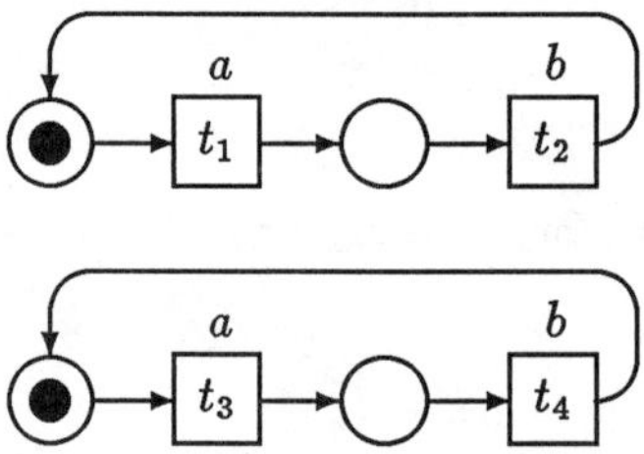

and in some contexts may be modelled by the following COSY processes:

$$\underline{process}\ a; b\ \underline{end}\ \underline{process}\ a; b\ \underline{end}.$$

Full synchrony means that $P||_{Ev}Q$ must always behave like P and Q, thus for instance:

$$\mu X.a \longrightarrow X \, ||_{Ev} \, \mu X.b \longrightarrow X$$

generates an empty behaviour. This case has no general COSY equivalent, however, if we assume additionally that events occurring in P and Q are identical, a case when the concept of full synchrony really makes sense, the operator $||_{Ev}$ can be modelled quite easily in COSY. For instance the TCSP process:

$$\begin{array}{ll} & \mu X.((a \longrightarrow c \longrightarrow X) \,\Box\, (\mu X.b \longrightarrow d \longrightarrow X)) \\ ||_{Ev} & \mu X.((c \longrightarrow a \longrightarrow X) \,\Box\, (\mu X.b \longrightarrow d \longrightarrow X)) \end{array}$$

corresponds to:

$$\underline{path}\ (a; c), (b; d)\ \underline{end}\ \underline{path}\ (c; a), (b; d)\ \underline{end}.$$

The fact that A may vary from $\emptyset$ to Ev gives the operator $||_A$ a lot of flexibility, which cannot always be achieved in COSY without introducing some transformations and additional compositions. For example, the TCSP process:

$$(\mu X.a \longrightarrow b \longrightarrow X)||_{\{b\}}(\mu X.a \longrightarrow b \longrightarrow X)$$

corresponds to the labelled Petri net below:

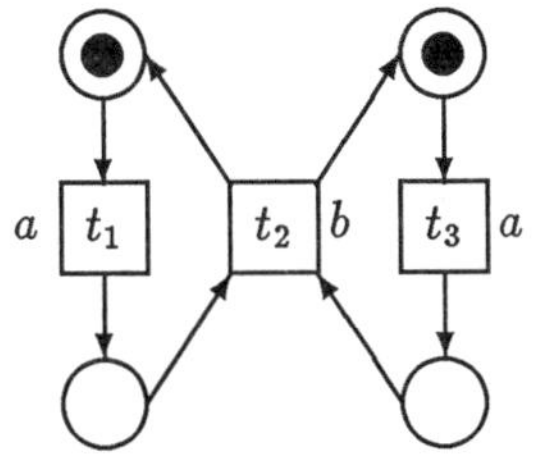

and to some degree of accuracy may be represented by the following paths:

$$\underline{path}\ a_1; b\ \underline{end}\ \underline{path}\ a_2; b\ \underline{end}.$$

Generally, it might be said that if the COSY path P_P models a TCSP process P, and the COSY path P_Q models a TCSP process Q, and $A = Ev(P) \cap Ev(Q)$, where $Ev(P)$, $Ev(Q)$ denote events occurring in P and Q respectively, then the paths $P_P P_Q$ model the TCSP process $P||_A Q$. The first path program we considered in this book, namely:

$$\underline{path}\ a; b, c\ \underline{end}\ \underline{path}\ (d; e)*; b\ \underline{end}$$

can be regarded as the COSY equivalent of the TCSP process:

$$\begin{array}{ll} & (\mu X.a \longrightarrow (b \longrightarrow skip \,\Box\, c \longrightarrow skip); X) \\ ||_{\{b\}} & \mu Y.(((\mu X.d \longrightarrow e \longrightarrow X) \,\Box\, skip); b \longrightarrow skip); Y) \end{array}$$

We now will show the distinction between $P \sqcap Q$ and $P \Box Q$. This distinction is quite subtle since both $P \sqcap Q$ and $P \Box Q$, considered apart from any environment, behave identically. However, it is possible to embed them in an environment in which $P \sqcap Q$ can deadlock at the first step, but $P \Box Q$ cannot. For example, let $P = \mu X.a \rightarrow X$, $Q = \mu X.b \rightarrow X$ and $Q_1 = \mu X.a \rightarrow b \rightarrow X$. Then:

$$(P \Box Q)||Q_1 = ((\mu X.a \longrightarrow X)\Box(\mu X.b \longrightarrow X))||_{\{a,b\}}(\mu X.a \longrightarrow b \longrightarrow X)$$

corresponds to:

$$\underline{path}\ a, b\ \underline{end}\ \underline{path}\ a; b\ \underline{end}$$

and clearly it does not deadlock. On the other hand:

$$(P \sqcap Q)||Q_1 = ((\mu X.a \longrightarrow X) \sqcap (\mu X.b \longrightarrow X))||_{\{a,b\}}(\mu X.a \longrightarrow b \longrightarrow X)$$

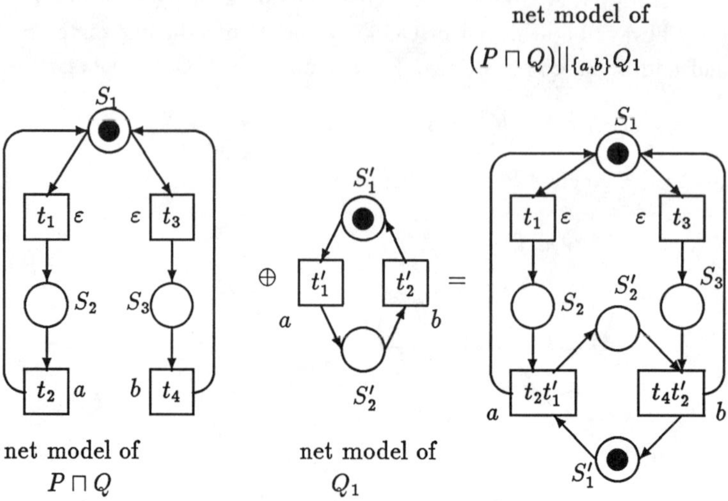

Figure 5.3. Net models of $P \sqcap Q$, Q_1, and $(P \sqcap Q)||_{\{a,b\}}Q_1$ with $P = \mu X.a \rightarrow X$, $Q = \mu X.b \rightarrow X$, and $Q_1 = \mu X.(a \rightarrow b \rightarrow X)$

may deadlock immediately if in $P \sqcap Q$ a nondeterministic internal choice is Q. Intuitively the TCSP process $P \sqcap Q$, Q_1 and $(P \sqcap Q)||_{\{a,b\}}Q_1$ correspond to the labelled Petri nets in Fig. 5.3.

Consequently $(P \sqcap Q)||_{\{a,b\}}Q_1$ might be modelled by the following e-path:

$$\underline{epath}\ ((nil;a),(nil;b))^*\ \underline{end}\ \underline{epath}\ (a;b)^*\ \underline{end}$$

which, when Petri net semantics is used, clearly may deadlock after performing *nil* which must be followed by b.

The *hiding* operation $P \backslash A$ conceals all occurrences of all events from the set A in P, and $P[b/a]$ is a *relabelling* where the TCSP process $P[b/a]$ represents the behaviour of P with all occurrences of the event a renamed to b. The semantics of TCSP may be defined in various ways. The original approach was the *sequence semantics* ([Hoa80], called *trace semantics* by the author, but we shall use the name *sequence* semantics, to distinguish it from the *traces* of Mazurkiewicz considered in Sect. 5.4), where the behaviour of any process is simply described by the set of all possible *sequences* of event occurrences it generates. We shall show how the sequence semantics can be introduced for some simplified version of TCSP. Let $\surd$ denote the event of successful termination, and let $Ev(P)$, the set of events of the process P be defined as follows:

1. $Ev(stop) = \emptyset$, $Ev(skip) = \{\surd\}$
2. $Ev(a \longrightarrow P) = \{a\} \cup Ev(P)$

3. $Ev(P;Q) = Ev(P\Box Q) = Ev(P \sqcap Q) = Ev(P||_A Q) = Ev(P) \cup Ev(Q)$
4. $Ev(P \setminus A) = Ev(P) \setminus A$
5. $Ev(P[b/a]) = (Ev(P) \setminus \{b\}) \cup \{a\}$
6. $Ev(\mu X.P) = Ev(P)$

We assume that $P||_A Q$ is allowed only if $A = Ev(P) \cap Ev(Q)$, and we shall write $P||Q$ in such a case. We also assume that if $\surd \in Ev(P)$ and $\surd \notin Ev(Q)$ then $P||Q$ is allowed only if $Ev(Q) \subseteq Ev(P)$. This will ensure that successful termination of P automatically cuts short any further activity of Q. We also assume that $\mu X.P$ is correct only if the equation $X = P$ has an unique solution (the least fixed point) in the domain of sets of sequences of event occurrences (i.e., if P is guarded [Hoa85]). We shall denote this solution by $fix_{sequences}(X = P)$. For more details the reader should refer to, for instance ([Hoa85]), here we mention only that if $P = a \longrightarrow b \longrightarrow X$, then we have an equation $X = a \longrightarrow b \longrightarrow X$ and

$$fix_{sequences}(X = a \longrightarrow b \longrightarrow X) = \{a,b\}^*\{\varepsilon, a\} = \{\varepsilon, a, ab, aba, abab, \ldots\}$$

For every set of sequences of event occurrences L, let

$$\mathcal{S}_{\surd}(L) = \{x \in L \mid \#_{\surd}(x) = 0\} \cup \{x\surd \in L | \#_{\surd}(x) = 0\}$$

The elements of $\mathcal{S}_{\surd}(L)$ belong to L and either contain no $\surd$ at all or the event $\surd$ occurs only once at the very end.
Let $seq(P)$ denote the set of all sequences of event occurrences generated by a TCSP process P. Under the above assumptions we may define $seq(P)$ as follows:

1. $seq(stop) = \{\varepsilon\}$, $seq(skip) = \{\varepsilon, \surd\}$
2. $seq(a \longrightarrow P) = \{\varepsilon\} \cup \{a\}seq(P)$
3. $seq(P;Q) = \{x \mid x \in seq(P) \wedge \#_{\surd}(x) = 0\} \cup \{xy \mid x\surd \in seq(P) \wedge y \in seq(Q)\}$
4. $seq(P\Box Q) = seq(P \sqcap Q) = seq(P) \cup seq(Q)$
5. $seq(P||Q) = \mathcal{S}_{\surd}(L_{P||Q})$, where $L_{P||Q} = \{x \in (Ev(P) \cup Ev(Q))^* \mid h_{Ev(P)}(x) \in seq(P) \wedge h_{Ev(Q)}(x) \in seq(Q)\}$.
6. $seq(P \setminus A) = h_{Ev(P\setminus A)}(seq(P))$
7. $seq(P[b/a]) = \{x[b/a] \mid x \in seq(P)\}$, where $x[b/a]$ is the string derived from x by replacing all occurrences of a by occurrences of b.
8. $seq(\mu X.P) = fix_{sequences}(x = P)$.

The sequence semantics as defined above has one advantage, it is relatively easy to define and to handle. However, it can distinguish neither between "external" and "internal" nondeterminism, nor between nondeterminism and concurrency. For instance the processes:

$$\begin{aligned} P_1 &= a \longrightarrow skip || b \longrightarrow skip, \\ P_2 &= a \longrightarrow b \longrightarrow skip \Box b \longrightarrow a \longrightarrow skip, and \\ P_3 &= a \longrightarrow b \longrightarrow skip \sqcap b \longrightarrow a \longrightarrow skip \end{aligned}$$

are indistinguishable since:

$$seq(P_1) = seq(P_2) = seq(P_3) = \{\varepsilon, a, b, ab, ba, ab\surd, ba\surd\}.$$

To distinguish formally between "internal" and "external" nondeterminism, as well as to support a more formal treatment of such properties as *deadlock* or *divergence*[1] the *failure* model of processes was introduced. Formally, the failures of a process are defined as a relation:

$$failures(P) = \{(s, X) | s \in seq(P) \wedge X \in refusals(P/s)\},$$

where for $X \subseteq Ev(P)$, P/s denotes the process P/s after engaging in events of sequence s, and $refusals(P/s)$ is the set of all refusals of the process P. Suppose that X is a set of events which are offered initially by the environment of a process P. If it is possible for P to deadlock on its first step when placed in this environment, we say that X is a *refusal* of P. Thus, if (s, X) is a failure of P, this means that P can engage in the sequence of events recorded by s, and then refuse to do anything more, in spite of the fact that its environment is prepared to engage in any of the events from X. We will not analyze the failure semantics to any great detail, but only show how refusals are defined for some simple processes, namely:

$$\begin{aligned} refusals(stop) &= \{X | X \subseteq Ev \wedge X \textit{ is finite}\}, \\ refusals(c \longrightarrow P) &= \{X | X \subseteq Ev(P) \setminus \{c\}\}, \\ refusals(P \sqcap Q) &= refusals(P) \cup refusals(Q), \\ refusals(P \Box Q) &= refusals(P) \cap refusals(Q). \\ refusals(P || Q) &= \{X \cup Y | X \in refusals(P) \wedge Y \in refusals(Q)\} \end{aligned}$$

Again, although COSY and TCSP use the notion of *uninterpreted atomic events* which is capable of occurring whenever the synchronization constraints of its immediate context allows, as the most primitive concept, their main concerns are different. Besides providing theoretical support for the message oriented

[1] A process diverges when it is engaged in an infinite unbroken sequence of internal actions invisible to its environment, and as a result leaves its environment waiting forever for a response (see [Hoa85, BR85] for details).

programming language CSP, TCSP, like CCS, provides insights into the aspects of system structure, abstraction and equivalence problems. In fact, although the basic tools of CCS and TCSP are different, bisimulation for CCS and failures for TCSP, there are a lot of obvious similarities between them. Recently even the notion of CCSP, a common approach to both CCS and TCSP has been developed ([Old87]).

As we pointed out in the section on COSY and CCS, the former emphasises the treatment of "true" concurrency (as Petri nets do), and it focusses on the expression and consequent proving of properties of non-sequential systems such as absence of deadlock, adequacy, starvation, etc.

The failure semantics for TCSP was introduced in [BHR84], the improved version capable of dealing with divergence problems was given in [BR85]. The main stream of CSP theory is presented in the monograph [Hoa85]. The operational semantics of CSP, similar in the style to that presented for CCS in the previous subsection, can be found in [Plo82],[Old86]. Petri net semantics for CSP was considered in [GR85],[Old87], and approaches to defining partial order semantics were made in [BD85], [Rei84],[HK88],[Bes87].

5.3 Full COSY and Petri Nets

In Chap. 2.3 we have already discussed the Petri net semantics for path programs. In this section we show how processes and priorities correspond to some classes of nets.

5.3.1 Uniquely Named Comma&Star-free Programs and Asymmetric Choice Nets

We show that Petri nets corresponding to uniquely named comma&star-free programs can be regarded as "essentially equivalent" to asymmetric choice nets.

A uniquely named comma&star-free program is a COSY program $Pr = P_1 \ldots P_n Q_1 \ldots Q_m$ where $Pr = P_1 \ldots P_n$ are paths, and $Qr = Q_1 \ldots Q_m$ are processes, and for every P_i, $i = 1, \ldots, n$, and Q_j, $j = 1, \ldots, m$, neither P_i nor Q_j involve occurrences of "," or "*", and no event name occurs more than once within P_i or Q_j.

As we have shown in Chap. 3.2, the semantics of a COSY program $Pr = P_1 \ldots P_n Q_1 \ldots Q_m$ is described by the semantics of a path program $Path(Pr)$. The rules for transforming Pr into $Path(Pr)$ are defined in detail also in Chap. 3.2. For uniquely named comma&star-free program Pr, the structure of $Path(Pr)$ is very simple. We shall say that a path $P = \underline{path}\ seq\ \underline{end}$ is **conflict-restricted** if *seq* can be derived from the nonterminal *crsequence* of the following grammar:

$$crsequence = crorelement\{;crorelement\}^*$$
$$crorelement = event\{,event\}^*$$

A path program $Pr = P_1 \dots P_n$ is said to be **syntactically-conflict-restricted** (abbv. SCR-path program) if every P_i is conflict restricted. Clearly every SCF-path program from Chap. 2.4.1 is also a SCR-path program. A path program Pr is uniquely named if no event name occurs more than once within P_i for $i = 1, \dots, n$. Directly from the definition of $Path(Pr)$ for a basic program $Pr = P_1 \dots P_n Q_1 \dots Q_m$ we have the following result.

Corollary 5.1 *For every uniquely named comma&star-free program Pr, the path program $Path(Pr)$ is an uniquely named SCR-path program.* □

The construction of a labelled Petri net $LN(Pr)$ corresponding to an uniquely named SCR-path program Pr is also easy, as it is the simplest case illustrated in Fig. 2.27 (see Section 2.3 for details). The following Corollary follows immediately from the construction of $LN(Pr)$.

Corollary 5.2 *Let Pr be an uniquely named SCR-path program, and let $LN(Pr) = (S, T, F, M_0, \nu, Ev)$ be a labelled net corresponding to Pr. Then the labelling mapping $\nu : T \longrightarrow Ev$ is one-to-one.* □

A labelled Petri net corresponding to a COSY program $Pr = P_1 \dots P_n Q_1 \dots Q_m$ is exactly the net corresponding to $Path(Pr)$, i.e. $LN(Path(Pr))$. Fig. 5.4 illustrates the notions discussed above.

A marked net $N = (S, T, F, M_0)$ is called an **asymmetric choice net**[2] if and only if:

$$(\forall s_1, s_2 \in S) s_1^\bullet \cap s_2^\bullet = \emptyset \vee s_1^\bullet \subseteq s_2^\bullet \vee s_2^\bullet \subseteq s_1^\bullet.$$

The major property and importance of asymmetric choice nets is that we can formulate static sufficient (but not necessary!) conditions for their liveness. The asymmetric choice nets were probably the most complex nets for which liveness properties can be characterized by H-deadlocks and H-traps (see free-choice nets and Theorem 2.9 in Chap. 2.3).

Theorem 5.1 (Liveness of Asymmetric Choice Nets)
If $N = (S, T, F, M_0)$ is an asymmetric choice net, and if every non-empty H-deadlock $S' \subseteq S$ contains a non-empty H-trap S'' such that $S'' \cap M_0 \neq \emptyset$, then N is live. □

[2] These nets were introduced under the name "extended simple", and this name occurs frequently in the Petri net literature. However, the recent edition of *Notation and Terminology on Petri Net Theory* ([BF86]) calls them asymmetric choice nets, and reserves the name "simple" for another, really much simpler class of nets.

$$
\begin{array}{lll}
Pr = P_1 : \underline{path}\ a; b\ \underline{end} & \Leftrightarrow & \underline{path}\ a\&1, a\&2; b\ \underline{end} \\
\quad Q_1 : \underline{process}\ a; c\ \underline{end} & & \underline{path}\ a\&1; c\&1\ \underline{end} \\
\quad Q_2 : \underline{process}\ a; c\ \underline{end} & & \underline{path}\ a\&2; c\&2\ \underline{end} \\
 & & \Updownarrow
\end{array}
$$

$LN(Path(Pr)) =$

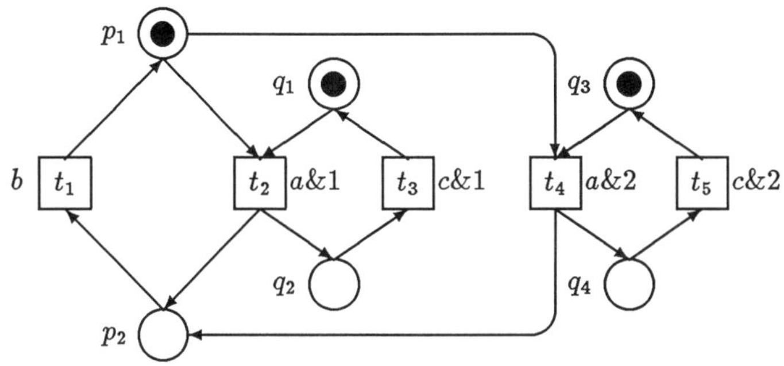

Figure 5.4. Uniquely named Comma&star-free program represented as SCR-path program and net with one-to-one labelling.

We shall now show that nets generated by comma&star-free programs are a proper subset of asymmetric choice nets.

Theorem 5.2
Let $Pr = P_1 \ldots P_n Q_1 \ldots Q_m$ be an uniquely defined comma&star-free program, and let $LN(Path(Pr)) = (S, T, F, M_0, \nu, Ev)$ be a labelled net corresponding to Pr. Then $N = (S, T, F, M_0)$ satisfies the following property:
$(\forall s_1, s_2 \in S) s_1^\bullet \cap s_2^\bullet = \emptyset \vee s_1^\bullet = s_2^\bullet \vee card(s_1^\bullet) = 1 \vee card(s_2^\bullet) = 1.$

Proof. From Corollaries 5.1 and 5.2 it follows that $\nu : T \longrightarrow Ev$ is one-to-one. Since, by the definition $\nu(T) = Ev$, we may assume that ν is the identity mapping. We shall say that a place of a net N belongs to a path (a process) if a path (a process) gave rise to its existence, by the rules for constructing $LN(Path(Pr))$ (see Sections 2.3.4–2.3.7 and 3.2 for details). Thus in Fig. 5.4, p_1 and p_2 belong to a path, while q_1, q_2 (and q_3, q_4) belong to a process.

Let $s_1, s_2 \in S$, and $s_1^\bullet \cap s_2^\bullet \neq \emptyset$. From the rules for constructing $Path(Pr)$ it follows that we have to consider two cases: both s_1 and s_2 belong to paths, and one, say s_1, belongs to a path, while the other, say s_2, belongs to a process. If s_1, s_2 belong to paths, then by the rules for constructing $LN(Path(Pr))$, $s_1^\bullet = s_2^\bullet$, and, if s_1 belongs to a path and s_2 to a process, then $card(s_1^\bullet) = 1$. □

Corollary 5.3 *The marked net MN from Theorem 5.2 is an asymmetric choice net.* □

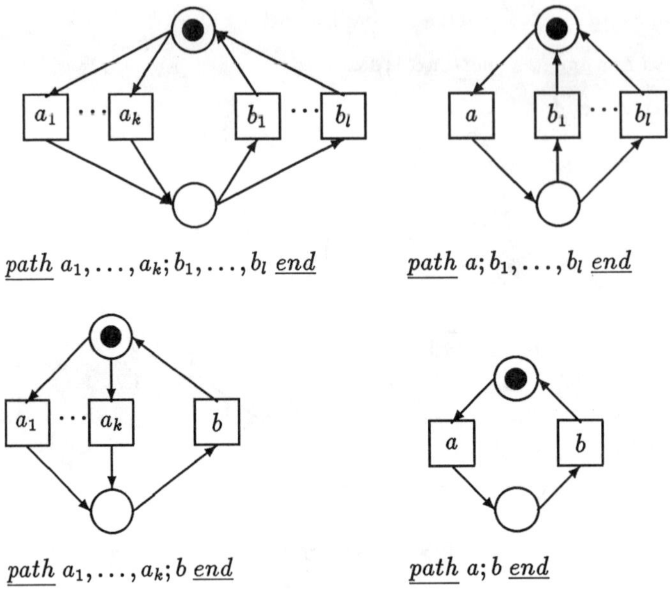

Figure 5.5. The structure of cyclic atoms and corresponding paths.

We shall say that a marked net $N = (S, T, F, M_0)$ is a **cyclic atom** if and only if the following conditions are satisfied:

(1) $card(S) = 2 \wedge card(M_0) = 1$

(2) $(\forall t \in T) card({}^\bullet t) = card(t^\bullet) = 1$

(3) $(\forall s \in S)\ s^\bullet \neq \emptyset \wedge {}^\bullet s \neq \emptyset$.

In other words, one may say that a *cyclic atom* is a 2-place strongly connected state machine net. Let $ATOMS^{cyc}$ denote the family of cyclic atoms. An important property of cyclic atoms is that every cyclic atom corresponds to a uniquely named conflict-restricted path of one of the following four forms:

$$\underline{path}\ a_1, \ldots, a_k; b_1, \ldots, b_l,\ \underline{end}$$
$$\underline{path}\ a; b_1, \ldots, b_l,\ \underline{end}$$
$$\underline{path}\ a_1, \ldots, a_k; b\ \underline{end}$$
$$\underline{path}\ a; b\ \underline{end}$$

Figure 5.5 shows all forms of cyclic atoms and their corresponding paths.

In what follows we shall refer to a marked net being "essentially equivalent" to another marked net. Intuitively, two marked nets are "essentially equivalent" if all of the relevant behavioural properties of the two nets can be regarded as the same. In particular, the second net is live if and only if the first one is live, and so on. More precisely, for "essentially equivalent" nets, their marking sequences and processes should correspond uniquely to each other, except for "unimportant details", where these details may involve some additional places and transitions. We appeal to the reader to convince himself that indeed such a correspondence can be found for all the constructions which follow. If the reader refuses to go along with this he or she may safely interpret the results involving net transformations as follows: formally, they state that liveness is preserved, and informally, they imply that the resulting net is in some sense "related" to the original net. The concept of "essential equivalence" has been in use since the late seventies, however, its satisfactory formal definition has not been given yet. Fig. 5.6 and 5.7 present some examples of transformation preserving "essential equivalence". For more details the reader is advised to refer to [BS83].

Theorem 5.3 (Representation of Asymmetric Choice Nets)
[Bes82],[LSB79],[BS83]
Every asymmetric choice net $N = (S, T, F, M_0)$ can be transformed into an "essentially equivalent" smd-net $N' = (S', T', F', M_0')$ such that,
(1) $(\forall s_1, s_2 \in S')\ s_1^\bullet \cap s_2^\bullet = \emptyset \vee card(s_1^\bullet) = 1 \vee card(s_2^\bullet) = 1.$ [3].
(2) $SMD(N') \subseteq ATOMS^{cyc}$. □

The full construction used in the proof of Theorem 5.3 is sophisticated and quite complex. The basic idea of the construction which guarantees the condition (1) of Theorem 5.3 is illustrated in Fig. 5.6. The net N of Fig. 5.6 is an asymmetric choice net but it does not satisfy condition (1) of Theorem 5.3. It is transformed into the net N' in which the original marking is retained but two new places and four new transitions are added. The net N' is "essentially equivalent" to N and satisfies the condition(1) of Theorem 5.3.
To prove (2) of Theorem 5.3 various transpositions are used, among others all those of Fig. 5.7.
We shall now define a subclass of uniquely named SCR-path programs that can be used to represent nets satisfying (1) and (2) of Theorem 5.3. Let $Pr = P_1 \ldots P_n$ be an uniquely named SCR-path program. Let Ev_i^{nc} denote the following set of events:

$$a \in Ev_i^{nc} :\Leftrightarrow \quad P_i = \underline{path}\ seq_1; a; seq_2\ \underline{end} \vee P_i = \underline{path}\ a; seq\ \underline{end} \vee$$
$$P_i = \underline{path}\ seq; a\ \underline{end} \vee P_i = \underline{path}\ a\ \underline{end}.$$

Define $Ev_i^c = Ev_i \setminus Ev_i^{nc}$. Thus, $a \in Ev_i^{nc}$ means that *a is never involved in any local conflict* in P_i, while $a \in Ev_i^c$ means that *a is involved in some local conflict* in P_i.

---
[3] The nets satifying this property are frequently called "simple nets", however the recent edition of *Notation and Terminology on Petri Net Theory* ([BF86]) reserves the name "simple" for a different class of nets, and does not provide a name for those in Theorem 5.3(1)

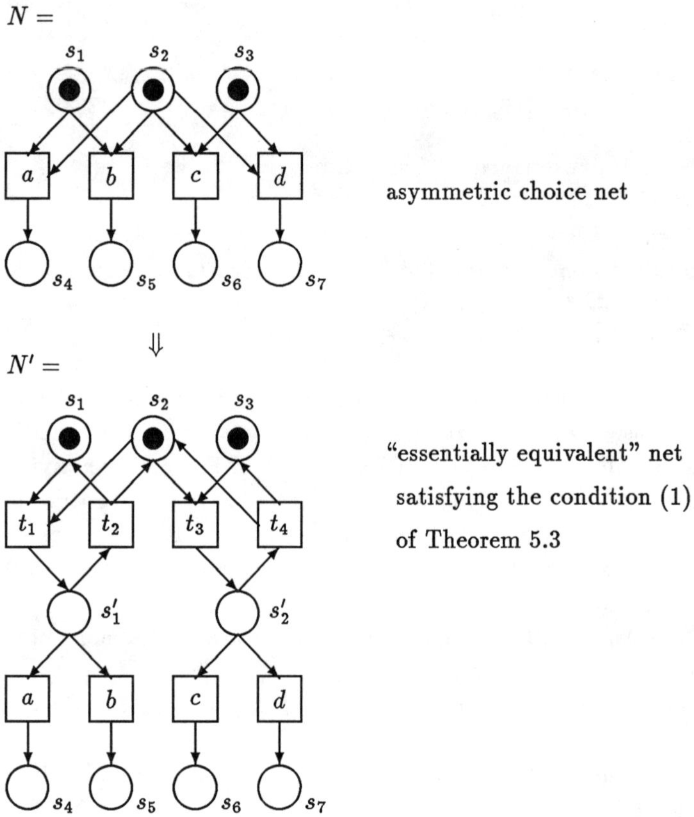

Figure 5.6. Illustration of one of the basic constructions from Theorem 5.3

Transforming a net to "essentially equivalent" one with property ${}^\bullet s \cap s^\bullet = \emptyset$, all $s \in S$

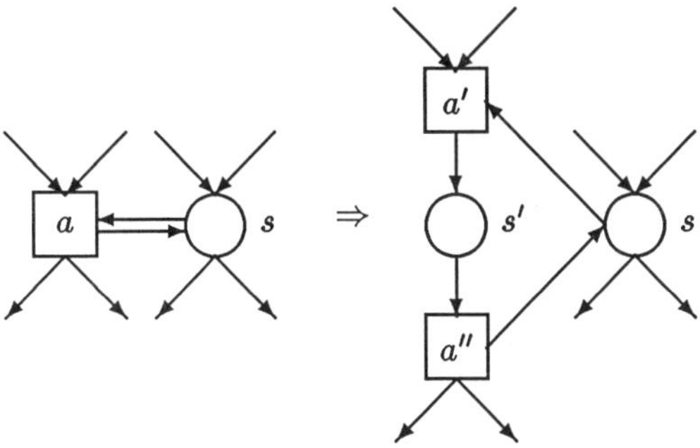

Transforming a net into "essentially equivalent" smd-net

$SMD(N') = \{N_1', N_2', N_3'\}$

$N =$ $\qquad$ $N' =$

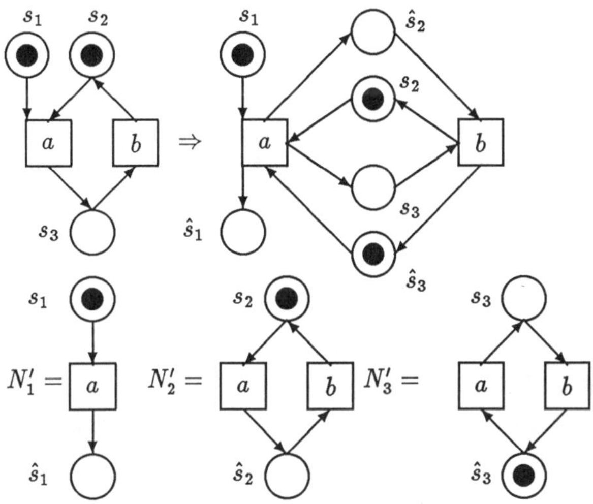

Transforming a state machine into "essentially equivalent" strongly connected and smd-net

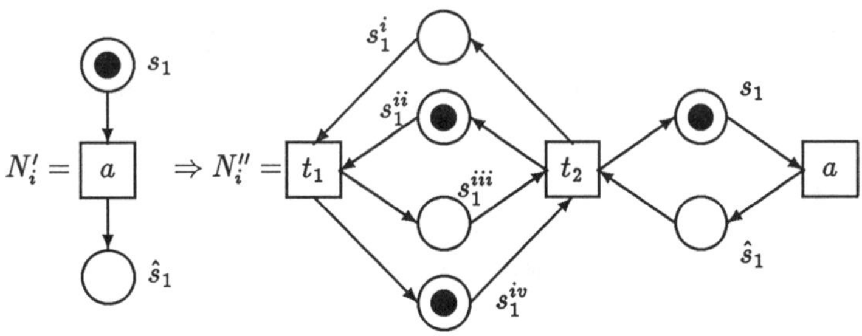

Figure 5.7. Examples of transformations preserving "essentially equivalence".

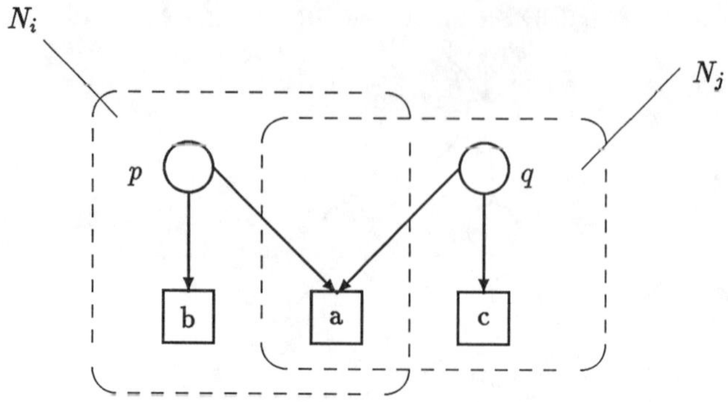

Figure 5.8.

A uniquely named SCR-path program $Pr = P_1 \dots P_n$ is called a **reduced SCR-path program** if and only if:

$$(\forall a \in Ev)(a \in Ev_i \cap Ev_j \wedge i \neq j) \Rightarrow a \in Ev_i^{nc} \vee a \in Ev_j^{nc}$$

In other words if an event a belongs to more than one path, either it is not involved in any local conflict at all, or it is involved in local conflict in exactly one path.

Lemma 5.1
Let $N = (S, T, F, M_0)$ *be a smd-net satisfying:*
(1) $(\forall s_1, s_2 \in S) s_1^\bullet \cap s_2^\bullet = \emptyset \vee card(s_1^\bullet) = 1 \vee card(s_2^\bullet) = 1$
(2) $SMD(N) \subseteq ATOMS^{cyc}$
Then there is a reduced SCR-path Pr *such that* $LN(Pr) = (S, T, F, M_0, \nu, Ev)$ *where* $\nu : T \longrightarrow Ev$ *is a one-to-one function.*

Proof. Suppose that $SMD(N) = \{N_1, \dots, N_k\}$. For $i = 1, \dots, k$, let P_i be the path corresponding to N_i as shown in Fig. 5.5. Clearly $Pr = P_1 \dots P_k$ is an uniquely named SCR-path program. Suppose that Pr is not reduced. Hence, there is $a \in Ev$ such that $a \in Ev_i^c \cap Ev_j^c$ for some $i \neq j$. This means that we have the situation as depicted in Fig. 5.8, i.e. the condition (1) is not satisfied. □

The transformation of Pr into $Path(Pr)$ is also correctly defined if Pr is just a path program. In such a case $Path(Pr) = Pr$. We shall say that two basic COSY programs (a path program is a special case of basic COSY programs) Pr_1 and Pr_2 are "essentially equivalent" if the corresponding nets, $LN(Path(Pr_1))$ and $LN(Path(Pr_2))$ are "essentially equivalent".

Now we show that reduced SCR-path programs can be transformed into some class of "essentially equivalent" uniquely named comma&star-free programs. But first we shall introduce some auxiliary transformations.

$Pr =$
path $a, b; c$ end
path $c, d; b$ end
path $a; b; e$ end

$Ev^c = \{a, b, c, d\}$

$a - loosening$ →

$Pr^1 =$
path $a, b; c$ end
path $c, d; b$ end
path $(\alpha_a; \beta_a); b; e$ end
path $\alpha_a; a; \beta_a$ end

↓ $b - loosening$

$Pr^2 =$
path $a, b; c$ end
path $c, d; (\alpha_b; \beta_b)$ end
path $(\alpha_a; \beta_a); (\alpha_b; \beta_b); e$ end
path $\alpha_a; a; \beta_a$ end
path $\alpha_b; b; \beta_b$ end

$c - loosening$ ←

$Pr^3 =$
path $a, b; (\alpha_c; \beta_c)$ end
path $c, d; (\alpha_b; \beta_b)$ end
path $(\alpha_a; \beta_a); (\alpha_b; \beta_b); e$ end
path $\alpha_a; a; \beta_a$ end
path $\alpha_b; b; \beta_b$ end
path $\alpha_c; c; \beta_c$ end

↓ $d - loosening$

$Pr^4 =$
path $a, b; (\alpha_c; \beta_c)$ end
path $c, d; (\alpha_b; \beta_b)$ end
path $(\alpha_a; \beta_a); (\alpha_b; \beta_b); e$ end
path $\alpha_a; a; \beta_a$ end
path $\alpha_b; b; \beta_b$ end
path $\alpha_c; c; \beta_c$ end
path $\alpha_d; d; \beta_d$ end

$Pr^{loosen} = Pr^4 \quad Confsets = \{\{a, b\}, \{c, d\}\}$

$\gamma_{ab} = \gamma_{\{a,b\}}$ → $\gamma_{cd} = \gamma_{\{c,d\}}$

$Pr' =$
path $\gamma_{ab}; (\alpha_c; \beta_c)$ end
path $\gamma_{cd}; (\alpha_b; \beta_b)$ end
path $(\alpha_a; \beta_a); (\alpha_b; \beta_b); e$ end
process $\alpha_a; \gamma_{ab}; \beta_a$ end
process $\alpha_b; \gamma_{ab}; \beta_b$ end
process $\alpha_c; \gamma cd; \beta_c$ end
process $\alpha_d; \gamma_{cd}; \beta_d$ end

↙ $Path$

$Path(Pr') =$
path $\gamma_{ab}\&1, \gamma_{ab}\&2; (\alpha_c\&3; \beta_c\&3)$ end
path $\gamma_{cd}\&3, \gamma_{cd}\&4; (\alpha_b\&2; \beta_b\&2)$ end
path $(\alpha_a\&1; \beta_a\&1); (\alpha_b\&2; \beta_b\&2); e$ end
path $\alpha_a\&1; \gamma_{ab}\&1; \beta_a\&1$ end
path $\alpha_b\&2; \gamma_{ab}\&2; \beta_b\&2$ end
path $\alpha_c\&3; \gamma_{cd}\&3; \beta_c\&3$ end
path $\alpha_d\&4; \gamma_{cd}\&4; \beta_d\&4$ end

Figure 5.9. Illustration of Lemma 5.2

Let $Pr = P_1 \dots P_n$ be a reduced SCR-path program. Define $Ev^c = Ev_1^c \cup \dots \cup Ev_n^c$. Let $a \in Ev^c$. Define $c(a) = j$ if $a \in Ev_j^c$. Since Pr is a reduced SCR-path program then $c(a)$ is unambiguous. Let $\alpha_a, \beta_a \notin Ev$. We say that $Pr^a = P_1' \dots P_n' P^a$ was derived from Pr by **a-loosening** if and only if:

(1) $P'_{c(a)} = P_{c(a)} \wedge (\forall i \notin i(a)) P_i' = P_i$

(2) For every $i \in i(a) \setminus \{j\}$ P_i' is derived from P_i by replacing each occurrence of a in P_i by an occurrence of $(\alpha_a; \beta_a)$.

(3) $P^a = \underline{path}\ \alpha_a; a; \beta_a\ \underline{end}$

For example:

$$\begin{aligned} Pr_{49} = \quad & P(1) : \underline{path}\ a, b; c, d\ \underline{end} \\ & P(2) : \underline{path}\ a; b\ \underline{end} \end{aligned}$$

after a-loosening is transformed into:

$$\begin{aligned} Pr_{49}^a = \quad & \underline{path}\ a, b; c, d\ \underline{end} \\ & \underline{path}\ (\alpha_a; \beta_a); b\ \underline{end} \\ & \underline{path}\ \alpha_a; a; \beta_a\ \underline{end} \end{aligned}$$

The same Pr_{49} after c-loosening is transformed into:

$$\begin{aligned} Pr_{49}^c = \quad & \underline{path}\ a, b; c, d\ \underline{end} \\ & \underline{path}\ a; b\ \underline{end} \\ & \underline{path}\ \alpha_c; c; \beta_c\ \underline{end} \end{aligned}$$

Note that Pr^a obtained from a reduced SCR-path program is also a reduced SCR-path program. We shall consider Pr^a and Pr as "essentially equivalent".

Lemma 5.2 *For every reduced SCR-path program Pr, there is an uniquely named comma&star-free program Pr' such that $LN(Pr)$ and $LN(Path(Pr'))$ are "essentially equivalent".*

Proof. Assume that $Pr = P_1 \dots P_n$. If $Ev^c = \emptyset$ then Pr is already an uniquely named comma&star-free program. Suppose that $Ev^c = \{a_1, \dots, a_k\}$, $k \geq 1$. Let $Pr^0, \dots, Pr^k$ be path programs such that $Pr^0 = Pr$, and for $i = 1, \dots, k$, Pr^i is derived from Pr^{i-1} by the a_i-loosening. Define $Pr^{loosen} = Pr^k$. Clearly Pr and Pr^{loosen} are "essentially equivalent", and Pr^{loosen} is of the form $Pr^{loosen} = P_1' \dots P_n' P^{a_1} \dots P^{a_k}$, where $P^{a_i} = \underline{path}\ \alpha_{a_i}; a_i; \beta_{a_i}\ \underline{end}$ for $i = 1, \dots, k$. Note that $Ev^c(Pr^{loosen}) = Ev^c(Pr) = \{a_1, \dots, a_k\}$, for every $a_j \in Ev^c(Pr^{loosen}) card(i(a_j)) = 2 \wedge a_j \in Ev(P^{a_j})$.
Let $Confsets$ be the following family of sets of events:

$$\begin{aligned} \{b_1, \dots, b_k\} \in Confsets \Leftrightarrow \quad & (\exists P_j') P_j' = \underline{path}\ seq_1; b_1, \dots, b_l; seq_2\ \underline{end} \\ & \vee P_j' = \underline{path}\ seq; b_1, \dots, b_l\ \underline{end} \\ & \vee P_j' = \underline{path}\ b_1, \dots, b_l; seq\ \underline{end} \\ & \vee P_j' = \underline{path}\ b_1, \dots, b_l\ \underline{end} \end{aligned}$$

Note that: $A, B \in Confsets \Rightarrow (A \cap B = \emptyset \vee A = B)$, and $\bigcup_{A \in Confsets} A = \{a_1, \ldots, a_k\}$, so the family $Confsets$ is a partition of $\{a_1, \ldots, a_k\}$. For every $A \in Confsets$, let γ_A be an event such that $\gamma_A \notin Ev(Pr^{loosen})$. Assume also that $\gamma_A \neq \gamma_B$ if $A \neq B$. Define Pr' as a COSY program derived from Pr^{loosen} by the following transformation:

(1) every path $\alpha_a; a; \beta_a$ end is replaced by process $\alpha_a; \gamma_A; \beta_a$ end, where $A \in Confsets$ and $a \in A$.

(2) in every P'_i, $i = 1, \ldots, n$, every *corelement* $b_1, \ldots, b_l$ is replaced by the event γ_A, where $A = \{b_1, \ldots, b_l\}$.

The correctness of the above transformation follows from the fact that $Confsets$ is a partition of $\{a_1, \ldots, a_k\}$ and Pr^{loosen} is uniquely named. As one may see $Path(Pr')$ and Pr^{loosen} are identical up to renaming of event names. Fig.5.9 illustrates the above constructions. □

Now we are able to formulate the major result of this section.

Theorem 5.4 (Asym. Choice Nets and Comma&Star-free Programs)

Asymmetric choice nets and nets corresponding to uniquely named comma&star-free programs can be regarded as "essentially equivalent".

Proof.
($\Leftarrow$) Follows immediately from Corollary 5.3.
($\Rightarrow$) Follows from Theorem 5.3, Lemma 5.1 and Lemma 5.2 in this order. □

The "essential equivalence" between comma&star-free programs and asymmetric choice nets was discovered by Eike Best and published in [LSB79], [Bes82]. The term "essentially equivalent" is taken from [BS83] where liveness preserving transformations among various classes of nets are defined and discussed. In [Bes82] this relation is called "behavioural equivalence". The major difference between our solution and the previous ones is that we use pure COSY to describe the construction from Lemma 5.2, while in the previous presentations, this part was done in terms of Petri Nets. The concept of asymmetric choice nets and Theorem 5.1 follows from [Hol74]. A more general proof of Theorem 5.1 can also be found in [JV80] where it is Theorem 4.4.

5.3.2 Priority Path Programs, Priority Nets and Inhibitor Nets

In this section we show how the concept of priorities introduced and analyzed for path programs in Chap. 2.6 can be extended to Petri nets. It turns out that this new kind of priority differs from and is more powerful than the existing concept of priority for nets (as defined in [Pet81]). We also show that a widespread belief that the power of standard priority nets and so called inhibitor nets is the same (compare [Pet81]), is true only as far as Petri net languages are concerned. In terms of multiple firing sequences the standard priority nets are weaker than inhibitor nets, while our new kind of priority nets are equivalent to inhibitor nets. We start with a formal definition of multiple firing sequence semantics for nets.

Let $N = (S, T, F, M_0)$ be a marked Petri net. Define:

$$\begin{array}{lcl} MTS(N) & = & \{A_1 \ldots A_k | (\exists M \subseteq S)\ M_0[A_1 \ldots A_k\rangle M\} \cup \{\lambda\} \\ TS(N) & = & \{t_1 \ldots t_r | (\exists M' \subseteq S)\ M_0[t_1 \ldots t_r\rangle M\} \cup \{\varepsilon\} \end{array}$$

where λ denotes the empty multiple sequence, ε denotes the empty string $A_i \subseteq T$, $t_i \in T$. The sets $MTS(N)$, $TS(N)$ are called **multiple transition sequences**, and **transition sequences** respectively.

In Chap. 2.6 we gave a definition of labelled state machine decomposible nets. We now extend this definition to all marked nets considered in this book. One of the principal assumptions characterizing the COSY approach is the assumption that **no event can occur concurrently with itself.** In Petri net terminology this means that a labelling function must satisfy some conditions.

By a **labelled net** we mean a 6-tuple:

$LN = (S, T, F, M_0, \nu, Ev)$, where
$N = (S, T, F, M_0)$ is a **marked net**,
Ev is a finite set of **events (labels)**, $\varepsilon \notin Ev$.
$\nu : T \longrightarrow Ev$ is a mapping (**labelling**) satisfying:
(1) $\nu(T) = Ev$
(2) $(\forall t_1, t_2 \in T)(\exists M \in [M_0\rangle)\ M$ enables $\{t_1, t_2\} \Rightarrow \nu(t_1) \neq \nu(t_2)$.

For every labelled net $LN(S, T, F, M_0, \nu, Ev)$, N will denote the net (S, T, F, M_0). The labelling ν can be standardly extended to $\nu : T^* \longrightarrow Ev^*$, by $\nu(\varepsilon) = \varepsilon$, $\nu(xt) = \nu(x)\nu(t)$, for $x \in T^*$, $t \in T$.

The set of strings $ES(LN) = \{\nu(x) | x \in TS(N)\}$ is called the set of **event sequences** of LN (or **language** of LN). We can extend ν to $\nu : 2^T \longrightarrow 2^{Ev} \cup \{\lambda\}$ by $\nu(\emptyset) = \lambda$, $\nu(\{t_1, \ldots, t_k\}) = \{\nu(t_1), \ldots, \nu(t_k)\}$, $t_i \in T$, as well as to $\nu : (2^T)^* \longrightarrow (2^{Ev})^*$, in a standard way by $\nu(\lambda) = \lambda$, $\nu(xA) = \nu(x)\nu(A)$ for $x \in (2^T)^*$, $A \subseteq T$.

The set of multiple sequences $MES(LN) = \{\nu(x) | x \in MTS(N)\}$ is called the set of **multiple event sequences** of LN (or **subset languages** of LN). The reader may wonder if the definition of MES really achieves what it is supposed to, since if $\nu(t_1) = \nu(t_2) = a$ then $\nu(\{t_1, t_2\}) = \{a\}$, instead of the more intuitive *multi set* $\{a, a\}$ (see for instance Appendix A in [Pet81]), but as the lemma below asserts such situations never happen for multiples from $MTS(N)$.

Lemma 5.3
Let $A_1 \dots A_k \in MTS(N)$. *Then* $(\forall i = 1, \dots, k)\ card(A_i) = card(\nu(A_i))$.

Proof. Immediate from (2) of the definition of labelling and the firing rule for nets. □

By a **general labelled priority net** we mean a 7-tuple:

$$LPN = (S, T, F, M_0, \nu, Ev, <)$$

where $LN = (S, T, F, M_0, \nu, Ev)$ is a labelled net,

$<\subseteq T \times T$ is a **priority relation** satisfying : $t_1 < t_2 \Leftrightarrow \neg t_2 < t_1$.

We want to stress that asymmetry is the only property required of the priority relation, which in general **is not required to be transitive.**

The firing rule for general priority nets is modified from the conventional firing rule for nets to the extent that if several transitions are enabled at a given marking, only those whose priority is no less than any other enabled transition can fire. Formally this rule may be defined as follows. For every marking M, let $enabled(M)$ denote the set of all sets of transitions simultaneously enabled by M without priority taken into account, i.e.:

$$(\forall M \subseteq S)\ enabled(M) = \{A | A \subseteq T \wedge M \text{ enables } A\}.$$

For every marking M we shall say that the set of transitions $A \subseteq T$ is **simultaneously priority enabled** by M (or M **priority enables** A) if and only if:

1. M enables A (i.e. $A \in enabled(M)$).
2. $(\forall B \in enabled(M))(\forall a \in A)(\forall b \in B)\ \neg(b < a)$.

We shall write $M[A_1 \dots A_k\rangle_< M'$ to denote that marking M' was obtained from M by consecutively firing sets of transitions $A_1, \dots, A_k$. The formal definition of $[\dots\rangle_<$ is exactly the same as that for $[\dots\rangle$ (see Chap. 2.3.1) except that we have to replace each "M enables A" by "M priority enables A".

The sets of *event sequences and multiple event sequences for labelled priority nets* are defined as:

$$ES(LPN) = \{\nu(t_1 \dots t_r) | (\exists M \subseteq S)\ M_0[t_1 \dots t_r\rangle_< M\} \cup \{\varepsilon\},$$
$$MES(LPN) = \{\nu(A_1 \dots A_k) | (\exists M \subseteq S)\ M_0[A_1 \dots A_k\rangle_< M\} \cup \{\lambda\}.$$

For the general labelled priority nets defined above the relation $t_1 < t_2$ may be defined for **any** pair of transitions, even if they are not involved in any conflict and are independent otherwise. This might cause some interpretational problems since intuitively priority seems to contradict independence. In practice, priority is usually used to resolve a conflict in one or the other direction.

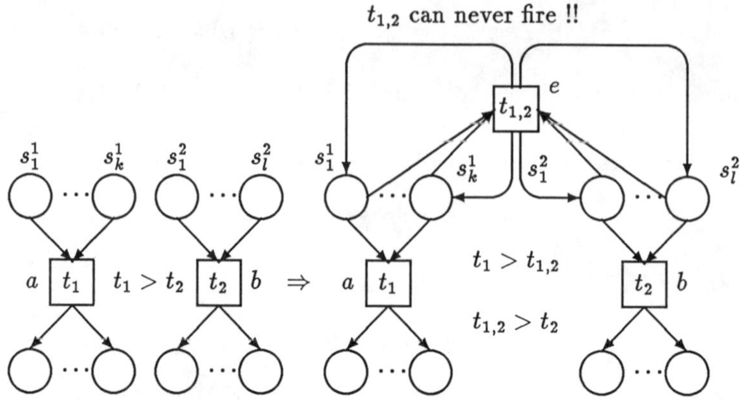

Figure 5.10. Transformation of general priority nets into conflict sensitive priority nets (Theorem 5.5)

A general labelled priority net $LPN = (S, T, F, M_0, \nu, Ev, <)$ is said to be **conflict sensitive** if and only if:

$$(\forall t_1, t_2 \in T)\ t_1 < t_2 \Leftrightarrow {}^\bullet t_1 \cap {}^\bullet t_2 \neq \emptyset.$$

The next theorem says that, in terms of multiple sequence semantics, the conflict sensitive labelled priority nets have the same power as general labelled priority nets.

Theorem 5.5

For every general labelled priority net LPN there is a conflict sensitive labelled priority net LPN' such that $MES(LPN) = MES(LPN')$.

Proof. First let us add to Ev a new symbol (event) e where $e \notin Ev$. Suppose that $t_1 > t_2$ and ${}^\bullet t_1 \cap {}^\bullet t_2 = \emptyset$. Let us add a new transition $t_{1,2}$, change the priority relation to $t_1 > t_{1,2}$, $t_{1,2} > t_2$, define $\nu(t_{1,2}) = e$, and modify the flow realtion F to:

$$\text{new } F := \text{ old } F \cup \{(s, t_{1,2}) | s \in {}^\bullet t_1 \cup {}^\bullet t_2\} \cup \{(t_{1,2}, s) | s \in {}^\bullet t_1 \cup {}^\bullet t_2\}.$$

This construction is illustrated in Fig. 5.10.

Note that the event $t_{1,2}$ can *never* be fired, it plays the same role as "ghost" events in Sect. 2.6.5, it "moves" priority through a net. Now the priority relation $<$ is no longer defined between t_1 and t_2, and of course ${}^\bullet t_i \cap {}^\bullet t_{1,2} = {}^\bullet t_i \neq \emptyset$, $i = 1, 2$. After repeating this construction for all t', t'' with $t' > t''$ and ${}^\bullet t' \cap {}^\bullet t'' = \emptyset$, we get a new labelled priority net LPN', which is conflict sensitive and $MES(LPN) = MES(LPN')$. □

$Pr =$ $\underline{priority}\ b > c\ \underline{end}$ $\Leftrightarrow$ $LPN(Pr) =$
$\underline{path}\ a; b\ \underline{end}$
$\underline{path}\ b, c\ \underline{end}$

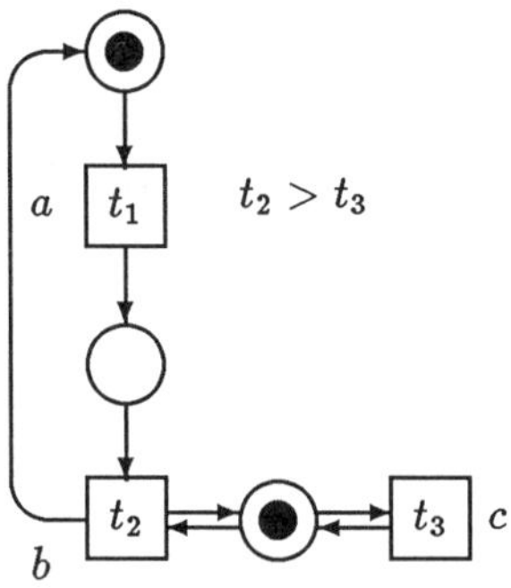

Figure 5.11. Relationship between priority path programs and conflict sensitive priority nets

We shall now show that priority path programs can be modelled by conflict sensitive labelled priority nets. Let $Pr =< P_1 \ldots P_n$ be a given priority path program.

We define $LPN(Pr) = (S, T, F, M_0, \nu, Ev, <_N)$, where

$$LN(P_1 \ldots P_n) = (S, T, F, M_0, \nu, Ev), \text{ and}$$
$$(\forall t_1, t_2 \in T)\ t_1 <_N t_2 \ :\Leftrightarrow\ \nu(t_1) < \nu(t_2).$$

The net $LPN(Pr)$ is said to be **generated by the priority path program** Pr. This construction is illustrated in Fig. 5.11.

Theorem 5.6
For every priority path program Pr, $MFS(Pr) = MES(LPN(Pr))$.

Proof. This straightforward proof (although not so short) is left to the reader. □

A general labelled priority net $LPN = (S, T, F, M_0, \nu, Ev, <)$ is said to be a **Hack labelled priority net** if and only if the relation $<$ is transitive. In popular Petri net literature (as for instance [Pet81]) the name "priority nets" usually corresponds to "Hack labelled priority nets" with additional assumptions that the relation $<$ is a total order. Hack labelled priority nets may not in general be conflict sensitive. Directly from Theorem 5.5 we have the following result.

Corollary 5.4
For every Hack labelled priority net LPN, there is a conflict sensitive labelled priority net LPN' such that $MES(LPN) = MES(LPN')$. □

We shall now show that as far as multiple sequence semantics is concerned, the Hack labelled priority nets are weaker than conflict sensitive priority nets.

Lemma 5.4
Let $LPN = (S, T, F, M_0, \nu, Ev, <)$ *be a Hack labelled priority net. Then* $(\forall M \subseteq S)$ *enabled*$(M) \neq \emptyset \Rightarrow (\exists t \in T)$ M *priority enables* $\{t\}$.

Proof. Let $A = \{t | \{t\} \in enabled(M)\}$, and let $\leq_A \subseteq A \times A$ be defined as:

$$t_1 \leq_A t_2 \Leftrightarrow t_1, t_2 \in A \wedge (t_1 < t_2 \vee t_1 = t_2).$$

Since $<$ is transitive, $\leq_A$ is a finite partial order. Each finite partial order must have at least one maximal element, so let t be a maximal element of $\leq_A$. This means that for every $t' \in A$, either $t' \leq_A t$ or t' and t are incomparable, in any case $\neg(t < t')$ is always satified. Thus M priority enables $\{t\}$. □
The above lemma follows entirely from the transitivity of $<$.

Theorem 5.7 (Limitations of Hack Labelled Priority Nets)
The multiple sequence set $\{\lambda, \{a, b\}, \{a\}, \{b\}\}$ *cannot be generated by any Hack labelled priority net.*

Proof. Suppose that $LPN = (S, T, F, M_0, \nu, Ev, <)$, $<$ is transitive, and $MES(LPN) = \{\lambda, \{a, b\}, \{a\}, \{b\}\}$. Since $\{a, b\} \in MES(LPN)$, then there is $A \subseteq T$ such that $A = \{t_1, t_2\}$, $\nu(t_1) = a$, $\nu(t_2) = b$, and M_0 priority enables A. This means that M_0 also priority enables $\{t_1\}$. Let $M_1 \subseteq S$ be such that $M_0[\{t_1\}\rangle_< M_1$. By Corollary 2.8 we have that M_1 enables t_2, i.e. $enabled(M_1) \neq \emptyset$. From Lemma 5.4 it follows that $(\exists t \in T) M_1$ priority enables $\{t\}$, which means that $\{a\}\{\nu(t)\} \in MES(LPN)$, a contradiction. □

A conflict sensitive labelled priority net generating $\{\lambda, \{a, b\}, \{a\}, \{b\}\}$ is shown in Fig. 5.12.

Thus, the generating capabilities of Hack labelled priority nets are weaker than that of conflict sensitive labelled priority nets.

The last class of Petri nets we shall now consider are the **labelled inhibitor nets.** These are labelled nets with a special kind of arc called inhibitor arcs. An inhibitor arc is only directed from place to transition, and graphically it has a small circle instead of an arrow head. This notation is borrowed from switching theory where the small circle means "not". The firing rule for a transition having incident inhibitor arcs is modified as follows: the transition is firable if and only if it was firable in the old sense and all inhibiting places have zero tokens. The same applies to sets of transitions. Fig. 5.13 shows an labelled inhibitor net generating the set of multiple sequences $\{\lambda, \{a, b\}, \{a\}, \{b\}\}$.

Formally, the labelled inhibitor nets can be defined as follows:

A labelled inhibitor net LIN is a 7-tuple

$$LIN = (S, T, F, M_0, \nu, Ev, IA)$$

where $LN = (S, T, F, M_0, \nu, Ev)$ is a labelled net, and $IA \subseteq S \times T$ is the set of **inhibitor arcs.**

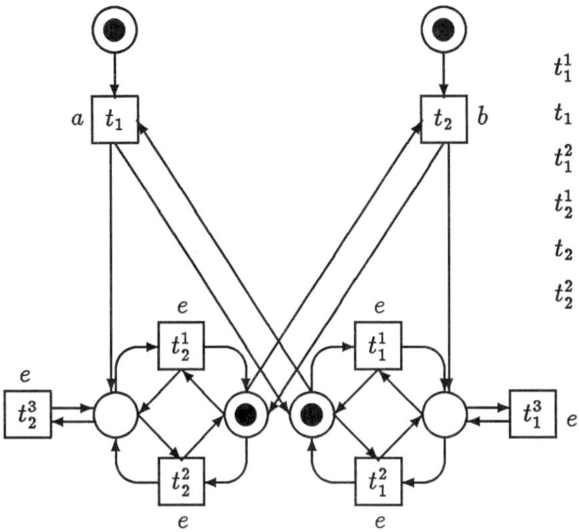

$t_1^1 > t_1,\ t_1^1 > t_1^3,$
$t_1 > t_1^2,\ t_1^3 > t_1^2,$
$t_1^2 > t_1^1,$
$t_2^1 > t_2,\ t_2^1 > t_2^3,$
$t_2 > t_2^2,\ t_2^3 > t_2^2,$
$t_2^2 > t_2^1.$

Figure 5.12. Conflict sensitive labelled priority net generating $\{\lambda, \{a, b\}, \{a\}, \{b\}\}$

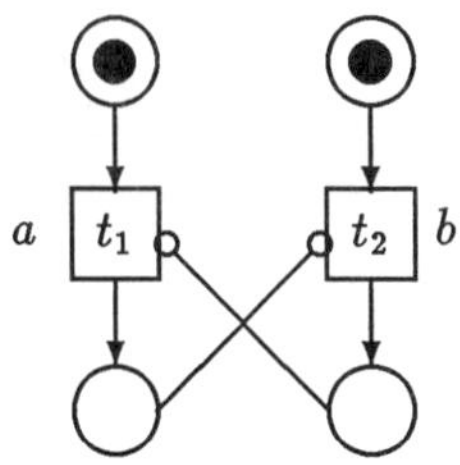

Figure 5.13. A labelled inhibitor net generating $\{\lambda, \{a, b\}, \{a\}, \{b\}\}$

For every marking M we say that the set of transitions $A \subseteq T$ is **simultaneously inhibitor enabled** by M (or M **inhibitor enables** A) if and only if:

1. *M enables A*
2. $M \cap \{s|(s,t) \in IA \wedge t \in A\} = \emptyset$.

We shall write $M[A_1 \dots A_k\rangle_{inh} M'$ to denote that the marking M' was obtained from M by consecutively firing sets of transitions $A_1, \dots, A_k$. The formal definition of $[\dots\rangle_{inh}$ is exactly the same as that of $[\dots\rangle$ (see Chap. 2.3.1) except that we have to replace each "*M enables A*" by "*M inhibitor enables A*".

The sets of **event sequences**, and **multiple event sequences for labelled inhibitor nets** are defined as:

$$\begin{aligned} ES(LIN) &= \{\nu(t_1 \dots t_r)|(\exists M \subseteq S)\ M_0[t_1 \dots t_r\rangle_{inh} M\} \cup \{\varepsilon\}, \\ MES(LIN) &= \{\nu(A_1 \dots A_k)|(\exists M \subseteq S)\ M_0[A_1 \dots A_k\rangle_{inh} M\} \cup \{\lambda\}. \end{aligned}$$

We shall show that labelled inhibitor nets and conflict sensitive labelled priority nets can be regarded as equivalent (with respect to multiple sequence semantics).

Theorem 5.8 (Equivalence of Inhibitor Nets and Priority Nets)
(1) For every labelled inhibitor net LIN there is a conflict sensitive labelled priority net LPN′ such that $MES(LIN) = MES(LPN')$.
(2) For every labelled conflict sensitive priority net LPN there is a labelled inhibitor net LIN′ such that $MES(LPN) = MES(LIN')$.

Proof. (1) First we add to Ev a new symbol (event) e. We replace every inhibitor arc $(s,t) \in IA$ by three additional transitions t^1, t^2, t^3 and an additional place s' such that:

$$\begin{aligned} &{}^\bullet(s') = (s')^\bullet = \{t, t^1, t^2\},\ {}^\bullet(t^1) = (t^1)^\bullet = {}^\bullet(t^2) = (t^2)^\bullet = \{s, s'\}, \\ &{}^\bullet(t^3) = (t^3)^\bullet = \{s\}, \end{aligned}$$

we place a token in s', set up: $\nu(t^1) = \nu(t^2) = \nu(t^3) = e$ and we define a priority: $t^1 > t, t > t^2, t^2 > t^1, t^1 > t^3, t^3 > t^2$. The construction is illustrated in Fig. 5.14. When applied to the net from Figure 5.13, it results in the net from Figure 5.12.

Note that the transitions t^1, t^2, t^3 can never be fired (like "ghost" events in Section 2.6.5). Clearly the new net LPN' is a labelled conflict sensitive net. We leave it to the reader to verify that $MES(LIN) = MES(LPN')$.
(2) Let $t_1 > t_2$, ${}^\bullet(t_1) \cap {}^\bullet(t_2) \neq \emptyset$ and $\nu(t_1) = a$, $\nu(t_2) = b$. The transformation is based on the following condition: $(\neg({}^\bullet t_1 \subseteq M)) \Rightarrow$ (M may priority enable $\{t_2\}$). This leads us to the following construction (which is illustrated in Fig. 5.15).

We replace t_2 by the set of transitions $\{t_2^A | A \neq \emptyset \wedge A \subseteq {}^\bullet t_1 \setminus {}^\bullet t_2\}$, with $\nu(t_2^A) = \nu(t_2)$, $t' < t_2^A$ if $t' < t_2$, $t' > t_2^A$ if $t' > t_2 \wedge t' \neq t_1$, ${}^\bullet(t_2^A) = {}^\bullet t_2$, $(t_2^A)^\bullet = t_2^\bullet$, for all nonempty $A \subseteq {}^\bullet t_1 \setminus {}^\bullet t_2$. Since t_2 disappears, so does $t_1 > t_2$. We also add

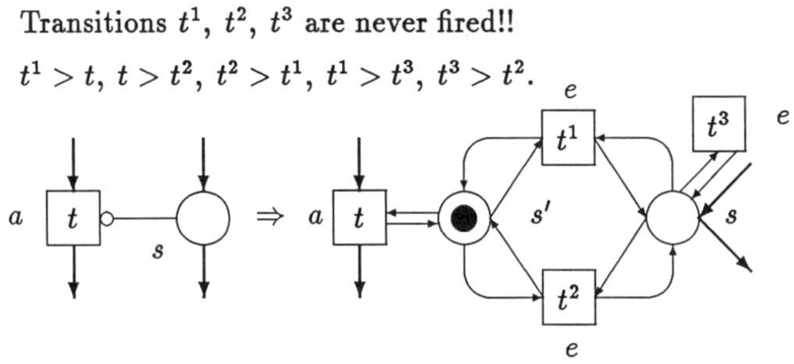

Figure 5.14. Transforming labelled inhibitor nets into equivalent labelled conflict sensitive priority nets

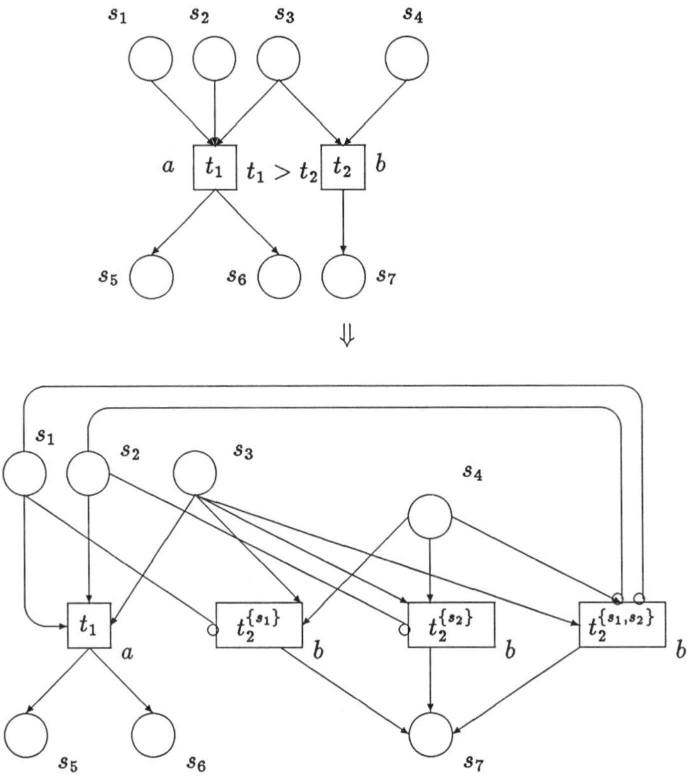

Figure 5.15. Transforming labelled conflict sensitive nets into equivalent labelled inhibitor nets

new inhibitor arcs, namely: $newIA := oldIA \cup \{(s, t_2^A) | s \in A \neq \emptyset \wedge A \subseteq {}^\bullet t_1 \setminus {}^\bullet t_2\}$. We repeat the above transformation until the relation $<$ completely disappears. The new net is a labelled inhibitor net, and generates the same set of multiple event sequences. □

Summarizing this section we can state that the general labelled priority nets, and labelled conflict sensitive priority nets, which come into existence as a net model for priority path programs, and labelled inhibitor nets are all equivalent in terms of multiple sequence semantics. The Hack labelled priority nets, i.e. the priority nets in standard text books (e.g. [Pet81]), are weaker.

Hack labelled priority nets were introduced in [Hac75] and labelled inhibitor nets in [FA73]. In both [Hac75] and [FA73] place capacities were infinite and weighted arcs were allowed. The paper [Hac75] is a landmark work in Petri net language theory. Among other things, the author shows that labelled inhibitor nets and Hack labelled priority nets can be transformed into each other in a language preserving manner. One of his transformations, slightly modified, was used in our proof of Theorem 5.8 (2). The second transformation, from labelled inhibitor nets into Hack labelled priority nets, does not preserve the standard properties of nets, since it introduces additional sequentialization (and because of this sequentialization it cannot preserve multiple sequences). For nets considered in this book (place capacity equals 1, no weighted arcs), the language (event sequence) semantics does not seem to be very exciting. One can easily show that all classes of nets considered here generate the same class of languages, namely prefix-closed regular languages. Recently, a brand new concurrent semantics for Petri nets with priorities has been proposed by E. Best and M. Koutny [BK].

5.4 Vector Sequences and Mazurkiewicz Traces

The main goal of this section is to show a close resemblance between the algebra of vector sequences and the algebra of (Mazurkiewicz) traces. The basic properties of the former are described in detail in Chap. 2.2.1 and 2.2.2, so we will here only define properties of the latter.

By a **concurrent alphabet** we mean any pair (Ev, ind) where Ev is a finite set of **symbols (event names)**, and $ind \subseteq Ev \times Ev$ is a symmetric and irreflexive relation called **independency**.

Let $\hat{-} \subseteq Ev^* \times Ev^*$ be the following relation:

$$(\forall x, y \in Ev^*) x \hat{-} y :\Leftrightarrow$$
$$(\exists x_1, x_2 \in Ev^*)(\exists a, b \in Ev)(a, b) \in ind \wedge x = x_1 ab x_2 \wedge y = x_1 ba x_2.$$

Let $\hat{=} = (\hat{-})^*$, i.e. $x \hat{=} y :\Leftrightarrow x = y \vee ((\exists x_1, \ldots, x_k) x \hat{-} x_1 \hat{-} \ldots \hat{-} x_k = y)$. It can be easily seen that $\hat{=}$ is an equivalence relation, which is usually referred to as the **trace equivalence** over (Ev, ind). In fact, we have already defined and used this relation as an auxiliary notion to characterize the equality $\underline{x} = \underline{y}$ in Theorem 2.3.

Equivalence classes of the $\hat{=}$ relation are called (Mazurkiewicz) **traces** over (Ev, ind). A trace generated by a string $x \in Ev^*$ is an equivalence class of $\hat{=}$ containing x, and is denoted by $[x]_{\hat{=}}$.

The **concatenation of traces** is defined by:

$$(\forall x, y \in Ev^*)\ [x]_{\hat{=}}[y]_{\hat{=}} \ = \ [xy]_{\hat{=}}.$$

The set of all traces over (Ev, ind) is the set of all equivalence classes of $\hat{=}$, i.e. the set $Ev^*/\hat{=}$.

Let $Ev = Ev_1 \cup \ldots \cup Ev_n$, and let $ind \subseteq Ev \times Ev$ be standardly defined as: $(a, b) \in ind :\Leftrightarrow (\forall Ev_i) a \notin Ev_i \vee b \notin Ev_i$. Clearly (Ev, ind) is a concurrent alphabet. Let Vev be the set of vector events generated by distribution $Ev_1, \ldots, Ev_n$. The theorem below shows the basic relationship between Vev^* and $Ev^*/\hat{=}$.

Theorem 5.9 (Equivalence of Traces and Vector Sequences)
(1) $(\forall x, y \in Ev^*)\ [x]_{\hat{=}} = [y]_{\hat{=}} \Leftrightarrow \underline{x} = \underline{y}$.
(2) $(\forall x \in Ev^*)\ [x]_{\hat{=}} \ = \ lin(\underline{x})$.

Proof. (1) From Theorem 2.3 it immediately follows that $\underline{x} = \underline{y} \Leftrightarrow x \hat{=} y$.
(2) From the definition of lin and Theorem 2.3. □

Let $tv : Ev^*/\hat{=} \longrightarrow Vev^*$, $vt : Vev^* \longrightarrow Ev^*/\hat{=}$ be mappings defined by:

$$(\forall [x]_{\hat{=}} \in Ev^*/\hat{=})\ tv([x]_{\hat{=}}) = \underline{x},\ (\forall \underline{x} \in Vev^*)\ vt(\underline{x}) = [x]_{\hat{=}}.$$

The correctness of the definition of tv and vt is guaranteed by Theorem 5.9(1), and their basic properties are the following:

Corollary 5.5
(1) $(\forall x, y \in Ev^*) tv([x]_{\hat{=}}[y]_{\hat{=}}) = tv([x]_{\hat{=}} tv([y]_{\hat{=}}),$
$vt(\underline{xy}) = vt(\underline{x}) vt(\underline{y})$.
(2) $(\forall x \in Ev^*) vt(tv([x]_{\hat{=}})) = [x]_{\hat{=}},\ tv(vt(\underline{x})) = \underline{x}$. □

For every trace $[x]_{\hat{=}}$ we can define the canonical partial order $PO_{[x]_{\hat{=}}}$ by $PO_{[x]_{\hat{=}}} = \bigcap_{y \in [x]_{\hat{=}}} TO_y$ (compare Chap. 2.2.2).

Corollary 5.6
$(\forall x \in Ev^*)\ PO_{\underline{x}} = PO_{[x]_{\hat{=}}}$. □

Thus the algebra of (Mazurkiewicz) traces, and the algebra of vector sequences may be regarded as isomorphic. The basic difference is that traces use the relation ind, while vector sequences require the distribution of the set of events onto $Ev_1, \ldots, Ev_n$. The approach based on the independency relation as primary concept makes, for instance, the analysis of equivalence between different behaviours, more convenient. On the other hand the explicit event distribution makes contributions of sequential subsystems to global behaviour more visible.

Traces (of Mazurkiewicz) were introduced in [Maz77] as a tool to define concurrent semantics of Petri nets (vector sequences originate from [Shi79b]). Both Mazurkiewicz traces and vector sequences can be regarded as two different representations of **partially commutative monoids** (see [Lal79]). Partially commutative monoids were introduced in [CF69] for combinatorial applications without any reference to concurrency theory (see [Lal79] Chap.11). The use of this concept to model concurrency begins with [Maz77]. The results concerning the connections between partial orders and Mazurkiewicz traces can be found in [Maz86], [CM85], [Zie87]. Some influence of [CM85] can be found in Sect. 2.2.1 and 2.2.2. During the last decade the theory and application of Mazurkiewicz traces have been developed substantially (see [Maz86] and [AR88] for reviews and bibliography). The relationship between Mazurkiewicz traces and vector sequences is analyzed in [Shi85]. We use the name "Mazurkiewicz trace" since the name "trace" is also used, in particular by the CSP-related school, to denote a sequence of event occurrences (see [Hoa85]).

5.5 COSY and Synchronized Behaviours

Let $S_1, \ldots, S_n$ be sequential systems. For every $i = 1, \ldots, n$, let Ev_i be the set of events of Ev_i, $B_i \subseteq Ev_i^*$ be the sequential behaviour generated by S_i. Let CS be a concurrent system built from $S_1, \ldots, S_n$ and let $Ev = Ev_1 \cup \ldots \cup Ev_n$. To find the concurrent behaviour of CS, B_{CS}, we have to take into account both $B_1, \ldots, B_n$ and the way they are synchronized. The vector firing sequence approach provides the following formula for B_{CS}^{VFS}:

$$B_{CS}^{VFS} = B_1 \times \ldots \times B_n \cap Vev^*$$

We assume additionally, as we did throughout the rest of this book, that each B_i is regular. Such an assumption has allowed us to focus on synchronization problems rather than own complexity of sequential behaviours. The above formula for B_{CS} assumes a "simple handshake" kind of synchronization. We may allow more sophisticated synchronization means by replacing Vev^* by more complex structures. Define: $GVev = (Ev_1 \cup \{\varepsilon\}) \times \ldots \times (Ev_n \cup \{\varepsilon\}) \setminus \{\underline{\varepsilon}\}$.

Recall that concatenation of vectors of strings is defined componentwise, i.e., $(x_1, \ldots, x_n)(y_1, \ldots, y_n) = (x_1y_1, \ldots, x_ny_n)$, for every $(x_1, \ldots, x_n), (y_1, \ldots, y_n) \in Ev_1^* \times \ldots \times Ev_n^*$.

Clearly $Vev \subseteq GVev$, but if $n > 1$ then always $Vev \neq GVev$.

Let $V \subseteq GVev$ be the set of vectors which *must be performed simultaneously* in CS.

Let $ind_V \subseteq V \times V$ be the relation defined by:

$$(\forall v_1, v_2 \in V)(v_1, v_2) \in ind_V \Leftrightarrow v_1 v_2 = v_2 v_1 \in GVev \setminus V$$

The relation ind_V may be interpreted as **independency imposed by** V. As one may easily observe the relation ind_{Vev} corresponds to ind.

Corollary 5.7

$$(\forall a, b \in Ev)\ (a, b) \in ind \Leftrightarrow (\underline{a}, \underline{b}) \in ind_{Vev}.$$

□

The set of V^*-**synchronized behaviours of** CS is defined as:

$$B_{CS}^{V^*} = B_1 \times \ldots \times B_n \cap V^*$$

If $V = Vev$, then $B_{CS}^{Vev^*} = B_{CS}^{VFS}$, so we have our standard vector sequence semantics. In general there may be no relationship between V and Vev, except that they are both subsets of $GVev$.
Consider the path program Pr_4:

$$Pr_4 = \begin{cases} P(1) & \underline{path}\ a; b\ \underline{end} \\ P(2) & \underline{path}\ b; a\ \underline{end} \end{cases}$$

and

$$V = \left\{ \begin{bmatrix} a \\ b \end{bmatrix}, \begin{bmatrix} b \\ a \end{bmatrix} \right\} \subset GVev, \textit{ where}$$

$$GVev = \left\{ \begin{bmatrix} a \\ a \end{bmatrix}, \begin{bmatrix} b \\ b \end{bmatrix}, \begin{bmatrix} a \\ b \end{bmatrix}, \begin{bmatrix} b \\ a \end{bmatrix}, \begin{bmatrix} a \\ \varepsilon \end{bmatrix}, \begin{bmatrix} b \\ \varepsilon \end{bmatrix}, \begin{bmatrix} \varepsilon \\ a \end{bmatrix}, \begin{bmatrix} \varepsilon \\ b \end{bmatrix} \right\}.$$

Note that in this case

$$Vev = \left\{ \begin{bmatrix} a \\ a \end{bmatrix}, \begin{bmatrix} b \\ b \end{bmatrix} \right\} \text{ and } Vev \cap V = \emptyset.$$

Clearly $B_{Pr_4}^{VFS} = VFS(Pr_4) = \{\underline{\varepsilon}\}$ as this is the simplest example of total deadlock when "handshaking" synchronization is used, but

$$B_{Pr_4}^{V^*} = FS_1 \times FS_2 \cap V^* = \left(\begin{bmatrix} a \\ b \end{bmatrix} \begin{bmatrix} b \\ a \end{bmatrix} \right)^* .$$

For every $A \subseteq V^*$, $V \subseteq GVev$, we define:

$$Pref_{V^*}(A) = \{v \in V^* \mid (\exists w \in V^*) vw \in A\}.$$

Clearly $Pref_{Vev^*}(A) = Pref(A)$, where $Pref(A)$ is as defined in Section 2.2.1. We shall say that A is *prefix-closed* in V^* if $A = Pref_{V^*}(A)$.

Let $V \subseteq GVev$ and $Contr \subseteq V^*$, $Contr \neq \{\varepsilon\}$, $Contr \neq \emptyset$ and $Contr$ is prefix-closed. Every such set of vectors $Contr$ can be treated as a **synchronization controller.** If the synchronization is controlled by $Contr$, then the behaviour of CS is defined as:

$$B_{CS}^{Contr} = B_1 \times \ldots \times B_n \cap Contr.$$

The set V is now interpreted as the set of atomic vectors of events. The sets V^* and Vev^* are special cases of synchronization controllers.

If $Contr$, treated as a language over the alphabet $GVev$, is regular, we have a **finite state controller.** One may show that in such a case, at every state of the evolution of a controlled system, the control allows a vector of events to be fired depending on its own state, and change its state depending on the vector just fired. Both $V^*(V \subseteq GVev)$ and Vev^* are finite state controllers.

The concept of V^*-synchronized behaviours and finite state controller originates from [Niv79] and was subsequently developed in [AN80], [Arn82]. In [Arn82] it is shown that behaviours defined by finite state controllers correspond to rational relations in $Ev_1^* \times \ldots \times Ev_n^*$. In the original approach of [Niv79], [AN80] and [Arn82] it was assumed that if the ith component of $v \in V$ is the empty string, it means that S_i *must wait while* v *is performed.* Thus the original synchronized behaviours semantics was an interleaving semantics, while vector firing sequence semantics was always truly concurrent semantics. On the other hand the mathematical model of synchronized behaviours semantics is also correct without this assumption, so it can be used to build a general truly concurrent vector-oriented semantics.

Chapter 6

Historical Perspective

6.1 Introduction

This historical perspective of the development of the COSY approach was written particularly to compensate for omissions we have had to make to obtain a text of reasonable size. There are several kinds of omissions we have had to make:

1. Omission of results in COSY theory which exist in the published literature, but which are based on different techniques which would have to be explained to make the book self contained. Their inclusion would have lengthened the present text considerably.
2. Omission of discussion of other important formal approaches to concurrency.
3. Omission of discussion of other important path notations based on different models of concurrency.
4. Omission of discussion of path or COSY like models which have largely appeared independently of knowledge of the similarity with our approach.

We will however repeat as little as possible from the biographical remarks at the end of individual chapters and sections. This historical sketch is solely intended to give a general background for our work, and to mention important related work which would otherwise have been completely omitted from the book, and hence cause the reader misunderstandings and concerns about possible lack of awareness of this related work on the part of the authors.

The particular combination of results about COSY presented in this book is the result of the development of a uniform method for representing as many results from COSY theory as possible, and the form of the presentation is very much determined by the intellectual background and experience of the two authors.

To indicate the ideas which have influenced our presentation we will briefly indicate the development of the authors' ideas up to the point at which they began to co-operate consciously on further development of COSY. This part of the historical perspective is a more detailed discussion of the overall perspective describing the development of the COSY theory within the framework of related work in concurrency theory.

6.2 Conceptual and Methodological Framework of COSY Approach

One of the main characteristics of the COSY approach is that it is **behaviour and history oriented.** This means that the semantics of expressions in the COSY language is formally defined in terms of its corresponding "execution" histories representing their corresponding behaviours consisting of elementary operations occurring in the expression.

The origins of this approach can be found in the *Graph-Shemata* of Kaluzhnin [Kal59] and the *Program Schemata* of Yanov [Yan58], both based on *Markov Algorithms* [Mar51, Mar54], and all are studied in [Lau67]. We will omit the discussion of the work of Markov and Khaluznin here and refer the reader to [Lau67] or the cited original papers. We will start from some of the ideas of Yanov relevant to the present discussion.

Based on Kaluzhnin's work, Yanov introduced Program Schemata after recognizing two crucial facts:

1. It is a general characteristic of every type of algorithm that they assign to an object a sequence of elementary operations, and usually different sequences to different objects. Each member of such a sequence depends on properties of the object that has been transformed in a manner completely determined by the preceding members of the sequence.
2. A **finite** set of predicates is sufficient to express properties of objects (before and after transformations), so that for a given algorithm, the dependence of the sequence of elementary operations on the object undergoing transformations is uniquely determined by such finitely many predicates. That is, which objects are channeled to which operators and in what order, if at all, depends on the built in sensitivity of test-operators with respect to a finite number of properties of objects.

This dependence can be expressed by means of a finite sequence of **operators** (symbols for elementary actions) $A_1, \ldots, A_n$, **predicates**, and certain auxiliary symbols $|_i$, $_i|$ $(i = 1, 2, \ldots)$ called left- and right-halfbrackets, respectively.

The sequence $A_{i_1} A_{i_2} \ldots A_{i_m}$ indicates that the elementary operations denoted by these operators are to be executed in that order, and is called an **A-sequence**. The sequence:

$$A_{i_1} \alpha |_k A_{i_2} \ldots {}_k| A_{i_3} \ldots$$

where α is a predicate, indicates the following: after the application of operator A_{i_1} to the object at hand, consider if the object transformed by A_{i_1} is such as to satisfy condition α or not. If yes, $\alpha = 1$. If no, $\alpha = 0$. If $\alpha = 1$ apply the next operator on the right of the left-halfbracket following α, i.e. the operator A_{i_2}. If $\alpha = 0$ disregard symbols within the halfbrackets with subscript k and apply A_{i_3} immediately. Such a sequence is called a **schematically represented algorithm** by Yanov.

If we abstract from the significance of the operators in schematic representations of algorithms, by regarding the operators as elementary symbols and predicates as independent propositional variables, the resulting expressions are called **program schemata**. If we assume the properties that qualify as interpretations for propositional variables as $p_1, p_2, \ldots p_n$ fixed, and before an object is submitted to the algorithm we assign to it an **evaluation** consisting of n-tuples of zeros and ones, where a 0 indicates that the object does not have the corresponding p_i, and a 1 indicates that the object has the corresponding property p_i. Furthermore, assume that an evaluation may change *only* as a result of the application of an operation. Given a program schema and a sequence of evaluations, one can obtain a unique A-sequence. To rephrase this, let $\mathcal{A}(p_1, \ldots, p_k, A_1, \ldots, A_n)$ be a program schema, then given a sequence of sets of values

$$\Delta_1, \Delta_2, \ldots, \Delta_m, \Delta_{m+1}, \ldots$$

of propositional variables $p_1, \ldots, p_k$, an A-sequence

$$A_{i_1}, A_{i_2}, \ldots, A_{i_m}, A_{i_{m+1}}, \ldots$$

is determined by the schema. So the semantics of the schema is formalized in terms of control predicates which are assigned values by the given sequences of deltas and in terms of the A-sequence or history of operation executions.

Both the level of abstraction for operations, namely uninterpreted elementary actions, and the history orientation of program schemata was chosen as the starting point of the COSY approach, as we shall see in more detail later.

Several notions of history were developed soon after by Thiele in his **Graph Schemata Calculi** [Thi66],[Thi65]and also [Lau68]. What is of concern here are the various notions of equivalence which Thiele develops on the basis of the notions of operation-decision-process, operation-process, result-process, and result function. In Yanov's approach two program schemata are equivalent if and only if they generate the same A-sequences for all admissible sequences of evaluations. Let us introduce the necesary notions from Thiele to discuss his notions of process.

Given a specific algorithmic language $\mathcal{L}$ over some operation-decision-structure $[J, O, T]$, J an arbitrary domain of individuals, the domain of (transformable) objects; O an arbitrary domain of n-place partial functions over J called basic operations; T a domain of n-place (two valued) test attributes over J, an **operation-decision-process** Π is for $\xi \in J$:

$$\Pi(\xi) = [\Pi_{0\xi}, \Pi_{1\xi}, \ldots]$$

where $[\Pi_{0\xi}, \Pi_{1\xi}, \ldots]$ is a (possibly infinite) sequence of $\Pi_{i\xi} \in O \cup T$ $(i \geq 0)$.

An **operation-process** Ω for $\xi \in J$ is:

$$\Omega(\xi) = [\omega_{0\xi}, \omega_{1\xi}, \ldots]$$

where $[\omega_{0\xi}, \omega_{1\xi}, \ldots]$ is a (possibly infinite) sequence of $\omega_{i\xi} \in O$ $(i \geq 0)$.

A **result-process** P for $\xi \in J$ is:

$$P(\xi) = [\rho_{0\xi}, \rho_{1\xi}, \ldots]$$

where $[\rho_{0\xi}, \rho_{1\xi}, \ldots]$ is a (possibly infinite) sequence of $\rho_{i\xi} \in J$ $(i \geq 0)$.

A **result-function** φ_P associated with the result-process P over J for $\xi \in J$ is:

$$\varphi_P(\xi) = \eta_\xi$$

if $\xi \in J$, $P(\xi)$ has a last element, and the last element is $\eta_\xi \in J$.

Note that one has increasingly less information as one moves from operation-decision-processes to result-functions.

Note also that these different notions are not to be thought of as mutually exclusive alternatives, but as mutually consistent and complementary notions at differing levels of abstraction. This notion of complementarity will be elaborated further in the sequel.

The monograph [LLS68] explains the foundations of the Vienna Definition Language VDL used in the classical formal definition of PL/I [WAF$^+$69], and makes use of a notion of computation as a (possibly infinite) sequence of states, however, because of non-deterministic aspects of PL/I (e.g. the order of elaboration of declarations, and the order of evaluation of arguments to procedures and functions is left open in the language definition), programs usually correspond to computations which are sets of sequences of states. These sets really represent trees, since all sequences in the set have some identical prefixes which would lead to tree structure if their common prefixes were collapsed. Such "trees" and their use resemble the computation trees used in many approaches to the semantics of non-determinism.

The states used in this approach were however much more complex than the evaluation sequences of Yanov, but included a store component which can be seen as playing a similar role as evaluation sequences in program schemata where dependence on these "states" was expressed in terms of the test predicates of the schemata. In its simplest form such a state was a function from the free individual variables of a program to values from the individual domain, that is, a semantic valuation as used in the semantic theory predicate logic. But appart from this difference, these state sequences directly correspond to result-sequences in the sense of Thiele.

The notion of **complementary formal definitions** of programming languages was introduced in [Lau71a, Lau71b] and [HL74] which also introduce the notion of **interpretive model** which was used in Lauer's proof of soundness of the Hoare proof system for partial correctness [Lau71a, Lau71b], and was later developed further and used in Cook's proof of completeness of the Hoare proof system [Coo78]. Subsequently these models have become standard in work on program verification (cf. e.g. [deB80, pp.146 ff.]).

The idea of complementary definitions was exemplified in [Lau71a, Lau71b] by the construction of a hierarchy of formal systems at differing levels of abstraction. At the bottom was a hierarchy of history (trace) oriented operational definitions called **Interpretive Models**, which represented the levels of operation-decision-process and result-process from the Thiele hierarchy. However, operation-decision-processes now consisted of sequences of pairs, the first member of which is the program (the control) and the second of which is a state, defined as a function from individual variables of the program to values in the individual domains. A computation is then considered to be a sequence:

$$(1)\ (P_0, s_0)(P_1, s_1)\ldots(P_m, s_m)\ldots$$

where P_{i+1} is a simple syntactic transformation of P_i representing the change of control due to the execution of P_i in state s_i, and s_{i+1} is the effect (if any) of updating values of free variables of the program. The style of this simple model originated from [dS69], and the use of a function **comp** to specify the computations corresponding to an arbitrary program in the interpretive model was followed in [Coo78] and [deB80]. These interpretive models were given a formulation in terms of labelled transition systems by Plotkin [Plo81], and have subsequently been used extensively for specifying operational semantics of non-determinism and concurrency (cf. e.g. [Plo81], [Mil88]). This structural operational semantics was also extended in [LL85] by using transitions labelled by vector firing sequences to allow the specification of true concurrency in this approach, which had previously been reduced to non-deterministic interleaving by proponents of this approach.

Result processes consisted simply of the sequences of states corresponding to the operation-decision-processes e.g.:

$$(2)\ s_0 s_1 \ldots s_m \ldots$$

in the above case. Note that (1) retains a history of all control decisions and operation executions, as well as all intermediate results and their order, whereas

(2) only retains a history of all intermediate results and their order. Both models are used to define interpretation functions which define the set of computations, in either of the two senses above, corresponding to an arbitrary program and initial state.

In [Lau71a, Lau71b] two additional formal definitions are introduced at even higher abstract levels, respectively. At the very top program semantics was specified in a Hoare calculus for partial correctness. Intermediate between the top and bottom levels program semantics was defined by an axiom system for binary relations over states in the sense used above. The style of this relational theory also originated in [dS69], and was extensively developed in [deB80],[deR76] and [Bes83]. In terms of levels of abstraction, the relational theory abstracts from the notion of computation which is present in the interpretive (computational) models at the bottom level, but it keeps the notion of state from these computational models. The Hoare calculus further abstracts from the notion of states present in the relational theory, and only works with predicate formulas, leaving the states implicit.

Our experience with the use of Hoare calculi and relational theories to define the semantics of sequential and deterministic languages suggested that it was perhaps premature to work in the area of concurrency theory on the basis of such calculi and theories, since the concept of concurrency needs first to be studied in terms of more concrete (computational, behavioural) models which have been abstracted from real situations involving concurrency, or some other well developed theory of time or causality from e.g. Physics. Only then will it be possible to identify the properties of concurrency and non-determinacy which are important in the practical case, which should then be axiomatized and formalized at more abstract levels. Otherwise we obtain formalizations which may be theoretically tractable but practically infeasible as has proven to be the case for most approaches to concurrency which have moved to the more abstract levels too soon or from the very outset. We will discuss the import of these statements for the initial decisions taken in determining the COSY approach as characterized in this book.

In [Lau71a, Lau71b] we also introduced the non-deterministic choice operator at the level of statements having the form $(Q_1 \mathbf{or} Q_2)$, where Q_1, Q_2 are arbitrary statements of the programming language. This meant that the computations corresponding to programs had to be computation trees rather than sequences, but again represented as sets of sequences representing trees. The semantic interpreter produced a set corresponding to all possible non-deterministic choices.

The history or trace based approach used in [Lau71a, Lau71b] was considerably developed by Hoare in [Hoa78b] and has found many applications since in specifying models of Hoare logics, particularly for the case of CSP and TCSP.

The notions of computations in the style of Lauer's interpretive models underwent further development in [Bes81] to give a formal semantics of atomicity in concurrent systems, and by Hillen [Hil85] to specify the semantics of CSP.

At the inception of the original COSY effort [Lau75a], which arose out a suggestion of Brian Randall [Lau74], we were faced with the task of formally specifying the semantics of Path Expressions [CH74] in a way that would allow the expression of true concurrency. This meant we had to decide how to define true concurrency first. There were several alternatives open to us, such as, defining concurrency in terms of the notion of time (as done for instance in temporal logic), or to adapt an existing theory of concurrency to the needs at hand. We decided against the use of a notion of time since time is not an objective phenomenon for various philosophical reasons. The notion of causality is however an more objective phenomenon and there existed the notion of concurrency in physics which can be defined in terms of causality. Thus, necessary sequentiality of event occurrences is identified with causal dependence between the events, and concurrency is identified with the absence of causal dependence between event occurrences. It turned out that there existed a theory of concurrency based on the primitive notion of causality in the for of Petri net theory [Pet62] which had been developed to a considerable degree by the time we began the COSY effort. Hence, this theory was our basis for examining the expression of concurrency in the the path notation, provided we could specify the behavioural semantics of this notation by corresponding Petri nets. [Lau74] introduces the syntax directed style of defining the semantics of non-deterministic and concurrent in terms of labelled condition event nets which was used throughout the development of the COSY notation [LC75], [LBS77], and in most subsequent work using nets to define semantics of programming languages [Old87], [Bes88].

As a result of the specification of the concurrency operator of the original path notation [CH74] in terms of Petri nets in [Lau74] it was seen that the construct had a very complex semantics, which indicated itself in the complexity of the corresponding labelled net. The path notation used a single path expression to specify the synchronization of an abstract data type in terms of the names of procedures associated with the abstract data type. In order to express necessary sequentialization of invocations of procedures p and q, one would write: $\underline{path}\ p; q\ \underline{end}$. In order to express necessary non-deterministic exclusive choice between the procedure invocations one would write: $\underline{path}\ p, q\ \underline{end}$. To express the possibility of unbounded concurrent (possibly simultaneous) invocations of the procedure p mutually exclusive with invocations of procedure q one would write:

$$\underline{path}\ \{p\}, q\ \underline{end}$$

which becomes a solution to the reader/writer problem [CHP71] in the form:

$$\underline{path}\ \{read\}, write\ \underline{end}.$$

The net semantics for this concurrency construct turned out to be complex and the corresponding net was large, and furthermore, the correctness of the construct presupposed certain priorities between events in the system which could only be modelled by using priority nets as discussed in Sect. 5.3.2. This suggested to us that path notation be modified to allow simpler and more controlled parallelism to be expresses and studied. Furthermore, it was decided to extract path expressions from their embedding in programming notation, and to use uninterpreted basic events in place of procedure names. This eventually led to the Path-Process notation of [LC75], [Cam76], which is essentially the basic COSY notation presented in this book.

This level of abstractness with regard to the basic actions of the formalism also followed from the use of net semantics (condition event nets are at this level), and it is the same level that was later used in CCS [Mil80] and TCSP [Hoa80]. We believe that the fact that all these approaches found this particular level of abstraction convenient for the discussion of concurrency in general tends to ensure that we made a good choice of level of abstraction at the outset.

Returning to our consideration of the various notions of process or computation earlier in this section, we realized that the notion of (labelled) causal net from net theory, which define the standard semantics of condition event nets in Petri net theory, were precisely the generalization of computations as sequences (to express deterministic sequential processes), sets of sequences or trees (to express non-deterministic sequential processes, or concurrent processes reduced to non-deterministic processes), to certain sets of labelled acyclic directed graphs (to express the concurrent aspects and the non-deterministic aspects of processes orthogonally). The fact that interleaving models of concurrency reduce concurrency to non-deterministic interleaving means that the distinction between these two aspects disappears, and choices now indicate either real choices necessitated by the nature of the processes being described or the choices necessitated by the limitations of the model. This is probably one of the most telling reasons against the use of interleaving models to study concurrency.

In [HIL72] we report on the development of algorithms which determine the absence of deadlock in condition event nets from their initial markings and the structure of the net, without the need for generating all possible behaviours of the net. This gave rise to another important idea at the base of the COSY approach, namely to provide a suitable setting for determining the extent to which properties of dynamic (global) behaviours of a program (e.g. absence of deadlock or starvation) could be expressed and analyzed in terms of the static structure of the program. We can rephrase this by saying that we are concerned with the reduction of certain aspects of semantics to syntax to the greatest degree possible.

Let us reiterate that we felt that it was too early to move to higher levels of abstraction than operational (computational) models permitting the explicit expression and study of concurrency structure of processes, and hence we decided against the use of, for example, a relational theory or Hoare logic in the study of concurrency. Hence our concentration on the study of structural properties of concurrent computations, rather than a concern for the immediate construction of axiom systems or logics for concurrency, since the general nature of concurrency was not well enough understood. Of course it was always our intent to develop appropriate logics eventually, which are consistent with the more explicit conctrete model of concurrency represented by Petri net theory and COSY theory. More on this later.

Our position at the outset can be described as follows:

1. Use net-semantics to understand what concurrency means – especially since it is independent of the notion of time, a notoriously difficult concept to formalize, and since it depends solely on the notion of causality, which is less arbitrary than the notion of time.
2. Redesign the COSY-path notation to obtain a convenient linguistic tool for expressing behaviours corresponding to nets (usually presented as graphs), and to obtain a more structured method of composing (or decomposing) nets and expressing net structure (e.g. nets corresponding to COSY specifications are state-machine decomposable). Furthermore, try to define sub-languages of COSY which correspond to known subclasses of nets for which the structure of the net and its initial marking guarantee e.g. absence of deadlock. This will enable the proof of such properties of programs by determining if the program can be stated in the language of the appropriate sub-language.
3. Development of an independent direct semantics for COSY based on language or automata theoretic methods. Very soon after the start of the COSY effort we recognized that our attempt to reduce synchronization semantics to syntactic criteria would be facilitated by a more direct language theoretic approach to the semantics of COSY programs in terms of histories which would be equivalent to causal-net semantics for the corresponding marked condition event nets for the programs.

   This first led to the use of **firing sequences** and **projection functions** to define an interleaving semantics for COSY [Shi77], [LSB79], and later to the notion of **vector firing sequences** enabling us to express concurrency in processes in a manner consitent with the causal basis of concurrency in net theory. This gave us three standard complementary semantic techniques to work with for various purposes.

Another influence of work in net theory was the recognition of the need for generators for expressing complex paths in a compact and perspicuous form to be able to able to manage size and complexity, and to support abstraction. The macro COSY notation is a direct consequence of the recognition that generators used in net theory could be modified to generalize the notion of distributor introduced in the path-process notation [LC75]. The notion of replicator which was developed as part of this macro notation is a more general version of the replicators which were later incorporated in CSP and *occam* [INM84]. See [LJ87] for historical details of this development which was contributed to by H. Genrich, P.R. Torrigiani, R. Devillers, Y. Cotronis and R. Janicki.

While the COSY project was initiated by Peter Lauer, Ryszard Janicki joined in the effort in 1982. His prior research included coroutine theory ([Jan76],[Jan79b]), concurrency-like relations [Jan79a] and compositional and decompositional properties of Petri nets [Jan84b]. He was strongly influenced by the algebraic approach to the fundamentals of computer science represented by the "Warsaw school" of theoretical computer science (cf. [Paw68], [Bli74], [Maz71]). The reader might find traces of this influence, mainly in the sense of tools and methodology used, in many parts of this book.

The first two members of the COSY project were M.W. Shields and E. Best, and each was mainly responsible for one of the major concerns of the project. Best had worked in net theory [Bes74] was mainly responsible for developing the net semantics of COSY and characterizing corresponding sub-languages of COSY and net sub-categories that would allow transferral of results about COSY to net theory and vice versa. This work culminated in [LSB79], [Bes82], [Bes87].

Shields was mainly responsible for the development of complementary language and automata theoretic semantics for COSY consistent with the net semantics. This work culminated in [Shi79b], [LSB79] and [Shi85].

Since the early eighties others such as M. Koutny, T. Müldner, F.Okulicka became involved in the COSY oriented research. We hope that major contributions of all persons involved in the mainstream of the COSY approach are included in this book.

There have been a number of other path notations which were developed independently of our work [Cam76], and which we have had to ignore in this book due to space constraints. These omissions are perhaps justified by the fact that our path expressions were designed on the basis of net semantics and other path notations were designed on entirely different bases and for other, usually more practical programming purposes, which make them less directly comparable to our notation and largely irrelevant for our purposes in this book. For the sake of completeness however we list a fairly complete set of references here [And79], [Cam83], [CK79], [HO85], [Sch76], [Tam88].

Other obvious omissions are discussions of approaches to concurrency based on Hoare logic [Hoa85], Modal logic [Pnu86], Dynamic logic [Har79], Process logic [BK86] and [BBK87], Temporal logic [Kro87] and [OL80], etc. But as we have pointed out above, the book concentrates on history or trace based semantics of COSY semantics of true concurrency, and now that we have enough results in place we feel that we will be able to develop a (modal) logic based on the notion of causality and capable of treating true concurrency which will have, for example, the advantages of both linear time and branching time logic, each of which lacks expressive power relative to the other [EH86]. For these reasons we feel justified in omitting important work being done in these approaches in this volume. But we will rectify these omissions in future publications as we develop our causality based (modal) logics and contrast them with other logic based approaches.

For similar reasons, we are unable to present any discussion of Linear Logic [Gir87] and its relationship to theory of concurrency. Recently, a close relationship has been established between Petri nets and theories in the tensor fragment of Linear Logic [GG90].

Finally, we have almost entirely omitted reference to the considerable ammount of work that has recently been done in so called "trace-theory" [Kel73], [vdS83]. A thorough presentation and contrasting of these newer results would produce another volume the size of the present. Again we hope to present such comparisons in future publications now that we have a relatively uniform presentation of our results in a self-contained volume.

Similarly, we have had to omit discussion of the **Denotational Approach** to concurrency theory [Kah74], [Kel78], [Win82], [Hen83]; the work based on **Transition Systems** [Kel72], [Sif82], [Plo81], [Mil88]; Montanari **histories** [DM87]; Winskel's **Event Structures** [Win88]; **Actor Models** [Hew77], [Agh86]; **Parallel Schemata and Vector Addition Systems** [KM69]; **Data Flow** approach to concurrency [GWK84]; **Functional Programming** approach [DR81]; **Logic Programming** [ST83]; **Algorithmic Logic** [SM81]; **IO Automata** [LT87]; **Partially Ordered Multi Sets (pomsets)** [Pra86].

Bibliography

[Agh86] G. Agha. *Actors, A Model of Concurrent Computation in Distributed Systems.* MIT Press, 1986.

[AHU74] A.V. Aho, J.E. Hopcroft, and J.D. Ullman. *The Design and Analysis of Computer Algorithms.* Addison-Wesley, 1974.

[AN80] A. Arnold and M. Nivat. *Controlling behaviours of systems: Some basic concepts and some applications.* Lecture Notes in Computer Science 88. Springer Verlag, 1980. pp.113-122.

[And79] S. Andler. Predicate Path Expressions. In *6th ACM Symposium on Principles of Programming Languages*, 1979.

[AR88] I.J. Aalbersberg and G. Rozenberg. Theory of Traces. *Theoretical Computer Science*, 60:1–82, 1988.

[Arn82] A. Arnold. Synchronized Behaviours of Processes and Rational Relations. *Acta Informatica*, 17:21–29, 1982.

[Avi85] A. Avizienis. The N-version Approach to Fault-tolerant Software. *IEEE Trans. on Software Eng.*, pages 1491–1501, 1985.

[Bac89] R.J.R. Back. *A Method for Refining Atomicity in Parallel Algorithms.* Lecture Notes in Computer Science 366. Springer Verlag, 1989. pp.199-216.

[BBK87] J.C.M. Baeten, J.A. Bergstra, and J.W. Klop. Ready-Trace Semantics for Concrete Process Algebra with the Priority Operator. *The Computer Journal*, 30(6):498–506, 1987.

[BD85] E. Best and R. Devillers. Concurrent Behaviour : Sequences, Processes and Programming Languages. GMD-Studien 99, GMD Bonn, May 1985.

[BD87] E. Best and R. Devillers. Sequential and Concurrent Behaviours in Petri Net Theory. *Theoretical Computer Science*, 55:87–136, 1987.

[Bes74] E. Best. Beiträge zur Petrinetz-Theorie. Master's thesis, Universität Karlsruhe, 1974. Diplomarbeit, 94pp.

[Bes81] E. Best. *Semantics, Verification and Design of Concurrent Programs using Atomic Actions.* PhD thesis, University of Newcastle upon Tyne, 1981. 244pp.

[Bes82] E. Best. Adequacy Properties of Path Programs. *Theoretical Computer Science*, 18:149–171, 1982.

[Bes83] E. Best. Relational Semantics of Concurrent Programs (with some applications). In D. Björner, editor, *Formal Description of Programming Concepts-II.* North-Holland, 1983. pp.431-452.

[Bes87] E. Best. *COSY : Its Relation to Nets and to CSP.* Lecture Notes in Computer Science 255. Springer Verlag, 1987. pp.416–440.

[Bes88] E. Best. Kausale Semantik Nichtsequentieller Programme. In *Habilitationsschrift zur Erlangung der Venia Legendi*, 1988. 187pp.

[BF86] E. Best and C. Fernandez. Notation and Terminology on Petri Net Theory. *Petri Net Newsletter*, 23:21–46, 1986.

[BH73] P. Brinch Hansen. *Operating Systems Principles.* Prentice Hall, 1973.

[BHR84] S.D. Brookes, C.A.R. Hoare, and A.W. Roscoe. A Theory of Communicating Sequential Processes. *JACM*, 31(7):560–599, 1984.

[BK] E. Best and M. Koutny. Petri Net Semantics of Priority Systems. *Theoretical Computer Science.* (to appear).

[BK86] J.A. Bergstra and J.W. Klop. *Process Algebra: specification and verification in bisimulation semantics.* CWI Monograph 4. North Holland, 1986. pp.61–94.

[Bli74] A. Blikle. *Proving programs by sets of computations.* Lecture Notes in Computer Science 28. Springer Verlag, 1974. pp.313-358.

[Bos88] P. Bose. Heuristic rule-based transformations of enhanced vectorization. In *Proc. 17th Internat. Conf. on Parallel Processing, Vol. II.* Penn State Press, 1988. pp.63-66.

[BR85] S.D. Brookes and A.W. Roscoe. *An improved failures model for Communicating Sequential Processes.* Lecture Notes in Computer Science 197. Springer Verlag, 1985. pp.281-306.

[BRR86a] W. Brauer, W. Reisig, and G. Rozenberg, editors. *Petri Nets: Control Models and their Properties, Part I.* Lecture Notes in Computer Science 254. Springer Verlag, 1986.

[BRR86b] W. Brauer, W. Reisig, and G. Rozenberg, editors. *Petri Nets: Control Models and their Properties, Part II.* Lecture Notes in Computer Science 255. Springer Verlag, 1986.

[BS83] E. Best and M.W. Shields. *Some Equivalence Results on Free Choice Nets and Simple Nets, and on the Periodicity of Live Free Choice Nets.* Lecture Notes in Computer Science 159. Springer Verlag, 1983. pp.141-159.

[BS86] G. Berry and R. Sethi. From Regular Expressions to Deterministic Automata. *Theoretical Computer Science*, 48:117–126, 1986.

[Cam76] R.H. Campbell. *Path Expressions: A technique for process synchronization.* PhD thesis, University of Newcastle upon Tyne, 1976.

[Cam83] R.H. Campbell. Distributed Path Pascal. In Y. Paker and J.R. Verjus, editors, *Distributed Computer Systems: Synchronization, Control and Communication.* Academic Press, 1983. pp.191-223.

[CF69] P. Cartier and D. Foata. *Problèmes combinatoires de commutation et rearrangements.* Lecture Notes in Mathematics 85. Springer Verlag, 1969.

[CH74] R.H. Campbell and A.N. Habermann. *The specification of process synchronization by path expressions.* Lecture Notes in Comper Science 16. Springer Verlag, 1974. pp.89-102.

[CHP71] P.J. Courtois, F. Heymans, and D.L. Parnas. Concurrent Control with "Readers" and "Writers". *CACM*, 14:667–668, 1971.

[CK79] R.H. Campbell and R.B. Kolstadt. Path expressions in Pascal. In *4th Int. Conf. on Software Engineering*, Munich, 1979. pp.212-219.

[CL75] R.H. Campbell and P.E. Lauer. A hierarchy of buffer systems in a parallel processing environment. Technical Report MRM/88, Computing Laboratory, University of Newcastle upon Tyne, 1975.

[CL81] J.Y. Cotronis and P.E. Lauer. A syntax for the macro notation. Technical Report ASM/83, Computing Laboratory, University of Newcastle upon Tyne, 1981.

[CL83] J.Y. Cotronis and P.E. Lauer. *Two Way Channel with Disconnect.* Lecture Notes in Computer Science 207. Springer Verlag, 1983. pp.184-198.

[CL84] R.H. Campbell and P.E. Lauer. RECIPE : Requirements for an Evolutionary Computer-based Information Processing Environment. In *Proceedings of the Software Process Workshop*, Egham, U.K., 1984. IEEE Computer Society Press, 1984, pp.67-76.

[CM85] R. Cori and Y. Metivier. Recognizable Subsets of Partially Abelian Monoids. *Theoretical Computer Science*, 35:179–189, 1985.

[Coo78] S.A. Cook. Soundness and completeness of an axiom system for program verification. *SIAM Journal on Computing*, 7(1), 1978.

[Cot82] J.Y. Cotronis. *Programming and Verifying Asynchronous Systems.* PhD thesis, University of Newcastle upon Tyne, 1982.

[deB80] J. deBakker. *Mathematical Theory of Program Correctness.* Prentice Hall, 1980.

[deR76] W.P. deRoever. *Recursive Program Schemes: Semantics and Proof Theory.* PhD thesis, University of Amsterdam, 1976. Mathematical Centre Tracts 70, 1976, pp.112.

[Dev79a] R. Devillers. A simplified hyper fast Banker. Technical Report ASM/56, Computing Laboratory, University of Newcastle upon Tyne, 1979.

[Dev79b] R. Devillers. On the Banker with Several Currencies. Technical Report ASM/57, Computing Laboratory, University of Newcastle upon Tyne, 1979.

[Die88] H.G. Dietz. Finding large-grain parallelism in loops with serial control dependencies. In *Proc. 17th Internat. Conf. on Parallel Processing Vol.II.* Penn. State Press, 1988. pp.114-121.

[Dij68] E.W. Dijkstra. Co-operating Sequential Processes. In F. Genuys, editor, *Programming Languages.* Academic Press, 1968.

[Dij75] E.W. Dijkstra. Guarded commands, nondeterminacy, and a calculus for the derivation of programs. *CACM*, 18(8):453–457, 1975.

[Dij76] E.W. Dijkstra. *A Discipline of Programming.* Prentice Hall, 1976.

[DJW85] B.T. Denvir, M.I. Jackson, and M.I. Wray, editors. *The Analysis of Concurrent Systems.* Lecture Notes in Computer Science 207. Springer Verlag, 1985.

[DM87] P. Degano and U. Montanari. A model for distributed systems based on graph rewriting. *JACM*, 34:411–449, 1987.

[DR81] J. Darlington and M. Reeve. Alice - a multiprocessor reduction machine for the parallel evaluation of applicative languages. In *Proc. ACM Conf. on Functional Programming Languages and Computer Architecture*, 1981. pp.65-75.

[dS69] J. deBakker and D. Scott. A Theory of Programs. An outline of joint work by the authors in response to a letter from H. Bekic and P.E. Lauer, 1969.

[EH86] E.A. Emerson and J.Y. Halpern. "Sometimes" and "Not Never" Revisited: On Branching versus Linear Time Temporal Logic. *JACM*, 33:151–178, 1986.

[FA73] M. Flynn and T. Agerwala. Comments on capabilities, limitations and "correctness" of Petri nets. *Computer Architecture News*, 2(4), 1973.

[GG90] V. Gehlot and G. Gunter. Normal Process Representatives. In *5th IEEE Symposium on Logic in Computer Science*, Philadelphia, 1990. pp.200-207.

[Gir87] J.-Y. Girard. Linear Logic. *Theoretical Computer Science*, 50:1–102, 1987.

[GL73] S.J. Garland and D.C. Luckham. Program Schemes, Recursion Schemes, and Formal Languages. *J. Comp. System Sci.*, 7:119–160, 1973.

[GL81] H. Genrich and K. Lautenbach. System Modelling with High-level Petri Nets. *Theoretical Computer Science*, 13:109–136, 1981.

[Gol80] J. Goldberg. SIFT, A provable fault-tolerant computer for aircraft control. In *Proceedings of IFIP'80 Congress, Tokyo, Japan*, North-Holland, 1980.

[GR85] U. Goltz and W. Reisig. *CSP-programs as nets with individual tokens.* Lecture Notes in Computer Science 188. Springer Verlag, 1985. pp.169-196.

[GWK84] J.R. Gurd, I. Watson, and C. Kirkham. The Manchester dataflow project. In D.A. Duce, editor, *Distributed computing systems programme.* Peter Peregrinus Ltd., 1984. pp.270-289.

[Hab69] A.N. Habermann. Prevention of system deadlocks. *CACM*, 12(7), 1969.

[Hab73] A.N. Habermann. On a solution and a generalization of the cigarette smokers problem. Technical report, Carnegie Mellon University, 1973.

[Hac75] M. Hack. Petri Net Languages. Computation Structures Group Memo 127, MIT, 1975. Project MAC.

[Ham83] B.C. Hamshere. A Computer Based Environment for the Design and Analysis of Concurrent Systems: SIMULA Implementation of the COSY Notation. In *11th Annual Conf. of SIMULA Users*, Paris, 1983.

[Har79] D. Harel. *First Order Dynamic Logic.* Lecture Notes in Computer Science 68. Springer Verlag, 1979.

[Hen83] M.C.B. Hennessy. Modelling Finite Delay Operators. CSR Report 153-83, Dept. of Computer Science, University of Edinburgh, 1983. 58pp.

[Hew77] C.E. Hewitt. Viewing control structures as patterns of passing messages. *Journal of Artificial Intelligence*, 8(3):323–364, 1977.

[Hey89] T. Hey. *Experiments in MIMD Parallelism.* Lecture Notes in Computer Science 366. Springer Verlag, 1989. pp.28-42.

[HIL72] A. Hansal, H. Izbicki, and P.E. Lauer. Some Algorithms on Graph-theoretical Matrices. Lab. Report LB 25.6.034, IBM Laboratory Vienna, February 1972.

[Hil85] D. Hillen. Ein nebenläufiges Kalkül für CSP-ähnliche Sprachen. (Diplomarbeit), Bonn University, 1985. also: Arbeitspapiere der GMD Nr.222, 1986.

[HK88] D.V. Hung and E. Knuth. A noninterleaving semantics for Communicating Sequential Processes: a fixed-point approach. *Acta Cybernetica*, 8(3):293–311, 1988.

[HL74] C.A.R Hoare and P.E. Lauer. Consistent and Complementary Formal Theories of the Semantics of Programming Languages. *Acta Informatica*, 3:135–153, 1974.

[HO85] M.R. Headington and A.E. Oldehoeft. Open Predicate Path Expressions and their implementation in highly parallel computing environments. In *Int. Conf. on Parallel Processing.* IEEE Computer Society Press, 1985. pp.239-246.

[Hoa78a] C.A.R Hoare. Communicating Sequential Processes. *CACM*, 21(8):666–697, 1978.

[Hoa78b] C.A.R Hoare. Some properties of predicate transformers. *Journal of the ACM*, 25(3):461–480, 1978.

[Hoa80] C.A.R. Hoare. A Model for Communicating Sequential Processes. In R.M. McKeag and A.M. Macnaghten, editors, *On the Construction of Programs.* Cambridge University Press, 1980. pp.229-254.

[Hoa85] C.A.R. Hoare. *Communicating Sequential Processes.* Prentice Hall, 1985.

[Hol74] A.W. Holt. Final Report for the Project "Development of the Theoretical Foundations for Description and Analysis of Discrete Information Systems". CADD-7405-2011, Applied Data Research, Wakefield Mass., 1974.

[Hov68] J.W. Hovender. Avoiding Deadlock in Multitasking Systems. *IBM System Journal*, 7(2):74–84, 1968.

[INM84] INMOS. *OCCAM Programming Manual.* Prentice Hall, 1984.

[Jan76] R. Janicki. *Vectors of Coroutines.* Lecture Notes in Computer Science 45. Springer Verlag, 1976. pp.377-384.

[Jan79a] R. Janicki. *A Characterization of Concurrency-Like Relations.* Lecture Notes in Computer Science 70. Springer Verlag, 1979. pp.109-122.

[Jan79b] R. Janicki. Analysis of coroutines by means of vectors of coroutines. *Fundamenta Informaticae*, 2(3):289–316, 1979.

[Jan81a] R. Janicki. *A construction of concurrent schemes by means of sequential solution and concurrency relation.* Lecture Notes in Computer Science 107. Springer Verlag, 1981. pp.327-337.

[Jan81b] R. Janicki. On the Design of Concurrent Schemes. In *Proc. 2nd IEEE Conf. on Distributed Computing Systems*, Paris, 1981. IEEE Press. pp.455-466.

[Jan84a] R. Janicki. *A method for developing concurrent systems.* Lecture Notes in Computer Science 167. Springer Verlag, 1984. pp.155-166.

[Jan84b] R. Janicki. Nets, Sequential Components and Concurrency Relations. *Theoretical Computer Science*, 29:155–166, 1984.

[Jan85a] R. Janicki. An Equivalence Notion for Path Expressions Systems. *Elektron. Inf.verarb. Kybern. EIK*, 21(6):283–295, 1985.

[Jan85b] R. Janicki. Transforming sequential systems into concurrent systems. *Theoretical Computer Science*, 36:27–58, 1985.

[Jan87] R. Janicki. A Formal Semantics for Concurrent Systems with Priorities. *Acta Informatica*, 24:33–55, 1987.

[Jan88] R. Janicki. How to relieve a programmer from synchronization details. In *Proc. 16th ACM Computer Science Conference*, Atlanta, Georgia, 1988. pp.438-446.

[Jen81] K. Jensen. Coloured Petri Nets and the Invariant Method. *Theoretical Computer Science*, 14:317–336, 1981.

[JK87] R. Janicki and M. Koutny. *On equivalent execution semantics of concurrent systems.* Lecture Notes in Computer Science 266. Springer Verlag, 1987. pp.89-103.

[JK89] R. Janicki and M. Koutny. *Towards a Theory of Simulation for Verification of Concurrent Systems.* Lecture Notes in Computer Science 366. Springer Verlag, 1989. pp.73-88.

[JK91a] R. Janicki and M. Koutny. *Invariants and Paradigms of Concurrency Theory*. Lecture Notes in Computer Science 506. Springer Verlag, 1991. pp.59-74.

[JK91b] R. Janicki and M. Koutny. On some implementation of optimal simulation. *DIMACS Series in Discrete Mathematics and Theoretical Computer Science*, 3:231–250, 1991.

[JL88] R. Janicki and P.E. Lauer. On the semantics of priority systems. In *Proc. of the 1988 International Conference on Parallel Processing*, Pheasant Run, Illinois, August 15-19 1988. pp.150-156.

[JLD85] R. Janicki, P.E. Lauer, and R. Devillers. *Maximal Concurrent Evolution of Non-sequential Systems*. Lecture Notes in Computer Science 197. Springer Verlag, 1985. pp.268-280.

[JLKD86] R. Janicki, P.E. Lauer, M. Koutny, and R. Devillers. Concurrent and maximally concurrent evolution of non-sequential systems. *Theoretical Computer Science*, 43:213–238, 1986.

[JM89] R. Janicki and T. Müldner. Complete Sequential Specification allows for Concurrent Execution. In *Proc. of 17th Ann. ACM Comp. Science Conference*, Louisville, Kentucky, 1989. pp.221-231.

[JM90] R. Janicki and T. Müldner. Transformations of sequential specifications into concurrent specifications. *Theoretical Computer Science*, 77:97–129, 1990.

[JV80] M. Jantzen and R. Valk. *Formal Properties of Place/Transition Nets*. Lecture Notes in Computer Science 84. Springer Verlag, 1980. pp.165-212.

[Kah74] G. Kahn. The semantics of a simple language for parallel programming. In *IFIP Congress 1974*. North Holland, 1974. pp.471-475.

[Kal59] L.A. Kaluzhnin. Ob Algoritmizatsii Matematicheskikh Zadach (On Algorithmization of Mathematical Problems). *Problemy Kibernetiki*, 2:51–67, 1959.

[Kel72] R.M. Keller. Vector replacement systems: a formalism for modelling asynchronous systems. Technical Report 117, Princeton University, 1972.

[Kel73] R.M. Keller. Parallel Program Schemata and Maximal Parallelism I: Fundamental Results. Parallel Program Schemata and Maximal Parallelism II: Construction of Closures. *JACM*, 20(3, 4):514–537, 696–710, 1973.

[Kel78] R.M. Keller. Denotational models for parallel programs with indeterminate operators. In E.J. Neuhold, editor, *Formal Descriptions of*

Programming Languages. North Holland, 1978. pp.337-363.

[KM69] R.M. Karp and R.E. Miller. Parallel Program Schemata. *JCSS*, 3:147–195, 1969.

[KM89] M. Koutny and L. Mancini. Synchronizing events in replicated systems. *Journal of Systems and Software*, 9:183–190, 1989.

[Kou] M. Koutny. Adequacy-preserving Transformations of COSY Path Programs. *Theoretical Computer Science.* (to appear).

[Kou84] M. Koutny. *On the Merlin-Randell Problem of Train Journeys.* Lecture Notes in Computer Science 167. Springer Verlag, 1984. pp. 179-190.

[Kou86] M. Koutny. The Merlin-Randell Problem of Train Journeys. *Acta Informatica*, 23:429–463, 1986.

[Kou88] M. Koutny. *Adequacy-preserving Transformations of COSY Path Programs.* Lecture Notes in Computer Science 335. Springer Verlag, 1988. pp.368-379.

[Kro87] F. Kroeger. *Temporal Logic of Programs.* EATCS Monographs on Theoretical Computer Science 8. Springer Verlag, 1987.

[Lal79] G. Lallement. *Semigroups and Combinatorial Applications.* J. Wiley and Sons, 1979.

[Lam85] L. Lamport. What it means for a concurrent program to satisfy a specification: why no one has specified priority. In *12th ACM Symposium on Principles of Programming Languages*, New Orleans, Louisiana, 1985. pp.78-83.

[Lau67] P.E. Lauer. The formal explicates of the notion of algorithm: An introduction to the theory of computability with special emphasis on the various formalisms underlying the alternative explicates. Technical Report TR25.072, IBM Laboratory Vienna, 1967. 297pp.

[Lau68] P.E. Lauer. An introduction to H. Thiele's notions of algorithm, algorithmic process and graph-schemata calculus. Technical Report TR25.079, IBM Laboratory Vienna, 1968. 34pp.

[Lau71a] P.E. Lauer. Consistent and Complementary Formal Theories of the Semantics of Programming Languages. In *IVth Int. Congress for Logic, Methodology and Philosophy of Science*, 1971.

[Lau71b] P.E. Lauer. *Consistent Formal Theories of the Semantics of Programming Language.* PhD thesis, Queen's University, Belfast, 1971. Also Technical Report TR25.121, IBM Laboratory Vienna, (1971), 122pp.

[Lau74] P.E. Lauer. Path Expressions as Petri Nets, or Petri Nets with fewer Tears. Technical Report MRM/70, Computing Laboratory, University of Newcastle upon Tyne, 1974.

[Lau75a] P.E. Lauer. A Project to Investigate a Design Technique for Asynchronous Systems of Processes. Technical Report Memo MRM/89, Computing Laboratory, University of Newcastle upon Tyne, 1975.

[Lau75b] P.E. Lauer. On the logical description of simple conventional hardware configurations. Technical Report TR80, Computing Laboratory, University of Newcastle upon Tyne, 1975.

[Lau83] P.E. Lauer. Computer System Dossiers. In P. Paker and J.P. Verjus, editors, *New Advances in Distributed Computing Systems: Synchronization, Control and Communication.* Academic Press, 1983. pp.109-138.

[Lau85] P.E. Lauer. *A Simple Railway System.* Lecture Notes in Computer Science 207. Springer Verlag, 1985. pp.271-292.

[LBS77] P.E. Lauer, E. Best, and M.W. Shields. On the problem of achieving adequacy of concurrent programs. In E.J. Neuhold, editor, *Formal Description of Programming Concepts.* North Holland, 1977. pp.301-334.

[LC75] P.E. Lauer and R.H. Campbell. Formal Semantics of a Class of High-Level Primitives for Coordinating Concurrent Processes. *Acta Informatica*, 5:297–332, 1975.

[Len82] C. Lengauer. A methodology for programming with concurrency. *Science of Computer Programming*, 2:19–52, 1982.

[LH81] P.E. Lauer and B.C. Hamshere. A computer based environment for the design and analysis of highly parallel and distributed systems. Technical Report ASM/88, Computing Laboratory, University of Newcastle upon Tyne, 1981.

[LH82] C. Lengauer and E.C.R. Hehner. A methodology for programming with concurrency. *Sci. of Computer Programming*, 2:1–18, 1982.

[LJ84] P.E. Lauer and J. Just. Modelling the Control Structures for Microprocessor Systems by means of Petri-Nets. In *3rd International Association of Science an Technology for Development, International Symposium on Modelling, Identification and Control MIC'84*, Innsbruck, Austria, February 14-17 1984.

[LJ87] P.E. Lauer and R. Janicki. An Introduction to the MACRO COSY Notation. In K. Voss, H.J. Genrich, and G. Rozenberg, editors, *Concurrency and Nets: Advances in Petri Nets.* Springer Verlag, 1987. pp.284-314.

[LJJ83] P.E. Lauer, J.R. Just, and R. Janicki. On the Description of Simple Microprocessor Configurations by means of Petri-Nets. In *International Conference on Microprocessors and Microprogramming*, Budapest, Hungary, 1983.

[LL85] W. Li and P.E. Lauer. Using the structural operational approach to express true concurrency. In E.J. Neuhold and G. Chroust, editors, *Formal Models in Programming*. North-Holland, 1985. pp.373-397.

[LLS68] P. Lucas, P.E. Lauer, and H. Stigleitner. Method and notation for the formal definition of programming languages. Technical Report TR25.087, IBM Laboratory Vienna, 1968.

[LS78] P.E. Lauer and M.W. Shields. Abstract specification of resource accessing disciplines: adequacy, starvation, priority and interrupts. *SIGPLAN Notices*, 13(12):41–59, 1978.

[LS79] A. Van Lamsverde and M. Sintzoff. Formal derivation of strongly correct concurrent programs. *Acta Informatica*, 12(1):1–31, 1979.

[LS81] P.E. Lauer and M.W. Shields. Interpreted COSY programs: Programming and Verification. In E. Gelenbe, editor, *Proceedings 2nd International Conference on Distributed Computing Systems*, Paris, 1981. IEEE Computer Society Press.

[LSB79] P.E. Lauer, M.W. Shields, and E. Best. Formal Theory of the Basic COSY Notation. Technical Report TR143, Computing Laboratory, University of Newcastle upon Tyne, 1979.

[LSC81] P.E. Lauer, M.W. Shields, and J.Y. Cotronis. *Formal Behavioural Specification of Concurrent Systems without Globality Assumptions.* Lecture Notes in Computer Science 107. Springer Verlag, 1981. pp.115–150.

[LT78] P.E. Lauer and P.R. Torrigiani. Towards a System Specification Language Based on Paths and Processes. In *Proceedings of a course: Le Traitement Parallele*, IRIA, Lille, France, 1978. pp.469–527.

[LT87] N.A. Lynch and M.R. Tuttle. Hierarchical Correctness Proofs for Distributed Algorithms. In *Proc. 6th Ann. ACM Symposium on Principles of Distributed Computing*, Vancouver, 1987. ACM. pp.137-151.

[LTD80] P.E. Lauer, P.R. Torrigiani, and R.E. Devillers. *A COSY Banker : Specification of Highly Parallel and Distributed Resource Management.* Lecture Notes in Computer Science 83. Springer Verlag, 1980. pp.223-239.

[LTS79] P.E. Lauer, P.R. Torrigiani, and M.W. Shields. COSY: A System Specification Language based on Paths and Processes. *Acta Informatica*, 12:109–158, 1979.

[Mar51] A.A. Markov. Teoriya Algorifmov (Theory of Algorithms). *Trudy Mat. Inst. Steklov*, 38:176–189, 1951. English translation, Am. Math Soc. Translations Ser.2 15, (1960), pp.1-14.

[Mar54] A.A. Markov. Teoriya Algorifmov (Theory of Algorithms). *Trudy Mat. Inst. Steklov*, 42:1–444, 1954. English translation, Israel Program for Scientific Translation, (1961).

[Maz71] A. Mazurkiewicz. Proving algorithms by tail functions. *Information and Control*, 18:220–226, 1971.

[Maz72] A. Mazurkiewicz. Recursive Algorithms and Formal Languages. *Bull. Acad. Polon. Sci., Ser. Sci. Math. Astron. Phys.*, 20:119–160, 1972.

[Maz77] A. Mazurkiewicz. Concurrent Program Schemes and their Interpretations. Technical Report DAIMI-PB-78, Aarhus University, 1977.

[Maz86] A. Mazurkiewicz. *Trace Theory.* Lecture Notes in Computer Science 255. Springer Verlag, 1986. pp.297-324.

[Mil80] R. Milner. *A Calculus of Communicating Systems.* Lecture Notes in Computer Science 92. Springer Verlag, 1980. 171pp.

[Mil83] R. Milner. Calculi for Synchrony and Asynchrony. *Theoretical Computer Science*, 25:267–310, 1983.

[Mil84] R. Milner. *Lectures on a Calculus of Communicating Systems.* Lecture Notes in Computer Science 197. Springer Verlag, 1984. pp.197-220.

[Mil86] M. Milanetti. Project to port the COSY Environment to UNIX: Final Report. Technical Report ASM/137, Dept. of Computer Science and Systems, McMaster University, 1986.

[Mil88] R. Milner. Operational and Algebraic Semantics of Concurrent Processes. LFCS Report Series 88-46, University of Edinburgh, 1988. 42pp.

[Mil89] R. Milner. *Communication and Concurrency.* Prentice Hall, 1989.

[MK86] L. Mancini and M. Koutny. Formal specification of n-modular redundancy. In *1986 ACM Computer Science Conference*, pages 192–204, Cincinnati, Ohio, February 1986.

[Mos85] B. Moszkowski. *A Temporal Analysis of Some Concurrent Systems.* Lecture Notes in Computer Science 207. Springer Verlag, 1985. pp.359-364.

[MR78] P. Merlin and B. Randell. Notes of Deadlock Avoidance on the Train Set. Technical Report MRM/144, Computing Laboratory, University of Newcastle upon Tyne, 1978.

[MY60] R. McNaughton and H. Yamada. Regular expressions and state graphs for automata. *IRE Trans. on Electronic Computers*, 9(1):38–47, 1960.

[Nie86] M. Nielsen. *CCS and its Relationship to Net Theory*. Lecture Notes in Computer Science 255. Springer Verlag, 1986. pp.393-415.

[Niv79] M. Nivat. Sur la synchronisation des processus. L.I.T.P. 81-64, University Paris VII, 1979.

[Oke91] S.J. Okech. BCS: A Computer Based Tool for the Design and Analysis of Concurrent Systems Specified in the Basic COSY Notation – C^{++} Implementation. Master's thesis, McMaster University, 1991. , 313pp.

[Oku86] F. Okulicka. *The Semantics of Path Expressions with Priorities*. PhD thesis, Warsaw Technical University, 1986.

[Oku90] F. Okulicka. On priority in COSY. *Theoretical Computer Science*, 74:199–216, 1990.

[OL80] S. Owicki and L. Lamport. Proving liveness properties of concurrent programs. Technical Report S&L 1 Op.57, Computer Systems Laboratory, Stanford University, 1980. 48pp.

[Old86] E.-R. Olderog. *TCSP: Theory of Communicating Sequential Processes*. Lecture Notes in Computer Science 255. Springer Verlag, 1986. pp.441-465.

[Old87] E.-R. Olderog. *Operational Petri Net Semantics for CCSP*. Lecture Notes in Computer Science 266. Springer Verlag, 1987. pp.196-223.

[Par75] D.L. Parnas. On a solution of the cigarette smoker's problem (without conditional statements). *CACM*, 18(3), 1975.

[Par81] D. Park. *Concurrency and Automata on Infinite Sequences*. Lecture Notes in Computer Science 104. Springer Verlag, 1981.

[Pat71] S.S. Patil. Limitations and capabilities of Dijksta's semaphore primitives for co-ordination among processes. Computation Structures Group Memo 57, Project MAC, MIT, 1971.

[Paw68] Z. Pawlak. On the notion of a computer. In *Logic, Meth. and Phil. of Science III*. North Holland, 1968. pp.255-267.

[Pet62] C.A. Petri. *Kommunikation mit Automaten*. PhD thesis, Schriften des Rheinisch-Westfälischen Institutes für Instrumentelle Mathematik an der Universität Bonn, 1962.

[Pet81] J.L. Peterson. *Petri Net Theory and the Modelling of Systems*. Prentice Hall, 1981.

[Plo81] G.D. Plotkin. A structural approach to operational semantics. Lecture Notes, Aarhus University, 1981. 172pp.

[Plo82] G.D. Plotkin. An operational semantics for CSP. In D. Björner, editor, *Formal Description of Programming Concepts.* North Holland, 1982. pp.199-225.

[Pnu86] A. Pnueli. *Applications of Temporal Logic to the Specification and Verification of Reactive Systems.* Lecture Notes in Computer Science 224. Springer Verlag, 1986. pp.510-584.

[Pra86] V. Pratt. Modelling concurrency with partial orders. *Int. Journal of Parallel Programming*, 15(1):33–71, 1986.

[Pró90] P. Prószyński. *The Core Concurrency.* Lecture Notes in Computer Science 468. Springer Verlag, 1990. pp.336–345.

[Rei84] W. Reisig. *Partial order semantics versus interleaving semantics for CSP-like languages and its impact on fairness.* Lecture Notes in Computer Science 172. Springer Verlag, 1984. pp.403-413.

[Rei85] W. Reisig. *Petri Nets.* Springer Verlag, 1985.

[Rei87] W. Reisig. *A Strong Part of Concurrency.* Lecture Notes in Computer Science 266. Springer Verlag, 1987. pp.238–272.

[Ros85] A.W. Roscoe. *A CSP solution to the "trains" problem.* Lecture Notes in Computer Science 207. Springer Verlag, 1985. pp.384-388.

[SA88] K. Smith and W.F. Appelbe. FAT-An interactive FORTRAN parallelizing tool. In *Proc. 17th Internat. Conf. on Parallel Processing.* The Pennsylvania State University Press, 1988. pp.58-62.

[Sch76] E.A. Schneider. *Synchronization of Finite State Shared Resources.* PhD thesis, Carnegie-Mellon University, 1976. 125pp.

[Shi77] M.W. Shields. A new class of adequate paths. Technical Report ASM/20, University of Newcastle upon Tyne, 1977.

[Shi79a] M.W. Shields. *Adequate Path Expressions.* Lecture Notes in Computer Science 70. Springer Verlag, 1979. pp.249-265.

[Shi79b] M.W. Shields. COSY Train Journeys. Technical Report ASM/65, University of Newcastle upon Tyne, 1979.

[Shi81] M.W. Shields. On the non-sequential behaviour of a class of systems satisfying a generalized free-choice property. Technical Report CSR-92-81, University of Edinburgh, 1981.

[Shi85] M.W. Shields. Concurrent Machines. *The Computer Journal,* 28(5):449–466, 1985.

[Shi89] M.W. Shields. *Behavioural Presentations.* Lecture Notes in Computer Science 354. Springer Verlag, 1989. pp.673–689.

[Sif82] J. Sifakis. A unified approach for studying the properties of transition systems. *Theoretical Computer Science*, 18:227–258, 1982.

[Siu90] S. Siu. Proving properties of non-sequential systems by means of the COSY model. Master's thesis, McMaster, 1990.

[Skl76] J.R. Sklaroff. Redundancy management techniques for space shuttle computers. *IBM Journal Res. and Develop.*, 20:20–28, 1976.

[SL80] M.W. Shields and P.E. Lauer. *Programming and verifying concurrent systems in COSY.* Lecture Notes in Computer Science 88. Springer Verlag, 1980. pp.576-586.

[SM30] E. Szpilrajn-Marczewski. Sur l'extension de l'ordre partiel. *Fundamenta Mathematicae*, 16:386–389, 1930.

[SM81] A. Salwicki and T. Müldner. *On the Algorithmic Properties of Concurrent Programs.* Lecture Notes in Computer Science 125. Springer Verlag, 1981. 169-197.

[ST83] E. Shapiro and A. Takeuchi. Object Oriented Programming in Concurrent Prolog. *New Generation Computing*, 1(1):25–48, 1983.

[Tam88] C.C. Tam. *OPE_q^+–an extended path expression language.* PhD thesis, Imperial College of Science and Technology, London, 1988. 225pp.

[Thi65] H. Thiele. Zur Bewertung von Algorithmen. *Zeitschrift für Psychologie (Kybernetik Sonderband)*, 171:275–285, 1965.

[Thi66] H. Thiele. Wissenschaftstheoretische Untersuchungen in Algorithmischen Sprachen I: Theorie der Graphschemata Kalküle. *VEB Deutscher Verlag der Wissenschaften Berlin*, 1966. 219pp.

[TL77] P.R. Torrigiani and P.E. Lauer. An Object Oriented Notation for Paths and Processes. In *AICA Annual Conference Pisa Vol.3*, 1977. pp.349–371.

[Tor78] P.R. Torrigiani. *Synchronic aspects of data types: construction of a non-algorithmic solution of the Banker's problem.* Lecture Notes in Computer Science 65. Springer Verlag, 1978.

[vdS83] J.L.A. van de Snepscheut. *Trace Theory and VLSI Design.* PhD thesis, Technische Hogeschool Eindhoven, 1983.

[WAF+69] K. Walk, K. Alber, M. Fleck, H. Goldmann, P.E. Lauer, E. Moser, P. Oliva, H. Stigleitner, and G. Zeisel. Abstract Syntax and Interpretation of PL/I. Technical Report TR25.098, IBM Laboratory Vienna, 1969.

[Win82] G. Winskel. *Event Structure Semantics for CCS and Related Languages.* Lecture Notes in Computer Science 140. Springer Verlag, 1982. pp.561-577.

[Win88] G. Winskel. *An introduction to event structures.* Lecture Notes in Computer Science 354. Springer Verlag, 1988. pp.364–397.

[Won85] P. Wong. Realization of a high-level formalism in a computer based environment for designing distributed systems. Technical Report ASM/130, Computing Laboratory, University of Newcastle upon Tyne, 1985.

[Yan58] Y.I. Yanov. O Logicheskikh Skhemakh Algoritmov. *Problemy Kibernetiki*, 1:75–127, 1958. English translation, *Problems of Cybernetics 1*, (1958), pp.75-127. German translation, *Probleme der Kybernetik 1*, (1962), pp.87-144.

[Zie87] W. Zielonka. Notes on Finite Asynchronous Automata. *Informatique Theorique et Applications*, 21(2):99–135, 1987.

Appendix A

Algebra of Relations

Let X be a set. By 2^X we shall denote the "power set" of X, i.e., $2^X = \{Y \mid Y \subseteq X\}$. By $card(X)$ we denote the **cardinality** of of X. If $X = \{a_1, \ldots, a_m\}$, then $card(X) = m$, if X is infinite but countable, $card(X) = \infty$. Cardinalities of uncountable sets are not considered in this book. Any subset R of $X \times X$ is called a *relation*. We shall sometimes write aRb instead of $(a,b) \in R$.
Let $R, Q \subseteq X \times X$.

Definition A.1 (Composition) *The relation RQ, called* **composition** *of R and Q is defined as :*
$(a,b) \in RQ :\Leftrightarrow (\exists c \in X)(a,c) \in R \wedge (c,b) \in Q.$

Definition A.2 (Powers) *For every $R \subseteq X \times X$, let*
1. $R^0 = \{(a,a) | a \in X\}$,
2. $R^1 = R$ *and* $R^{i+1} = R^i R$ *for* $i = 1, 2, \ldots$,
3. $R^* = \bigcup_{i=0}^{\infty} R^i$.

Definition A.3 (Equivalence Relation) *A relation $R \subseteq X \times X$ is called an* **equivalence relation** *if and only if :*
1. $(\forall a \in X)(a,a) \in R$
2. $(\forall a,b \in X)(a,b) \in R \Rightarrow (b,a) \in R$
3. $(\forall a,b,c \in X)(a,b) \in R \wedge (b,c) \in R \Rightarrow (a,c) \in R$

Let R be an equivalence relation.

Definition A.4 (Equivalence Class) *For every $a \in X$, let $[a]_R$ denote the set $[a]_R = \{b | (a,b) \in R\}$. Clearly $a \in [a]_R$ and $[a]_R$ is called an* **equivalence class** *of R containing a.*

Theorem A.1 *For every equivalence relation R :*

1. $X = \bigcup_{a \in X}[a]_R$
2. $(\forall a, b \in X)[a]_R = [b]_R \Leftrightarrow (a, b) \in R$
3. $(\forall a, b \in X)(a, b) \notin R \Leftrightarrow [a]_R \cap [b]_R = \emptyset$ □

Definition A.5 (Equivalence Class Family) *For any equivalence relation $R \subseteq X \times X$, X/R will denote the* **family** *of all equivalence classes of R, i.e., $X/R = \{[a]_R | a \in X\}$.*

For every relation $R \subseteq X \times X$:

Definition A.6 (Domain, Inverse and Complement)
$Dom(R) = \{a \in X \,|\, (\exists b \in X)(a, b) \in R \vee (b, a) \in R\}$,
$R^{-1} = \{(a, b) \in X \times X \,|\, (b, a) \in R\}$
$\overline{R} = Dom(R) \times Dom(R) \setminus (R \cup R^{-1})$.
In other words :
$(a, b) \in \overline{R} :\Leftrightarrow a, b \in Dom(R) \wedge (a, b) \notin R \wedge (b, a) \notin R$.

Definition A.7 (Partial Order) *A relation $R \subseteq X \times X$ is called a* **partial order** *if and only if :*

1. $(\forall a \in Dom(R))(a, a) \in R$
2. $(\forall a, b \in Dom(R))(a, b) \in R \wedge (b, a) \in R \Rightarrow a = b$
3. $(\forall a, b, c \in Dom(R))(a, b) \in R \wedge (b, c) \in R \Rightarrow (a, c) \in R$

Definition A.8 (Total Order) *A relation $R \subseteq X \times X$ is called a* **total order** *if and only if R is a partial order and :*
$(\forall a, b \in Dom(R))(a, b) \in R \vee (b, a) \in R$.

Definition A.9 *For every* finite *partial order $R \subseteq X \times X$, $\hat{R} \subseteq X \times X$ is the relation defined as follows :*
$a\hat{R}b :\Leftrightarrow aRb \wedge a \neq b \wedge ((\forall c \in X)(aRc \wedge cRb) \Rightarrow (a = c \vee b = c))$.

Definition A.10 (Antichain) *Let $R \subseteq X \times X$ be a partial order.*
A set $A \subseteq X$ is called an **antichain** *of R if and only if $(\forall a, b \in A)$ $a\overline{R}b$ or $a = b$.*

Definition A.11 (Labelled partial order) *By a* **labelled partial order** *we we mean a triple $LPO = (R, \nu, \Sigma)$ where $R \subseteq X \times X$ is a partial order, Σ is an alphabet, and $\nu : Dom(R) \longrightarrow \Sigma$ is a labelling function.*

Lemma A.1 *For every finite partial order R :*
$R = (\hat{R})^*$.

Proof. Since $\hat{R} \subseteq R$ and $R^* = R$, then $(\hat{R})^* \subseteq R$. Let aRb and $a \neq b$. Let C be the set of all c such that aRc and cRb. Because R is finite and $a \neq b$ there is $C' = \{a_0, \ldots, a_k\} \subseteq C$ such that $a_0 = a$, $a_k = b$, $i \neq j \Rightarrow a_i \neq a_j$, and $a_{i-1}Ra_i$ for $i = 1, \ldots, k$. But this means that $a(\hat{R})^*b$. □

Lemma A.2 *For all relations $R, Q \subseteq X \times X$:*
$(R \cup Q)^* = (R^* \cup Q^*)^*$

Proof. Since $R \cup Q \subseteq R^* \cup Q^*$ then $(R \cup Q)^* \subseteq (R^* \cup Q^*)^*$. Since $R \subseteq R \cup Q$ and $Q \subseteq R \cup Q$, we have $R^* \subseteq (R \cup Q)^*$ and $Q^* \subseteq (R \cup Q)^*$. Thus $R^* \cup Q^* \subseteq (R \cup Q)^*$. Hence : $(R^* \cup Q^*)^* \subseteq ((R \cup Q)^*)^* = (R \cup Q)^*$. □

Lemma A.3 *Let X be a countable set and let $R \subseteq X \times X$ be a partial order. There is a total order $T \subseteq X \times X$ such that $R \subseteq T$ and $Dom(R) = Dom(T)$.*

Proof. Since X is countable there is $f : X \longrightarrow \{1, 2, \ldots\}$ such that $f(x) \neq f(y)$ for $x \neq y$. Define $T = \{(x, y) \mid xRy \cup (x\overline{R}y \cap f(x) < f(y))\}$. □

Lemma A.3 holds also for uncountable partial orders. It is then usually referred as Szpirlajn-Marczewski Extension Theorem [SM30]. In the uncountable case the proof uses Kuratowski-Zorn Lemma.

Appendix B

Automata and Formal Language Theory

B.1 Strings

By **alphabet** we mean **any** finite set of elements. Let A be an alphabet. By a **string** (word) over A we mean *any finite sequence* of elements of A juxtaposed. For example if $a, b, c \in A$ then $abcb$ is a string over A. The **length** of a string x, denoted by $length(x)$, is the number of elements composing x. For instance, $length(abcb) = 4$. The string consisting of zero elements is called the **empty string** and denoted by ε. Thus $length(\varepsilon) = 0$. The **concatenation** of two strings x and y is the string xy. For every string x we have $x\varepsilon = \varepsilon x = x$. For every string x, $x^0 = \varepsilon$ and $x^n = x^{n-1}x$ for $n \geq 1$. For example $(aab)^3 = aabaabaab$. The set of all strings over A, including ε, is denoted by A^*. For every string $x \in A^*$ and every $a \in A$, $\#_a(x)$ denotes the number of occurrences of a in x. For example, $\#_a((aab)^3) = 6$, $\#_b((aab)^3) = 3$.

B.2 Languages

Every subset L of A^* is called a **language** over A.

Definition B.1 (Concatenation of Languages) *For every $L_1, L_2 \subseteq A^*$, the language L_1L_2 is called a* **concatenation** *of L_1 and L_2, and defined :*
$L_1L_2 = \{xy | x \in L_1 \wedge y \in L_2\}$.

Definition B.2 (Prefixes of Languages) *For every language $L \subseteq A^*$,*
$Pref(L) = \{x | (\exists y) xy \in L\}$.

We shall write $Pref(x)$ instead of $Pref(\{x\})$ for every string x.

Definition B.3 (Powers of Languages) *For every* $L \subseteq A^*$,
$L^0 = \{\varepsilon\}$ *and*
$L^n = L^{n-1}L$ *for* $n \geq 1$, *and*
for every $L \subseteq A^*$, $L^* = \bigcup_{i=0}^{\infty} L^i$.

B.3 Regular Expressions and Languages

Definition B.4 (Regular Expressions) *The set of all* **regular expressions** *over an alphabet* A, *denoted* $REX(A)$, *is the least language over the alphabet* $A \cup \{(,), \cup, *, \varepsilon, \emptyset\}$ *satisfying the conditions below.*

1. $\varepsilon \in REX(A), \emptyset \in REX(A)$ *and* $(\forall a \in A) A \in REX(A)$
2. $(\forall R_1, R_2 \in REX(A))(R_1 \cup R_2) \in REX(A), (R_1 R_2) \in REX(A)$
3. $(\forall R \in REX(A))\ R^* \in REX(A)$

For every $R \in REX(A)$, *the language* $L(R)$ *defined by* R *is the least subset of* A^* *satisfying the following conditions:*

1. $(\forall a \in A) L(a) = \{a\}, L(\varepsilon) = \{\varepsilon\}, L(\emptyset) = \emptyset$
2. $L((R_1 \cup R_2)) = L(R_1) \cup L(R_2)$
3. $L((R_1 R_2)) = L(R_1) L(R_2)$
4. $L((R)) = L(R)$
5. $L(R^*) = L(R)^*$

Two regular expressions R_1 *and* R_2 *are said to be* **equivalent**, $R_1 =_E R_2$, *if and only if* $L(R_1) = L(R_2)$. *Usually, by slight abuse of language, we write* $R_1 = R_2$ *instead of* $R_1 =_E R_2$.

Definition B.5 (Regular Languages) *A language* L *is* **regular** *if and only if there is a regular expression* R *such that* $L = L(R)$.

Definition B.6 (Occurrence Count) *For every string* $x \in A^*$, *and every* $a \in A$, *let* $\#_a(x)$ *denote the number of occurrences of* a *in* x.

For instance $\#_a(abcaab) = 3$.

Lemma B.1 (Existence of Non-regular Languages) *The languages*
$\{a^n b^n | n \geq 0\}$
$\{x \in A^* | \#_a(x) = \#_b(x)\}$
are not regular □

Definition B.7 (Homomorphism) *Let* A_1, A_2 *be alphabets. A mapping*

$h : A_1^* \longrightarrow A_2^*$ is called a **homomorphism** *if for every $x, y \in A_1^*$, $h(xy) = h(x)h(y)$.*

Theorem B.1 (Homomorphisms and Regular Languages)
Let $L_1, L_2, L \subseteq A_1^*, L' \subseteq A_2^*, h : A_1^* \longrightarrow A_2^*$ be a homomorphism, and let L, L', L_1, L_2 be regular languages. Then:
$L_1L_2, L_1 \cup L_2, L_1 \cap L_2, L^*, h(L), Pref(L)$, and $h^{-1}(L')$
are all regular, where $h^{-1}(L') = \{x \in A_1^* \mid h(x) \in L'\}$ □

Definition B.8 (Substitution) *Let A, B be alphabets. A* **substitution** *f is any mapping $f : A \longrightarrow 2^{B^*}$. Thus f associate a language with each symbol of A. The mapping f is extended to $f : A^* \longrightarrow 2^{B^*}$ as follows:*

(1) $f(\varepsilon) = \varepsilon$

(2) $f(xa) = f(x)f(a)$ for $x \in A^*, a \in A$

The mapping f is extended to $\hat{f} : 2^{A^*} \longrightarrow 2^{B^*}$ by defining:

$$(\forall L \subseteq A^*)\ \hat{f}(L) = \bigcup_{x \in L} f(x).$$

The language $\hat{f}(L)$ is regarded as the result of applying the substitution f to the language L. Ususally, the mappings $\hat{f}$ and f are denoted by the same symbol f.

Lemma B.2 *For every substitution f and language L_1, L_2, L:*

(1) $f(L_1 \cup L_1) = f(L_1) \cup f(L_2)$

(2) $f(L_1, L_2) = f(L_1)f(L_2)$

(3) $f(L^*) = f(L)^*$ □

B.4 Finite State Automata and Grammars

Definition B.9 (Finite State Automaton) *By a* **finite state automaton** *we mean a 5-tuple*
$M = (Q, A, \delta, q_0, \mathcal{F})$
where :
Q is a finite set called the set of **states**

A is an **alphabet**
$q_0 \in Q$ *is the* **initial state**
$\mathcal{F} \subseteq Q$ *is the set of* **final states**
$\delta : Q \times A \longrightarrow 2^Q$ *is a* **transition function.**
The function δ *is extended to* $\hat{\delta} : Q \times A^* \longrightarrow 2^Q$ *as follows :*
1. $\hat{\delta}(q, \varepsilon) = \{q\}$
2. $\hat{\delta}(q, xa) = \{p | (\exists r \in \hat{\delta}(q, x)) p \in \delta(r, a)\}$, *for* $a \in A$ *and* $x \in A^*$.

Definition B.10 (Languages Accepted by Finite State Automata) *The* **language** $L(M)$ *accepted by finite state automaton* M *is defined as :* $L(M) = \{x | \hat{\delta}(q_0, x) \in F\}$.

Definition B.11 (Deterministic Automaton) *A finite state automaton M is called* **deterministic** *if for every* $q \in Q$ *and* $a \in A$, $card(\delta(q, a)) \leq 1$.

Theorem B.2 (Equivalences of Languages) *Let* $L \subseteq A^*$. *Then the following are equivalent :*
1. *L is regular.*
2. *There is a finite state automaton M such that $L = L(M)$.*
3. *There is a deterministic finite state automaton M such that $L = L(M)$.*
4. *Let equivalence relation* $R_L \subseteq A^* \times A^*$ *be defined by :*
$(x, y) \in R_L :\Leftrightarrow ((\forall z \in A^*) xz \in L \Leftrightarrow yz \in L)$.
Then $card(A^*/R_L) < \infty$, *i.e. the number of equivalence classes of* R_L *is finite.*

□

Definition B.12 (Minimal Automaton)
A finite automaton $M = (Q, A, \delta, q_0, \mathcal{F})$ *is called* **minimal** *if for every automaton* $M' = (Q', A', \delta', q_0', \mathcal{F}')$, $L(M) = L(M')$ *implies* $card(Q) \leq card(Q')$.

Definition B.13 (Quotient Automata) *Let* $L \subseteq A^*$ *be a regular language and let M be the following finite automaton:*

$$M = (A^*/R_L, A, \delta, [\varepsilon]_{R_L}, \mathcal{F})$$

where:

R_L *is the relation from Theorem B.2 (4),*
$\delta([x]_{R_L}, a) = [xa]_{R_L}$ *for every* $x \in A^*, a \in A$,
$\mathcal{F} = \{[x]_{R_L} | x \in L\}$.

M is a finite automaton called a **quotient automaton of** L.

Theorem B.3 (Quotient Automata)

(1) If M is a quotient automaton of L, then $L(M) = L$.

(2) Every quotient automaton is a minimal automaton. □

Definition B.14 (Grammar)

By a **grammar** *we mean a 4-tuple $G = (Non, Ter, Pro, \sigma)$ where:*

Non *is a finite set of* **nonterminals**,

Ter *is a finite set of* **terminals**,

$Non \cap Ter = \emptyset$,

Pro *is a finite set of productions,*

$\sigma \in Non$ *is the* **start symbol**.

Each production is of the form $x \longrightarrow y$ where $x \in (Non \cup Ter)^+$ and $y \in (Non \cup Ter)^*$.

Definition B.15 (Derivation)

The relation $\mapsto \subseteq (Non \cup Ter)^+ \times (Non \cup Ter)^*$ defined by

$x \mapsto y \;:\Leftrightarrow$

$$(\exists x_1, x_2, u, v \in (Non \cup Ter)^*)\; x = x_1 u x_2 \wedge y = x_1 v x_2 \wedge u \longrightarrow v \in Pro,$$

is called a **direct derivation**.

The relation $\mapsto^*$, i.e.,

$x \mapsto^* y \Leftrightarrow x = y$ *or* $x \mapsto x_1 \mapsto \ldots \mapsto x_k = y$, $k \geq 1$, *is called a* **derivation**.

Definition B.16 (Languages Generated by Grammars)

The **language** $L(G)$ *generated by grammar G is defined as:*

$$L(G) = \{x \in Ter^* \mid \sigma \mapsto^* x\}.$$

Definition B.17 (Context-Free Grammar and Language)

A grammar $G = (Non, Ter, Pro, \sigma)$ is called **context-free** *if all productions are of the form $A \longrightarrow x$ where $A \in Non$, $x \in (Non \cup Ter)^*$. For context-free grammars we usually write:*

$$A := x_1 \mid \ldots \mid x_k$$

instead of:

$$A \longrightarrow x_1,\ A \longrightarrow x_2,\ \ldots,\ A \longrightarrow x_k.$$

A language is called context-free if there is a context-free grammar which generates it.

Comment B.1.

Languages which are not context-free are usually defined in two steps. First a context-free grammar generating a superlanguage is provided. Next, a set of context-sensitive restrictions, usually in a semi-formal manner, is added. In such a case the derivation means a derivation in the context-free grammar which additionally satisfies all context-sensitive restrictions.

Appendix C

Elements of Graph Theory

Definition C.1 (Undirected Graphs) *An* **undirected graph**, *denoted* $G = (V, E)$, *consists of a set of* **vertices** V *and a set of pairs of vertices* E *called* **edges**. *More formally,*
$e \in E \Rightarrow (e \subseteq V \wedge card(e) = 2)$.

Definition C.2 (Paths in Graphs) *A sequence of vertices* $v_1, \ldots, v_k, k \geq 2$, *is called a* **path** $v_1, \ldots, v_k$ **in** G, *if* $\{v_i, v_{i+1}\} \in E$ *for each* $i = 1, \ldots, k-1$.

Definition C.3 (Connected Graph) *An undirected graph* $G = (V, E)$ *is* **connected** *if and only if for every two* $v, v' \in V$:
$v \neq v' \Rightarrow$ *there is a path* $v_1, \ldots, v_k$ *in* G *and* $v = v_1, v' = v_k$.

Definition C.4 (Undirected Tree) *An undirected connected graph* G *is an* **undirected tree** *if for every path* $v_1, \ldots, v_k$, *in* G : $i \neq j \Rightarrow v_i \neq v_j$ *for* $i, j = 1, \ldots, k$.

Definition C.5 (Undirected Spanning Tree) *An undirected tree* $G_T = (V_T, E_T)$ *is an* **undirected spanning tree** *of the graph* $G = (V, E)$, *if and only if:*
$V_T = V \wedge E_T \subseteq E$.

Lemma C.1 *If* $G = (V, E)$ *is an undirected tree and* $G' = (V', E')$ *is a subgraph of* G, *i.e.* $V' \subseteq V$ *and* $E' \subseteq E$, *then:*
$E' \neq E \Rightarrow V' \neq V$ □

Definition C.6 (Directed Graph) *A* **directed graph**, *denoted* $DG = (V, A)$, *consists of a set of vertices* V *and a set of* **directed pairs** *of vertices, called* **arcs** *or* **edges**, *i.e.* $A \subseteq V \times V$. *We assume that directted graphs do not have self-arcs, i.e.,* $(a, b) \in A \Rightarrow a \neq b$.

Every directed graph $DG = (V, A)$ defines an undirected graph $G_{DG} = (V, E_A)$, where $E_A = \{\{v, v'\} | (v, v') \in A\}$.
A directed graph DG is **connected** if G_{DG} is connected.

Definition C.7 (Directed Tree) *A connected directed graph* $DG = (V, A)$ *is called a* **directed tree** *if:*

1. $(\forall v \in V)(v', v) \in A \wedge (v'', v) \in A \Rightarrow v' = v''$
2. $(\exists v_0 \in V)(\forall v \in V)(v, v_0) \notin A$

One may prove that v_0 is defined in an unambiguous way, and v_0 is called the **root** of the tree DG.

Lemma C.2 (Directed and Undirected Trees)

1. *For every directed tree* DG, *the graph* G_{DG} *is an undirected tree.*
2. *For every undirected tree* $G = (V, E)$, *and every* $v \in E$, *there is only one directed tree* $DG = (V, A)$ *such that* v *is the root of* DG *and* $G_{DG} = G$ □

Definition C.8 (Directed Spanning Tree) *A directed tree* $DG_T = (V_T, A_T)$ *is a* **directed spanning tree** *of the directed graph* $DG = (V, A)$, *if and only if:* $V_T = V \wedge A_T \subseteq A$.

Appendix D

Proofs of Theorems 2.25, 2.26 and 2.28

Lemma D.1
$(\forall i = 1, \ldots, n)\ Pr = P_1 \ldots P_n \in \Gamma_0 \Rightarrow \varepsilon \notin Cyc(P_i)$.

Proof. Suppose that $\varepsilon \in Cyc(P_i)$ for some $i \leq n$. Then $pr_i(\varepsilon) \in FS(P_i)$ has cyclic decompositions of different lengths, e.g. $< \varepsilon, \varepsilon >$ and $< \varepsilon, \varepsilon, \varepsilon >$, contradicting (3) of Γ_0. □

For every $\underline{x} \in VFS(Pr)$, where $Pr = P_1 \ldots P_n \in \Gamma_0$, and every $\delta =< v_1, \ldots, v_k > \in cd_{P_1}(pr_1(\underline{x})) \cup \ldots \cup cd_{P_n}(pr_n(\underline{x}))$, let $nc_{Pr}(\underline{x}) = k$. From (3) of Γ_0 it follows that $nc_{Pr}(\underline{x})$ is correctly defined for every $Pr \in \Gamma_0$.

Lemma D.2 *Let $Pr = P_1 \ldots P_n \in \Gamma_0, \underline{x} \in VFS(Pr)$ and $k = nc_{Pr}(\underline{x}) \geq 1$.*

1. $(\forall i = 1, \ldots, n) card(cd_{P_i}(pr_i(\underline{x}))) = 1$
2. *If $cd_{P_i}(pr_i(\underline{x})) =< v_{i1}, \ldots, v_{ik} >$, for $i \leq n$, then:*
   - *(a) for every $j \leq k, v_{1j}, \ldots, v_{nj}$ differ from ε and begin with the same event,*
   - *(b) for every $j \leq k-1, \underline{v}_j = (v_{1j}, \ldots, v_{nj}) \in Vcyc(Pr)$ and $\underline{v}_k = (v_{1k}, \ldots, v_{nk}) \in Pref(Vcyc(Pr))$.*
   - *(c) there are $z_1, \ldots, z_n$ such that $(v_{1k}z_1, \ldots, v_{nk}z_n) \in Vcyc(Pr)$, and $(\forall i = 1, \ldots, n) z_i = \varepsilon \Leftrightarrow v_{ik} \in Cyc(P_i)$.*

Proof. For every $i \leq n$, let $< v_{i1}, \ldots, v_{ik} >$ be any cyclic decomposition of $pr_i(\underline{x})$. Also, for every $j \leq k$, let $\delta_j = (v_{1j}, \ldots, v_{nj})$. So far we do not know if $\delta_j \in Vev(Pr)^*$. We shall first show that $\delta_1 \ldots \delta_j \in VFS(Pr)$ for all $j \leq k-1$. There is a minimal prefix $\underline{y}$ of $\underline{x}$ such that $v_{i1} \ldots v_{ij} \in Pref(pr_i(\underline{y}))$ for all $i \leq n$. From the minimality of $\underline{y}$ it follows that there is $l \leq n$ such that $v_{l1} \ldots v_{lj} = pr_l(\underline{y})$. Suppose now that $\underline{y} \neq \delta_1 \ldots \delta_j$. Then there must be $r \leq n$ such that $v_{r1} \ldots v_{rj} \neq pr_r(\underline{y})$. Hence we can denote $pr_r(\underline{y}) = v_{r1} \ldots v_{rj}u$, where u is a non-empty prefix of $v_{rj+1} \ldots v_{rk}$. Thus $pr_r(\underline{y})$ has a cyclic decomposition of length greater than j. On the other hand, $pr_l(\underline{y})$ has a cyclic decomposition of length j which produces a contradiction to (3) of Γ_0 since $\underline{y} \in VFS(Pr)$. Hence $\delta_1 \ldots \delta_j = \underline{y} \in VFS(Pr)$, for $j \leq k-1$. Clearly $\delta_1 \ldots \delta_k = \underline{x} \in VFS(Pr)^*$, thus from Lemma 2.3 we have that for every $j \leq k, \delta_j \in Vev(Pr)^*$, so we can write $\delta_j = \underline{v}_j$ for some $v_j \in Ev(Pr)^*$. Consequently, since $v_{ij} \in Cyc(P_i)$ for $j < k$ and $v_{ik} \in Pref(Cyc(P_i))$ we obtain:

$$\underline{v}_1, \ldots, \underline{v}_{k-1} \in Vcyc(Pr) \textit{ and } \underline{v}_k \in Pref(Vcyc(Pr)),$$

so we have proven (2.b).

Suppose now that $a \in Ev(Pr)$ is such that $\underline{a} \in VFS(Pr)$. By (3) of Γ_0, we must have $pr_i(\underline{a}) = a$ for all $i \leq n$. This, (2.b) and Lemma D.1 imply that for every $j \leq n$, the sequences $v_{1j}, \ldots, v_{nj}$ differ from ε and begin with the same event (recall that $\underline{v}_j = (v_{1j}, \ldots, v_{nj})$), so we have proven (2.a).

We can now prove (1). Suppose that (1) is not satisfied for $i = 1$. Then $k \geq 2$ and there is $l \leq k-1$ such that $(v_{11}, \ldots, v_{1l-1}, t_{1l}, t_{1l+1}, \ldots, t_{1k})$ is a cyclic decomposition of $pr_i(\underline{x})$ and $t_{1l} \neq v_{1l}$. Let a be the first event of v_{2l+1}. As we have already shown, a is the first event in both v_{1l+1} and t_{1l+1}. Also, we have $\underline{u} = \underline{v}_1 \ldots \underline{v}_l \underline{a} \in VFS(Pr)$, since $\underline{v}_1 \ldots \underline{v}_l \underline{v}_{l+1} \in VFS(Pr)$ as we demonstrated at the beginning of this proof, and from the same part of the proof : $\underline{w} = \underline{v}_1 \ldots \underline{v}_{l-1} \tau \underline{a} \in VFS(Pr)$, where $\tau = (t_{1l}, v_{2l}, \ldots, v_{nl})$. Furthermore from the same part of this proof we have that $\underline{u} \in Pref(\underline{x})$ and $\underline{w} \in Pref(\underline{x})$. Since a is the first event in both v_{2l+1} and v_{1l+1}, we have $a \in Ev(P_1) \cap Ev(P_2)$. But $\underline{u}, \underline{w} \in Pref(\underline{x})$ implies : $pr_i(\underline{u}), pr_i(\underline{w}) \in Pref(pr_i(\underline{x}))$ for $i = 1, \ldots, n$, which means that either $pr_i(\underline{u}) \in Pref(pr_i(\underline{w}))$ or $pr_i(\underline{w}) \in Pref(pr_i(\underline{u}))$, for $i = 1, \ldots, n$. Consider $pr_1(\underline{u})$ and $pr_1(\underline{w})$. Both of them end with a, and both contain the same number of occurrences of a, thus $pr_1(\underline{u}) = pr_1(\underline{w})$. Hence $v_{11} \ldots v_{1l-1} v_{1l} = v_{11} \ldots v_{1l-1} t_{1l}$, contradicting $t_{1l} \neq v_{1l}$. This completes the proof of (1).

What still remains to be shown is (2.c). This, however, follows directly from (2.b) and (2) of Γ_0. □

Lemma D.3 *Let* $Pr \in \Gamma_0$ *and let* $\underline{x}, \underline{y} \in VFS(Pr)$ *be such that* $\underline{x} \in Pref(\underline{y})$. *Suppose that* $i \leq n$, $cd_{P_i}(pr_i(\underline{x})) = < v_1, \ldots, v_r >$ *and* $cd_{P_i}(pr_i(\underline{y})) = < t_1, \ldots, t_r, s >$, *where* $r, s \geq 0$. *Then :*

$r \leq s, < v_1, \ldots, v_{r-1} > = < t_1, \ldots, t_{r-1} >$ *if* $r > 1$, *and* $v_r \in Pref(t_r) \setminus \{\varepsilon\}$ *if* $r > 0$.

Proof. Observe that the lemma holds for $r = 0$. Let $r > 0$. By Lemma D.1 and the fact that $pr_i(\underline{x}) \in Pref(pr_i(\underline{y}))$, we have $pr_i(\underline{x}) = t_1 \ldots t_{q-1} t$, where $1 \leq q \leq s$ and t is a nonempty prefix of t_q. Hence $< t_1, \ldots, t_{q-1}, t >$ is a cyclic decomposition of $pr_i(\underline{x})$. Thus by Lemma D.2 (1), $< v_1, \ldots, v_r > = < t_1, \ldots, t_{q-1}, t >$, which completes the proof. □

Let $Pr = P_1 \ldots P_n \in \Gamma$ and $Qr = Q_1 \ldots Q_m \in \Omega$. For every $\underline{x} \in VFS(Pr \longrightarrow Qr)$, let $\underline{x}|_{Pr} = (x_1, \ldots, x_n)$, where each x_i is obtained from $pr_i(\underline{x})$ by deleting the events which do not belong to $Ev(Pr)$, i.e. $x_i = h_{Ev(P_i)}(pr_i(\underline{x}))$ for $i = 1, \ldots, n$. Thus, $\underline{x}|_{Pr}$ projects histories of $Pr \longrightarrow Qr$ onto Pr. One can easily see that $\underline{x}|_{Pr} \in VFS(Pr)$.

Lemma D.4 *Let $Qr = Q_1 \ldots Q_m \in \Omega, \underline{x} \in Vev(Qr)^*$ and $\#_\omega(\underline{x}) = 0$. Suppose $k \geq 1$ and $y_1, \ldots, y_n \in Ev(Qr)^*$ are such that $y_i = y_{i1}\omega y_{i2}\omega \ldots y_{ik}\omega y_{ik+1}$, where*
1. $pr_i(\underline{x}) = y_{i1}y_{i2} \ldots y_{ik}y_{ik+1}$ for all $i \leq n$;
2. if $i, j \leq n$ and $a \in Ev(Q_i) \cap Ev(Q_j)$ then
$$\#_a(y_{i1} \ldots y_{is}) = \#_a(y_{j1} \ldots y_{js}), \text{ for all } s = 1, \ldots, k.$$
Then : $(y_1, \ldots, y_n, pr_{n+1}(\underline{x}), \ldots, pr_m(\underline{x})) \in Vev(Qr)^*$.

Proof. This straightforward proof is left to the reader. □

Lemma D.5 *Let $Pr = P_1 \ldots P_n \in \Gamma_0, Qr = Q_1 \ldots Q_m \in \Omega$ and $\underline{x} \in VFS(Pr \longrightarrow Qr)$. For every $i \leq n$ there is exactly one $t = t_1\omega \ldots t_k\omega t_{k+1}$ and :*
1. $k = nc_{Pr}(\underline{x}|_{Pr})$
2. $\#_\omega(t) = k$
3. $pr_i(\underline{x}) = t_1v_1 \ldots t_kv_kt_{k+1}$, where $< v_1, \ldots, v_k >= cd_{P_i}(pr_i(\underline{x}|_{Pr}))$
4. if $v_k \notin Cyc(P_i)$ then $t_{k+1} = \varepsilon$.

This unique t will be denoted by $\underline{x}|_{Q_i}$.

Proof. If $k = 0$ then the lemma is true since $\varepsilon \notin Cyc(P_i)$ (Lemma D.1). Suppose that $k \geq 1$. Clearly, there is $t \in FS(Q_i)$ such that $t = t_1\omega \ldots t_l\omega t_{l+1}, l \geq 1$, and :

$$\#_\omega(t) = l;$$
$$pr_i(\underline{x}) = t_1u_1 \ldots t_lu_lt_{l+1}, \text{ where } u_1, \ldots, u_{l-1} \in Cyc(P_i) \text{ and}$$
$$u_L \in Pref(Cyc(P_i) \setminus \{\varepsilon\}.$$

Thus, $< u_1, \ldots, u_{l-1}, u_l >$ is a cyclic decomposition of $pr_i(\underline{x}|_{Pr})$. Hence by Lemma D.2 (1), $l = k$ and $< u_1, \ldots, u_l >=< v_1, \ldots, v_k >$. The unicity of t follows from $\varepsilon \notin Cyc(P_i)$. Suppose that $t_{k+1} \neq \varepsilon$. Then $pr_i(\underline{x}|_{Pr}) \in Cyc(P_i)^*$. This and Lemma D.2 (1) yields $v_k \in Cyc(P_i)$, which completes the proof. □

Lemma D.6 *If $Pr = P_1 \ldots P_n \in \Gamma_0$ and $Qr = Q_1 \ldots Q_m \in \Omega$ then :*
$Pr \longrightarrow Qr$ is adequate $\Leftrightarrow$ Qr is adequate.

Proof. First we shall show (1) of Corollary 2.14. Let $\underline{x} \in VFS(Pr \longrightarrow Qr)$ and let $k = nc_{Pr}(\underline{x}|_{Pr})$. If $k = 0$ then $\underline{x} \in VFS(Qr) \subseteq Insert(Pr, Qr)$. Suppose that $k \geq 1$. Let us define : $\underline{x}|_{Qr} = (\underline{x}|_{Q_1}, \ldots, \underline{x}|_{Q_n}, pr_{n+1}(\underline{x}), \ldots, pr_m(\underline{x}))$. We shall show that $\underline{x}|_{Qr} \in VFS(Qr)$. Clearly, for every $i \leq m$, $pr_i(\underline{x}|_{Qr}) \in FS(Q_i)$. Thus it suffices to show that $\underline{x}|_{Qr} \in Vev(Qr)^*$. Let $\underline{x}|_{Q_i} = w_{i1}\omega \ldots w_{ik}\omega w_{ik+1}$, for all $i \leq n$, and $\omega \notin Ev(w_{ij})$. Define $v = h_{Ev(Qr)\setminus\{\omega\}}(x)$. Clearly $\underline{v} \in Vev(Qr)^*$ and $\underline{v} = (v_1, \ldots, v_n, pr_{n+1}(\underline{x}), \ldots, pr_m(\underline{x}))$, where $v_i = w_{i1} \ldots w_{ik} w_{ik+1}$ for $i = 1, \ldots, n$. Consequently, by Lemma D.4, it suffices to show that if $i, j \leq n$ and $a \in Ev(Q_i) \cap Ev(Q_j)$ then for all $s = 1, \ldots, k, \#_a(w_{i1} \ldots w_{is}) = \#_a(w_{j1} \ldots w_{js})$. Since $\#_\omega(w_{i1} \ldots w_{is}) = \#_\omega(w_{j1} \ldots w_{js}) = 0$, we may assume $a \neq \omega$. Let us consider :

$$pr_i(\underline{x}) = w_{i1} v_{i1} \ldots w_{is} v_{is} \ldots w_{ik+1}$$

$$pr_j(\underline{x}) = w_{j1} v_{j1} \ldots w_{js} v_{js} \ldots w_{jk+1}$$

where for $l = 1, \ldots, n, < v_{l1}, \ldots, v_{lk} >= cd_{P_l}(pr_l(\underline{x}|_{P_r}))$. Let b be the first event in $v_{is} \neq \varepsilon$. By Lemma D.2 (2.a), b is also the first event of v_{js}. This and the fact that $(v_{1r}, \ldots, v_{nr}) \in VFS(Pr)$ for all $r \leq k$ (see Lemma D.2 (2.b)) implies (by Lemma 2.2 (1)) that

$$\#_b(w_{i1} v_{i1} \ldots w_{is} b) = \#_b(w_{j1} v_{j1} \ldots w_{js} b) \stackrel{\text{df}}{=} k_s.$$

Let x' be a string such that $\underline{x}' = \underline{x} \in VFS(Pr \longrightarrow Qr)$ and in x' all b's are pushed to the right as far as possible. Let x'_s be a string derived from x' by removing all event occurrences that occur in x' after k_s occurrences of b. Note that $h_i(x'_s) = w_{i1} v_{i1} \ldots w_{is} b$ and $h_j(x'_s) = w_{j1} v_{j1} \ldots w_{js} b$. But this means that for every $a \in Ev(Q_i) \cap Ev(Q_j), \#_a(w_{i1} v_{i1} \ldots w_{is} b) = \#_a(w_{j1} v_{j1} \ldots w_{js} b)$. Consequently, $\#_a(w_{i1} v_{i1} \ldots w_{is}) = \#_a(w_{j1} v_{j1} \ldots w_{js})$ for $a \in Ev(Q_i) \cap Ev(Q_j)$. Thus, by Lemma D.4, $\underline{x}|_{Q_r} \in Vev(Qr)^*$.
Hence $\underline{x}|_{Q_r} \in VFS(Qr)$ and, furthermore, $\#_\omega(\underline{x}|_{Q_r}) = k$. By Lemma D.2 (2.c), there are $z_1, \ldots, z_n$ such that $\underline{\tilde{u}} = (v_{1k} z_1, \ldots, v_{nk} z_n) \in Vcyc(Pr)$ and $z_i = \varepsilon \Leftrightarrow v_{ik} \in Cyc(P_i)$ for $i = 1, \ldots, n$. Also, by Lemma D.2 (2.b), we have $\underline{v}_j = (v_{1j}, \ldots, v_{nj}) \in Vcyc(Pr)$ for $j = 1, \ldots, k-1$. Thus, $< \underline{v}_1, \ldots, \underline{v}_{k-1}, \underline{\tilde{u}}, \underline{x}|_{Q_r} >\in insert_{P_r,Q_r}$. By Lemma D.5 (4), we have that $\underline{x} \in Pref(< \underline{v}_1, \ldots, \underline{v}_{k-1}, \underline{\tilde{u}} >\longrightarrow \underline{x}|_{Q_r})$, and, since $< \underline{v}_1, \ldots, \underline{v}_{k-1}, \underline{\tilde{u}} >\longrightarrow \underline{x}|_{Q_r} \in Insert(Pr, Qr), \underline{x} \in Pref(Insert(Pr, Qr))$, thus $VFS(Pr \longrightarrow Qr) \subseteq Pref(Insert(Pr, Qr))$. The opposite inclusion is, by the definition of $Insert(Pr, Qr)$, true for any $Pr \in \Gamma$ and $Qr \in \Omega$; hence (1) of Corollary 2.14 is satisfied.
The condition (2) of Corollary 2.14 is just (1) of Γ_0, so it is satisfied as well. Suppose now that $\underline{x}, \underline{y} \in Insert(Pr, Qr)$ and that $\underline{x} \in Pref(\underline{y})$. We first prove that here $image_{Q_r}(\underline{x}) = \{\underline{x}|_{Q_r}\}$ and $image_{Q_r}(\underline{y}) = \{\underline{y}|_{Q_r}\}$. Of course, $\underline{x}|_{Q_r} \in image_{Q_r}(\underline{x})$. Suppose that $\underline{x} =< \underline{t}_1, \ldots, \underline{t}_r >\longrightarrow \underline{u} = < \underline{t}'_1, \ldots, \underline{t}'_r >\longrightarrow \underline{u}'$. Clearly $\underline{t} = \underline{t}_1 \ldots \underline{t}_r = \underline{t}'_1 \ldots \underline{t}'_r \in VFS(Pr)$, and $< pr_i(\underline{t}_1), \ldots, pr_i(\underline{t}_r) >$, $< pr_i(\underline{t}'_1), \ldots, pr_i(\underline{t}'_r) >\in cd_{P_i}(pr_i(\underline{t}))$, for $i \leq n$. So from Lemma D.2 (1), $< \underline{t}_1, \ldots, \underline{t}_r >=< \underline{t}'_1, \ldots, \underline{t}'_r >$. This and Lemma D.1 yields $\underline{u} = \underline{u}'$, indeed $image_{Q_r}(\underline{x}) = \{\underline{x}|_{Q_r}\}$, and similarly for $\underline{y}$. Now, to prove (3) of Corollary 2.14, it suffices to show that $\underline{x}|_{Q_r} \in Pref(\underline{y}|_{Q_r})$. For $i > n, pr_i(\underline{x}|_{Q_r}) = pr_i(\underline{x})$,

$pr_i(\underline{y}|_{Qr}) = pr_i(\underline{y})$, so clearly, $pr_i(\underline{x}|_{Qr}) \in Pref(pr_i(\underline{y}|_{Qr}))$. For $i \leq n, pr_i(\underline{x}|_{Qr}) = \underline{x}|_{Q_i}$ and $pr_i(\underline{y}|_{Qr}) = \underline{y}|_{Q_i}$, where $\underline{x}|_{Q_i}$ and $\underline{y}|_{Q_i}$ are defined as in Lemma D.5. One may check that from Lemma D.1 and Lemma D.3 it follows that $\underline{x}|_{Q_i} \in Pref(\underline{y}|_{Q_i})$ for $i = 1,\ldots,n$. Hence $\underline{x}|_{Qr} \in Pref(\underline{y}|_{Qr})$ and consequently (3) of Corollary 2.14 is satisfied which completes the proof. □

Lemma D.7

$$Smooth(\Omega_0) \subseteq \Gamma_0 \ and\ Smooth(\Omega_1) \subseteq \Gamma_0.$$

Proof. Let $Pr = P_1 \ldots P_n \in Smooth(\Omega_0)$. We first show that Pr must satisfy (1) of Γ_0 and (2) of Γ_0. Let Qr_0 denote the path program defined by (1) in the definition of Ω_0, i.e.

$$Qr_0 = \begin{cases} P(1): & \underline{path}\ \omega; a_1; a\ \underline{end} \\ & \cdots \\ P(n): & \underline{path}\ \omega; a_n; a\ \underline{end} \end{cases}$$

Qr_0 is adequate since $VFS(Qr_0) = Pref(\{\underline{\omega a_1} \ldots \underline{a_n a}\}^*)$. One can easily see that for every $\underline{x} \in VFS(Pr \longrightarrow Qr_0), \underline{x}|_{Pr} \in Vcyc(Pr)^* Pref(Vcyc(Pr))$. Hence, since $Pr \longrightarrow Qr_0$ is adequate, (1) of Γ_0, i.e. $Ev(Vcyc(Pr)) = Ev(Pr)$, must be satisfied.

Let $\underline{x} \in Pref(Vcyc(Pr))$. Then, $\underline{x}\underline{a_{i_1}} \ldots \underline{a_{i_s}} \in VFS(Pr \longrightarrow Qr_0)$, where : $s \geq 0, 1 \leq i_1 < \ldots < i_s \leq n$, and for all $j \leq n, pr_j(\underline{x}) \in Cyc(P_j) \Leftrightarrow j \in \{i_1,\ldots,i_s\}$. Since $Pr \longrightarrow Qr_0$ is adequate, there is $\underline{z} \in Vev(Pr \longrightarrow Qr_0)^*$ such that $\underline{x}\underline{a_{i_1}} \ldots \underline{a_{i_s}}\underline{z}\underline{a} \in VFS(Pr \longrightarrow Qr_0)$ and $\#_a(\underline{z}) = 0$. Clearly, $\underline{x}\underline{a_{i_1}} \ldots \underline{a_{i_s}}\underline{z}|_{Pr} \in Vcyc(Pr), \underline{x} \in Pref(\underline{x}\underline{a_{i_1}} \ldots \underline{a_{i_s}}\underline{z}|_{Pr})$, and for all $j = \{i_1,\ldots,i_s\}, pr_i(\underline{x}) = pr_j(\underline{x}\underline{a_{i_1}} \ldots \underline{a_{i_s}}\underline{z}|_{Pr})$. Hence (2) of Γ_0 is satisfied. Suppose that Pr does not satisfy (3) of Γ_0. Without loss of generality we can assume that there is $\underline{x} \in VFS(Pr)$ such that there are $< u_1,\ldots,u_k > \in cd_{P_1}(pr_1(\underline{x}))$ and $< v_1,\ldots,v_l > \in cd_{P_2}(pr_2(\underline{x}))$ such that $k > l$. Let $Qr_k = \psi_k\phi_k\xi \ldots \xi \in \Omega_0$. Observe that Qr_k is adequate since $VFS(Qr_k) = Pref((\{\underline{bac}\}\{\underline{\omega},\underline{a}\})^*)$ if $k = 1$, and $VFS(Qr_k) = Pref((\{\underline{\omega},\underline{a}\}^{k-1}\{\underline{bac}\})^*)$ if $k > 1$. Since $Ev(Pr) \cap Ev(Qr_k) = \emptyset$ and Qr_k consists of n single paths, $\underline{x} \in Vev(Pr)^* \subseteq Vev(Pr \longrightarrow Qr_k)^*$. Since $b \in Ev(\psi_k)$ only, $\underline{x}_1\underline{b}\underline{x}_2\underline{b} \ldots \underline{x}_s\underline{b}\underline{x}_{s+1} \in Vev(Pr \longrightarrow Qr_k)$ where $\underline{x} = \underline{x}_1\underline{x}_2 \ldots \underline{x}_s\underline{x}_{s+1}, \underline{x}_i \in Vev(Pr \longrightarrow Qr_k)^*$ for any $s \geq 0$. This means that the definition of $\underline{y}$ given below is a correct definition of an element of $Vev(Pr \longrightarrow Qr_k)^*$:

$$\underline{y} = \begin{cases} (u_1 \ldots u_{k-1}bu_kb, pr_2(\underline{x}),\ldots,pr_n(\underline{x})) & \text{if } u_k = \varepsilon \\ (u_1 \ldots u_{k-1}bu_k, pr_2(\underline{x}),\ldots,pr_n(\underline{x})) & \text{if } u_k \neq \varepsilon \end{cases}$$

where, we set $u_1 \ldots u_{k-1} = \varepsilon$ if $k = 1$. If $k = 1$ then $l = 0$ and $u_k \neq \varepsilon$, hence $\underline{y} = (bu_1, \varepsilon, pr_3(\underline{x}), \ldots, pr_n(\underline{x}))$, and clearly $\underline{y} \in VFS(Pr \longrightarrow Qr_1)$. If $k \geq 2$ then

$$\underline{y} = (u_1 \ldots u_{k-1}bb, v_1 \ldots v_l, pr_3(\underline{x}), \ldots, pr_n(\underline{x}))\ if\ u_k = \varepsilon,\ or$$

$$\underline{y} = (u_1 \ldots u_{k-1}bu_k, v_1 \ldots v_l, pr_3(\underline{x}), \ldots, pr_n(\underline{x}))\ if\ u_k \neq \varepsilon$$

but this means (see the shape of Qr_k) that $\underline{y} \in VFS(Pr \longrightarrow Qr_k)$. Let $Pr \longrightarrow Qr_k = Q'_1 \ldots Q'_n$. In all cases $\underline{y}$ has the following two properties :

- if $pr_1(\underline{y})z_1 \in FS(Q'_1)$ then any occurrence of a in z_1 is preceded by an occurrence of c;
- if $pr_2(\underline{y})z_2 \in FS(Q'_2)$ then any occurrence of c in z_2 is preceded by an occurrence of a;

which imply that for every $\underline{yz} \in VFS(Pr \longrightarrow Qr_k), \#_a(\underline{z}) = \#_c(\underline{z}) = 0$. Hence $Pr \longrightarrow Qr_k$ is not adequate, contradicting $Pr \in Smooth(\Omega_0)$. Thus, $Smooth(\Omega_0) \subseteq \Gamma_0$.
In exactly the same way we may prove $Smooth(\Omega_1) \subseteq \Gamma_0$. In particular, every $Pr \in Smooth(\Omega_1)$ satisfies (1) and (2) of Γ_0 since $Qr_0 \in \Omega_1$. □

Proof of Theorems 2.25 and 2.26. From Lemma D.6 we have $\Gamma_0 \subseteq Smooth(\Omega)$. This and Lemma D.7 give us: $Smooth(\Omega_0) \subseteq \Gamma_0 \subseteq Smooth(\Omega)$ and $Smooth(\Omega_1) \subseteq \Gamma_0 \subseteq Smooth(\Omega)$. On the other hand : $\Omega_0 \subseteq \Omega$ and $\Omega_1 \subseteq \Omega$, then $Smooth(\Omega) \subseteq Smooth(\Omega_0)$ and $Smooth(\Omega) \subseteq Smooth(\Omega_1)$, hence $Smooth(\Omega_1) = Smooth(\Omega_0) = \Gamma_0 = Smooth(\Omega)$. □

Proof of Theorem 2.28 (Sketch). Clearly $Smooth(\Omega_{\omega CF}) \subseteq Smooth(\Omega_3)$ since $\Omega_3 \subseteq \Omega_{\omega CF}$. Let $Pr = P_1 \ldots P_n \in Smooth(\Omega_3)$. To show that Pr satisfies (1) of Γ_1, we observe the path program R_1 defined by (1) of Ω_3 is equal to Qr_0 from the proof of Lemma D.7, so we can repeat the first part of that proof. To show that (2) of Γ_1 we proceed similarly to the proof of the second part of the proof of Lemma D.7. Suppose now that Pr does not satisfy (3) of Γ_1. Without loss of generality we may assume that there is $\underline{x} \in VFS(Pr)$ such that $pr_1(\underline{x}) \in Cyc(P_1)^k$ and $pr_2(\underline{x}) \in Cyc(P_2)^m$, where $k > m \geq 0$. Let $Qr = \psi_k\phi_{k,m}\xi \ldots \xi$. Clearly $Qr \in \Omega_3$, and Qr is adequate since $VFS(Qr) = Pref((\{\underline{\omega}\}^m\{\underline{a}\}\{\underline{\omega}\}^{k-m}\{\underline{bdc}\})^*)$. On the other hand one can easily see that $\underline{xc} \in VFS(Pr \longrightarrow Qr)$ and $\underline{xc}$ leads to deadlock, contradicting $Pr \in Smooth(\Omega_3)$, which completes the proof. □

Appendix E

Proofs of Theorems 2.37 and 2.38

Let $\underline{x} \in VFS$. We say that:

Definition E.1 (Greatest Prefix) *$\underline{y} \in Pref(\underline{x}) \cap MAX^*(\underline{\varepsilon})$ is the* **greatest prefix** *of $\underline{x} \in VFS$ if*
$(\forall \underline{z} \in Pref(\underline{x}) \cap MAX^*(\underline{\varepsilon}))\underline{z} \in Pref(\underline{y})$.

We say that:

Definition E.2 (Minimal Extension) *Let $\underline{x} \in Scope_{MAX}$. $\underline{y}' \in MAX^*(\underline{\varepsilon})$ is the* **minimal extension** *of $\underline{x}$ if $\underline{x} \in Pref(\underline{y}')$ and*
$(\forall \underline{z}' \in MAX^*(\underline{\varepsilon}))\underline{x} \in Pref(\underline{z}') \Rightarrow length(\underline{y}') \leq length(\underline{z}')$.

From Theorem 2.31 (3) it follows that $\underline{y} \in Pref(\underline{z})$ or $\underline{z} \in Pref(\underline{y})$, for all $\underline{y}, \underline{z} \in Pref(\underline{x}) \cap MAX(\underline{\varepsilon})$. Hence a greatest prefix always exists. Its unicity follows directly from Definition E.1. Concerning minimal extensions, it may happen that there is no unique minimal extension of $\underline{x}$ in $MAX^*(\underline{\varepsilon})$. For

$$Pr = \begin{cases} P(1): & \underline{path}\ b, c\ \underline{end} \\ P(2): & \underline{path}\ a\ \underline{end} \end{cases}$$

and for $\underline{x} = \underline{a}$, there are two minimal extensions of $\underline{x}$ in $MAX^*(\underline{\varepsilon}) : \underline{ab}$ and $\underline{ac}$. We shall now prove two lemmas characterizing the greatest prefixes and minimal extensions.

Lemma E.1 *Let $\underline{x} \in Scope_{CON} \setminus Scope_{MAX}$ and $\underline{y}$ be the greatest prefix of $\underline{x}$ in $MAX^*(\underline{\varepsilon})$, i.e.*
$\underline{y} \in Pref(\underline{x}) \cap MAX^*(\underline{\varepsilon})$ and $(\forall \underline{z} \in Pref(\underline{x}) \cap MAX^*(\underline{\varepsilon}))\underline{z} \in Pref(\underline{y})$. Then:
$(\exists A_1, \ldots, A_k \in Ind)\underline{x} = \underline{y}\underline{A_1} \ldots \underline{A_k} \wedge \underline{A_1} \ldots \underline{A_k} \doteq \underline{A_1} \ldots \underline{A_k} \wedge k \geq 2$.

Proof. Let $\underline{x} \doteq \underline{B_1} \ldots \underline{B_m}$. Suppose that $\underline{y} \neq \underline{\varepsilon}$ and $\underline{y} \doteq \underline{C_1} \ldots \underline{C_p}$. By Theorem 2.31(3) we have $p \geq m$ and $B_i = C_i$ for $i = 1, \ldots, m$. Let us set $k = m - p$ (if $\underline{y} = \underline{\varepsilon}$, we set $k = m$ and $p = 0$) and $A_i = C_{p+i}$ for $i = 1, \ldots, k$. Thus, $\underline{x} = \underline{y}\underline{A_1} \ldots \underline{A_k}$ and $\underline{A_1} \ldots \underline{A_k} \doteq \underline{A_1} \ldots \underline{A_k}$. Because $\underline{x} \notin MAX^*(\underline{\varepsilon})$, we have $k \geq 1$.

Suppose $k = 1$. Then $\underline{x} = \underline{y}\underline{A_1}, A_1 \in Ind$ and $\underline{y} \in MAX^*(\underline{\varepsilon})$. But this means that $A_1 \in enabled_{CON}(\underline{y}) \setminus enabled_{MAX}(\underline{y})$, and there is a $B \in enabled_{MAX}(\underline{y})$ such that $A_1 \subset B$, $A_1 \neq B$, and $\underline{y}\underline{B} \in MAX^*(\underline{\varepsilon})$. Of course, $\underline{y}\underline{A_1}$ is a prefix of $\underline{y}\underline{B}$, so $\underline{x} = \underline{y}\underline{A_1} \in Pref(MAX^*(\underline{\varepsilon})) = Scope_{MAX}$, which is a contradiction. □

Lemma E.2 *Let* $\underline{x} \in Scope_{MAX} \setminus MAX^*(\underline{\varepsilon}), \underline{y}$ *be its greatest prefix in* $MAX^*(\underline{\varepsilon})$ *and* $\underline{y}'$ *be one of its least extensions in* $MAX^*(\underline{\varepsilon})$. *Then there is* $k \geq 1$ *and* $A_i, B_i, C_i, i = 1, \ldots, k$, *such that the following hold:*

(1) $\underline{y}' = \underline{y}\underline{A_1} \ldots \underline{A_k} \wedge \underline{A_1} \ldots \underline{A_k} \doteq \underline{A_1} \ldots \underline{A_k}$

(2) $\underline{x} = \underline{y}\underline{B_1} \ldots \underline{B_k} \wedge \underline{B_1} \ldots \underline{B_k} \doteq \underline{B_1} \ldots \underline{B_k}$

(3) $\underline{y}' = \underline{x}\underline{C_1} \ldots \underline{C_k}$ *(Note: if* $C_i = \emptyset$ *then* $\underline{C_i} = \underline{\varepsilon}$*)*

(4) $A_i = B_i \cup C_i$ *for* $i = 1, \ldots, k$

(5) $C_1 \neq \emptyset$

(6) $(\forall i = 1, \ldots, k)(\forall j = i, \ldots, k) B_j \cup C_i \in Ind$ *and* $B_j \cap C_i = \emptyset$.

Proof. Of course, there is a sequence $A_1, \ldots, A_k$ such that $\underline{y}' = \underline{y}\underline{A_1} \ldots \underline{A_k}, A_1 \in enabled_{MAX}(\underline{y})$ and $A_i \in enabled_{MAX}(\underline{y}\underline{A_1} \ldots \underline{A_{i-1}})$ for $i = 2, \ldots, k$. Note that $A_1, \ldots, A_k$ satisfies (1) (see Theorem 2.31 (2)).
If $k = 1$, the lemma simply expresses that $\underline{y} \neq \underline{x} \neq \underline{y}'$ and that we have $\underline{y} MAX \underline{y}', \underline{y} CON \underline{x} CON \underline{y}'$. For $k \geq 2$, let $\underline{x} = \underline{y}\underline{z_0}, \underline{y}' = \underline{x}\underline{w}$, and let $B_1, \ldots, B_k, C_1, \ldots, C_k$ be the sets constructed by:

(1) $B_1 = A_1 \cap Ev(\underline{z_0}), C_1 = A_1 \setminus B_1$,

(2) for $i = 1, \ldots, k-1, \underline{z_i}$ is constructed from $\underline{z_{i-1}}$ by dropping the first occurrence of each operation in B_i, and $B_{i+1} = A_{i+1} \cap Ev(\underline{z_i}), C_{i+1} = A_{i+1} \setminus B_{i+1}$.

From the fact that $\underline{y}' = \underline{y}\underline{z_0}\underline{w} = \underline{x}\underline{w} = \underline{y}\underline{A_1} \ldots \underline{A_k}$ and the above construction, it follows: $A_i = B_i \cup C_i, \underline{x} = \underline{y}\underline{B_1} \ldots \underline{B_k}, \underline{y}' = \underline{x}\underline{C_1} \ldots \underline{C_k}$, i.e. the points (3),(4), and the first half of (2) are satisfied. Property (5) results from the fact that if $C_1 = \emptyset$, then $\underline{y}\underline{B_1} = \underline{y}\underline{A_1}$ would be a greater prefix of $\underline{x}$ in $MAX^*(\underline{\varepsilon})$ than $\underline{y}$. Property (6) results from the fact that since $\underline{y}' = \underline{y}\underline{A_1} \ldots \underline{A_k} = \underline{y}\underline{B_1} \ldots \underline{B_k}\underline{C_1} \ldots \underline{C_k}$, each event $c \in C_1$ has to commute with all the events in $B_2, \ldots, B_k$; moreover, if $c \in B_j$, from the construction of B_1, c would belong to B_1 and not to C_1 since that means that $c \in Ev(\underline{z_{i-1}})$ and thus $c \in Ev(\underline{z_0})$; the same argument may then be resumed for $C_2, \ldots, C_k$ in that order.
To prove that $\underline{B_1} \ldots \underline{B_k} \doteq \underline{B_1} \ldots \underline{B_k}$ we have to show that:

(i) $(\forall i = 1, \ldots, k) B_i \neq \emptyset$,

(ii) $(\forall b \in B_i)(\exists c \in B_{i-1})(b, c) \notin ind$ for $i = 2, \ldots, k$.

Property (i) results from:

(a) if $B_k = \emptyset$, then $\underline{y}\underline{A}_1 \ldots \underline{A}_{k-1}$ would be an extension of $\underline{x}$ in $MAX^*(\underline{\varepsilon})$ smaller than $\underline{y}'$;

(b) Let us suppose that i is the greatest index in $\{1, \ldots, k-1\}$ such that $B_i = \emptyset$, from (6) we know that each $c \in C_i = A_i$ is independent of each operation in B_{i+1} ($\neq \emptyset$); but then A_i would not be maximally concurrent at $\underline{y}\underline{A}_1 \ldots \underline{A}_{i-1}$ since we may add $B_{i+1} \subseteq A_{i+1}$ to it.

Property (ii) results from an argument similar to the ones used above. Suppose $(\exists i \in \{2, \ldots, k\})(\exists b \in B_i)(\forall b' \in B_{i-1})(b, b') \in ind$. From (6) we have $(\forall c \in C_{i-1})(b, c) \in ind$. Hence $A_{i-1} = B_{i-1} \cup C_{i-1}$ would not be maximally independent at $\underline{y}\underline{A}_1 \ldots \underline{A}_{i-2}$ since one might add b to it. □

Now, we are able to prove a result very similar to Theorem 2.37.

Theorem E.1 *The following properties are equivalent:*

(1) $CON \not\approx MAX$

(2) there are $\underline{x} \in MAX^*(\underline{\varepsilon})$ *and* $A_1, \ldots, A_k \in Ind, k \geq 2$, *such that:*

(a) $\underline{y} = \underline{x}\underline{A}_1 \ldots \underline{A}_k \in VFS \wedge \underline{x}\underline{A}_1 \ldots \underline{A}_{k-1} \in Scope_{MAX}$,

(b) $(\forall a \in Ev)\underline{y}\underline{a} \in VFS \Rightarrow (\exists b \in A_{k-1} \cup A_k)(a, b) \notin ind$,

(c) $< A_1, \ldots, A_k >$ *is the left-normal form of* $\underline{z} = \underline{A}_1 \ldots \underline{A}_k$, *i.e.* $\underline{z} \stackrel{\bullet}{=} \underline{A}_1 \ldots \underline{A}_k$.

(d) $(\forall a \in A_k)(\exists b \in Ev)\underline{x}\underline{b} \in VFS \wedge (\forall c \in A_1 \cup \ldots \cup A_{k-1} \cup (A_k \setminus \{a\}))(b, c) \in ind$.

Proof. (1 ⇒ 2): For every $\underline{x} \in VFS$, let $gp(\underline{x})$ denote its greatest prefix in $MAX^*(\underline{\varepsilon})$. We shall also denote $d(\underline{x}) = \kappa(\underline{x}) - \kappa(gp(\underline{x}))$ and $T = Scope_{CON} \setminus Scope_{MAX} = VFS \setminus Pref(MAX^*(\underline{\varepsilon}))$. Let $V = \{\underline{x} \in T | (\forall \underline{y} \in T) d(\underline{x}) \leq d(\underline{y})\}$. Note that $(\forall \underline{x}, \underline{y} \in V) d(\underline{x}) = d(\underline{y})$. Let $k = d(\underline{x})$, where $\underline{x} \in V$. From Lemma E.1 we have $k \geq 2$. Let $\underline{y} \in V$. Clearly $V \subseteq VFS$, so the first part of (2a) is satisfied. We have $m = \kappa(\underline{y}) - k = \kappa(gp(\underline{y})) \geq 0$ and $\underline{y}$ has the following left-normal form:

$$\underline{y} \stackrel{\bullet}{=} \underline{B}_1 \ldots \underline{B}_m \underline{A}_1 \ldots \underline{A}_k \text{ where } gp(\underline{y}) \stackrel{\bullet}{=} \underline{B}_1 \ldots \underline{B}_m$$

(it may happen that $m = 0$, then $gp(\underline{y}) = \underline{\varepsilon}$).

Let $f, g : V \longrightarrow \{0, 1, 2, \ldots\}$ be functions defined as follows:

$$\textit{if } \underline{y} \stackrel{\bullet}{=} \underline{B}_1 \ldots \underline{B}_m \underline{A}_1 \ldots \underline{A}_k, \textit{ then } f(\underline{y}) = length(\underline{A}_k), \textit{ where } gp(\underline{y}) \stackrel{\bullet}{=} \underline{B}_1 \ldots \underline{B}_m$$

$$\textit{and } g(\underline{y}) = length(\underline{A}_1 \ldots \underline{A}_{k-1}),$$

Let us define $V_1 = \{\underline{y} \in V | (\forall \underline{z} \in V) f(\underline{y}) \leq f(\underline{z})\}, V_2 = \{\underline{y} \in V_1 | (\forall \underline{z} \in V_1) g(\underline{y}) \geq g(\underline{z})\}$. We may notice that since $(\forall \underline{y} \in V) g(\underline{y}) \leq (k-1) \cdot card(Ev)$,

V_2 is a nonempty set (in fact, we have strict inequality, as A_1 may not be in $enabled_{MAX}(\underline{B_1}\dots\underline{B_m})$). Let $\underline{y} \in V_2$. Since $\underline{y} \in V$ and $\underline{y} \doteq \underline{B_1}\dots\underline{B_m}\underline{A_1}\dots\underline{A_k}$, we set $\underline{x} = \underline{B_1}\dots\underline{B_m} = gp(\underline{y})$ and we have $\underline{A_1}\dots\underline{A_k} \doteq \underline{A_1}\dots\underline{A_k}$, i.e. (2.c) is satisfied.

Suppose now that $\underline{ya} \in VFS$ for some $a \in Ev$. Note that $\underline{ya} \in T$ (since $\underline{ya} \in Scope_{MAX} \Rightarrow \underline{y} \in Scope_{MAX}$). By Theorem 2.2, we have $\underline{ya} \doteq \underline{B_1}\dots\underline{B_m}\underline{C_1}\dots\underline{C_p}$, where $p \geq k, A_i \subseteq C_i$ for $i = 1,\dots,k$. Two cases may happen: $A_i = C_i$ for $i = 1,\dots,k$ and $p = k+1$ and $C_p = \{a\}$ or $p = k$ and $C_1 = A_1,\dots,C_{k-1} = A_{k-1}$ and $C_k = A_k \cup \{a\}$; and: there is only one j, such that $1 \leq j < k$ and $C_j = A_j \cup \{a\}$ and $C_i = A_i$, for $i \neq j, k = p$.

Now, suppose that $(\forall a' \in A_{k-1} \cup A_k)(a,a') \in ind$. In the first case, since $< B_1,\dots,B_m,C_1,\dots,C_p >$ is the left-normal form, we have $(\exists b \in C_{p-1})(a,b) \notin ind$ and $C_{p-1} = A_k$ or $C_{p-1} = A_{k-1}$, which is a contradiction. Let us consider the second case. Note that $gp(\underline{y})$ is a prefix of $gp(\underline{ya})$, thus $d(\underline{ya}) \leq d(\underline{y}) = k$, but this means that $\underline{ya} \in V$. Next, $j < k$ and $f(\underline{ya}) = f(\underline{y})$, so $\underline{ya} \in V_1$. But $g(\underline{ya}) = 1 + g(\underline{y})$, a contradiction since $\underline{y} \in V_2$! Therefore, we have $(\exists a' \in A_{k-1} \cup A_k)(a,a') \notin ind$, and the condition (2.b) is satisfied.

To prove (2d) and the second part of (2a), let us suppose that $a \in A_k$ and $A = A_k \setminus \{a\}$. We have to consider two cases:

Case 1: $A = \emptyset$. Since $d(\underline{x}\underline{A_1}\dots\underline{A_{k-1}}) = k-1$ and $(\forall \underline{x'} \in T)d(\underline{x'}) \geq k$, then $\underline{z} = \underline{x}\underline{A_1}\dots\underline{A_{k-1}} \in Scope_{MAX} \setminus MAX^*(\underline{\varepsilon})$.

Case 2: $A \neq \emptyset$. Suppose that $\underline{z} = \underline{x}\underline{A_1}\dots\underline{A_{k-1}}\underline{A} \in T$. Note that $\underline{x}\underline{A_1}\dots\underline{A_{k-1}} \doteq \underline{B_1}\dots\underline{B_m}\underline{A_1}\dots\underline{A_{k-1}}$; hence $\underline{z} \doteq \underline{B_1}\dots\underline{B_m}\underline{A_1}\dots\underline{A_{k-1}}\underline{A}$. This means that $\underline{z} \in V$, but $f(\underline{z}) = f(\underline{y}) - 1$, a contradiction since $\underline{y} \in V_2 \subseteq V_1$.

Thus in both cases, $\underline{z} = \underline{x}\underline{A_1}\dots\underline{A_{k-1}}\underline{A} \in Scope_{MAX} \setminus MAX^*(\underline{\varepsilon})$ (we set $\underline{\emptyset} = \underline{\varepsilon}$) which means that the second part of (2a) is satisfied. By Lemma E.2 we have $(\exists C_1 \subseteq Ev)C_1 \neq \emptyset, \underline{z}\underline{C_1} \in VFS, A_1 \cup C_1 \in enabled_{MAX}(\underline{x})$ and $(\forall b \in C_1)(\forall b' \in A_1 \cup \dots \cup A_{k-1} \cup A_k)(b,b') \in ind$, so that any $b \in C_1$ satisfies (2d).

$(2 \Rightarrow 1)$: Let $\underline{x} \doteq \underline{B_1}\dots\underline{B_m}$. Then, from (2c) and from the fact that, by Theorem 2.31 (2), $B_m \in maxenabled_{CON}(\underline{B_1}\dots\underline{B_{m-1}})$, we have $\underline{y} \doteq \underline{B_1}\dots\underline{B_m}\underline{A_1}\dots\underline{A_k}$. Since $A_k \neq \emptyset$, there is at least one $b \in Ev$ which satisfies (2.d). Thus, $b \in enabled_{CON}(\underline{x}\underline{A_1}\dots\underline{A_{k-1}})$ and $(\forall c \in A_{k-1})(b,c) \in ind$. Hence, $A_{k-1} \notin maxenabled_{CON}(\underline{x}\underline{A_1}\dots\underline{A_{k-2}})$. Let A be such that $A \cup A_k \in maxenabled_{CON}(\underline{x}\underline{A_1}\dots\underline{A_{k-1}})$. By (2.b) we have: $\underline{z} = \underline{x}\underline{A_1}\dots\underline{A_{k-1}}\underline{A_k}\underline{A} \Rightarrow \underline{z} \doteq \underline{B_1}\dots\underline{B_m}\underline{A_1}\dots\underline{(A_k \cup A)}$. Thus, by Theorem 2.36, $CON \not\approx MAX$. □

The next lemma shows that the condition of the above theorem can be strengthened.

Lemma E.3 *Let $\underline{x} \in MAX^*(\underline{\varepsilon})$ and let $A_1,\dots,A_k \in Ind$, where $k \geq 2$, be those from Theorem E.1 (2). Then:*

(a) $A_1 \notin maxenabled_{CON}(\underline{x})$
(b) $(\forall j < k) A_j \cap A_k = \emptyset$
(c) $(\exists b \in Ev)\underline{x}\underline{A_1} \ldots \underline{A_{k-1}}\underline{b} \in VFS \wedge (\forall c \in A_{k-1})(b, c) \in ind.$

Proof. (a) Let $b \in Ev$ satisfy Theorem E.1 (2.d). Then $b \in enabled_{CON}(\underline{x})$ and $(\forall b' \in A_1)(b, b') \in ind$, hence $A_1 \notin maxenabled(\underline{x})$.
(b) Let $a \in A_j \cap A_k$, where $j < k$. From Theorem E.1 (2.d) it follows that $(\exists b \in Ev)\underline{xb} \in VFS$ and $(\forall b' \in A_1 \cup \ldots A_{k-1} \cup A)(b, b') \in ind$, where $A = A_k \setminus \{a\}$. Since $a \in A_j$ and $j < k$, we have $(\forall b' \in A_1 \cup \ldots \cup A_k)(b, b') \in ind$. But this means that $\underline{yb} \in VFS$, and $(\forall b' \in A_{k-1} \cup A_k)(b, b') \in ind$, a contradiction with Theorem E.1(2.b).
(c) From Theorem E.1 (2d) we have $(\exists b \in Ev)\underline{xb} \in VFS$ and $(\forall c \in A_1 \cup \ldots \cup A_{k-1})(b, c) \in ind$. But this means that $\underline{x}\underline{A_1} \ldots \underline{A_{k-1}}\underline{b} \in VFS$ and $(\forall c \in A_{k-1})(b, c) \in ind$. □

Proof of Theorem 2.37. (1 ⇒ 2): The condition (2.d) of Theorem 2.37 follows immediately from (2.d) of Theorem E.1. The remaining conditions of Theorem 2.37 and Theorem E.1 are identical.
(2 ⇒ 1): Let $\underline{x} \doteq \underline{B_1} \ldots \underline{B_m}$. From (2c) and from the fact that, by Theorem 2.31 (2), $B_m \in maxenabled_{CON}(\underline{B_1} \ldots \underline{B_{m-1}})$, we have $\underline{y} \doteq \underline{B_1} \ldots \underline{B_m}\underline{A_1} \ldots \underline{A_k}$. By (2d) we have $A_{k-1} \notin maxenabled_{CON}(\underline{x}\underline{A_1} \ldots \underline{A_{k-2}})$. Let A be such that $A \cup A_k \in maxenabled_{CON}(\underline{x}\underline{A_1} \ldots \underline{A_{k-1}})$. By (2.b) we have:

$$\underline{z} = \underline{x}\underline{A_1} \ldots \underline{A_{k-1}}\underline{A_k}\underline{A} \Rightarrow \underline{z} \doteq \underline{B_1} \ldots \underline{B_m}\underline{A_1} \ldots \underline{A_{k-1}}\underline{(A_k \cup A)}.$$

Thus by Theorem 2.36, $MAX \not\approx CON$. □

Proof of Theorem 2.38. Let us take $\underline{x}, A_1, \ldots, A_k$, from the proof of Theorem E.1. Since $\underline{A_1} \ldots \underline{A_k} \doteq \underline{A_1} \ldots \underline{A_k}$, there are $a \in A_{k-1}$ and $b \in A_k$ (from Lemma E.3 (b) it follows that $A_{k-1} \cap A_k = \emptyset$, i.e. $a \neq b$), such that $a \neq b$ and $(a, b) \notin ind$. From the proof of Theorem E.1, we also have $\underline{x}\underline{A_1} \ldots \underline{A_{k-1}}\underline{(A_k \setminus \{b\})} \in Scope_{MAX}$ (provided $\underline{\emptyset} = \underline{\varepsilon}$), so that by Lemma E.2, there exists a $C \subseteq Ev$ such that $C \neq \emptyset, \underline{x}\underline{A_1} \ldots \underline{A_{k-1}}\underline{(A_k \setminus \{b\})}\underline{C} \in VFS, A_1 \cup C \in maxenabled_{CON}(\underline{x}), (\forall c \in C)((\forall c' \in A_k \setminus \{b\})(c, c') \in ind$, and $(\forall i = 1, \ldots, k - 1)(\forall c \in C)(\forall c' \in A_i)(c, c') \in ind$. Let $c \in C$. Suppose that $(b, c) \in ind$. In such a case $(\forall c' \in A_{k-1} \cup A_k)(c, c') \in ind$ and $\underline{x}\underline{A_1} \ldots \underline{A_k}\underline{c} \in VFS$, which contradicts Theorem E.1 (2.b). Thus, $(b, c) \notin ind$. Of course, $(a, c) \in ind$ since $a \in A_{k-1}$ and $c \in C$. Let $\underline{z} = \underline{x}\underline{A_1} \ldots \underline{A_{k-2}}\underline{(A_{k-1} \setminus \{a\})}$; we have $\underline{z} \in Scope_{MAX}, \underline{zab} \in VFS$ and $\underline{zac} \in VFS$. □

Appendix F

Proof of Theorem 2.29

In this appendix we shall use the following notation, for every $x, y \in Ev^*$ we shall write $x \leq y$ if $x \in Pref(y)$, and $x < y$ if $x \in Pref(y) \setminus \{y\}$, and similarly, $\underline{x} \leq \underline{y}$ if $\underline{x} \in Pref(\underline{y})$ and $\underline{x} < \underline{y}$ if $\underline{x} \in Pref(\underline{y}) \setminus \{\underline{y}\}$.

Lemma F.1 *Let $Pr = P_1 \ldots P_n$ be a connected FC-path program. Then the following are equivalent:*
(1) Pr is adequate
(2) Pr is deadlock-free
(3) $(\forall \underline{x} \in VFS)(\forall i = 1, \ldots, n)(\forall y_i \in Ev_i^*) pr_i(\underline{x}) y_i \in FS_i \Rightarrow$
$(\exists \underline{y} \in CON^*(\underline{x})) pr_i(\underline{y}) = pr_i(\underline{x}) y_i$.

Proof. $(1 \Rightarrow 2)$: Obvious.
$(2 \Rightarrow 3)$: In fact we are going to prove that $\neg(3) \Rightarrow \neg(2)$. The idea is that if $\neg(3)$ then we gradually block the system, coordinate by coordinate. Let $\underline{x} \in VFS$, $I \subseteq \{1, \ldots, n\}$, $I \neq \emptyset$. We say that $\underline{x}$ is I-blocked if and only if:

$$(\forall \underline{y} \in VFS) \underline{x} < \underline{y} \Rightarrow (\forall i \in I) pr_i(\underline{x}) = pr_i(\underline{y}).$$

We shall show that $\neg(3)$ implies we may construct a vector which is $\{1, \ldots, n\}$-blocked. First we shall show that $\neg(3)$ implies that for some $\underline{x}$ and some i, $\underline{x}$ is $\{i\}$-blocked (step 1). Then we shall show that blocking is transmitted accross connections between connected paths (step 2 and step 3).
Step 1: Let $\underline{x} \in VFS$, $i \in \{1, \ldots, n\}$, $y_i \in Ev_i^+$ be such that $pr_i(\underline{x}) y_i \in FS_i$ and $(\forall \underline{y} \in CON^*(\underline{x})) pr_i(\underline{y}) \neq pr_i(\underline{x}) y_i$. Let z_i be a maximal prefix of y_i such that

$$(\exists \underline{z} \in CON^*(\underline{x}))\ pr_i(\underline{z}) = pr_i(\underline{x}) z_i.$$

Assume that $y_i = z_i b v$, $b \in Ev_i$, $v \in Ev_i^*$. We claim that $\underline{z}$ is $\{i\} - blocked$. Suppose that it is not, i.e., $(\exists \underline{w} \in CON^*(\underline{z}))\ pr_i(\underline{w}) \neq pr_i(\underline{z})$. This means that there are $a \in Ev_i$, $u \in Ev_i^*$ such that $pr_i(\underline{w}) = pr_i(\underline{z}) a u$. If $a = b$ then $\underline{z}\ \underline{a} \in CON^*(\underline{x})$, $pr_i(\underline{z}\ \underline{a}) = pr_i(\underline{z}) a = pr_i(\underline{x}) z_i a_j$ and $z_i a \leq y_i$, a contradiction to the maximality of z_i.

If $a \neq b$, from the fact that Pr is a FC-path program it follows that $\underline{z}\,\underline{a} \in CON^*(\underline{x}) \Rightarrow \underline{z}\,\underline{b} \in CON^*(\underline{x})$. Thus we have:

$$pr_i(\underline{z}\,\underline{b}) = pr_i(\underline{z})b = pr_i(\underline{x})z_ib_j \text{ and } z_ib \leq y_i,$$

a contradiction to the maximality of z_i. Hence $\underline{z}$ is $\{i\}$-blocked.
Step 2: Suppose $j \in \{1,\dots,n\}, i \in I, Ev_i \cap Ev_j \neq \emptyset$ and $\underline{x} \in VFS$ is $I-blocked$. We show that $(\exists \underline{z} \in CON^*(\underline{x})) \in z$ is $(\{j\} \cup I) - blocked$. Let $a \in Ev_i \cap Ev_j$, and let $y_j \in Ev_j^*$ be such that $pr_j(\underline{x})y_j \in FS_j$ and $\#_a(y_j) > 0$. Since $a \in Ev_i$, $i \in I$ and $\underline{x}$ is I-blocked then $(\forall \underline{y} \in CON^*(\underline{x}))pr_j(\underline{y}) \neq pr_j(\underline{x})y_j$. Thus, from Step 1 it follows that there is $\underline{z} \in CON^*(\underline{x})$ which is $\{j\} - blocked$. Clearly, $\underline{z}$ is also $I - blocked$ and have $(\{j\} \cup I) - blocked$.
Step 3: We may now conclude our proof of: $\neg(3) \Rightarrow \neg(2)$. Note first that, since we have the logical rule: $\neg(\alpha \Rightarrow \beta) \equiv \neg(\neg\alpha \vee \beta) \equiv \alpha \wedge \neg\beta$, $\neg(3)$ is equivalent to:

$$(\exists \underline{x} \in VFS)(\exists i \in \{1,\dots,n\})(\exists y_i \in FS_i)pr_i(\underline{x})y_i \in FS_i \wedge$$

$$(\forall \underline{y} \in CON^*(\underline{x}))pr_i(\underline{y}) \neq pr_i(\underline{x})y_i.$$

Thus, if $\neg(3)$, then, by the assertion from Step 1, we may find $\underline{x}_1 \in VFS$ and $i_1 \in \{1,\dots,n\}$ such that $\underline{x}_1$ is $\{i_1\}$-blocked. Since Pr is connected, we may find distinct $i_2,\dots,i_n \in \{1,\dots,n\}$ such that if we set $I_1 = \{i_1\}, I_2 = \{i_1,i_2\},\dots,I_n = \{i_1,\dots,i_n\}$, then $I_n = \{1,\dots,n\}$, and $(\forall j \in \{2,\dots,n\})(\exists k \in I_{j-1})Ev_k \cap Ev_{i_j} \neq \emptyset$. We may now apply the assertion from Step 2 repeatedly to obtain vectors $\underline{x}_1 \leq \underline{x}_2 \leq \dots \leq \underline{x}_n$ such that $\underline{x}_j \in VFS$ and $\underline{x}_j$ is I_j-blocked for each $j = 1,\dots,n$. In particular $\underline{x}_n$ is $\{1,\dots,n\}$-blocked, which is clearly equivalent to P being in deadlock.
$(3 \Rightarrow 1)$: Let $\underline{x} \in VFS, a \in Ev$. We shall show that

$$(\exists \underline{xy} \in CON^*(\underline{x}))\#_a(\underline{y}) > 0$$

which implies adequacy. Let $a \in Ev_i$. Since $pr_i(\underline{x}) \in FS_i$ we may find $z \in Ev_i^*$ such that $\#_a(z) > 0$, and $pr_i(\underline{x})z \in FS_i$, so by (3) we may find $\underline{xy} \in CON^*(\underline{x})$ with $pr_i(\underline{y}) = z$. Clearly $\#_a(\underline{y}) = \#_a(z) > 0$, which ends the proof of this implication.
Finally, note that the implications: $(3) \Rightarrow (1) \Rightarrow (2)$ hold in general, for every path program $Pr = P_1 \dots P_n$, since we used neither the FC-property nor connectedness in their proof. □
The path program Pr_{22} from Chap. 2.4.4 shows that neither $(2) \Rightarrow (3)$ nor $(1) \Rightarrow (3)$ are true in general, and Pr_{23} from Chap. 2.4.4 shows that $(2) \Rightarrow (1)$ is also not true in general.

Proof of Theorem 2.29. This is a trivial consequence of Lemma F.1. To strengthen the intuition behind condition (3) of Lemma F.1, we show that, in fact this condition is equivalent to adequacy in the case of FC-path programs.

Lemma F.2 *Let $Pr = P_1 \ldots P_n$ be a FC-path program (not necessarily connected). Then the following are equivalent:*
(1) Pr is adequate
(2) $(\forall \underline{x} \in VFS)(\forall i = 1, \ldots, n)(\forall y_i \in Ev_i^*) pr_i(\underline{x}) y_i \in FS_i \Rightarrow$
$(\exists \underline{y} \in CON^*(\underline{x})) pr_i(\underline{y}) = pr_i(\underline{x}) y_i$.

Proof. $(1 \Rightarrow 2)$: If Pr is connected this is a trivial consequence of Lemma F.1. Assume that Pr is not connected. Without loss of generality we may assume that $Pr = Pr_1 \ldots Pr_k$ where $k \leq n$, each $Pr_j = P_{j1} \ldots P_{jk_j}$ is a connected path program, and $Ev(Pr_i) \cap Ev(Pr_j) = \emptyset$ if $i \neq j$. Thus,by the definition of $VFS, VFS = VFS(Pr_1) \times \ldots \times VFS(Pr_k)$, by the definition of the FC-property, each Pr_j is also a FC-path program, and by the definition of adequacy, each Pr_j is adequate for $j = 1, \ldots, k$. To distinguish between elements of Vev^* and $Vev(Pr_j)^*$ for $j = 1, \ldots, k$, the latter ones will be denoted by $\underline{x}_{Pr_j}$, that is to say, for every $x \in Ev(Pr_j)^*$ we have $\underline{x} \in Vev^*$, and $\underline{x}_{Pr_j} \in Vev(Pr_j)^*$.
Let $x \in Ev^*, \underline{x} \in VFS, i \in \{1, \ldots, n\}, y_i \in Ev(P_i)^*, pr_i(\underline{x}) y_i \in FS(P_i)$. Let $Pr_j = P_{j1} \ldots P_{jk_j}$ be such that $(\exists r) P_{jr} = P_i$. Let $x' = h_{Ev(Pr_j)}(x)$. Clearly $Ev(P_i) \subseteq Ev(Pr_j), pr_i(\underline{x}) = pr_r(\underline{x}'_{Pr_j})$ and $\underline{x}'_{Pr_j} \in VFS(Pr_j)$. Hence, by Lemma F.1, $(\exists y \in Ev(Pr_j)^*) \underline{y}_{Pr_j} \in CON^*_{Pr_j}(\underline{x}'_{Pr_j}) \wedge pr_r(\underline{y}_{Pr_j}) = pr_r(\underline{x}'_{pr_j}) y_i$. But since $VFS(Pr) = VFS(Pr_1) \times \ldots \times VFS(Pr_k)$, this also means that $\underline{xy} \in CON^*_{Pr}(\underline{x})$, which ends the proof.
$(2 \Rightarrow 1)$: In the same way as $(3) \Rightarrow (1)$ in the proof of Lemma F.1, where we did not use the assumption of connectedness. □
Finally, we show that for disconnected FC-path programs, Lemma F.2 cannot be replaced by Lemma F.1. Let Pr be the following path program:

$$Pr = \begin{cases} P(1): & \underline{path}\ a\ \underline{end} \\ P(2): & \underline{path}\ b; c\ \underline{end} \\ P(3): & \underline{path}\ c; b\ \underline{end} \end{cases}$$

Pr is a SCF-path program, so by Corollary 2.16 it is also free-choice. Since $Ev(P_1) \cap Ev(P_2 P_3) = \emptyset$, Pr is not connected. Finally, $VFS(Pr) = \{\underline{a}\}^*$, so it is deadlock free but not adequate.

Appendix G

Proof of Theorem 4.3

Let $Ev = Ev_1 \cup \ldots \cup Ev_n$, Vev be a set of event vectors and $ind \subseteq Ev \times Ev$ be an independency relation generated by $\{Ev_1, \ldots, Ev_n\}$. Let $vect, \overline{vect} : 2^{Ev^*} \longrightarrow 2^{Vev^*}$ be the following mappings:

$$(\forall L \subseteq Ev^*) vect(L) = \{\underline{x} | x \in L\},\ \overline{vect}(L) = h_1(L) \times \ldots \times h_n(L) \cap Vev^*.$$

Corollary G.1
1. $\overline{vect}(L) = \{\underline{x} \in Vev^* | h_i(x) \in h_i(L),\ i = 1, \ldots, n\}$
2. $vect(L) \subseteq \overline{vect}(L)$
3. $vect(Pref(L)) \subseteq Pref(vect(L))$
4. $Pref(\overline{vect}(L)) \subseteq \overline{vect}(Pref(L))$
5. $\overline{vect}(L_1)\overline{vect}(L_2) \subseteq \overline{vect}(L_1 L_2)$
6. $\overline{vect}(L_1) \cup \overline{vect}(l_2) \subseteq \overline{vect}(L_1 \cup L_2)$ □

Now, we rephrase Theorem 4.3 in an equivalent form which is easier to prove.

Theorem G.1 *Let* $CP = (P, Ev_1, \ldots, Ev_n) \in SCEPATH$*, i.e.,* $P \in SEPATH$*, and let* $L = RFS(P)$*. Then:*
(1) $Pref(vect(L)) = \overline{vect}(Pref(L))$
(2) $vect(L) = \overline{vect}(L)$

Proof. Since P is a synchronized $e - path$, then $P = \underline{epath}\ E\ \underline{end}$ where E is any expression derived from the nonterminal $sequence_\nabla$ of the grammar for synchronized $e - paths$. Clearly, E can be treated as a regular expression, and $RFS(P) = L(E)$, where $L(E)$ is a language defined by E (see Appendix B). The proof of Theorem G.1 is by induction on the structure of E. Clearly, for $E = a$, where a is an event, both (1) and (2) of Theorem G.1 are true. In the proof of the inductive step, we often identify regular expressions with languages defined by them. In what follows E, E_i will denote expressions derived from the nonterminal $sequence_\nabla$ of the grammar which defines synchronized $e - paths$. First we prove the following lemma:

Lemma G.1
(1) $\overline{vect}((\bigtriangledown;E)^*) = (\{\bigtriangledown\}\overline{vect}(E))^*$
(2) $\overline{vect}((E_1,(\bigtriangledown;E_2))) = \overline{vect}(E_1) \cup \{\bigtriangledown\}\overline{vect}(E_2)$.

Proof. (1) From Corollary G.1(1) we have $\overline{vect}((\bigtriangledown;E)^*) = \{\underline{x} \in Vev^* | h_i(x) \in h_i((\bigtriangledown;E)^*)\}$. For $i \notin i(\bigtriangledown)$ we have $h_i((\bigtriangledown;E)^*) = \{\varepsilon\}$, thus

$$\begin{aligned}&\{\underline{x} \in Vev^*|h_i(x) \in h_i((\bigtriangledown;E)^*)\} = \{x \in Vev^*|h_i(x) \in h_i((\bigtriangledown;E)^*), i \in i(\bigtriangledown)\} = \\ &\{\underline{x} \in Vev^*|h_i(x) \in (\{\bigtriangledown\}h_i(E)^*), i \in i(\bigtriangledown)\} = \\ &\{\underline{x} \in Vev^*|h_i(x) \in (\{\bigtriangledown\}h_i(E)^{k_i}), i \in i(\bigtriangledown), k_i \geq 0\}.\end{aligned}$$

Since $(\forall a \in Ev(E))i(a) \subseteq i(\bigtriangledown)$ we have: $(\forall i,j \in i(\bigtriangledown))k_i = k_j$, so

$$\begin{aligned}&\{\underline{x} \in Vev^*|h_i(x) \in (\{\bigtriangledown\}h_i(E)^{k_i}), i \in i(\bigtriangledown), k_i \geq 0\} = \\ &= \{\underline{x} \in Vev^*|h_i(x) \in (\{\bigtriangledown\}h_i(E)^{k}), i \in i(\bigtriangledown), k_i \geq 0\} = \\ &= \{\underline{\bigtriangledown\, y_1 \bigtriangledown\, y_2 \cdots \bigtriangledown\, y_k} | h_i(y_j) \in h_i(E), i \in i(\bigtriangledown), j = 1, \ldots, k, k \geq 0\} = (\{\bigtriangledown\}\overline{vect}(E))^*.\end{aligned}$$

(2) From Corollory G.1(1) we have $\overline{vect}((E_1,(\bigtriangledown;E_2))) = \{\underline{x} \in Vev^*|h_i(x) \in h_i((E_1,(\bigtriangledown;E_2)))\}$. For $i \notin i(\bigtriangledown)$ we have $h_i((E_1,(\bigtriangledown);E_2))) = \{\varepsilon\}$, thus

$$\begin{aligned}&\{\underline{x} \in Vev^*|h_i(x) \in h_i((E_1,(\bigtriangledown;E_2)))\} = \\ &= \{\underline{x} \in Vev^*|h_i(x) \in h_i((E_1,(\bigtriangledown;E_2))), i \in i(\bigtriangledown)\} = \\ &= \{\underline{x} \in Vev^*|h_i(x) \in h_i(E_1) \cup \{\bigtriangledown\}h_i(E_2), i \in i(\bigtriangledown)\}.\end{aligned}$$

Since $(\forall a \in Ev(E_1) \cup Ev(E_2))i(a) \in i(\bigtriangledown)$ and $\bigtriangledown$ is unique, we have

$$\{x \in Ev^*|h_i(x) \in h_i(E_1)\} \cap \{z \in Ev^*|h_j(x) \in \{\bigtriangledown\}h_j(E_2)\} = \emptyset \text{ for } i,j \in i(\bigtriangledown).$$

Therefore

$$\begin{aligned}&\{\underline{x} \in Vev^*|h_i(x) \in h_i(E_1) \cup \{\bigtriangledown\}h_i(E_2), i \in i(\bigtriangledown)\} = \\ &= \{\underline{x} \in Vev^*|h_i(x) \in h_i(E_1), i \in i(\bigtriangledown)\} \cup \{\underline{\bigtriangledown y}|h_i(y) \in h_i(E_2), i \in i(\bigtriangledown)\} = \\ &= \overline{vect}(E_1) \cup \{\bigtriangledown\}\overline{vect}(E_2).\end{aligned}$$ □

From Lemma G.1, it immediately follows that if $vect(E) = \overline{vect}(E)$, and $vect(E_i) = \overline{vect}(E_i)$, $i = 1,2$, then $vect((\bigtriangledown;E)^*) = \overline{vect}((\bigtriangledown;E)^*)$ and $vect((E_1,(\bigtriangledown;E_2)) = \overline{vect}((E_1,(\bigtriangledown;E_2))$.

Lemma G.2
(1) $\overline{vect}(Pref((\bigtriangledown;E)^*)) = \{\underline{\varepsilon}\} \cup \overline{vect}((\bigtriangledown;E)^*)\{\underline{\bigtriangledown}\}\overline{vect}(Pref(E))$.
(2) $\overline{vect}(Pref((E_1,(\bigtriangledown;E_2)))) = \overline{vect}(Pref(E_1)) \cup \{\underline{\bigtriangledown}\}\overline{vect}(Pref(E_2))$.

Proof. The proof is similar to that of Lemma G.1. Since $i(a) \subseteq i(\bigtriangledown)$ we conclude that $h_i(x)$ belongs to the image under h_i of expressions that contain $\bigtriangledown$'s, which implies that $\underline{x}$ belongs to the image under $\overline{vect}$ of such expressions. □

From the definition of the mapping $vect$, it follows that:

$$Pref(vect((\nabla; E)^*)) = \{\underline{\varepsilon}\} \cup vect((\nabla; E)^*)\{\underline{\nabla}\}Pref(vect(E)),$$
$$Pref(vect((E_1, (\nabla; E_2)))) = Pref(vect(E_1)) \cup \{\underline{\nabla}\}Pref(vect(E_2)).$$

From Lemmas G.1 and Lemma G.2 it follows that if, for $i = 1, 2$,

$$vect(E_i) = \overline{vect}(E_i) \text{ and } Pref(vect(E_i)) = \overline{vect}(Pref((E_i)),$$
$$\text{then } Pref(vect((\nabla; E)^*)) = \overline{vect}(Pref((\nabla; E)^*)) \text{ and}$$
$$Pref(vect((E_1, (\nabla; E_2)))) = \overline{vect}(Pref((E_1, (\nabla; E_2)))).$$

Thus, Theorem G.1 is true for $E = a, E = (\nabla; E')^*, E = (E_1, (\nabla; E_2))$.

Lemma G.3
For every regular expression E (not necessarily derived from sequence$_\nabla$): $\overline{vect}((E; a)) = \overline{vect}(E)\{\underline{a}\}$.

Proof. From Corollary G.1(5) we have $\overline{vect}(E)\{\underline{a}\} \subseteq \overline{vect}((E; a))$. Let $\underline{x} \in \overline{vect}((E_i; a)) \setminus \overline{vect}(E)\{\underline{a}\}$. Hence, for some $\underline{y} \notin \overline{vect}(E)$ we have $\underline{x} = \underline{ya}$. Since $\underline{y} \notin \overline{vect}(E)$, we have $(\exists i_0)h_{i_0}(y) \notin h_{i_0}(E)$, which implies that $h_{i_0}(ya) \notin h_{i_0}((E; a))$ and so $\underline{x} = \underline{ya} \notin \overline{vect}((E; a))$, which gives a required contradiction. □

Lemma G.4 *Let E be an expression derived from the nonterminal sequence$_\nabla$. If $\overline{vect}(Pref(E)) = Pref(vect(E))$, then for every $a \in Ev$,*

$$\overline{vect}(Pref((E; a))) = Pref(vect((E; a))).$$

Proof. Suppose that $\overline{vect}(Pref(E)) = Pref(vect(E))$. From Corollary G.1 we have $Pref(vect((E; a)) \subseteq \overline{vect}(Pref((E; a)))$. Let $\underline{x}$ be a shortest element (it may not be unique) of $\overline{vect}(Pref((E; a))) \setminus Pref(vect((E; a)))$. From the assumption of this Lemma it follows that $\underline{x} = \underline{ya}$, where $\underline{x} \notin Pref(vect(E)), \underline{y} \in Pref(vect(E))$ and $(\forall \underline{v} \in Vev^* \setminus \{\underline{\varepsilon}\})\underline{yav} \notin Pref(vect((E; a)))$. This would mean that $\overline{vect}(Pref((E; a))) \setminus Pref(vect((E; a))) = \overline{vect}(Pref((E; a))) \setminus vect((E; a))$, and so the following conditions hold:

(1) $(\forall i)h_i(ya) \in h_i(Pref((E; a)))$,

(2) $(\forall w \in (E; a))(\exists i)h_i(w) \neq h_i(ya)$.

For every $L \subseteq Ev^*, h_i(L) = \{h_i(z) | z \in L\}$, hence $(\forall i)(\exists w_i \in Pref((E; a)))h_i(w_i) = h_i(ya)$. Since $\underline{ya} \notin Pref(vect(E))$ we can assume that $w_i \in (E; a)$ for $i = 1, \ldots, n$.
The condition (2) implies that $(\exists i_1, i_2)w_{i_1} \neq w_{i_2}$. From the definition of $SEPATH$, it follows that $w_{i_1} = z \nabla v_1$, $w_{i_2} = zv_2$, $v_1, v_2 \in Ev^*$, $\#_\nabla(z) = 0$. Note that for the expressions derived from $sequence_\nabla$, conflicts may occur only

due to syntactic structures of the form $(E_1, (\nabla; E_2))$ or $((\nabla; E_1)^*; E_2)$.
Suppose that the conflict occurs due to the syntactic structure $((\nabla; E_1)^*; E_2)$. Thus, $(\exists k \geq 1)(\forall i \in i(\nabla))h_i(ya) = z_i \nabla t_1^i \nabla t_2^i \dots \nabla t_k^i r_i z_i'$, where $\#_\nabla(z_i) = 0$, $\nabla t_j^i \in h_i((\nabla; E_1))$ for $i = 1, 2, \dots, k-1$, $\nabla t_k^i \in h_i(Pref((\nabla; E_1)))$, $r_i \in h_i(Pref(E_2))$, $\#_\nabla(h_i(ya)) = k$. Let $A = \bigcup_{i\in i(\nabla)} \bigcup_{j=1}^{k} Ev(t_j^i)$.
Assume that $A = \{a_1, \dots, a_r\}$ and let $s_j = \#_{a_j}(\nabla t_1^i \nabla t_2^i \dots \nabla t_k^i)$, where i is any element from $i(a_j)$, for $j = 1, \dots, r$ (note that s_j is well-defined). Clearly, $\underline{ya} = \underline{x} = \underline{x_1 \nabla x_2}$, where $\#_\nabla(x_1) = 0$. Let x' be a string obtained from x by erasing first s_j occurrences of a_j from x_2, for $j = 1, \dots, r$, and k occurrences of ∇ from ∇x_2. Of course, $\underline{x}'$ is of the form $\underline{x}' = \underline{y}'\underline{a}$, and $\underline{x}' \in \overline{vect}(Pref((E; a))) \setminus Pref(vect((E; a)))$, which is a contradiction because $\underline{x}'$ is shorter than $\underline{x}$ (as $k \geq 1$) and $\underline{x}$ has been chosen to be the shortest element.
Suppose now that the conflict is due to the structure $(E_1, (\nabla; E_2))$. We consider the two cases: $v_2 = \varepsilon$ and $v_2 \neq \varepsilon$. Recall that $w_{i_1} = z \nabla v_1$ and $w_{i_2} = z v_2$.
Case 1: $v_2 = \varepsilon$. In this case, $\varepsilon \in E_1$, for example $E_1 = (\nabla_1; E_{11})^*; \dots; (\nabla_k; E_{1k})^*$. Using the same arguments as for the case $((\nabla; E_1)^*; E_2)$, we can erase from $\underline{ya} = \underline{x}$ both ∇ and all symbol occurrences from $(\nabla; E_2)$ in $(E_1, (\nabla; E_2))$. The resulting $\underline{x}' = \underline{y}'\underline{a}$ is shorter than $\underline{x}$ and belongs to the set $\overline{vect}(Pref((E; a))) \setminus Pref(vect((E; a)))$, which gives the required contradiction.
Case 2: $v_2 \neq \varepsilon$. Here $w_{i_2} = zcv_2'$, where $c \in Ev, cv_2' \in Pref(E_1)$, and $\nabla \notin i(c)$, which is in contradiction with the definition of $SEPATH$. □
In Lemma G.4 the assumption that E is an expression derived from $sequence_\nabla$ satisfying the conditions from the definition of $SEPATH$ is essential. Consider for example $E = (a, b)$, $Ev_1 = \{a, b\}$, $Ev_2 = \{a, c\}$, $Ev_3 = \{b, c\}$. Clearly E does not satisfy the conditions from $SEPATH$ definition and $\overline{vect}(Pref((a, b))) = Pref(vect((a, b))) = \{\underline{\varepsilon}, \underline{a}, unb\}$, but
$\overline{vect}(Pref((a, b; c))) = \{\underline{\varepsilon}, \underline{a}, \underline{b}, \underline{c}, \underline{ac}, \underline{bc}\} \neq \{\underline{\varepsilon}, \underline{a}, \underline{b}, \underline{ac}, \underline{bc}\} = Pref(vect((a, b; c)))$.
From Lemmas G.3 and G.4 it follows that for an expression E satisfying the conditions of $SEPATH$ definition, if $Pref(vect(E)) = \overline{vect}(Pref(E))$ and $vect(E) = \overline{vect}(E)$, then $Pref(vect((E; a))) = \overline{vect}(Pref((E; a)))$ and $vect((E; a)) = \overline{vect}((E; a))$, so Theorem G.1 is true for $E = (E'; a)$. To complete the proof of Theorem G.1 we need two more lemmas.

Lemma G.5 *Let $(E; (\nabla; E_1)^*)$ and $(E; E_1, (\nabla; E_2))$ satisfy the conditions of $SEPATH$. Then*
(1) $\overline{vect}((E; (\nabla; E_1)^*)) = \overline{vect}(E)\overline{vect}((\nabla; E_1)^*)$
(2) $\overline{vect}((E; E_1, (\nabla; E_2))) = \overline{vect}((E_1, (\nabla; E_2)))$.

Proof.
(1) From Corollary G.1(5) we know that $\overline{vect}(E)\overline{vect}((\nabla; E_1)^*) \subseteq \overline{vect}((E; (\nabla; E_1)^*))$. Let $\underline{x} \in \overline{vect}((E; (\nabla; E_1)^*))$, so $(\forall i)\ h_i(x) \in h_i((E; (\nabla; E_1)^*))$. Recall that for every $A \subseteq Ev$, $h_A : Ev^* \longrightarrow A^*$ is an erasing

homomorphism. Define $h^A : Ev^* \longrightarrow (Ev \setminus A)^*$ as $(\forall z \in Ev^*) h^A(z) = h_{Ev\setminus A}(z)$. If $\#_\nabla(x) = 0$ then $\underline{x} \in \overline{vect}(E) \subseteq \overline{vect}(E)\overline{vect}((\nabla; E_1)^*)$, so suppose that $\#_\nabla(x) > 0$. Clearly $x = x' \nabla x''$, $\#_\nabla(x') = 0$ and $x' \in Pref(\overline{vect}(E))$. Define $A = \{\nabla\} \cup \{a | i(a) \subseteq i(\nabla)\}$ and $y = x'h^A(\nabla x'')h_A(\nabla x'')$. Note that $\underline{x} = \underline{y}$ and $\underline{y} \in \overline{vect}(E)\overline{vect}((\nabla; E_1)^*)$.
(2) Again, from Corollary G.1(5) we have $\overline{vect}(E)\overline{vect}((E_1, (\nabla; E_2)) \subseteq \overline{vect}((E; E_1, (\nabla; E_2)))$. Let $\underline{x} \in \overline{vect}((E; E_1, (\nabla; E_2))$. We have to consider two cases: $\#_\nabla(x) = 1$ and $\#_\nabla(x) = 0$.
Case 1, $\#_\nabla(x) = 1$: Here $x = x' \nabla x''$ and $x' \in Pref(\overline{vect}(E))$. Let $A = \{\nabla\} \cup \{a | i(a) \subseteq i(\nabla)\}$ and $y = x'h^A(x'')h_A(\nabla x'')$. Clearly $\underline{x} = \underline{y}$ and $\underline{y} \in \overline{vect}(E)\overline{vect}((\nabla; E_2)) \subseteq \overline{vect}(E)\overline{vect}((E_1, (\nabla; E_2)))$.
Case 2, $\#_\nabla(x) = 0$: Here, $(\forall i) h_i(x) \in h_i(E)h_i(E_1)$. For every i, let $h_i(x) = y_i z_i$, where $z_i \in h_i(E_1)$ and z_i is the longest such suffix of $h_i(x)$. Define $A = \bigcup_{i \in i(\nabla)} Ev(z_i)$. Assume that $A = \{a_1, \ldots, a_r\}$. For every $j = 1, 2, \ldots, r$, let $k_j = min\{\#_{a_j}(z_i) | i \in i(\nabla)\}$. Now, let x' be the string obtained from x by erasing all last k_j occurrences of a_j, for all $a_j \in A$, and let x'' be the string obtained from x by erasing all symbol occurrences except the last k_j occurrences of a_j, for all $a_j \in A$. Once can easily verify that both x' and x'' are well defined and $\underline{x} = \underline{x}'\underline{x}''$, while $\underline{x}' \in \overline{vect}(E)$ and $\underline{x}'' \in \overline{vect}(E_1)$. Hence $\underline{x} \in \overline{vect}(E)\overline{vect}(E_1) \subseteq \overline{vect}(E)\overline{vect}((E_1, (\nabla; E_2)))$. □

Lemma G.6
Let $(E; (\nabla; E_1)^*)$ and $(E; E_1, (\nabla; E_2))$ satisfy the conditions of $SEPATH$, and let $\overline{vect}(Pref(E)) = Pref(vect(E))$. Then:
(1) $\overline{vect}(Pref((E; (\nabla; E_1)^*))) = Pref(vect((E; (\nabla; E_1)^*)))$,
(2) $\overline{vect}(Pref((E; E_1, (\nabla; E_2)))) = Pref(vect((E; E_1, (\nabla; E_2))))$.

Proof. (1) From Corollary G.1 it follows that

$$Pref(vect((E; (\nabla; E_1)^*))) \subseteq \overline{vect}(Pref((E; (\nabla; E_1)^*))).$$

Let $\underline{x}$ be a shortest (not necessarily unique) element of the set $\overline{vect}(Pref((E; (\nabla; E_1)^*))) \setminus Pref(vect((E; (\nabla; E_1)^*)))$.
Since $\overline{vect}(Pref(E)) = Pref(vect(E))$, we have $\underline{x} = \underline{y\nabla}$, $\underline{x} \notin Pref(vect(E))$, $\underline{y} \in Pref(vect(E))$, thus the following conditions hold:

(1) $(\forall i) h_i(y\nabla) \in h_i(Pref((E; \nabla))) \subseteq h_i(Pref((E; (\nabla; E_1)^*)))$,

(2) $(\forall w \in (E; \nabla) \subseteq (E; (\nabla; E_1)^*)) \; (\exists i) \; h_i(w) \neq h_i(y\nabla)$.

The remaining part of the proof of this case is almost identical to the proof of Lemma G.4.
(2) From Corollary G.1 it follows that

$$Pref(vect((E; E_1, (\nabla; E_2)))) \subseteq \overline{vect}(Pref((E; E_1, (\nabla; E_2)))).$$

Let $\underline{x}$ be a shortest element of the set $\overline{vect}(Pref((E; E_1, (\nabla; E_2)))) \setminus Pref(vect((E; E_1, (\nabla; E_2))))$. Since $\overline{vect}(Pref(E)) = Pref(vect(E))$, either $\underline{z} = \underline{y\nabla}$, $\underline{x} \notin Pref(vect(E))$, and $\underline{y} \in Pref(vect(E))$, or $\underline{x} = \underline{ya}$ for some $a \in Ev(E_1)$. But if $\underline{x} = \underline{ya}$ then one can easily prove that $\underline{x}' = \underline{y\nabla}$ also belongs to the set $\overline{vect}(Pref((E; E_1, (\nabla; E_2)))) \setminus Pref(vect((E; E_1, (\nabla; E_2))))$, the length of x and x' are the same, and $x' \notin Pref(vect(E))$. The remaining part of the proof is almost the same as the proof of Lemma G.4. □

From Lemmas G.5 and G.6 it follows that Theorem G.1 is true for $E = (E'; (\nabla; E_1)^*)$ and $E = (E'; E_1, (\nabla; E_2))$, which completes the proof of the theorem. □

Appendix H

Basic COSY Notations and Macro COSY Notation

Syntax of Basic COSY Path Programs.

B1: $\mathit{basicpathprogram} = \underline{\mathit{program}}\ \mathit{programbody}\ \underline{\mathit{endprogram}}$

B2: $\mathit{programbody} = \{\mathit{path}\}+$

B3: $\mathit{path} = \underline{\mathit{path}}\ \mathit{sequence}\ \underline{\mathit{end}}$

B4: $\mathit{sequence} = \mathit{orelement}\ \{;\ \mathit{orelement}\}^*$

B5: $\mathit{orelement} = \mathit{starelement}\ \{,\ \mathit{starelement}\}^*$

B6: $\mathit{starelement} = \mathit{element} \mid \mathit{element}*$

B7: $\mathit{element} = \mathit{event} \mid (\mathit{sequence})$

B8: $\mathit{event} = \mathit{name}$

B9: $\mathit{name} = \mathit{letter}\{\mathit{letter} \mid \mathit{digit}\}^*$

B10: $\mathit{letter} = a|\ldots|z|A|\ldots|Z$

B11: $\mathit{digit} = 0|\ldots|9$

Assume additionally that we may write some comments immediately before $\underline{\mathit{program}}, \underline{\mathit{path}}, \underline{\mathit{end}}$. We shall use this possibility to write names of programs and paths.

Syntax of Basic COSY Programs.

B1: *basicprogram* = <u>*program*</u> *programbody* <u>*endprogram*</u>

B2: *programbody* = $\{path\}^+\{process\}^* \mid \{path\}^*\{process\}^+$

B3: *path* = <u>*path*</u> *sequence* <u>*end*</u>

B4: *process* = <u>*process*</u> *sequence* <u>*end*</u>

B5: *sequence* = *orelement* $\{;$ *orelement*$\}^*$

B6: *orelement* = *starelement* $\{,$ *starelement*$\}^*$

B7: *starelement* = *element* | *element*∗

B8: *element* = *event* | (*sequence*)

B9: *event* = *name*

B10: *name* = *letter*$\{$*letter* | *digit*$\}^*$

B11: *letter* = $a \mid \ldots \mid z \mid A \mid \ldots \mid Z$

B12: *digit* = $0 \mid \ldots \mid 9$

Assume additionally that we may write some comments immediately before <u>*program*</u>, <u>*path*</u> and <u>*process*</u>. We shall use this possibility to write names of programs, paths and processes.

To shorten some otherwise lengthy expressions in the following rules, we adopt the convention:

$$\{item\ separator\ \underline{@}\} = item\ \{separator\ item\}^*.$$

Syntax of General COSY Macro Notation

M1. *mprogram* = program *mprogrambody* endprogram

M2. *mprogrambody* = {{*collectivisor*}*{*mpath*|*mprocess*|*bodyreplicator*}}+

M2.1. *collectivisor* = array {{*simpleardecl* | *replardecl*} *space* @}+ endarray

M2.2. *simpleardecl* = {*name space* @}+({{*iexpr* : |}*iexpr* @, }+)

M2.3. *replardecl* = *range_spec*[{*replardecl*|*subsevents*}]

M2.4. *subsevents* = {{*name space* @}+({*iexpr* , @}+)*space* @}+

M2.5. *range_spec* = #*indexvariable* : *iexpr*, *iexpr*, *iexpr*

M2.6. *iexpr* = *integer*|*rangevariable*|*idexvariable*|*iexpr* + *iexpr*|*iexpr* * *iexpr*
|*iexpr* ÷ *iexpr*|*iexpr* * **iexpr*|(*iexpr*)

M2.7. *space* = *the space symbol*

M2.8. *indexvariable* = *name*

M2.9. *rangevariable* = *name*

M2.10. *bodyreplicator* = *range_spec*[{*mpath*|*mprocess*|*bodyreplicator*}+]

M3.1. *mpath* = path *msequence* end

M3.2. *mprocess* = process *msequence* end

M4. *msequence* = {*morelement*; @}+

M5. *morelement* = {*gelement*, @}+

M5.1. *gelement* =
{*rreplicator*}*{*mstarelement*|*sreplicator*|*distributor*}{*lreplicator*}*

M6. *mstarelement* = *melement*|*melement**

M7. *melement* = *mevent*|(*msequence*)

M8. *mevent* = *name*{ |({{*iexpr*| },@}+)}

M9. *distributor* = *sep dim range* [*msequence*]

M9.1 *sep* = , | ;

M9.2 *dim* = | *iexpr*

M9.3 *range* = | # *iexpr*, *iexpr*, *iexpr*

M10. *sreplicator* = *range_spec*[{*concseq*|*imbrseq*}]

M11. *concseq* = *msequence sep* @

M12. *imbrseq* = *msequence sep* @ *sep* { |*msequence*} @ *sep msequence*|
(@ { |*msequence sep*} @ *sep msequence* {)/)*}|
(*msequence sep* @ { |*sep msequence*} @ {)/)*}|
(@ *msequence* @ {)|)*}|
({ |*msequence sep*} *imbrseq* { |*sep msequence*}{)/)*}|
msequence sep imbrseq | *imbrseq sep msequence*

M13. *lreplicator* = *range_spec*[{; |, }|{*concseq*|*imbrseq*}]

M14. *rreplicator* = *range_spec*[{*concseq*|*imbrseq*}|{; |, }]

M15. *integer* = { |+}{*digit*}+

M16. *name* = *letter*{*letter*|*digit*}*

M17. *letter* = *a*|...|*z*|*A*|...|*Z*

M18. *digit* = 0|...|9

Context-Sensitive Restrictions:

Crest1: If the upperbound and lowerbound of dimensions of the collective names are integers then the upperbound has to be greater than or equal to their corresponding implicit or explicit lowerbound.

Crest2: An array identifier may occur only once in collectivisors.

Crest3: Each replicator without range variables must specify a non-empty range for its index.

Irest1: Identifiers for replicator indices and range variables should be distinct from any identifiers used for simple events. Identifiers for replicator indices and range variables should also be distinct.

Irest2: Replicator index variables are only defined inside the [ ] of the replicators with which they are associated. In the scope of a replicator index variable no other replicator index variable having the same name is permitted.

BRrest: The range of the bodyreplicator indices should be non-empty.

Rrest1.1: For every replicator if inc is an integer then $inc \neq 0$.

Rrest1.2: For every concatenator and bodyreplicator if in, fi, inc are integers then $(fi - in) \div inc + 1 > 0$.

Rrest1.3: For every imbricator, if in, fi, inc are integers then $(fi = in) \div inc + 1 \leq 0 \Rightarrow FSTRIP(LSTRIP(t)) \neq \epsilon$.

Rrest2: The replicators should generate indexed events permitted by the corresponding collectivisors.

Drest1: For every distributor *sep dim range*[*msequence*], *msequence* must contain at least one slice event.

Drest2.1: If both *range* and *dim* are blank then the number of sections generated by the rightmost distributable (current) dimension must be the same for all slice events occurring in *msequence*.

Drest2.2: If *range* is blank, $dim = j$ and j is an integer, then the number of sections generated by jth dimension must be the same for all slice events occurring in *msequence* and having the jth dimension distributable.

Drest3: Inside a k-nested distributor there must be at least one slice event with k distributable dimensions, and no array slices with more than k distributable dimensions.

Drest4: If $dim = j$ then every slice event with maximal number of distributable dimensions must have the jth dimension as a distributable dimension.

Drest5.1: If *range* $= \#i_1, i_2, i_3$, *dim* is blank and i_1, i_2, i_3 are integers, then $i_2 \geq i_1 \geq 1$, $i_3 \geq 1$ and i_2 must be smaller than or equal to the number of sections generated by the rightmost distributable (current) dimension of any slice event occurring in *msequence*.

Drest5.2: If *range* $= \#i_1$, i_2, i_3, *dim* $= j$ and i_1, i_2, i_3, j are integers, then $i_2 \geq i_1 \geq 1$, $i_3 \geq 1$ and i_2 must be smaller than or equal to the number of sections generated by the jth dimension of any slice event occurring in *msequence* which has the jth dimension distributable.

List of Figures

List of Algorithms

List of Definitions

List of Theorems

List of Corollaries

List of Lemmas

Index

EATCS Monographs on Theoretical Computer Science

Vol. 1: K. Mehlhorn
Data Structures and Algorithms 1:
Sorting and Searching

Vol. 2: K. Mehlhorn
Data Structures and Algorithms 2:
Graph Algorithms and NP-Completeness

Vol. 3: K. Mehlhorn
Data Structures and Algorithms 3:
Multidimensional Searching and Computational Geometry

Vol. 4: W. Reisig
Petri Nets
An Introduction

Vol. 5: W. Kuich, A. Salomaa
Semirings, Automata, Languages

Vol. 6: H. Ehrig, B. Mahr
Fundamentals of Algebraic Specification 1
Equations and Initial Semantics

Vol. 7: F. Gécseg
Products of Automata

Vol. 8: F. Kröger
Temporal Logic of Programs

Vol. 9: K. Weihrauch
Computability

Vol. 10: H. Edelsbrunner
Algorithms in Combinatorial Geometry

Vol. 11: J. L. Balcázar, J. Díaz, J. Gabarró
Structural Complexity I

Vol. 12: J. Berstel, C. Reutenauer
Rational Series and Their Languages

Vol. 13: E. Best, C. Fernàndez C.
Nonsequential Processes
A Petri Net View

Vol. 14: M. Jantzen
Confluent String Rewriting

Vol. 15: S. Sippu, E. Soisalon-Soininen
Parsing Theory
Volume I: Languages and Parsing

Vol. 16: P. Padawitz
Computing in Horn Clause Theories

Vol. 17: J. Paredaens, P. De Bra, M. Gyssens, D. Van Gucht
The Structure of the Relational Database Model

Vol. 18: J. Dassow, G. Păun
Regulated Rewriting in Formal Language Theory

Vol. 19: M. Tofte
Compiler Generators
What they can do, what they might do, and what they will probably never do

Vol. 20: S. Sippu, E. Soisalon-Soininen
Parsing Theory
Volume II: LR(k) and LL(k) Parsing

Vol. 21: H. Ehrig, B. Mahr
Fundamentals of Algebraic Specification 2
Module Specifications and Constraints

Vol. 22: J. L. Balcázar, J. Díaz, J. Gabarró
Structural Complexity II

Vol. 23: A. Salomaa
Public-Key Cryptography

Vol. 24: T. Gergely, L. Úry
First-Order Programming Theories

Vol. 25: W. Wechler
Universal Algebra for Computer Scientists

EATCS Monographs on Theoretical Computer Science

R. Janicki, P. E. Lauer
Specification and Analysis of Concurrent Systems
The Cosy Approach

K. Jensen
Coloured Petri Nets
Basic Concepts, Analysis Methods and Practical Use, Vol. 1